Heat Transfer

Other Books by Adrian Bejan

Entropy Generation Through Heat and Fluid Flow, Wiley, 1982.

Convection Heat Transfer, 1984, Wiley, Fourth Edition, 2013.

Advanced Engineering Thermodynamics, 1988, Wiley, Fourth Edition, 2016.

Convection in Porous Media, with D. A. Nield, 1992, Springer, Fifth Edition, 2017.

Heat Transfer, Wiley, 1993.

Thermal Design and Optimization, with G. Tsatsaronis and M. Moran, Wiley, 1996.

Entropy Generation Minimization, CRC Press, 1996.

Shape and Structure, from Engineering to Nature, Cambridge University Press, 2000.

Heat Transfer Handbook, with A. D. Kraus, eds., Wiley, 2003.

La loi constructale, with S. Lorente, L'Harmattan, Paris, 2005.

Constructal Theory of Social Dynamics, with G. W. Merkx, eds., Springer, 2007.

Design with Constructal Theory, with S. Lorente, Wiley, 2008.

DESIGN IN NATURE: How the Constructal Law Governs Evolution in Biology, Physics, Technology, and Social Organization, with J. P. Zane, Doubleday, 2012.

THE PHYSICS OF LIFE: The Evolution of Everything, St. Martin's Press, 2016.

FREEDOM AND EVOLUTION: Hierarchy in Nature, Society and Science, Springer Nature, 2020.

TIME AND BEAUTY: Why Time Flies and Beauty Never Dies, World Scientific, 2022.

Heat Transfer: Evolution, Design and Performance, Wiley, 2022.

Heat Transfer

Evolution, Design and Performance

Adrian Bejan
Duke University

This edition first published 2022

© 2022 John Wiley & Sons, Inc.

The right of Adrian Bejan to be identified as the author of this work has been asserted in accordance with law.

Registered Office
John Wiley & Sons, Inc., 111 River Street, Hoboken, NJ 07030, USA

Editorial Office
111 River Street, Hoboken, NJ 07030, USA

For details of our global editorial offices, customer services, and more information about Wiley products visit us at www.wiley.com.

Wiley also publishes its books in a variety of electronic formats and by print-on-demand. Some content that appears in standard print versions of this book may not be available in other formats.

Library of Congress Cataloging-in-Publication Data is applied for
Hardback: 9781119467403

Cover Design: Wiley
Cover Image: © Adrian Bejan

Set in 10.5/13.5pt TimesLTStd by Straive, Chennai, India

SKY10033386_022222

Contents

Preface

The discipline of heat transfer grew as a sequence of solved problems. From these roots the doctrine emerged as a predictive method, not as a random collection of solved problems. The first problems were the most fundamental and the simplest, and they bear the names of their creators: Fourier, Prandtl, Nusselt, Reynolds, and their contemporaries. As the field grew, the problems became more ad-hoc and applied (i.e. relevant to this, but not to that), more complicated and disunited, and much more numerous and forgettable.

Hidden in this voluminous stream are the fundamental principles that emerge. Identifying these and building with them the structure of the discipline is the main characteristic of the present book. I teach not only structure but also strategy:

The structure is drawn with sharp lines: heat transfer versus thermodynamics, conduction versus convection, convection versus radiation, external convection versus internal convection, forced convection versus natural convection, combined convection and conduction, phase change, and radiation.

The strategy is to start with the simplest method (scale analysis), and follow with more laborious and exact methods. Scale analysis is powerful because it teaches how to determine (on the back of an envelope) the proper orders of magnitude of all the physical features that matter (temperature difference, heat flux, fluid velocity, boundary layer thickness, lost power). It reveals the correct dimensionless groups, which are the fewest such numbers. With them, we learn how to predict and correlate in the most compact form the results obtained analytically, numerically, and experimentally.

This book is an *idea-book* that points toward the future of the discipline, in four ways:

(i) The relationship between configuration and performance. Traditionally, the fundamentals of heat transfer are taught by first postulating the configuration (the "boundary conditions") and then solving the governing equations. The resulting solution describes the flow fields (pressure, temperature, velocity) and the currents that flow on these fields in the assumed configuration. The solution permits the calculation of global features such as pressure drop and heat transfer rate, which are important in practice.

The key word is "describe". How these features affect the desired performance of the greater installation that uses that configuration is another matter. Addressed even less is how to discover the flow configuration in the first place. This is the new point of view from which this book teaches thermal sciences. It teaches how configuration affects performance, and how to configure the flows such that performance is enhanced. It establishes *"performance"* and *"evolution"* (the freedom to change[1]) as fundamental concepts in thermal sciences.

---

1 A. Bejan, *Freedom and Evolution: Hierarchy in Nature, Society and Science*, Springer Nature, New York, 2020.

(ii) New configurations are being developed, adopted, and joined by new arrivals. This is the universal phenomenon of evolution as physics, bio and non bio, which includes technology evolution. Chief among the growing population of designs are the tree-shaped vasculatures that bathe entire areas and volumes. They are transforming the field of smart, high-density and multi-functional materials, and bringing them close to animal design. This book shows how principles of physics such as the constructal law[2] predict a future with more degrees of freedom, economies of scale, multi-scale design, vascularization, and hierarchical distribution of many small features among the few large features.

(iii) The oneness of natural convection and forced convection is a central feature. Traditionally, natural convection is presented as being "free," i.e. unlike forced convection configurations for which people must constantly pay with power. In reality, natural convection too is driven by power from invisible "engines" that inhabit the flow field. These engines drive the flow, and because of fluid flow and heat transfer they dissipate their power output in internal "brakes" that account for the irreversibility of the relative motion between fluid layers, and the resistance overcome by heat currents.

(iv) Thermodynamics is an essential component in a heat transfer treatise (Figure 1). Why, because nothing moves and nothing flows (heat, fluid) unless it is driven. Power is dissipated throughout the flow and temperature fields. Flows do not happen unless differences and gradients (pressure, temperature) are sustained over time, so that power is generated in order to drive flows.

The book rests on material from *Heat Transfer* (1993), and it features new work (formulas, figures, problems, references) and the history of ideas and terminology. Words have meaning, and a very important word is optimization: the activity of making changes and choosing between the alternatives that emerge. To opt is to make a *choice*. To be able to choose, one must have *freedom* to change the existing configuration (design, organization) and to choose from the alternate configurations that emerge after the change. To optimize is a relentless activity of making choices when confronted with the endless number of forks in the road of evolution.

To choose is natural. Every moving thing, animate and inanimate, does it because of freedom. Every river alters its course and bed cross section to flow more easily. Every animal group varies its migration routes to

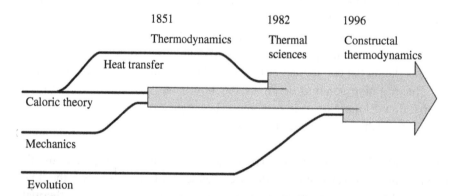

Figure 1 The evolution, spreading and merger of heat transfer with thermodynamics and evolution during the past two centuries.

---

2 For a finite-size flow system to persist in time (to live) it must evolve freely such that it offers greater access to what flows.

facilitate its movement, which is its life. Every wounded tissue heals itself in order to keep the whole body moving, which means to keep the whole alive.[3,4]

To change, choose and change again is the plot of the never ending movie called evolution. To find better choices is so good that it is addictive. The addiction reveals what "good" means. Freedom and design are concepts that belong in physics. They were placed firmly in physics as the constructal law (Figure 1).

With freedom new changes are made, more choices emerge, old bests die, and future bests are born. The best is short lived, precious today and derisory tomorrow. This truth is the mother of all evolution, including technology, and this book teaches it in terms of *evolutionary design* for heat and fluid flow and power.

January 2021   *Adrian Bejan*
Durham, NC

---

3 A. Bejan, *The Physics of Life: The Evolution of Everything*, St. Martin's Press, New York, 2016.
4 A. Bejan and J. P. Zane, *Design in Nature*, Doubleday, New York 2012.

About the Author

Adrian Bejan was awarded the 2018 Benjamin Franklin Medal "for his pioneering interdisciplinary contributions in thermodynamics and convection heat transfer that have improved the performance of engineering systems, and for constructal theory, which predicts natural design and its evolution in engineering, scientific, and social systems."

He received the 2019 Humboldt Research Award for lifetime achievement for "his pioneering interdisciplinary contributions to thermodynamics and Constructal Law – a law of physics that predicts natural design and its evolution in biology, geophysics, climate change, technology, social organization, evolutionary design and development, wealth and sustainability."

His degrees are from M.I.T.: BS (1971, Honors Course), MS (1972, Honors Course) and PhD (1975). He was a Fellow in the Miller Institute for Basic Research in Science at the University of California, Berkeley (1976–1978). At Duke University, he is the J. A. Jones Distinguished Professor. His research is in heat transfer, thermodynamics, and the physics that governs organization and evolution in nature.

Professor Bejan is the author of 700 peer-refereed journal articles and 30 books. His books are used worldwide in multiple editions and languages. He received the top international awards for thermal sciences. He is an honorary member of the American Society of Mechanical Engineers and a member of the Academy of Europe and the national academies of Mexico, Turkey, Moldova, and Romania.

He was awarded 18 honorary doctorates from universities in 11 countries, for example, the Swiss Federal Institute of Technology (ETH Zurich), the University of Rome I "La Sapienza," the National Institute of Applied Sciences (INSA) Lyon, and the University of Pretoria.

xvi

Acknowledgments

In this book, I benefited from the continuous creative support offered by Deborah Fraze and my most recent doctoral students and visiting professors: Umit Gunes, Abdulrahman Almerbati, Sinan Gucluer, Hitoshi Matsushima, and Hamad Almahmoud.

My spirits were kept high by Mary and our children Cristina, Teresa, and William. I am grateful for the continuous support from Wiley and my colleagues Sylvie Lorente, Pezhman Mardanpour, Tanmay Basak, Hooman Farzaneh, William Worek, Jose Lage, Jaime Cervantes, Abel Hernandez, Alexandru Morega, Heitor Reis, Bahri Sahin, Yousef Haseli, Adrian Sabau, and Shigeo Kimura.

List of Symbols

a_i	impurity-scattering resistivity coefficient (m·K^2/W), Eq. (1.26)
a_n	dimensionless characteristic values, Eq. (3.43) and Table 3.1
a_p	phonon-scattering resistivity coefficient (m/W·K), Eq. (1.26)
A	area (m^2)
A_c	cross-sectional area (m^2)
A_c	minimum free-flow area (m^2), Eq. (9.65)
A_{exp}	fin surface exposed to the fluid (m^2)
A_f	finned area (m^2), contributed by the exposed surfaces of the fins, Chapter 9
A_{fr}	frontal area (m^2), Eq. (9.65)
A_n	normal or projected area (m^2)
A_u	unfinned area (portions) of the wall (m^2), Chapter 9
A_0	bare surface, projected surface (m^2), Figure 2.10
$A_{0,f}$	finned portion of bare surface (m^2), Figure 2.10
$A_{0,u}$	unfinned portion of bare surface (m^2). Figure 2.10
b	thermal stratification parameter, Eq. (7.71)
b, b_T	transversal length scales (m), Eqs. (5.148)–(5.149)
b_n	dimensionless characteristic values, Table 4.1
B	cross-section shape number, Eq. (6.30)
B	driving parameter for film condensation, Eq. (8.26)
Be	Bejan number, $Be = \Delta P \cdot L^2 / \mu \alpha$, Appendix A
Bi	Biot number, Eq. (2.58)
Bo_y	Boussinesq number, $Bo_y = Ra_y Pr$
c	specific heat of incompressible substance (J/kg·K)
c	speed of light in vacuum, Appendix A
c_v	specific heat at constant volume (J/kg·K)
c_P	specific heat at constant pressure (J/kg·K)
C	capacity rate (W/K), or $\dot{m}c_P$
C_c	correction factor, Figure 10.27
C_D	drag coefficient, Eq. (5.138)
$C_{f,x}$	local skin friction coefficient, Eq. (5.39)
$\overline{C}_{f,x}$	average skin friction coefficient, Eq. (5.55)
C_{sf}	empirical constant for liquid–surface combination, Table 8.1

C_w	correction factor, Figure 10.29
d	diameter (m)
D	diameter (m)
D_h	hydraulic diameter (m), Eq. (6.28)
D_i	inner diameter (m)
D_o	outer diameter (m)
e	specific internal energy (J/kg), Eq. (5.12)
E	energy (J)
E	modulus of elasticity (N/m^2)
E	total hemispherical emissive power (W/m^2)
E_b	total hemispherical blackbody emissive power (W/m^2)
$E_{b,\lambda}$	monochromatic hemispherical blackbody emissive power (W/m·m^2)
Ec	Eckert number, Appendix A
E_λ	monochromatic hemispherical emissive power (W/m m^2)
f	factor, Figures 9.28 and 9.29
f	friction factor, Eq. (6.24)
f_v	vortex shedding frequency (s^{-1}), Eq. (5.137)
F	correction factor, Figures 9.10–9.14
F	force (N)
F	similarity streamfunction profile, Eq. (7.45)
F_D	drag force (N)
F_n	normal force (N)
Fo	Fourier number (dimensionless time), Eqs. (4.63)
F_r, F_θ, F_z	body forces per unit volume (N/m^3), Table 5.2
F_r, F_ϕ, F_θ	body forces per unit volume (N/m^3), Table 5.3
F_t	tangential force (N)
F_{12}	geometric view factor, Eq. (10.33)
g	gravitational acceleration (m/s^2)
G	mass velocity (kg/m^2·s), Eq. (9.58)
G	similarity vertical velocity profile, Eq. (7.43)
G	total irradiation (W/m^2)
Gr_y	Grashof number based on temperature difference and height y, $Gr_y = g\beta y^3\,\Delta T/\nu^2 = Ra_y/Pr$
Gz	Graetz number, Eq. (6.53)
$G_\mathcal{L}$	constant, Table 7.1
G_λ	monochromatic irradiation (W/m^2·m)
h	heat transfer coefficient for external flow (W/m^2·K), Eq. (1.50)
h	heat transfer coefficient for internal flow (W/m^2·K), Eq. (1.51)
h	Planck's constant (J·s), Appendix A
h	specific enthalpy (J/kg)
h_e	effective heat transfer coefficient (W/m^2·K), Eq. (9.7)
h_f	specific enthalpy of saturated liquid (J/kg)
h_{fg}	latent heat of condensation (J/kg), $h_g - h_f$

h'_{fg}	augmented latent heat of condensation (J/kg), Eqs. (8.10) and (8.17)
h''_{fg}	augmented latent heat of condensation (J/kg), Eq. (8.41)
h_g	specific enthalpy of saturated vapor (J/kg)
h_s	specific enthalpy of saturated solid (J/kg)
h_{sf}	latent heat of melting, or of solidification (J/kg), $h_f - h_s$
h_x	local heat transfer coefficient (W/m²·K) at position x
$\overline{h}_x$	average heat transfer coefficient (W/m²·K) averaged over length x
$\overline{h}_D$	heat transfer coefficient (W/m²·K) averaged over cylinder or sphere of diameter D
H	enthalpy (J)
H	enthalpy flowrate per unit length (W/m), Eq. (8.5)
H	height (m)
i	specific enthalpy (J/kg), Eq. (5.16)
I_b	total intensity of blackbody radiation (W/m²·sr)
$I_{b,\lambda}$	intensity of monochromatic blackbody radiation (W/m³·sr)
I_λ	intensity of monochromatic radiation (W/m³·sr)
j_H	Colburn j_H factor, Eq. (9.75)
J	electric current density (A/m²)
J	radiosity (W/m²)
Ja	Jakob number, Eq. (8.19)
J_0	zeroth-order Bessel function of the first kind, Figure 3.6 and Appendix E
J_1	first-order Bessel function of the first kind, Eq. (3.63)
k	Boltzmann's constant, Appendix A
k	thermal conductivity (W/m·K)
k_{avg}	average thermal conductivity (W/m·K), Eq. (2.50)
k_e	thermal conductivity due to conduction electrons (W/m·K)
k_l	thermal conductivity due to lattice vibrations (W/m·K)
k_i^{-1}	thermal resistivity due to impurity scattering (m·K/W)
k_p^{-1}	thermal resistivity due to phonon scattering (m·K/W)
k_s	sand roughness scale (mm), Figure 6.10
K	constant coefficient
K	permeability (cm²), Appendix B
K_c	contraction loss coefficient. Figures 9.23 and 9.24
K_e	enlargement loss coefficient, Figures 9.23 and 9.24
l	equivalent length (m), Eq. (7.84)
l	length (m)
l	mixing length (m), Eq. (5.109)
L	characteristic length (m), Eq. (7.76)
L	length (m)
$\mathcal{L}$	equivalent length (m), Eqs. (5.140) and (7.85)
L_c	corrected fin length (m), Eq. (2.95)
L_e	equivalent length (m), Table 10.5
L_0	Lorentz constant, 2.45×10^{-8} (V/K)²
m	fin parameter (m⁻¹), Eq. (2.75)

m	integer
m	mass (kg)
$\dot{m}$	mass flowrate (kg/s)
$\dot{m}'$	mass flowrate per unit length (kg/s·m)
$\dot{m}''$	mass flux (kg/m^2·s)
M	dimensionless factor, Eq. (6.84)
n	direction normal to the boundary (m), Table 3.2
n	integer
NTU	number of heat transfer units, Eq. (9.27)
Nu_x	Nusselt number based on the local heat transfer coefficient $h_x x/k$
$\overline{Nu}_D$	overall Nusselt number based on the surface-averaged heat transfer coefficient $\overline{h}_D D/k$, where D is the diameter
$\overline{Nu}_{\mathcal{L}}^0$	constant, Table 7.1
$\overline{Nu}_x$	overall Nusselt number based on the x-averaged heat transfer coefficient $\overline{h}_x x/k$
p	perimeter (m)
p	perimeter of contact with fluid (wetted perimeter) (m)
P	dimensionless parameter, Figures 9.10–9.14
P	pressure (Pa or N/m^2)
P	mechanical power (W)
Pe_D	Péclet number based on diameter, $U_\infty D/\alpha$, UD/α
Pe_x	Péclet number based on longitudinal length, $U_\infty x/\alpha$
Pr	Prandtl number, $Pr = v/\alpha$
Pr_t	turbulent Prandtl number, $Pr_t = \varepsilon_M/\varepsilon_H$
q	heat transfer rate (W)
q_b	total heat transfer rate through the fin (W)
q_{tip}	heat transfer through the tip of the fin (W)
q'	heat transfer rate per unit length (W/m)
q''	heat flux (W/m^2)
$\dot{q}$	volumetric rate of internal heat generation (W/m^3)
$q''_{w,x}$	local wall heat flux (W/m^2)
$\overline{q}''_{w,x}$	x-averaged wall heat flux (W/m^2), Eq. (5.80)
q_{1-2}	one-way heat current (W) from 1 to 2
q_{1-2}	net heat current (W) from 1 to 2
Q	heat transfer (J)
Q'	heat transfer interaction per unit length (J/m)
Q''	heat transfer interaction per unit area (J/m^2)
r	radial coordinate (m), Figures 1.7 and 1.8
r_i	inner radius (m)
r_o	outer radius (m)
$r_{o,c}$	critical outer radius (m), Eqs. (2.42) and (2.43)
r_s	thermal resistance of the scale (m^2·K/W), Eq. (9.5)
R	dimensionless parameter, Figures 9.10–9.14
R	radius (m)

R	function of r only, Chapter 3
R	ideal gas constant (kJ/kg·K), Appendix D
$\overline{R}$	universal ideal gas constant, Appendix A
Ra_y	Rayleigh number based on temperature difference and height y, $Ra_y = g\beta y^3\,\Delta T/\alpha v$
Ra_y^*	Rayleigh number based on heat flux and height y, $Ra_y^* = g\beta q_w'' y^4/\alpha v k$
Re	Reynolds number $V_{max}D_h/v$, Eq. (9.70)
Re_D	Reynolds number based on diameter, $U_\infty D/v$, UD/v
Re_l	local Reynolds number, Appendix F
Re_x	Reynolds number based on longitudinal length, $U_\infty x/v$
Re_y	condensate film Reynolds number, $4\Gamma(y)/\mu_l$, Eq. (8.22)
R_i	internal radiation resistance (m^{-2}), Eq. (10.78)
R_r	radition thermal resistance (m^{-2}), Eq. (10.45)
R_t	thermal resistance (K/W), Eq. (2.8)
s	empirical constant, Table 8.1
s_n	dimensionless characteristic values, Table 4.2
S	conduction shape factor, Eq. (3.33) and the header of Table 3.3
S	entropy (J/K)
St	x-independent Stanton number $h/\rho c_P U$
Ste	Stefan number, Eq. (4.119)
St_x	local Stanton number $h_x/\rho c_P U_\infty$
t	thickness (m)
t	time (s)
t_c	transition time scale (s), Eq. (4.9)
T	temperature (K or °C), Eqs. (1.5) and Figure 1.1
T_b	base temperature in fin analysis (K), Chapter 2
T_b	bulk, or mean temperature (K or °C)
T_c	center temperature (K), Chapter 4
T_i	initial temperature (K)
T_m	mean, or bulk temperature (K), Eq. (6.33)
T_m	melting point temperature (K)
T_{sat}	saturation temperature (K)
T_w	wall temperature (K or °C)
T_0	surface temperature (K), Chapter 4
T_0	reference temperature (K or °C)
T_∞	free-stream or reservoir temperature (K or °C)
u	specific internal energy (J/kg)
u	velocity component in the x direction (m/s)
U	average longitudinal velocity (m/s)
U	internal energy (J)
U	mean velocity (m/s), Eq. (6.1)
U	overall heat transfer coefficient (W/m^2·K)
U_∞	free stream velocity (m/s)
v	specific volume (m^3/kg)

v	velocity in the y direction (m/s)
v_n	normal velocity (m/s)
V	mean longitudinal velocity (m/s)
V	volume (m^3)
$\mathcal{V}$	volume (m^3), Chapter 9
w	mechanical transfer rate, or power (W)
W	width (m)
W	work transfer (J)
$\dot{W}$	work transfer rate, or power (W)
x	Cartesian coordinate (m), Figure 1.6
x_{tr}	transition length (m)
X	flow entrance length (m), Eqs. (6.4′) and (6.65)
X	function of x only, Chapter 3
X_l	longitudinal pitch (m)
X_t	transversal pitch (m)
X_T	thermal entrance length, Eqs. (6.32) and (6.65)
X_l^*	dimensionless longitudinal pitch X_l/D
X_t^*	dimensionless transversal pitch X_t/D
X, Y, Z	body forces per unit volume (N/m^3), Table 5.1
y	Cartesian coordinate (m), Figure 1.6
Y	function of y only, Chapter 3
Y_0	zeroth-order Bessel function of the second kind, Figure 3.6
z	axial position in cylindrical coordinates (m), Figure 1.7
z	Cartesian coordinate (m), Figure 1.6
Z	function of z only, Chapter 3

Greek Letters

α	heat transfer area density (m^2/m^3), Eq. (9.64)
α	thermal diffusivity (m^2/s), $\alpha = k/\rho c_P$
α	total absorptivity
α	total hemispherical absorptivity
α_0	temperature coefficient of electrical resistivity (°C^{-1}), Appendix B
α_λ	monocromatic hemispherical absorptivity
α_λ'	directional monochromatic absorptivity
β	coefficient of volumetric thermal expansion (K^{-1}), Eq. (5.18)
Γ	condensate mass flowrate per unit length (kg/s·m), Eq. (8.4)
δ	film thickness (m), Chapter 8
δ	skin thickness, boundary layer thickness (m)
δ	velocity boundary layer thickness (m), Eq. (5.25)
δ^*	displacement thickness (m), Eq. (5.58)
δ_s	thickness of shear layer (m), Eq. (7.38)
δ_T	thermal boundary layer thickness (m), Eq. (5.60)

δ_{99}	velocity boundary layer thickness (m), Eq. (5.57)
ΔP	pressure drop (N/m^2), Eq. (6.27)
ΔT	temperature difference (K)
ΔT_{lm}	log-mean temperature difference (K), Eqs. (6.105) and (9.22)
$\Delta \varepsilon$	correction term, Figure 10.30
ε	heat exchanger effectiveness, Eq. (9.29)
ε	overall surface efficiency, Eq. (9.4)
ε	total hemispherical emissivity
ε_f	fin effectiveness, Eq. (2.100)
ε_H	thermal eddy diffusivity (m^2/s), Eq. (5.100)
ε_M	momentum eddy diffusivity (m^2/s), Eq. (5.99)
ε_0	overall projected-surface effectiveness, Eq. (2.62)
ε_λ	monochromatic hemispherical emissivity
ε'_λ	directional monochromatic emissivity
η	fin efficiency, Eq. (2.98)
η	similarity variable
η_c	compressor isentropic efficiency
η_p	pump isentropic efficiency
θ	angular coordinate (rad), Figures 1.7 and 1.8
θ	excess temperature (K), Eq. (2.71)
θ	momentum thickness (m), Eq. (5.59)
θ	similarity temperature profile, Eq. (7.46)
θ	thermal potential function (W/m), Eq. (2.47)
θ_b	excess temperature of fin base (K)
κ	von Karman's constant, Eq. (5.112)
κ_λ	monochromatic extinction coefficient (m^{-1})
λ	characteristic value, Chapter 3
λ	dimensionless parameter in the Stefan solution, Eq. (4.118)
λ	wavelength (m)
μ	characteristic value, Chapter 3
μ	viscosity (kg/s·m)
ν	frequency (s^{-1})
ν	kinematic viscosity (m^2/s), $\nu = \mu/\rho$
ρ	density (kg/m^3)
ρ	total reflectivity
ρ_e	electrical resistivity (W·m/A^2)
σ	contraction ratio, Eq. (9.53)
σ	Stefan–Boltzmann constant, Appendix A
σ	surface tension (N/m), Table 8.2
σ_{xx}	normal stress (N/m^2), Eq. (5.8)
τ	angle of enclosure inclination (rad), Figure 7.18
τ	total transmissivity
$\tau_{w,x}$	local wall shear stress (N/m^2)

$\overline{\tau}_{w,x}$	x-averaged wall shear stress (N/m^2), Eq. (5.54)
τ_{xy}	tangential stress (N/m^2), Eq. (5.8)
τ_λ	monochromatic transmissivity
ϕ	angle of wall inclination (rad), Figure 7.7
ϕ	angular coordinate (rad), Figure 1.8
ϕ	dimensionless temperature profile, Eq. (6.45)
φ	porosity, Appendix B
Φ	viscous dissipation function (s^{-2}), Eq. (5.15)
Φ_i	mass fraction ρ_i/ρ
χ	correction factor, Figures 9.28 and 9.29
ψ	streamfunction (m^2/s), Eq. (5.45)
ω	solid angle (sr)

Subscripts

$()_a$	absorbed, Chapter 10
$()_{acc}$	acceleration
$()_{app}$	apparent
$()_{avg}$	average
$()_b$	base of the fin, Chapter 2
$()_b$	black, Chapter 10
$()_b$	bulk, mean
$()_c$	carbon dioxide, Chapter 10
$()_c$	centerline, center, midplane
$()_c$	cold
$()_c$	compressor
$()_{eddy}$	eddy transport
$()_f$	fluid
$()_g$	gas, Chapter 10
$()_h$	hot
$()_i$	initial
$()_i$	inner
$()_{in}$	initial
$()_{in}$	inlet
$()_l$	liquid
$()_{max}$	maximum
$()_{min}$	minimum
$()_{mol}$	molecular diffusion
$()_o$	outer
$()_{opt}$	optimal
$()_{out}$	outlet
$()_p$	pump
$()_r$	reflected, Chapter 10

$(\ )_{rad}$	radiation
$(\ )_{ref}$	reference
$(\ )_s$	shield, Chapter 10
$(\ )_s$	straight, Chapter 9
$(\ )_s$	surface, Chapter 10
$(\ )_{sat}$	saturation
$(\ )_v$	vapor
$(\ )_v$	water vapor
$(\ )_w$	wall
$(\ )_w$	water vapor, Chapter 10
$(\ )_0$	nozzle
$(\ )_0$	reference
$(\ )_0$	wall
$(\ )_\infty$	free stream

Superscripts

$(^-)$	time averaged, or volume averaged
$(\ )'$	fluctuation, Eq. (5.88)
$(\ )^+$	wall coordinates, Eq. (5.117)

About the Companion Website

This book is accompanied by a companion website:

www.wiley.com/go/bejan/heattransfer

This website includes:

- Solutions Manual

1

Introduction

1.1 Fundamental Concepts

1.1.1 Heat Transfer

Heat transfer Q (joules) is one kind of energy interaction between the system and its environment. The other kind, fundamentally different than Q, is work transfer W. The first-law statement for an infinitesimal process executed by a *closed system* is

$$\delta Q - \delta W = dE \tag{1.1}$$

where dE represents the change in the energy of the system. Energy is a thermodynamic property (function of state, path-independent quantity), whereas heat transfer and work transfer are not. The per-unit-time equivalent of Eq. (1.1) is

$$q - w = \frac{dE}{dt} \tag{1.2}$$

where q (watts) is the net *rate* of heat transfer experienced by the closed system. According to the thermodynamics sign convention [1], the value of q is positive when the heat transfer enters the system, that is, when the system is being heated by the other system (the environment) with which it communicates thermally.

The work transfer rate w is considered positive when the system does work on its environment. The work transfer can be of several origins: mechanical, electrical, and magnetic. For example, the electrical work per unit time (i.e. the power) supplied by an external battery for pushing an electric current through the system is represented by a negative w value.

The name of E is energy, not "thermal" energy as something supposedly different than "mechanical" energy. These are misnomers that betray the influence of caloric theory, which propagated to this day through the success of the science of heat transfer (cf. Preface, Figure 1).

The first law of thermodynamics does not make any distinction between heat transfer and work transfer – both are energy interactions (nonproperties) that must be distinguished from the energy change (property). The formal distinction between heat transfer and work transfer is made by the second law of thermodynamics, which for the same *closed system* and on a per-unit-time basis states that

$$\sum_i \frac{q_i}{T_i} \leq \frac{dS}{dt} \tag{1.3}$$

where S is the entropy inventory of the system (a property) and T_i is the Kelvin or Rankine temperature of the system boundary point i that is crossed by q_i. Each term of the type q_i/T_i represents the entropy transfer rate (watts/K) associated with the heat transfer rate q_i and the boundary point of temperature T_i.

Heat Transfer: Evolution, Design and Performance, First Edition. Adrian Bejan.
© 2022 John Wiley & Sons, Inc. Published 2022 by John Wiley & Sons, Inc.
Companion website: www.wiley.com/go/bejan/heattransfer

The inequality sign (1.3) accounts for the phenomenon of *irreversibility*: everything, by itself, flows from high to low. Heat flows from high temperature to low temperature. Cooling an object that is colder than the ambient requires refrigerator power. Fluid flows from high pressure to low pressure. Forcing a liquid to flow through a pipe from low pressure to high pressure requires pump power. Mixing goes one way, from unmixed to mixed. Separating the components that constitute the mixed requires external effort from the environment. Cleaning requires a greater effort than polluting.

Comparing Eqs. (1.2) and (1.3), we see the distinction between heat transfer and work transfer; only heat transfer appears in the second law. Work interaction carries zero entropy, whereas heat transfer is accompanied by entropy transfer q_i/T_i.

Historically, the heat transfer definition was based on a phenomenological description – it is the energy interaction driven by the temperature difference between the system and its environment. Work transfer was defined as in mechanics – it is equal to the force experienced by the system times the displacement of the point of application of that force.

1.1.2 Temperature

First, let's review what "equilibrium" means. Consider two closed systems whose boundaries are such that both systems cannot experience work transfer (e.g. two arbitrary amounts of air sealed in rigid containers), where "arbitrary" means that the mass, volume, and pressure of each system are not specified. If the systems are positioned sufficiently close to one another, we observe that changes occur in both systems. The changes can be documented by recording the air pressure versus time. The condition of the closed system is said to be one of *equilibrium* when, after a sufficiently long period, changes cease to occur inside the system. And, since this particular closed system is incapable of experiencing work transfer interactions, the long-time condition illustrated earlier is *thermal equilibrium.*

Let A and B be the closed systems that interact and reach thermal equilibrium in the preceding example. The same experiment can be repeated by using system A and a third system, C, which is also closed and unfit for work transfer. It is also a matter of common experience that if systems B and C are individually in thermal equilibrium with system A, then, when placed in direct communication, systems B and C do not undergo any changes as time passes. This second observation is summarized as follows. If systems B and C are separately in thermal equilibrium with a third system, then they are in thermal equilibrium with each other. This statement is recognized as the zeroth law of thermodynamics [1].[1]

Temperature T is the system property that determines whether the system is in thermal equilibrium with another system. Two systems, A and B, are in mutual thermal equilibrium when their temperatures are identical, $T_A = T_B$.

The temperature of a system is measured by placing the system in thermal communication with a special system (a test system) called *thermometer*. The thermometer must be sufficiently smaller than the system so that the heat transfer interaction with the thermometer is negligible from the point of view of the system. The thermometer, on the other hand, is designed so that the same heat transfer interaction leads to measurable effects such as changes in volume or electrical resistance.

The temperature scales that are currently being employed are displayed in Figure 1.1. Of practical importance are the relationships between the *thermodynamic* temperatures recorded on the Kelvin and Rankine

---

1 Numbers in square brackets indicate references at the end of each chapter.

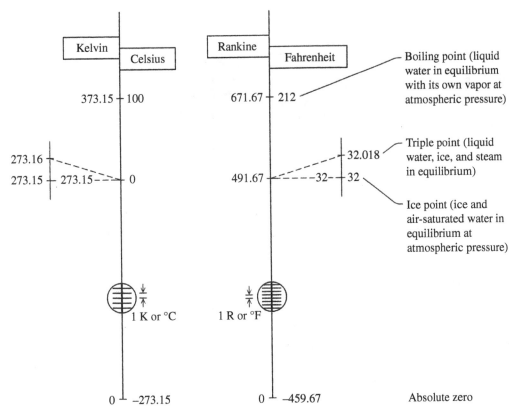

Figure 1.1 The relationships between the Kelvin, Celsius, Rankine, and Fahrenheit temperature scales. Source: Bejan [1].

scales and on the Celsius and Fahrenheit scales:

$$T(^{\circ}C) = T(K) - 273.15 \qquad T(^{\circ}F) = T(R) - 459.67$$

$$T(R) = \frac{9}{5}T(K) \qquad\qquad T(^{\circ}C) = \frac{5}{9}[T(^{\circ}F) - 32] \qquad\qquad (1.4)$$

Figure 1.1 shows that the Kelvin (or Celsius) degree is larger than the Rankine (or Fahrenheit) degree, $(1\,K,\ \text{or}\ 1\,^{\circ}C) = \frac{9}{5}(1\,R,\ \text{or}\ 1\,^{\circ}F)$. Note that 1 K is equal to 1 °C, yet most books of heat transfer and fluid mechanics report the problem answers for temperature in °C, not K. The reason is that most common temperature answers are for use in applications at (or above) room temperature, where thermometry is in °C, not K. These applications include combustion and metallurgy. In cryogenics, temperature answers to problems are for applications near liquid helium (4 K) and liquid nitrogen (77 K), because thermometry at low temperatures is in K, not negative °C. Deep cold did not exist when the °C and °F scales were put on thermometers. Furthermore, the temperature effect on thermophysical properties (e.g. Eq. (1.23)) is expressed as $T\,[K]$ raised to an exponent of order 1 and in cryogenics that temperature factor varies greatly when $T\,[K]$ varies just a little (that is not the case at room temperature).

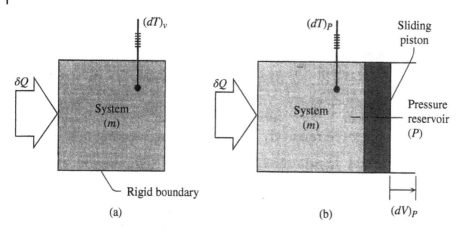

Figure 1.2 Two heating processes for measuring (a) the specific heat at constant volume and (b) the specific heat at constant pressure.

1.1.3 Specific Heats

Two thermodynamic concepts that are essential in the study of heat transfer are the specific heat at constant pressure c_P and the specific heat at constant volume c_v. In general, the two specific heats of a substance are different, the larger of the two being c_P, which can be measured during a process of (isobaric) heating in the apparatus shown in Figure 1.2b. One can measure the heat input δQ, the temperature rise of the material sample dT, and the volume expansion dV. The definition of c_P is

$$c_P = \left(\frac{\delta Q}{m \, dT} \right)_P \tag{1.5}$$

where m is the mass of the sample and, according to the first law of thermodynamics, $\delta Q = dU + \delta W$. Furthermore, since P is uniform throughout the system, the work transfer during the process is $\delta W = P \, dV$. Because P remains constant, the first law can be rewritten as $\delta Q = d(U + PV) = dH$, where dH is the enthalpy increase experienced by the sample. Recalling the specific enthalpy definition h, the c_P definition (1.5) becomes

$$c_P = \left(\frac{\partial h}{\partial T} \right)_P \tag{1.6}$$

The specific heat at constant volume is defined based on the isochoric heating process described in Figure 1.2a. Measuring the heat transfer into the sample (δQ) and the temperature rise (dT), c_v is calculated by

$$c_v = \left(\frac{\delta Q}{m \, dT} \right)_v \tag{1.7}$$

Invoking the first law for the heating process (and noting that the work transfer is zero), $\delta Q = dU$, we arrive at

$$c_v = \left(\frac{\partial u}{\partial T} \right)_v \tag{1.8}$$

where u is the specific internal energy of the sample.

For any pure substance, c_P and c_v are, in general, functions of both temperature and pressure. Two special limits of thermodynamic behavior are particularly important because in these domains the specific heats depend only on the temperature. The *ideal gas* domain is the highly compressible limit, and the functions $c_P(T)$ and $c_v(T)$ are related through Robert Mayer's equation [1]:

$$c_P(T) - c_v(T) = R \tag{1.9}$$

where R is the ideal gas constant of the substance that is being considered. In the *incompressible substance* limit, the specific heats are equal:

$$c_P(T) = c_v(T) = c(T) \tag{1.10}$$

and the symbol for their common value is c. Incompressible substances that occur frequently in this course are the solid bodies analyzed in Section 1.3 and the liquids whose flow is studied in Section 1.4.

1.2 The Objective of Heat Transfer

In the discipline of heat transfer, the objective is to describe the manner in which the dissimilarity between T_A and T_B governs the magnitude of the heat transfer *rate* between the system (A) and its environment (B). In general, heat transfer rate is influenced not only by the fact that T_A and T_B are different but also by the physical configuration formed by the heat-exchanging entities:

$$q = \text{function } (T_A, T_B, \text{ time, and, for both A and B, thermophysical properties, size, geometric shape,}$$

$$\text{relative movement, or flow}) \tag{1.11}$$

According to the definition of thermal equilibrium and temperature, the heat transfer rate function has the special value

$$q = 0 \quad \text{when} \quad T_A = T_B \tag{1.12}$$

Simpler approximations of the heat transfer function (1.11) will be developed by focusing on the three specific modes of heat transfer: conduction, convection, and radiation.

The discovery of the heat transfer rate function is just the starting point. The performance is the real objective. Even though the problems we encounter are extremely diverse, we can see the generality of Eq. (1.11) by focusing on three relatively large classes of performance:

1. **Thermal Insulations**. In this class the two temperatures (T_A, T_B) that drive the heat transfer rate are fixed. The key unknown is the heat transfer rate q, which is also named "heat loss" or "heat leak." In *thermal design* the objective is to minimize q, while keeping in mind certain economic and geometric constraints, such as the total cost of the heat transfer medium (the "insulation") and the total volume occupied by the system (the "size"). The thermal design activity consists of changing the constitution of the insulation (size, material, shape, structure, flow pattern) so that q decreases while T_A and T_B remain fixed. The behavior of the design (the configuration) is evolutionary.

2. **Heat Transfer Enhancement (Augmentation)**. In heat exchanger design, the heat transfer rate between the two streams is usually a prescribed quantity. Desirable from the thermodynamic performance standpoint is the flow of q across a smaller temperature difference $T_A - T_B$. This way the rate of entropy generation (or, proportionally, the destruction of useful mechanical power) is reduced [1]. The unknown in Eq. (1.11) is the temperature difference; the objective is to improve the thermal contact between the heat-exchanging entities, that is, to reduce the temperature difference $T_A - T_B$. This can be done by

changing the flow patterns of the two streams and the shapes of the solid surfaces bathed by the fluid streams (by using fins, for example). The behavior of the design (the configuration) is evolutionary.

3. **Temperature Control**. There are several other applications in which the main concern is the overheating of the hot surface (T_A) that produces the heat transfer rate q. In a dense package of electronics, q is generated by Joule heating, while the heat sink temperature T_B is provided by the ambient (e.g. a stream of atmospheric air). The temperature of the electrical conductors (T_A) cannot rise too much above the ambient temperature, because high temperatures threaten the error-free operation of the electrical circuitry. The heat transfer rate and flow configuration must vary in such a way that T_A does not exceed a certain ceiling value. Examples of temperature control applications include the cooling of nuclear reactor cores and the cooling of the outer skin of space vehicles during reentry. The high temperature T_A must be kept below that critical level where the mechanical strength characteristics of the hot surface deteriorate (e.g. the melting point).

1.3 Conduction

1.3.1 The Fourier Law

Consider the *solid* bar of material illustrated in Figure 1.3. The four long sides of this object are adiabatic (i.e. insulated perfectly) so that, if present, the transfer of heat will take place only in the longitudinal direction x. Any slice of thickness Δx in this bar constitutes a closed system in the thermodynamic sense used in the preceding section, because no mass is flowing through any of its six sides. Applied to the Δx system, the first law of thermodynamics, Eq. (1.2), states that

$$q_x - q_{x+\Delta x} - w = \frac{\partial E}{\partial t} \tag{1.13}$$

in which the heat transfer rate q_x points in the positive x direction. Equation (1.13) will be modified in three respects.

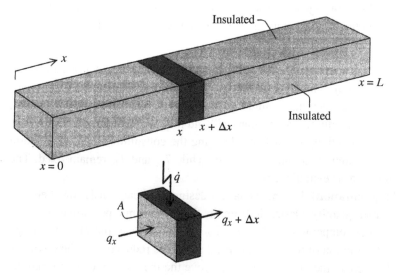

Figure 1.3 Unidirectional conduction through a solid body with internal heat generation.

First, the only energy inventory present in the stationary bar is the internal energy:

$$E = (\rho A \, \Delta x)u \qquad (1.14)$$

where u is the specific internal energy, ρ is the density of the solid material, and $\rho A \, \Delta x$ is the mass of the system. We recall that the internal energy change of an incompressible substance is proportional to its temperature change [1]:

$$du = c \, dT \qquad (1.15)$$

where c is the lone specific heat of the solid, Eq. (1.10). Combining Eqs. (1.14) and (1.15), we replace the right side of Eq. (1.13) with

$$\frac{\partial E}{\partial t} = \rho c A \, \Delta x \frac{\partial T}{\partial t} \qquad (1.16)$$

In this statement we assumed that the temperature variation along the bar is sufficiently small so that the specific heat c may be treated as a constant (in general, c is a function of temperature). The partial derivative sign is used because the bar temperature is generally a function of both t and x.

The second modification of Eq. (1.13) stems from the observation that in most cases the sign of the work transfer rate w is negative. For example, the rate of internal (volumetric) heating caused by an electric current that passes through the bar (i.e. the dissipation of electrical power) is represented by $-w$, not w. For this reason, we write

$$-w = (A \Delta x)\dot{q} \qquad (1.17)$$

and recognize $\dot{q}$ (W/m^3) as the volumetric rate of internal heat generation in the solid. Equation (1.17) applies to situations such as resistive (Joule) heating, where the mechanical power is transformed into a volumetric heating effect. In general, the work transfer rates that contribute to w in Eq. (1.2) can have additional effects such as the compression or stretching of the solid medium.

Third, guided by the observation that the function q vanishes when the medium is isothermal, Eq. (1.12), we *assume* that q_x is proportional to the local temperature difference in the x direction, $q_x = C(T_x - T_{x+\Delta x})$. It is found experimentally that the proportionality factor C is itself proportional to $A/\Delta x$, that is, $C = kA/\Delta x$, where k is the *thermal conductivity* coefficient and C is the thermal conductance. In the limit $\Delta x \to 0$, the assumed expression for the local heat current becomes

$$q_x = -kA \frac{\partial T}{\partial x} \qquad (1.18)$$

This assumption is recognized as the *Fourier law of heat conduction*[2] or the Fourier law of thermal diffusion. It is not a law of physics. It is an empirical relation (a correlation of measurements) that serves as definition for the thermal conductivity coefficient k, which must be measured experimentally. Equation (1.18) and the thermal conductivity coefficient were first introduced by Biot [2].

The positive values that are exhibited by k (e.g. Figure 1.5), coupled with the negative sign of the right side of Eq. (1.18), are the fingerprint of the second law of thermodynamics, which demands that q_x must always

---

2 Jean Baptiste Joseph Fourier (1768–1830) was a French mathematician and public servant (governor, prefect). He developed the general methodology (Fourier series, Chapters 3 and 4) for solving problems of heat conduction. Arguably the founder of the modern discipline of heat transfer, Fourier also had a great impact on the development of the field of applied mathematics.

flow in the direction of lower temperatures. Finally, the Fourier law can be used to rewrite the second term of Eq. (1.13) as

$$q_{x+\Delta x} = q_x + \frac{\partial q_x}{\partial x} \Delta x \quad \text{(Taylor series)}$$

$$= -A \left[k\frac{\partial T}{\partial x} + \frac{\partial}{\partial x}\left(k\frac{\partial T}{\partial x} \right) \Delta x \right] \tag{1.19}$$

After these simplifications, Eq. (1.13) becomes a partial differential equation (also called "conduction" equation) for the temperature function $T(x, t)$:

$$\underbrace{\frac{\partial}{\partial x}\left(k\frac{\partial T}{\partial x} \right)}_{\text{Net longitudinal conduction}} + \underbrace{\dot{q}}_{\text{Internal heat generation}} = \underbrace{\rho c\frac{\partial T}{\partial t}}_{\text{Thermal inertia}} \tag{1.20}$$

This equation shows that *three* effects compete in the energy balance at any point inside the bar: the internal heat generation, the retarding effect of thermal inertia, and the *net* transfer of heat by longitudinal conduction. The word "net" means the *difference* between the conduction current that arrives at a given x and the conduction current that leaves. The units of all the terms appearing in Eq. (1.20) are W/m^3. Thermal "inertia" means that a finite sample must be the recipient of net heat transfer if its temperature is to rise. When the net heat transfer input is fixed, the temperature rises faster when the group ρc is smaller. The group ρc represents the thermal inertia per unit of sample volume or the specific heat capacity of the medium.

If the variation of temperature along the bar is small enough so that the thermal conductivity may be treated as a constant, the one-dimensional conduction equation, Eq. (1.20), assumes the simpler form

$$\frac{\partial^2 T}{\partial x^2} + \frac{\dot{q}}{k} = \frac{1}{\alpha}\frac{\partial T}{\partial t} \tag{1.21}$$

The new coefficient here is the *thermal diffusivity* of the material:

$$\alpha = \frac{k}{\rho c} \tag{1.22}$$

In Eq. (1.21), the thermal diffusivity α is assumed constant, because ρ, c, and k have been approximated as temperature independent.

The conduction equation, Eq. (1.20), was developed with reference to the solid material shaped as a bar in Figure 1.3. The same equation applies to a column of incompressible and motionless liquid or gas. A completely analogous derivation of the conduction equation can be constructed for a one-dimensional column of fluid that is not incompressible and whose pressure is constant and uniform. When the temperature variation along the bar is sufficiently small so that the local dilation of the fluid does not induce a significant movement (flow) in the x direction, the conduction equation is the same as Eq. (1.20), except that c is replaced by c_P, which is also treated as a constant.

1.3.2 Thermal Conductivity

The thermal conductivity k can be determined by using the definition (1.18) and a conducting column such as in Figure 1.3, where one can measure both the heat transfer rate and the temperature gradient. In the most general case, the measured k value will depend not only on the local thermodynamic state of the material sample (e.g. temperature, pressure) but also on the orientation of the sample relative to the heat current q and on the point inside the sample where the k measurement is being performed (for example, the position x in

Figure 1.4 Classification of thermally conducting media in terms of their homogeneity and isotropy:
(a) nonhomogeneous and anisotropic,
(b) nonhomogeneous and isotropic, (c) homogeneous and anisotropic, and (d) homogeneous and isotropic.

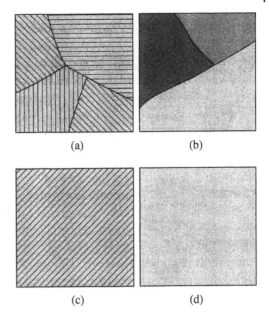

(a)   (b)

(c)   (d)

Figure 1.3). This general case is illustrated in Figure 1.4a, in which the conducting material is anisotropic and nonhomogeneous.

The remainder of Figure 1.4 shows three classes of materials for which the k function revealed by experiments is progressively simpler. In Figure 1.4b, the material is isotropic and nonhomogeneous. In this case the k value depends on the point where the measurement is made, but not on how the material is oriented relative to the heat current.

A homogeneous and anisotropic material is illustrated in Figure 1.4c. Crystalline solids, meat, wood, and the windings of electrical machines can be described in this manner, provided the distance between adjacent fibers or laminae is much smaller than the size of the conducting sample. In such cases the measured k value depends on the orientation of the sample and on thermodynamic properties such as the temperature, but not on the point of measurement.

Most of the thermal conductivity data stored in handbooks refer to homogeneous and isotropic materials (Figure 1.4d). The empirical relations between k and T are illustrated in Figure 1.5. Important is that the temperature can have a sizeable effect on k when the heat-conducting system occupies a wide range on the absolute temperature scale. The thermal conductivity differentiates between materials known as "good conductors" and "poor conductors"; from the top to the bottom of Figure 1.5, the k values decrease by 6 orders of magnitude.

Several theories account for the temperature trends exhibited in Figure 1.5. For low pressure (ideal) gases, kinetic theory [6] argues that the energy transport that is represented by the Fourier law, Eq. (1.18), has its origin in the collisions between gas molecules. In the case of monatomic gases, the thermal conductivity is expected to depend on T:

$$k = k_0 \left(\frac{T}{T_0} \right)^n \tag{1.23}$$

such that the exponent is $n = \frac{1}{2}$ and k_0 is the conductivity measured at a reference temperature, T_0. In reality, the n exponent of a curve such as that of helium in Figure 1.5 is somewhat larger than the theoretical value,

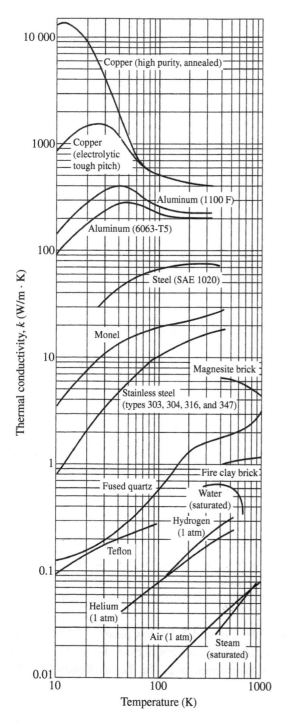

Figure 1.5 Dependence of thermal conductivity on temperature (the *k* data are from Refs. [3–5]). Source: Scott [3]; Rohsenow and Choi [4]; Bejan [5].

$n \simeq 0.7$. The merit of the power law expression (1.23) is that, with an appropriate exponent n, it can be fitted to the conductivity data of any other gas, so as to obtain a compact $k(T)$ formula that holds over a wide temperature range. In low-pressure monatomic gases, the gas density is proportional to P/T, while c_P is constant [1]. Consequently, the thermal diffusivity expression that corresponds to Eq. (1.23) is

$$\alpha = \alpha_0 \left(\frac{T}{T_0}\right)^{n+1} \left(\frac{P}{P_0}\right)^{-1} \tag{1.24}$$

showing that α depends on both T and P and increases more rapidly than k as T increases.

The thermal conductivity of metallic solids is attributed to the movement of conduction electrons (the "electron gas"), k_e, and the effect of lattice vibrations, k_l, energy quanta of which are called phonons, $k = k_e + k_l$. In metals, the electron movement plays the dominant role, such that as a very good approximation $k \cong k_e$. The movement of the conduction electrons is impeded by scattering, which is the result of the interactions between electrons and phonons and between electrons and impurities and imperfections (e.g. fissures, boundaries) that exist in the material. These electron-scattering mechanisms are accounted for in an additive-type formula for the *thermal resistivity* (the inverse of thermal conductivity):

$$\frac{1}{k_e} = \frac{1}{k_p} + \frac{1}{k_i} \tag{1.25}$$

in which, according to electron conduction theory [7], the phonon-scattering resistivity (k_p^{-1}) and the impurity-scattering resistivity (k_i^{-1}) depend on thermodynamic temperature:

$$\frac{1}{k_p} = a_p T^2 \quad \text{and} \quad \frac{1}{k_i} = \frac{a_i}{T} \tag{1.26}$$

The coefficients a_p and a_i are two characteristic constants of the metal. Equation (1.26) shows that at low temperatures the thermal resistivity is due primarily to impurity scattering, while the effect of phonon scattering plays an important role at high temperatures. Putting Eqs. (1.25) and (1.26) together, one can see that the thermal conductivity of a metal obeys the relation

$$k = \frac{1}{a_p T^2 + (a_i/T)} \tag{1.27}$$

which reveals a conductivity maximum at a characteristic temperature:

$$k_{\max} = \frac{3}{2^{2/3}} a_p^{1/3} a_i^{2/3} \quad \text{at} \quad T = \left(\frac{a_i}{2a_p}\right)^{1/3} \tag{1.28}$$

The k maximum shifts toward higher temperatures as the impurity-scattering effect a_i increases. These features are most evident in the shapes of the $k(T)$ curves of copper, Figure 1.5, in which the impurity content increases as we shift from the "high purity" curve to the "electrolytic tough pitch" curve. The maximum disappears entirely from the thermal conductivity curves of highly impure alloys such as stainless steel. In such cases the impurity-scattering resistivity overwhelms the phonon-scattering effect over a much wider temperature domain. Consequently, the simple formula

$$k \cong \frac{T}{a_i} \tag{1.29}$$

is a good fit for the conductivity data at temperatures below room temperature.

Another result of the electron conduction theory stems from analogies between the electron transport of energy (thermal diffusion) and the transport of electricity (electrical diffusion). This result is recognized as the *Wiedemann–Franz law* [7]:

$$k\frac{\rho_e}{T} = L_0 \quad \text{(constant)} \tag{1.30}$$

where ρ_e is the electrical resistivity of the metal. Recall that the electrical resistance of a conductor of length L and cross-sectional area A_c is $\rho_e L/A_c$, and that the units of ρ_e are [$W \cdot m/A^2$]. The Lorentz constant, $L_0 = 2.45 \times 10^{-8}(V/K)^2$, is a universal constant of all metals. Equation (1.30) holds particularly well in the two temperature extremes, at cryogenic temperatures and at temperatures well above room temperature. At intermediate temperatures, the k value calculated with Eq. (1.30) overestimates the measured thermal conductivity. The agreement between the Wiedemann–Franz law and k measurements at intermediate temperatures improves considerably as the impurity of the metal increases; therefore, Eq. (1.30) is a good approximation for the thermal conductivity of an impure metal over the entire temperature range in most applications.

The practical value of Eq. (1.30) is that it allows us to estimate the thermal conductivity based on an electrical resistivity measurement, that is, based on a considerably simpler measurement. Furthermore, the Wiedemann–Franz law shows that when the thermal conductivity is nearly independent of temperature (e.g. copper above room temperature, Figure 1.5), the electrical resistivity increases almost linearly with the temperature. This behavior has important consequences in the design of electrical cables that are thermally "stable," that is, safe against the threat of burn-up or thermal runaway instability.

The thermal conductivity coefficient k is a *macroscopic*, or aggregate, fingerprint of phenomena that occurs at the molecular level. In this sense, k is similar to the properties (u, h, s, T, P, etc.) encountered in thermodynamics; the sample of conducting material whose k value is measured contains an immense number of molecules so that the sample can be treated as a *continuum*. The heat transfer theory that stands behind the conduction and convection parts of the present treatment is based on the view that the medium is a continuum.

1.3.3 Cartesian Coordinates

The energy conservation statement, Eq. (1.20), can be generalized. Switching from Figure 1.3 to Figure 1.6, we recognize the Cartesian system of coordinates (x,y,z) attached to the body, which is the focus of this analysis. Note the agreement – the good geometric fit – between the overall shape of the body and the coordinate system. The shape of the body demands that we use a Cartesian system, not a cylindrical system (Figure 1.7) or a spherical system (Figure 1.8). The problem solver will learn to appreciate this observation, for the heat conduction analysis of a parallelepiped is the simplest in a Cartesian system. The first law for the infinitesimal closed system of size $\Delta x \Delta y \Delta z$ is

$$(q_x'' - q_{x+\Delta x}'')\Delta y \Delta z + (q_y'' - q_{y+\Delta y}'')\Delta x \Delta z$$
$$+ (q_z'' - q_{z+\Delta z}'')\Delta x \Delta y + \dot{q}\Delta x \Delta y \Delta z = \rho c \Delta x \Delta y \Delta z \frac{\partial T}{\partial t} \tag{1.31}$$

where each q'' term represents a *heat flux* or heat transfer rate per unit area (W/m^2). In Eq. (1.18), for example, the corresponding heat flux would have been $q_x'' = dq_x/dA$.

The Fourier law can be invoked for each of the heat fluxes shown in Figure 1.7:

$$q_x'' = -k\frac{\partial T}{\partial x}, \quad q_y'' = -k\frac{\partial T}{\partial y}, \quad q_z'' = -k\frac{\partial T}{\partial z} \tag{1.32}$$

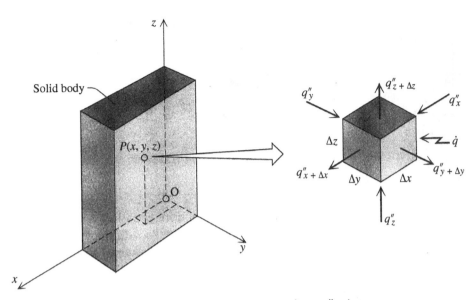

Figure 1.6 Three-dimensional Cartesian system of coordinates.

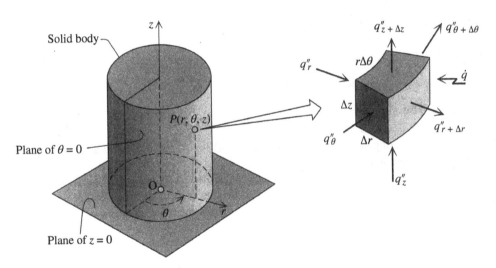

Figure 1.7 Cylindrical system of coordinates.

and then substituted into Eq. (1.35). Dividing the resulting equation by $\Delta x \Delta y \Delta z$ and invoking the limits $\Delta x \to 0$, $\Delta y \to 0$, and $\Delta z \to 0$, we obtain the equation for energy conservation at a point in a Cartesian frame:

$$\frac{\partial}{\partial x}\left(k\frac{\partial T}{\partial x}\right) + \frac{\partial}{\partial y}\left(k\frac{\partial T}{\partial y}\right) + \frac{\partial}{\partial z}\left(k\frac{\partial T}{\partial z}\right) + \dot{q} = \rho c \frac{\partial T}{\partial t} \tag{1.33}$$

This form is useful in applications where all of its terms have a meaning, namely, (i) the temperature is unsteady (time dependent), (ii) the internal heating effect $\dot{q}$ is present, and (iii) the thermal conductivity is not constant. Simpler versions of Eq. (1.33) apply to the key configurations covered by the present treatment:

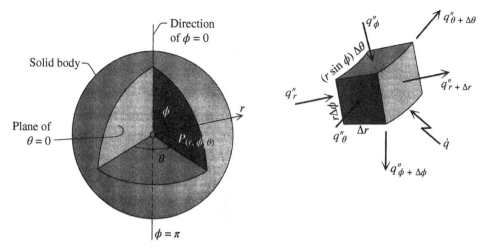

Figure 1.8 Spherical system of coordinates.

Unsteady, constant conductivity, with internal heat generation:

$$\frac{\partial^2 T}{\partial x^2} + \frac{\partial^2 T}{\partial y^2} + \frac{\partial^2 T}{\partial z^2} + \frac{\dot{q}}{k} = \frac{1}{\alpha}\frac{\partial T}{\partial t} \tag{1.34}$$

Unsteady, constant conductivity, without internal heat generation:

$$\frac{\partial^2 T}{\partial x^2} + \frac{\partial^2 T}{\partial y^2} + \frac{\partial^2 T}{\partial z^2} = \frac{1}{\alpha}\frac{\partial T}{\partial t} \tag{1.35}$$

Steady state, constant conductivity, without internal heat generation:

$$\frac{\partial^2 T}{\partial x^2} + \frac{\partial^2 T}{\partial y^2} + \frac{\partial^2 T}{\partial z^2} = 0 \tag{1.36}$$

The sum of the first three terms appearing on the left side of Eq. (1.34) is often abbreviated as $\nabla^2 T$, where the $\nabla^2(\ )$ notation is the Laplacian operator:

$$\nabla^2 = \frac{\partial^2}{\partial x^2} + \frac{\partial^2}{\partial y^2} + \frac{\partial^2}{\partial z^2} \tag{1.37}$$

1.3.4 Cylindrical Coordinates

The terminology for the heat conduction in the cylindrical coordinate system (r,θ,z) is defined in Figure 1.7. Note again the good fit between the overall shape of the conducting body and the position of the coordinate system: the z axis coincides with the axis of symmetry of the body. The infinitesimal closed system that contains the arbitrary point P has the volume $r\Delta r\Delta\theta\Delta z$. In place of the Fourier law (1.32), we have the three heat fluxes in cylindrical coordinates:

$$q_r'' = -k\frac{\partial T}{\partial r}, \quad q_\theta'' = -\frac{k}{r}\frac{\partial T}{\partial \theta}, \quad q_z'' = -k\frac{\partial T}{\partial z} \tag{1.38}$$

The energy conservation equation for unsteady conduction in a body with internal heat generation and variable conductivity becomes

$$\frac{1}{r}\frac{\partial}{\partial r}\left(kr\frac{\partial T}{\partial r}\right) + \frac{1}{r^2}\frac{\partial}{\partial \theta}\left(k\frac{\partial T}{\partial \theta}\right) + \frac{\partial}{\partial z}\left(k\frac{\partial T}{\partial z}\right) + \dot{q} = \rho c\frac{\partial T}{\partial t} \tag{1.39}$$

The special forms of this conduction equation are the following:

Unsteady, constant conductivity, with internal heat generation:

$$\frac{1}{r}\frac{\partial}{\partial r}\left(r\frac{\partial T}{\partial r}\right) + \frac{1}{r^2}\frac{\partial^2 T}{\partial \theta^2} + \frac{\partial^2 T}{\partial z^2} + \frac{\dot{q}}{k} = \frac{1}{\alpha}\frac{\partial T}{\partial t} \tag{1.40}$$

Unsteady, constant conductivity, without internal heat generation:

$$\frac{1}{r}\frac{\partial}{\partial r}\left(r\frac{\partial T}{\partial r}\right) + \frac{1}{r^2}\frac{\partial^2 T}{\partial \theta^2} + \frac{\partial^2 T}{\partial z^2} = \frac{1}{\alpha}\frac{\partial T}{\partial t} \tag{1.41}$$

Steady state, constant conductivity, without internal heat generation:

$$\frac{1}{r}\frac{\partial}{\partial r}\left(r\frac{\partial T}{\partial r}\right) + \frac{1}{r^2}\frac{\partial^2 T}{\partial \theta^2} + \frac{\partial^2 T}{\partial z^2} = 0 \tag{1.42}$$

On the left side of Eq. (1.42) is the Laplacian operator in cylindrical coordinates:

$$\nabla^2 = \frac{\partial^2}{\partial r^2} + \frac{1}{r}\frac{\partial}{\partial r} + \frac{1}{r^2}\frac{\partial^2}{\partial \theta^2} + \frac{\partial^2}{\partial z^2} \tag{1.43}$$

such that the abbreviated version of Eq. (1.42) is $\nabla^2 T = 0$.

1.3.5 Spherical Coordinates

Recommended for the study of conduction in objects bounded by spherical surfaces is the spherical system of coordinates (r, ϕ, θ) defined in Figure 1.8. The relations between heat fluxes and temperature gradients are

$$q''_r = -k\frac{\partial T}{\partial r}, \quad q''_\phi = -\frac{k}{r}\frac{\partial T}{\partial \phi}, \quad q''_\theta = -\frac{k}{r\sin\phi}\frac{\partial T}{\partial \theta} \tag{1.44}$$

The resulting energy conservation equation for a body with time-dependent temperature, variable conductivity, and internal heat generation is

$$\frac{1}{r^2}\frac{\partial}{\partial r}\left(kr^2\frac{\partial T}{\partial r}\right) + \frac{1}{r^2\sin\phi}\frac{\partial}{\partial \phi}\left(k\sin\phi\frac{\partial T}{\partial \phi}\right)$$
$$+ \frac{1}{r^2\sin^2\phi}\frac{\partial}{\partial \theta}\left(k\frac{\partial T}{\partial \theta}\right) + \frac{\dot{q}}{k} = \rho c\frac{\partial T}{\partial t} \tag{1.45}$$

In the order of increasing simplicity, the special versions of the above equation are the following:

Unsteady, constant conductivity, with internal heat generation:

$$\frac{1}{r^2}\frac{\partial}{\partial r}\left(r^2\frac{\partial T}{\partial r}\right) + \frac{1}{r^2\sin\phi}\frac{\partial}{\partial \phi}\left(\sin\phi\frac{\partial T}{\partial \phi}\right)$$
$$+ \frac{1}{r^2\sin^2\phi}\frac{\partial}{\partial \theta}\left(\frac{\partial T}{\partial \theta}\right) + \frac{\dot{q}}{k} = \frac{1}{\alpha}\frac{\partial T}{\partial t} \tag{1.46}$$

Unsteady, constant conductivity, without internal heat generation:

$$\frac{1}{r^2}\frac{\partial}{\partial r}\left(r^2\frac{\partial T}{\partial r}\right) + \frac{1}{r^2\sin\phi}\frac{\partial}{\partial \phi}\left(\sin\phi\frac{\partial T}{\partial \phi}\right)$$
$$+ \frac{1}{r^2\sin^2\phi}\frac{\partial}{\partial \theta}\left(\frac{\partial T}{\partial \theta}\right) = \frac{1}{\alpha}\frac{\partial T}{\partial t} \tag{1.47}$$

Steady state, constant conductivity, without internal heat generation:

$$\frac{1}{r^2}\frac{\partial}{\partial r}\left(r^2\frac{\partial T}{\partial r}\right) + \frac{1}{r^2\sin\phi}\frac{\partial}{\partial\phi}\left(\sin\phi\frac{\partial T}{\partial\phi}\right) + \frac{1}{r^2\sin^2\phi}\frac{\partial}{\partial\theta}\left(\frac{\partial T}{\partial\theta}\right) = 0 \tag{1.48}$$

The last equation can also be written as $\nabla^2 T = 0$, in which ∇^2 is the Laplacian operator in spherical coordinates:

$$\nabla^2 = \frac{\partial^2}{\partial r^2} + \frac{2}{r}\frac{\partial}{\partial r} + \frac{1}{r^2\sin\phi}\frac{\partial}{\partial\phi}\left(\sin\phi\frac{\partial}{\partial\phi}\right) + \frac{1}{r^2\sin^2\phi}\frac{\partial^2}{\partial\theta^2} \tag{1.49}$$

1.3.6 Initial and Boundary Conditions

The problems addressed beginning with Chapter 2 will show that the heat conduction equation is not sufficient for determining the temperature distribution and heat flow through the conducting body. In problems where the temperature field is time dependent, in addition to recognizing the proper form of the energy equation, the problem solver must recognize the proper *initial condition* and *boundary conditions* that belong to the heat flow configuration.

For example, if the problem is to determine the temperature history inside a metallic bar that is quenched by sudden immersion in a pool of oil, the initial condition is the statement that in the beginning (i.e. at the time $t = 0$) the temperature throughout the bar is high and uniform. The boundary conditions are statements made with regard to the temperature, the heat flux, or the relationship between temperature and heat flux at each point around the periphery of the conducting medium. The boundary conditions are assumptions that the location and shape of the boundary is known. The writing of boundary conditions becomes clearer after we learn the concept of heat transfer coefficient.

In steady-state conduction problems, the temperature distribution depends only on the position inside the conducting body. The energy equation does not contain the $\partial T/\partial t$ term, and since this equation will not be integrated in time, the specification of an initial condition is not required. Boundary conditions, however, are required over the entire surface that defines (surrounds) the conducting medium.

Example 1.1 *Thermal Conductivity: Conduction in a Slab*

Figure E1.1 shows a device for measuring the thermal conductivity of polystyrene. A thin metallic plate heater is sandwiched between two polystyrene slabs. This arrangement is itself sandwiched between two copper plates. It is sealed along the top, bottom, and the sides to prevent the flow of water into the polystyrene.

The plate heater is powered by a battery and, acting as a resistance in the circuit, generates 0.1 W over each square centimeter of its extent. The temperature of the outer surfaces of the polystyrene layers is held at $T_s = 0\,°C$ by contact with an equilibrium mixture of ice and water. Note that the temperature drop across each copper plate is negligible. Thermocouples imbedded in the plate heater indicate that the plate temperature T_p is uniform and equal to 62.5 °C. Calculate the polystyrene thermal conductivity k measured in this experiment.

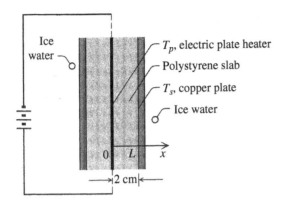

Figure E1.1

Solution

The geometry is one-dimensional because the heat generated by the metallic strip can only escape laterally, that is, unidirectionally. Since there are two sides to the metallic strip, the heat transfer rate per unit area (heat flux) conducted through one of the polystyrene layers is $q_x'' = 0.05 \text{ W/cm}^2 = 500 \text{ W/m}^2$. The heat flux is oriented in the positive x direction. Because q_x'' is known, we can calculate the thermal conductivity k if we know the temperature gradient that drives q_x'' away from one face of the plate heater:

$$q_x'' = -k\left(\frac{dT}{dx}\right)_{x=0} \tag{1}$$

In other words,

$$k = \frac{q_x''}{-(dT/dx)_{x=0}} \tag{2}$$

The problem reduces to finding the temperature gradient $(dT/dx)_x = 0$. For this we solve Eq. (1.21) in the polystyrene half-slab $0 \le x \le L$. Because in the present case the temperature is steady and $\dot{q} = 0$ in the polystyrene, Eq. (1.21) reduces to

$$\frac{d^2 T}{dx^2} = 0 \tag{3}$$

The general solution is $T = c_1 x + c_2$, in which the constants of integration (c_1, c_2) are determined from the boundary conditions:

$$T = T_p \text{ at } x = 0 \tag{4}$$

$$T = T_s \text{ at } x = L \tag{5}$$

The important observation is that the temperature distribution across the polystyrene is linear, which means that the temperature gradient dT/dx is the same at every x:

$$\frac{dT}{dx} = \left(\frac{dT}{dx}\right)_{x=0} = c_1 \tag{6}$$

In conclusion, we have to calculate only c_1, not c_1 and c_2. By using Eqs. (4) and (5),

$$T_p = c_2, \quad T_s = c_1 L + c_2 \tag{7}$$

we obtain $c_1 = (T_s - T_p)/L$, or

$$\frac{dT}{dx} = \frac{0° - 62.5\,°C}{0.02\,m} = -3125\,°C/m \tag{8}$$

By substituting this temperature gradient and $q_x'' = 500\,W/m^2$ in Eq. (2), we arrive at

$$k = \frac{q_x''}{-dT/dx} = 500\,\frac{W}{m^2}\,\frac{m}{3125\,°C} = 0.16\,W/m \cdot K \tag{9}$$

1.4 Convection

Convective heat transfer, or, simply, convection, is the heat transfer executed by the flow of a fluid. The fluid acts as carrier or conveyor belt for the energy that it draws from (or delivers to) a solid wall. The two most common entities that engage in convective-type heat transfer interactions – the entities that in Eq. (1.11) were labeled A and B – are the solid wall and the fluid stream with which the wall comes in direct contact. Convection is that special heat transfer mechanism in which the characteristics of the flow (e.g. configuration, velocity distribution, turbulence) affect greatly the heat transfer rate between the wall and the stream.

The geometric relationship between wall and stream is illustrated by the two convective configurations shown in Figures 1.9 and 1.10. In the external flow configuration, Figure 1.9, the flow engulfs the body with which it interacts thermally. The transition from the body surface temperature (T_w) to the fluid temperature far into the stream (T_∞) is made by a special region of the flow (called "boundary layer") that coats the solid wall. Across the same flow region, the configuration, velocity of the fluid decreases from its free stream value to zero at the wall.

The question is how the temperature extremes (T_w, T_∞) and the flow dictate the magnitude of the heat transfer rate between the body and the stream. The same question can be formulated on a per-unit-area basis: "What is the relationship between the local heat flux q'' and the temperature extremes and the flow?" For if we know the local heat flux q'', we can presumably integrate it over the entire surface of the body to determine the total body-to-stream heat transfer rate q.

The traditional approach consists of defining the external-flow *heat transfer coefficient h* by writing

$$q'' = h(T_w - T_\infty) \tag{1.50}$$

so that (cf. Eq. (1.12)) the heat flux vanishes when the stream is in thermal equilibrium with the wall. The h definition (1.50) does not mean that h is a constant. Furthermore, the heat transfer coefficient symbol h should not be confused with that of specific enthalpy.

The question formulated in the preceding paragraph reduces to finding the heat transfer coefficient and the manner in which the h value is influenced by the characteristics of the flow. For this, the analyst must know not only the temperature distribution in the near-wall fluid but also the flow (velocity) distribution. Out of necessity, then, convection analyses are based not only on fluid flow generalizations of the energy conservation statements of Section 1.3 but also on statements that account for the conservation of mass and momentum in the flow field.

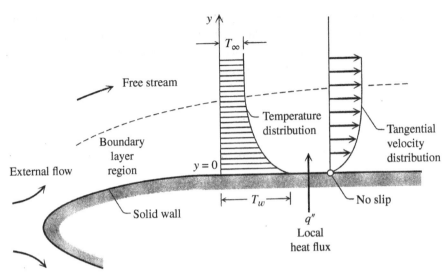

Figure 1.9 External flow configuration of convective heat transfer.

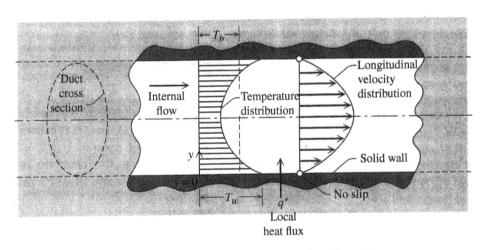

Figure 1.10 Internal flow configuration of convective heat transfer.

When the heat transfer surface surrounds the stream and guides it, the configuration is internal flow. Figure 1.10 shows the main features of the temperature and velocity distributions in the vicinity of the wall. The heat transfer question can be reduced again to the problem of finding a "heat transfer coefficient"; however, this time there is no free-stream temperature (T_∞) that can be identified in the stream. The internal flow alternative to define the heat transfer coefficient is

$$q'' = h(T_w - T_b) \tag{1.51}$$

in which the bulk temperature T_b is a weighted cross section-averaged temperature of the stream, which is also called *mean temperature*. The proper definition of this average will be discussed in connection with Eq. (6.33). The important thought to retain at this stage is that the definition of heat transfer coefficient in internal flow, Eq. (1.51), differs from that used in external flow, Eq. (1.50).

The path to evaluate h in external or internal flow can be seen by looking at the fluid side of the solid wall. Figures 1.9 and 1.10 show that the infinitesimally thin fluid layer situated at $y = 0^+$ is stuck to the wall; in fluid mechanics, this feature is recognized as the *no-slip condition* at a solid boundary. Since the $y = 0^+$ fluid layer is not moving, the wall heat flux that traverses it (q'') is ruled by pure conduction; the Fourier law, Eq. (1.32), applies to this layer as well:

$$q'' = -k \left(\frac{\partial T}{\partial y} \right)_{y=0} \tag{1.52}$$

where k is the thermal conductivity of the fluid and T is the temperature distribution in the fluid. Combining Eqs. (1.50)–(1.52), we find that

$$h = -\frac{k}{T_w - T_\infty} \left(\frac{\partial T}{\partial y} \right)_{y=0^+} \quad \text{(external flow)} \tag{1.53}$$

$$h = -\frac{k}{T_w - T_b} \left(\frac{\partial T}{\partial y} \right)_{y=0^+} \quad \text{(internal flow)} \tag{1.54}$$

In conclusion, to calculate h we must first determine the temperature distribution in the fluid situated close to the wall. The temperature distribution, in turn, depends on the velocity distribution. Therefore, a prerequisite for calculations of convective heat transfer is an understanding of the flow that makes contact with the wall, that is, an understanding of the fluid mechanics of the configuration.

Figure 1.11 exposes the dramatic effect that the fluid and the flow regime have on the order of magnitude of the heat transfer coefficient. These convection regimes are analyzed systematically beginning with Chapter 5. First will be the external flow configurations, in which the emphasis will fall on the flow region situated next to the wall – the boundary layer. Heat transfer to internal flows, or duct flows, will be discussed in Chapter 6. The flows of Chapters 5 and 6 together represent the larger class of forced convection, because in each of these configurations the fluid is forced (by a fan or a pump) to flow past the solid walls of interest. Flows that happen "naturally" are those driven by the buoyancy effect felt by the relatively warmer regions of a flow. These are examples of natural, or free, convection and form the subject of Chapter 7.

Historically, Eq. (1.50) was first written by Fourier [8], who introduced in this way the concept of heat transfer coefficient ("external conductivity," in his terminology), to which he gave the symbol h. Fourier emphasized the fundamental difference between h and the "proper" thermal conductivity k. More than 100 years earlier, Newton [9] had published an essay in which he reported measurements showing that the rate of temperature decrease (dT/dt) of a body immersed in a fluid is at all times proportional to the body–fluid temperature difference ($T - T_\infty$). This is why beginning with Fourier's contemporaries (e.g. Péclet [10]), Eq. (1.50) acquired the name "Newton's law of cooling." This is not correct because Newton's statement can be written [11] as $dT/dt = b(T - T_\infty)$, which is not the same as Eq. (1.50); furthermore, the b coefficient (assumed constant) accounts for the ratio h/c, that is, the heat transfer coefficient divided by the specific heat. The concepts of heat transfer coefficient and specific heat were unknown in Newton's time.

Convection currents in liquids were first discovered experimentally by Count Rumford [12, 13], who also visualized the flow by suspending neutrally buoyant particles in the liquid. The heat transfer effect of such currents was named "convection" by Prout [14].

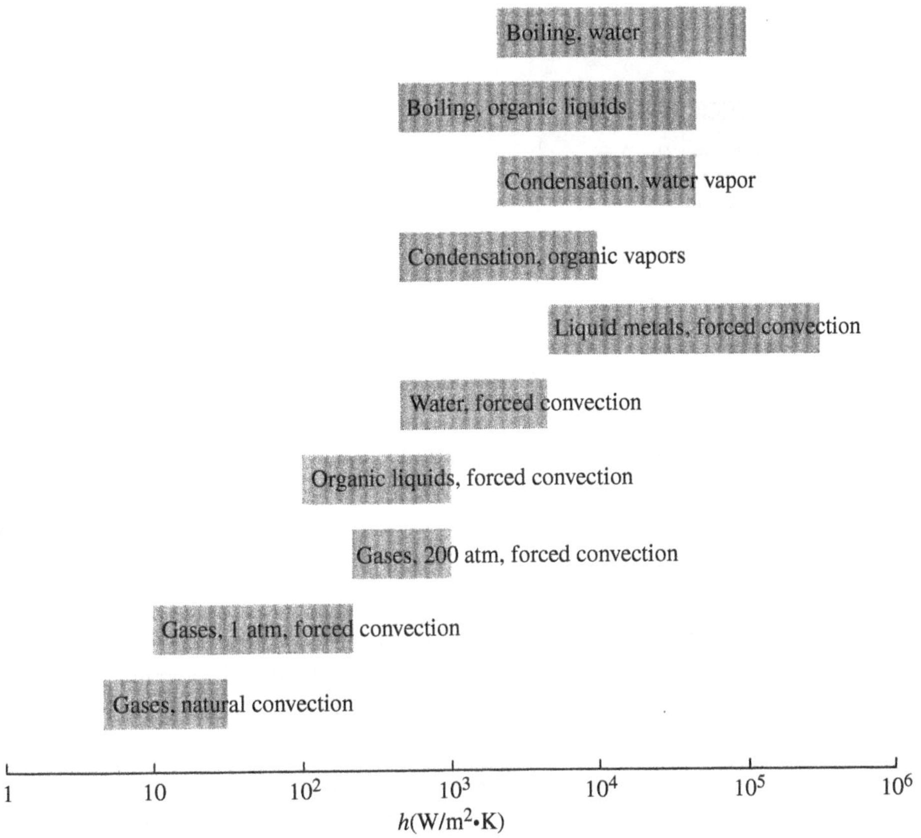

Figure 1.11 Effect of the fluid type and flow regime on the order of magnitude of the heat transfer coefficient.

Example 1.2 *External Flow*

An amount of fine coal powder has been stockpiled as the 2-m-deep layer shown in Figure E1.2. Due to a reaction between coal particles and atmospheric gases, the layer generates heat volumetrically at the uniform rate $\dot{q} = 20$ W/m^3. The coal temperature is in the steady state. The ambient air temperature is $T_\infty = 25\,°C$, and the heat transfer coefficient at the upper surface of the coal stockpile is $h = 5$ W/m$^2 \cdot$ K. Calculate the temperature of the upper surface, T_w.

Solution

Expressed per unit of surface area covered by the coal pile, the heat flux generated by the layer of coal is

$$q'' = \dot{q}H = 20\,\frac{\text{W}}{\text{m}^3}2\,\text{m} = 40\,\text{W/m}^2 \tag{1}$$

This heat current escapes through the upper surface of temperature T_w:

$$q'' = h(T_w - T_\infty) \tag{2}$$

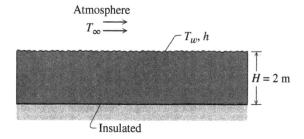

Figure E1.2

because the lower surface has been modeled as insulated (zero heat flux). Equation (2) delivers the unknown temperature of the upper surface of the coal stockpile:

$$T_w = T_\infty + \frac{q''}{h} = 25\,°C + 40\,\frac{W}{m^2}\,\frac{m^2 \cdot K}{5\,W}$$
$$= 25 + 8\,°C = 33\,°C$$

Example 1.3 *Conduction in Series with Convection*

Attached to a flat wall of temperature T_b is a plate of thickness b, length L, and width W (see Figure E1.3). This plate is made of a highly conductive metal and, as a consequence, its temperature is practically uniform. The plate is bathed on all its exposed sides by a fluid of temperature T_∞. The heat transfer coefficient has the same value h on all the surfaces wetted by the fluid.

The plate is attached to the wall with a layer of glue of thickness t and thermal conductivity k. Derive an expression for the heat transfer rate that passes from T_b to T_∞ through the glue-plate system. Under what conditions is this heat transfer rate controlled (impeded the most) by the layer of glue?

Solution

The temperature T of the isothermal plate is unknown. The heat transfer rate q is first conducted across the layer of glue:

$$q = k\frac{bW}{t}(T_b - T) \tag{1}$$

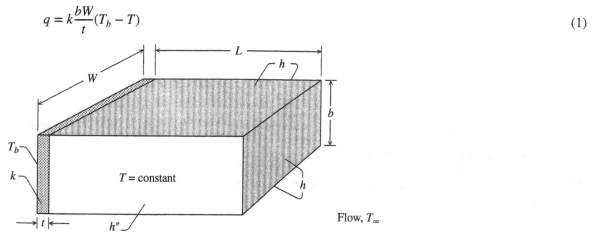

Figure E1.3

Later, the same q is convected away from the lateral surfaces of the plate:

$$q = h(2WL + 2Lb + Wb)(T - T_\infty) \qquad (2)$$

By eliminating q between Eqs. (1) and (2), we obtain an expression for the plate temperature:

$$T = \frac{1}{1+B}T_b + \frac{B}{1+B}T_\infty \qquad (3)$$

where B denotes the dimensionless group:

$$B = \frac{ht}{k}\left(2\frac{L}{b} + 2\frac{L}{W} + 1\right) \qquad (4)$$

Finally, by substituting the T expression (3) in Eq. (1) or Eq. (2), we obtain the relation for the heat transfer rate:

$$q = k\frac{bW}{t}(T_b - T_\infty)\frac{B}{1+B} \qquad (5)$$

This expression shows that when h, t, k, L, W, and b are such that the dimensionless B number is greater than 1, the heat transfer rate is "controlled" by the conduction across the glue layer:

$$q \cong k\frac{bW}{t}(T_b - T_\infty)\,(B \gg 1) \qquad (6)$$

Equation (3) indicates that in the same limit the plate temperature approaches the fluid temperature T_∞. This is why Eqs. (6) and (1) look similar.

1.5 Radiation

Next to conduction and convection, the third distinct mechanism for heat transfer is thermal radiation. This mechanism is covered in detail in Chapter 10. Here we identify the most essential aspect that distinguishes radiation from conduction and convection.

Consider the evacuated inner space that surrounds the spherical container B shown in Figure 1.12. The "vacuum jacket" is an insulation feature employed in the design of many devices, for example, storage vessels for cryogenic liquids, nitrogen, and helium. The evacuated space serves as insulation, precisely because the material (the continuum) that would have acted by contact for conduction and convection is absent. Had

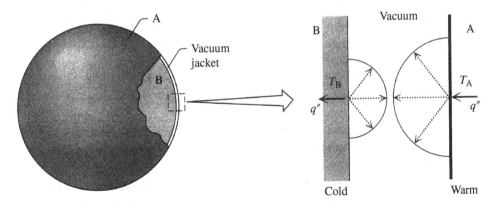

Figure 1.12 Thermal radiation across an evacuated space.

the annular gap been filled with a solid, even with one of low conductivity, it would have been penetrated radially by conduction heat fluxes similar to the q_r'' arrow shown in Figure 1.8. Had a fluid been present in the gap (e.g. air or the vapor of the paint that coats the two facing surfaces), buoyancy would have driven a closed-loop flow (a "circulation") in the gap. Riding on this flow would have been a convective heat current of the kind that will be analyzed in the natural convection segments of this book.

Experience shows that even when the annular gap is completely evacuated, a finite heat transfer current still leaves the outer (warm) shell and lands on the inner (cold) shell. This net heat transfer rate is due to thermal radiation, that is, to the interaction of two bodies that can affect one another *from a distance*.

The thermal radiation effect can be explained on the basis of electromagnetic wave theory. The net heat transfer rate by radiation (from A to B in Figure 1.12) represents the difference between the energy stream with which the warm surface "bombards" the cold surface and the weaker energy stream emanating from the cold surface. Under certain simplifying assumptions (explained in Chapter 10), the net radiation current becomes

$$q_{\text{A-B}} = \beta A(T_\text{A}{}^4 - T_\text{B}{}^4) \tag{1.55}$$

where A is the area of either shell and T_A and T_B are the absolute (Kelvin, or Rankine) temperatures of the two surfaces.

The proportionality factor labeled β depends in general on T_A and T_B, on the relative size of the two surfaces, on their degree of smoothness and cleanliness, and on the materials out of which they are made. Special forms of this proportionality factor will be discussed in Chapter 10. Until then, note that Eq. (1.55) is a special case of the function presented in Eqs. (1.11) and (1.12).

The feature that distinguishes radiation from conduction and convection is that radiation can proceed even in the absence of a continuous medium. Radiation heat transfer can also occur when a sufficiently transparent continuous medium (e.g. dry air) separates the heat-exchanging surfaces, as in the case of domestic and solar heating arrangements around us. Indeed, in many problems the challenge is to evaluate the total heat transfer rate when the radiation, convection, and conduction mechanisms participate simultaneously.

1.6 Evolutionary Design

Heat transfer happens. It is natural because temperature differences and gradients are everywhere. In the devices conceived by humans, just like in the design of the animal body, heat transfer is present because the bigger entity (device, animal, human, and machine specimen) represents a design with purpose.

In science "purpose" is not mentioned, yet, many scientific words and concepts account for the same irrefutable aspect of reality, for example, performance, function, objective, cause, self-organization, self-optimization, cost, good, better, easy, and difficult. This multitude of terms makes it difficult to address "performance" in a fundamental way. Here are a few examples.

To the biologist who studies the design of the fur of a warm-blooded animal, performance is the animal's ability to thrive in a cold environment, which depends on the performance of its fur as a thermal insulation. To the manufacturer of a heat exchanger, performance is the cost of materials, labor, and fuel used during manufacturing, which have everything to do with the performance of the whole heat exchanger as a facilitator of heat transfer. To the designer of a large-scale steam turbine power plant, in addition to the total cost, performance is the overall energy conversion efficiency of the whole plant, which is a global measure to which contributes the functioning of each of the heat exchangers through which the steam flows.

All these interpretations belong together in a most useful and tutorial way if approached from the most fundamental point of view available in science: thermodynamics [1, 15–21]. Heat transfer is one kind of energy transfer between two systems (the system and its environment), and this kind has "performance" that distinguishes it unmistakably from the performance of the other kind of energy transfer, which is work transfer. Performance is why heat transfer is eminently not analogous with work transfer (for details, study Refs. [15–17]).

In this section we follow the path of the concept of performance from its fundamental place in thermodynamics to the many and diverse counterparts of performance that are recognized in other scientific undertakings [17]. Along the way, we will discover opportunities to correct some of the erroneous interpretations of heat transfer performance that continue to propagate in the current literature.

1.6.1 Irreversible Heating

Heat transfer is a flow of energy (q) that proceeds by itself in one direction, from a high temperature (T_H) to a lower temperature (T_L). This feature of the flow is illustrated in Figure 1.13a. The thermodynamic system is traversed by q. The system occupies the square space shown between T_H and T_L. Said another way, the system is embedded in an environment that is not isothermal: the environment has a cold zone (T_L) and a hot zone (T_H). The system is sandwiched between those two zones.

Mass does not cross the system boundary: the system is closed. Changes do not occur in the system over time: the system is in steady state, or stationary state. The first law states that the energy inventory of the system (E) is constant because the net flow of energy into the system is zero. The rate of heat transfer into the system (q) is the same as the rate of heat transfer out of the system (q), and, in addition, the lateral portions of the system boundary (the vertical sides in Figure 1.13a) are not crossed by heat transfer and work transfer.

For the system shown in Figure 1.13a, the second law states that the net flow of entropy into the system cannot be greater than the rate of increase in the entropy inventory (S) of the system. The rate of increase (dS/dt) is zero because the system is in steady state. The net flow of entropy into the system is the difference between the entropy current that flows in (q/T_H) and the entropy current that flows out (q/T_L). Therefore, in

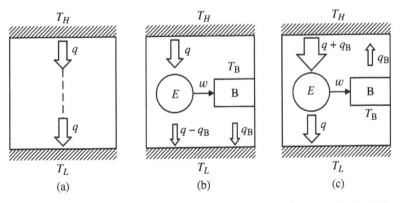

Figure 1.13 The destruction of useful power. Drawing from 1976 [22] and 1982 [23] for illustrating the performance of heat transfer across a finite temperature difference. Source: Bejan and Paynter [22]; Bejan [23].

this case the second law reduces to

$$0 \geq \frac{q}{T_H} - \frac{q}{T_L} \tag{1.56}$$

or to the inequality sign in

$$s_{\text{gen}} = q\left(\frac{1}{T_L} - \frac{1}{T_H}\right) \geq 0 \tag{1.57}$$

where the equal sign after s_{gen} is the definition of the rate of entropy generation of the system.

The inequality sign in either Eq. (1.56) or Eq. (1.57) accounts for the one-way nature of heat transfer. If the sign is ">," then $T_H > T_L$ and q flows from "high" to "low." The heat transfer is said to be *irreversible*, and the system generates entropy, $s_{\text{gen}} > 0$.

The limiting class of systems for which the $>$ sign is so weak that it resembles the equal sign, T_H equals T_L, and q flows across a space that is isothermal. The direction of q in Figure 1.13a (down or up) loses its meaning. In such systems the heat transfer is *reversible* because when $T_H = T_L$ the q arrow could be rotated 180° without violating the second law.

This concludes the thermodynamics of any closed system in steady state, which is traversed by a heat current. Think of this thermodynamic system as a "temperature gap" system. If this is all that thermodynamics teaches about Figure 1.13a, then what about performance?

To discover the physics meaning of performance, examine the rest of Figure 1.13. Panels (b) and (c) are two special cases of (a), because inside of panel (a) nothing was drawn. Panel (a) is the "any system" that obeys the first law and the second law. The number of systems that are thermodynamically equivalent to (a) is infinite.

In system (b), the entering heat current (q) drives an engine (any engine) that generates work per unit time, i.e. power (w). The remaining portion of the heat current ($q - w$) is rejected as heat transfer to the T_L side of the boundary. The power current (w) cannot leave the system because the only exit for energy flow is for heat transfer, not work transfer. To be equivalent to system (a), inside system (b) there must also exist a purely dissipative system that functions as a brake (B), which converts the power (w) into the heat current (q_B).

In sum, the heat current rejected by the dissipator ($q_B = w$) adds itself to the heat current rejected by the engine ($q - w$), and the result is that the original heat current (q) crosses undiminished the T_L portion of the boundary. This is why panel (b) is a special case of the general system (a).

If hot enough, which is questionable, the dissipator is eligible to reject its heat current to either side, T_L or T_H, that is if its temperature (T_B) is higher than both T_L and T_H. In panel (b) the heat current from the dissipator is rejected to the T_L zone of the environment, and the total heat current rejected to T_L is q, just like in panel (a). In panel (c) the q_B current is rejected to T_H which means that the net heat current that enters the temperature gap system from the T_H zone is equal to q, just like in panel (a).

The discussion of systems (b) and (c) in relation to the general system (a) does not depend on the efficiency of the imagined engine (E). The energy conversion efficiency of the engine, $\eta = w/q$, can have any value between 0 and the efficiency in the limit of reversible engine operation (the Carnot limit):

$$\eta_{\text{rev}} = 1 - \frac{T_L}{T_H} \tag{1.58}$$

In this limit, w attains its highest value:

$$w_{\text{rev}} = q\left(1 - \frac{T_L}{T_H}\right) \tag{1.59}$$

which according to the Gouy–Stodola theorem [1] is proportional to s_{gen} of Eq. (1.57):

$$w_{rev} = T_L s_{gen} \qquad (1.60)$$

In reality, all efficiencies (η) are below η_{rev}, and in the energy flows inside systems (b) and (c), this means that as η decreases from one engine model to another, the heat rejection rate from the dissipator ($q_B = w$) decreases, while the heat current rejected directly to the cold zone ($q - w$) increases. The total heat current rejected to T_L by either (b) or (c) is equal to q, i.e. it is independent of η.

The performance (or imperfection) of the heat flow q from T_H to T_L is measured by the ceiling value of the useful power (w_{rev}) that would have been available for harvesting but was lost immediately (destroyed, dissipated) because the "any" system (a) is a purely thermal system, i.e. a system incapable of work transfer with its environment.

According to Eq. (1.59) the destroyed power is proportional to the product of the heat current (q) and the temperature gap ($T_H - T_L$). Ideal performance happens in the limit where the product (current) × (difference) is zero. This is why in the examples with which this section began, the performance is consistently higher in two seemingly contradictory domains of design change (cf., Section 1.2):

- Thermal insulation, which is to diminish thermal contact, i.e. to decrease q when ($T_H - T_L$) is given.
- Heat transfer augmentation, which is to improve the thermal contact (i.e. to decrease $T_H - T_L$) when the heat current is given.

This way we see how the thermodynamics of heat transfer performance accounts for and unites the many and disparate activities of performance increase (i.e. design evolution) recognized at the start of this chapter. Cost, manufacturing, size, and transportation owe their existence to the fact that when heat transfer crosses a finite temperature difference, there is a penalty that the greater system (plant, vehicle, animal) pays in terms of useful power (exergy rate, fuel consumption rate) [24]. The oneness of these manifestations of evolutionary design for performance increase constitutes the core of thermodynamics (cf. Preface, Figure 1).

1.6.2 Reversible Heating

Consider the process in which a solid body of mass m and constant specific heat c is heated reversibly. At any time during the heating process, the body is isothermal, and its lone temperature (T) rises from the initial temperature $T_1 = T_L$ to the final temperature $T_2 = T_H$. The solid body is a closed system that can experience heat transfer, but not work transfer. During the heating process the internal energy inventory of the body increases by the amount

$$E_2 - E_1 = mc(T_H - T_L) \qquad (1.61)$$

and its entropy inventory increases by the amount

$$S_2 - S_1 = mc \ln \left(\frac{T_H}{T_L} \right) \qquad (1.62)$$

These quantities are the only changes in the extensive properties that the body has, unless we account for derived (compounded) extensive properties such as enthalpy, exergy, and Helmholtz and Gibbs free energies [1].

The preceding section concluded that if the heat transfer is to be *reversible*, then the temperature of the environment (from where the heat transfer originates) must be the same as the instantaneous system temperature T. Heating requires an environment that transfers heat to the body. We distinguish two possibilities:

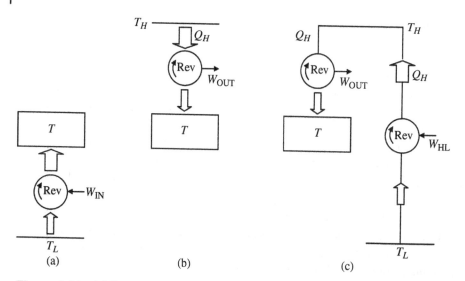

Figure 1.14 (a) Reversible heating from below, (b) reversible heating from above, and (c) the equivalence between (a) and (b).

Heating from Below. The temperature reservoir is at T_L, and the body (T) is heated by the heat current rejected to it by a reversible refrigerator placed between T and T_L (Figure 1.14a). In this case the environment consists of the T_L zone, the refrigerator, and the rest, which produces the power required by the refrigerator. The analysis of the heating process is left as an exercise. While the body temperature rises from T_L to T_H, the total work input required by the refrigerator amounts to

$$W_{IN} = mcT_L \left(\frac{T_H}{T_L} - 1 - \ln \frac{T_H}{T_L} \right) \tag{1.63}$$

The total heat transfer into the body is $Q_{IN} = mc\,(T_H - T_L) = E_2 - E_1$, cf. Eq. (1.61), while the heat transfer from the T_L reservoir to the refrigerator is $Q_L = mcT_L \ln(T_H/T_L)$.

Heating from Above. The heating originates from the T_H zone, and between T_H and the body (T), there is a reversible engine that delivers work while the body temperature rises from T_L to T_H (Figure 1.14b). The total work output is

$$W_{OUT} = mcT_H \left(\ln \frac{T_H}{T_L} - 1 + \frac{T_L}{T_H} \right) \tag{1.64}$$

The total heat transfer from the T_H zone into the body is

$$Q_H = mcT_H \ln \frac{T_H}{T_L} \tag{1.65}$$

The difference $Q_H - W_{OUT}$ is the total heat transfer to the body, which equals $mc(T_H - T_L) = E_2 - E_1$, in accord with Eq. (1.61).

Subtle is the observation that W_{OUT} and W_{IN} are related, and their relation is useful in order to check the validity of either Eq. (1.63) or Eq. (1.64). The relation is due to the fact that in the environment there is only one true "reservoir," which in Figure 1.14c is labeled T_L. The process of "heating from above" is possible provided that a refrigerator is placed between T_H and T_L, so that it can deliver the heat transfer Q_H from the level T_H. The total work input required by the refrigerator is

$$W_{HL} = mc(T_H - T_L) \ln \frac{T_H}{T_L} \tag{1.66}$$

The network input required (from the environment) to heat the body from above reversibly is $(W_{HL} - W_{OUT})$. It can be verified by inspection that $(W_{HL} - W_{OUT})$ is the same as the work required to heat the body reversibly from below (W_{IN}).

Equations (1.61)–(1.66) complete the thermodynamics of the heating process 1–2 executed reversibly by a solid body (m) with an assumed constant specific heat (c) and with one temperature (T) in every state between state 1 and state 2. On this background, the recent "entransy" literature based on the claim (reviewed in [15–17]) that heating reversibly a solid body is analogous to charging an electric capacitor is false. The error behind this claim has been identified by multiple independent authors listed chronologically in Refs. [25–32]. Increasing the internal energy of a solid body is by heat transfer, while charging an electric capacitor is by work transfer.

Heat transfer is not to be confused with work transfer. With work transfer, one system moves and deforms another system that resists change. With heat transfer, one system warms a cold system, or cools a warm system. Movement and morphing a body are identifiable macroscopically, while warming and cooling a body are not.

Publishing new science that respects the physics (i.e. nature) is hard. Publishing false science is easy, especially in the Internet era. Advice to the aspiring scientist: question authority and consensus, stay away from the group think, and bear in mind that "marching columns do not climb peaks" [33].

References

1 Bejan, A. (2016). *Advanced Engineering Thermodynamics*, 4e. Hoboken, NJ: Wiley.
2 Biot, J.B. (1816). *Traité de Physique*, vol. 4, 669. Paris.
3 Scott, R.B. (1959). *Cryogenic Engineering*, 344–345. Princeton, NJ: Van Nostrand.
4 Rohsenow, W.M. and Choi, H.Y. (1961). *Heat, Mass and Momentum Transfer*, 518. Englewood Cliffs, NJ: Prentice-Hall.
5 Bejan, A. (2013). *Convection Heat Transfer*, 4e. Hoboken, NJ: Wiley.
6 Tsederberg, N.V. (1965). *Thermal Conductivity of Gases and Liquids*, Chapter II. Cambridge, MA: MIT Press.
7 Scurlock, R.G. (1966). *Low Temperature Behavior of Solids: An Introduction*, 51–64. New York: Dover.
8 Fourier, J. (1878). *Analytical Theory of Heat* (transl. with notes, by A. Freeman). New York: G. E. Stechert & Co.
9 Newton, I. (1701). Scala graduum caloris, calorum descriptiones & signa. *Philos. Trans. R. Soc. London* 8: 824–829; translated from Latin in *Philos. Trans. R. Soc. London*, Abridged, Vol. IV (1694–1702), 1809, pp. 572–575.
10 Péclet, E. (1860). *Traité de la chaleur considérée dans ses applications*, 3e, vol. 1, 364. Paris: Victor Masson.
11 Bergles, A.E. (1988). Enhancement of convective heat transfer: Newton's legacy pursued. In: *History of Heat Transfer* (eds. E.T. Layton Jr., and J.H. Lienhard), 53–64. New York: American Society of Mechanical Engineers.

12 Count of Rumford (1798). *Essays, Political, Economical and Philosophical*, vol. II. London: T. Cadell and W. Davis, Essay VII.

13 Brown, S.C. (1947). The discovery of convection currents by Benjamin Thompson, Count of Rumford. *Am. J. Phys.* 15: 273–274.

14 Prout, W. (1834). *Bridgewater Treatises*, vol. 8, 65. Philadelphia, PA: Carey, Lea & Blanchard.

15 Bejan, A. (2017). Evolution in thermodynamics. *Appl. Phys. Rev.* 4: 011305.

16 Bejan, A. (2018). Thermodynamics today. *Energy* 160: 1208–1219.

17 Bejan, A. (2019). Thermodynamics of heating. *Proc. R. Soc. A* 475 https://doi.org/10.1098/rspa.2018.0820.

18 Pramanick, A. (2014). *The Nature of Motive Force*. Berlin: Springer-Verlag.

19 Dincer, I. (ed.) (2018). *Comprehensive Energy Systems*, vol. 1, Part A. Amsterdam: Elsevier.

20 Rocha, L. (2009). *Convection in Channels and Porous Media: Analysis, Optimization, and Constructal Design*. Saarbrücken: VDM Verlag.

21 Lorenzini, G., Moreti, S., and Conti, A. (2011). *Fin Shape Optimization Using Bejan's Constructal Theory*. San Francisco, CA: Morgan & Claypool Publishers.

22 Bejan, A. and Paynter, H.M. (1976). *Solved Problems in Thermodynamics*. Massachusetts Institute of Technology, Department of Mechanical Engineering.

23 Bejan, A. (1982). *Entropy Generation through Heat and Fluid Flow*. New York: Wiley.

24 Bejan, A., Lorente, S., Martins, L., and Meyer, J.P. (2017). The constructal size of a heat exchanger. *J. Appl. Phys.* 122: 064902.

25 Grazzini, G., Borchiellini, R., and Lucia, U. (2013). Entropy versus entransy. *J. Non-Equilib. Thermodyn.* 38: 259–271.

26 Herwig, H. (2014). Do we really need entransy? *J. Heat Transfer* 136: 045501.

27 Bejan, A. (2014). Entransy, and its lack of content in physics. *J. Heat Transfer* 136: 055501.

28 Awad, M.M. (2014). Entransy is now clear. *J. Heat Transfer* 136: 095502.

29 Oliveira, S.R. and Milanez, L.F. (2014). Equivalence between the application of entransy and entropy generation. *Int. J. Heat Mass Transfer* 79: 518–525.

30 Manjunath, K. and Kaushik, S.C. (2014). Second law thermodynamic study of heat exchangers: a review. *Renewable Sustainable Energy Rev.* 40: 348–374.

31 Grazzini, G. and Rochetti, A. (2014). Thermodynamic optimization of irreversible refrigerators. *Energy Convers. Manage.* 84: 583–588.

32 Sekulic, D., Sciubba, E., and Moran, M.J. (2015). Entransy: a misleading concept for the analysis and optimization of thermal systems. *Energy* 80: 251–253.

33 Bejan, A. (2020). *Freedom and Evolution: Hierarchy in Nature, Society and Science*, 139. New York: Springer Nature.

Problems

Conduction

1.1 The one-dimensional conductor shown in Figure 1.3 stretches from $x = 0$ to $x = L$, and the respective end temperatures are maintained at T_0 and T_L. The long sides of the conductor are insulated. The temperature distribution in the conductor is steady. Let q_0 represent the heat current that enters the conductor through the $x = 0$ cross section. Assume that the value of q_0 is positive, in other words, that heat is conducted in the positive x direction. Invoke the second law to prove that q_0 flows toward lower temperatures, for example, by showing that T_L cannot be greater than T_0.

1.2 Derive the heat conduction equation corresponding to the cylindrical coordinates shown in Figure 1.7. Consider the general case where the conduction phenomenon is unsteady, the thermal conductivity is a function of temperature, and the effect of internal heat generation is present.

1.3 Consider the statement of conservation of energy in the infinitesimal chunk of material shown on the right side of Figure 1.8. Derive from this statement the conduction equation in spherical coordinates, for the case of unsteady conduction with temperature-dependent thermal conductivity, and with internal heat generation.

1.4 Figure P1.4 shows the distribution of temperature with depth in the rock formation of a geothermal field. The vertical direction x points downward, and $x = 0$ represents the ground level. The average thermal conductivity of the rock formation near the Earth's surface is $k \cong 17.4 \, \text{W/m} \cdot \text{K}$. Calculate the heat flux that crosses the $x = 0$ surface into the atmosphere. Estimate also the heat transfer rate released by the rock formation into the atmosphere through a ground area of $1 \, \text{km}^2$.

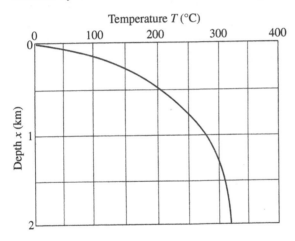

Figure P1.4

Convection

1.5 The plane heat exchanger surface shown in Figure P1.5 is bathed by a stream of city water. In time, the surface becomes covered with a 0.1-mm-thin solid layer consisting of an accumulation of solids collected from the water stream. This process is called "fouling" and is similar to the formation of plaque on teeth. The thermal conductivity of the accumulated solid is $0.6 \, \text{W/m} \cdot \text{K}$. The heat flux from the heat exchanger surface to the water stream is $0.1 \, \text{W/cm}^2$. Calculate the temperature difference across the accumulated layer; in other words, how much does the temperature of the wetted surface drop because of fouling?

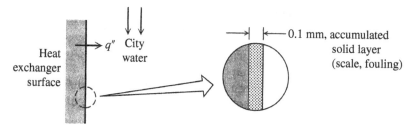

Figure P1.5

1.6 With reference to the illustration accompanying Example 1.2 in the text, consider the case where the layer of coal loses heat to the ground at the rate of $30 \, \text{W/m}^2$. Chemical reactions heat the layer of coal at the rate of $50 \, \text{W/m}^3$, which is distributed uniformly throughout the coal volume. The coal layer is 2 m deep, the ambient air temperature is $30 \, °\text{C}$, and the heat transfer coefficient at the upper surface is $h = 15 \, \text{W/m}^2 \cdot \text{K}$. Calculate the temperature of the upper surface of the layer of coal.

1.7 During a cold winter, the surface of a river is covered by a layer of ice of unknown thickness L. Known are the lake water temperature $T_w = 4 \, °\text{C}$, the atmospheric air temperature $T_a = -30 \, °\text{C}$, and the temperature of the underside of the ice layer $T_0 = 0 \, °\text{C}$. The thermal conductivity of ice is $k = 2.25 \, \text{W/m} \cdot \text{K}$. The convective heat transfer coefficients on the water and air sides of the ice layer are $h_w = 500 \, \text{W/m}^2 \cdot \text{K}$ and $h_a = 100 \, \text{W/m}^2 \cdot \text{K}$, respectively. Calculate the temperature of the upper surface of the ice layer, T_s, and the ice thickness, L (Figure P1.7).

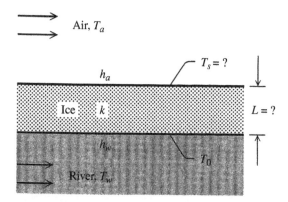

Figure P1.7

1.8 During cold weather, the temperature of the human skin ($30 \, °\text{C}$) is lower than the core temperature of the body ($36.5 \, °\text{C}$). The transition between the two temperatures occurs across a subskin layer with an approximate thickness of 1 cm, which acts as an insulating coat. The thermal conductivity of the living tissue in this layer is approximately $0.42 \, \text{W/m} \cdot \text{K}$.
a) Estimate the heat flux that escapes through the skin surface. Treat the subskin tissue as a motionless conducting medium.
b) The ambient air temperature is $20 \, °\text{C}$. Calculate the heat transfer coefficient between skin and air (Figure P1.8).

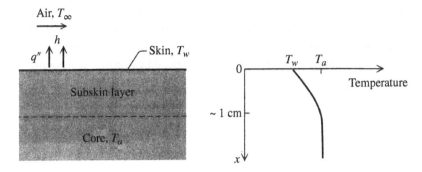

Figure P1.8

1.9 The electric oven shown in Figure P1.9 is used to heat a continuous sheet of metal from 25 to 1000 °C. The oven and sheet are two-dimensional, that is, sufficiently wide in the direction perpendicular to the figure. The metal speed is such that each point on the sheet spends two hours inside the oven. The sheet thickness is 1 cm, the ambient temperature is 25 °C, and the temperature of the outer surface of the oven (uniform all around) is 50 °C. The heat transfer coefficient between the outer surface and the surrounding air is 20 W/m² · K. The metal is 304 stainless steel. Calculate the electric power input to the oven, per unit oven width, q'. What percentage of this power input is lost by convection to the atmosphere?

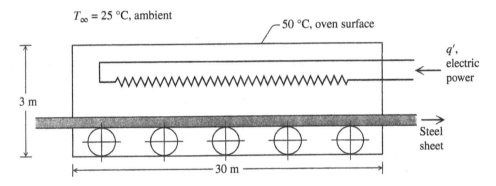

Figure P1.9

1.10 The oven described in the preceding problem has a constant-thickness wall made of firebrick. The temperature of the internal surface is approximately 900 °C, the temperature of the external surface is 50 °C, and the ambient temperature is 25 °C. The convection heat transfer coefficient at the external surface is 20 W/m² · K. Calculate the heat flux through the wall and the wall thickness.

Evolutionary Design

1.11 One way to establish the equivalence between panels (a) and (b) in Figure 1.13 is to analyze the most general (real, irreversible) heat engine shown on the left side of Figure P1.11. The engine is a closed system operating in steady state or in an integral number of cycles, receiving the heat current

q from the temperature reservoir T_H and rejecting a heat current to the temperature T_L. Show that the "any" irreversible heat engine is equivalent to the assembly shown on the right side of the figure; in the reversible limit, the "any" engine delivers w_{rev}, and one fraction of that power ($w_{rev} - w$) is dissipated as a heat current rejected to T_L. In this problem, the special value $w = 0$ accounts for each of the three drawings made in Figure 1.13.

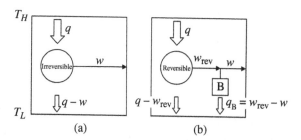

Figure P1.11

1.12 With reference to Figure 1.14a, show that the total work (W_{IN}) required to heat reversibly a solid body (m, c) from $T = T_L$ to $T = T_H$ is expressed by Eq. (1.63). As the body temperature rises from T to $T + dT$, the reversible refrigerator requires the work input δW, receives heat transfer δQ_L from T_L, and delivers heat transfer δQ to the body at T. Start with the first law and the second law to determine δW as a function of mc, T, T_L, and dT, and then integrate δW from $T = T_L$ to $T = T_H$.

1.13 Show that during reversible heating from above (Figure 1.14b), the total work output of the reversible engine is given by Eq. (1.64). Begin the analysis with the first law and the second law for the infinitesimal change of state of the system (m, c), from (E, T) to ($E + dE$, $T + dT$). Express the infinitesimal work output from the reversible engine (δW) as a function of mc, T_H, T, and dT. To determine W_{OUT}, integrate δW from $T = T_L$ to $T = T_H$.

1.14 The furnace that heats a house consumes fuel in proportion with the rate at which it generates heat, q_f. The house must be heated at the fixed rate q. A portion of q_f is carried by the exhaust into the ambient. The remainder is equal to q.

Concerned by the dumping of heating into the ambient, fuel waste, global warming, and carbon footprint, the home owner considers replacing the furnace with an electric heater. In modern designs, the electric heater is almost perfectly insulated, which means that its rate of electric power consumption equals q. Is the home owner right?

Compare the performance of the two designs on the same basis, namely, the rate of fuel consumption that is responsible for the heating that originally serves as source for the house heating requirement q. Note that the electric power dissipated in the electric heater is produced by a power plant that burns fuel, and that the power plant efficiency is 40%.

1.15 The hot components of a power plant must be fitted with thermal insulation so that they do not leak heat excessively to the ambient (T_0). The thermal conductivity of the insulation is known (k). The total volume of the insulation (V) is fixed.

A simple model of the hot components is the two-chamber model shown in Figure P1.15. The hottest is the furnace, which is enclosed by a surface of area A_H and high temperature T_H. The thickness of the insulation mounted on A_H is t_H. This thickness is sufficiently small so that the volume of the insulation on A_H is $A_H t_H$.

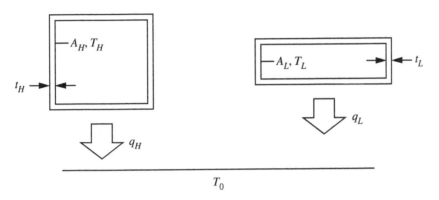

Figure P1.15

The rest of the hot components (pipes, feed water heaters, turbines) are not as hot as the furnace. They are inside an enclosure with area A_L, temperature T_L, insulation thickness t_L, and insulation volume $A_L t_L$.

The heat leaks q_H and q_L are by pure conduction and are driven by the temperature differences $\Delta T_H = T_H - T_0$ and $\Delta T_L = T_L - T_0$, respectively. These temperature differences are known. The problem is to determine t_H/t_L, that is, how to distribute the available insulation on A_H and A_L.

1. Minimize the total heat leak from A_H and A_L to the ambient, namely, $q_H + q_L$, and determine the optimal ratio t_H/t_L as a function of other parameters of the two-chamber model. Does t_H/t_L depend on A_H/A_L?
2. Explain why 1 W of heat leak from T_H is not the same as 1 W of heat leak from T_L. Which do you think is more damaging to the performance of the power plant?
3. Imagine that q_H can be intercepted outside A_H and used to run a Carnot engine between T_H and T_0. The power producible in this way (W_H) is lost because q_H is dumped straight into the ambient. Imagine the equivalent scenario for q_L, and derive a formula for the Carnot power W_L that is lost because of this second heat leak.
4. Minimize the total loss of power ($W_H + W_L$) and determine the optimal ratio t_H/t_L.
5. Compare the t_H/t_L results obtained at parts 1 and 4 above. Which ratio is larger? Which is more relevant for actual implementation? (Figure P1.15)

1.16 Here we explore the trade-off between two bars made of different materials for the purpose of facilitating thermal contact between two ends at different temperatures. Heat conduction is in the axial direction along each bar. The length, cross-sectional area, and thermal conductivity of the two bars are given, $(L, A, k)_1$ and $(L, A, k)_2$. The overall end-to-end distance is fixed, $L_1 + L_2 = L$. The objective is to find the ratio L_1/L_2 or A_1/A_2, such that the end-to-end thermal resistance is minimal. Show that this design solution is ruled by the dissimilarity ratio $k_1 V_1/(k_2 V_2)$, where V_1 and V_2 are the volumes of the two bars. Show further that the bar with the higher thermal conductivity should be thinner than the bar with lower thermal conductivity (Figure P1.16).

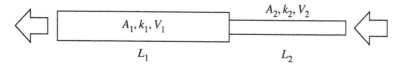

Figure P1.16

1.17 When not in tension, an elastic bar has the length L and cross-sectional area A. Material properties are the modulus of elasticity E, mass m, density ρ, and specific heat c. In state 1 the bar is in tension, and its elongation x is much smaller than L. The bar is isothermal, at temperature T_1. Consider the process 1–2, which is triggered by the fact that the bar snaps into two or more pieces. During this process the system (the bar) is perfectly insulated. In the final state (state 2), the tension is zero, and the temperature is uniform (T_2). Determine the temperature change ($T_2 - T_1$) as a function of the initial elongation (x) and other properties of the bar. Does the temperature change depend on the size and shape (the design) of the bar cross section?

2

Unidirectional Steady Conduction

2.1 Thin Walls

2.1.1 Thermal Resistance

Consider the insulation effect provided by a layer of material whose thermal conductivity k is constant and the layer thickness is L. Across the layer, the temperature varies from T_0 (the temperature of the surface of the enveloped body, Figure 2.1) to the temperature of the outer skin of the layer, T_L. The question is how the temperature difference drives the heat transfer through the layer.

The analysis is particularly simple when the layer looks thin compared with the dimensions of the covered body. For example, if the layer thickness L is much smaller than the radius of curvature of the surface, we can treat the layer as locally "plane," as is demonstrated on the right side of Figure 2.1. Such a description is, of course, ideal in the case of a layer that is plane throughout its extent, like a sheet of glass, or the plasterboard covering of the wall in a house. This is why it is sometimes referred to as the "plane wall," even though the analysis applies to other wall shapes in the thin-wall limit.

In the Cartesian coordinate system attached to the inner surface of the layer, the temperature of the material varies in the transversal direction x. The equation that governs this temperature variation is the one for steady conduction in a material with constant conductivity and no internal heat generation:

$$\frac{d^2T}{dx^2} = 0 \tag{2.1}$$

The boundary conditions that apply to the $T(x)$ distribution are

$$T = T_0 \quad \text{at} \quad x = 0 \tag{2.2}$$
$$T = T_L \quad \text{at} \quad x = L \tag{2.3}$$

The solution to Eq. (2.1) is $T = c_1 x + c_2$, in which the two constants of integration are determined from the boundary conditions (2.2) and (2.3). In conclusion, the solution is a *linear* temperature distribution that reaches T_0 and T_L at the two boundaries of the conducting medium:

$$T = T_0 + \left(T_L - T_0\right) \frac{x}{L} \tag{2.4}$$

This result is needed to calculate the heat flux:

$$q'' = -k\frac{dT}{dx} = \frac{k}{L}\left(T_0 - T_L\right) \tag{2.5}$$

Note that the heat flux q'' is defined positive when pointing in the positive x direction and that it is conserved from $x = 0$ to $x = L$. The gradient dT/dx independent of x.

Heat Transfer: Evolution, Design and Performance, First Edition. Adrian Bejan.
© 2022 John Wiley & Sons, Inc. Published 2022 by John Wiley & Sons, Inc.
Companion website: www.wiley.com/go/bejan/heattransfer

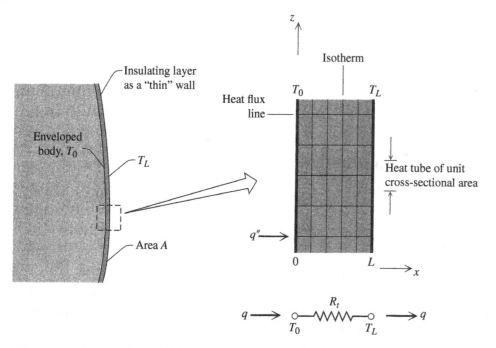

Figure 2.1 Thermal resistance posed by a sufficiently thin wall.

Figure 2.1 further shows that the *isotherms* (the constant-T lines) are equidistant and parallel to the two faces of the wall. Perpendicular to the family of isotherms are the *heat flux lines*, which indicate the path followed by q'' through the conducting medium. The space contained between two adjacent flux lines is the *heat tube* aligned with the direction of q''. The heat tubes have the same thickness because the heat flux q'' is distributed uniformly over the face of the wall.

If the surface covered by the layer is A, the total heat transfer rate is

$$q = q''A = \frac{kA}{L}\left(T_0 - T_L\right) \tag{2.6}$$

We learn from this result that the steady leakage of heat across the layer is proportional to the temperature difference that drives it, $T_0 - T_L$, and the special group of physical quantities kA/L. The heat transfer rate q increases as the cross-sectional area A of its flow increases, as the layer is made out of materials with greater k values, and as the layer becomes thinner. The kA/L group is the *thermal conductance* of the layer. The inverse of kA/L is the *thermal resistance* of the layer:

$$R_t = \frac{L}{kA} \tag{2.7}$$

Combining Eqs. (2.6) and (2.7), we can see that the "fall" of q across the temperature "drop" $(T_0 - T_L)$ and "through" the resistance R_t

$$q = \frac{T_0 - T_L}{R_t} \tag{2.8}$$

is completely analogous to that of an electrical current in a single-resistance circuit. This analogy, which also gives R_t its name, is stressed further by the electrical resistance symbol used on the right side of Figure 2.1.

2.1.2 Composite Walls

The thermal resistance concept is useful when estimating the heat transfer rate through a *composite wall* (Figure 2.2). Two or more sheets are sandwiched and, together, they bridge the temperature gap from T_0 to T_L. Each sheet has its own thermal resistance:

$$R_{t,i} = \frac{L_i}{k_i A} \quad (i = 1, 2, 3) \tag{2.9}$$

where k_i is the thermal conductivity of the sheet material and L_i is the sheet thickness. Assuming that two adjacent sheets have the same temperature at the interface (i.e. that the "thermal contact resistance" is zero), for the three sheets we write

$$\begin{aligned}
T_0 - T_1 &= q R_{t,1} \\
T_1 - T_2 &= q R_{t,2} \\
T_2 - T_L &= q R_{t,3}
\end{aligned} \tag{2.10}$$

where T_1 and T_2 are the interface temperatures and q is the heat transfer rate. Note that q remains constant as it traverses the entire sandwich. Summing up the three equations, we obtain the composite-wall equivalent of Eq. (2.6), namely, a relationship between the heat transfer rate and the *overall* temperature difference that drives it:

$$q = \frac{T_0 - T_L}{\dfrac{L_1}{k_1 A} + \dfrac{L_2}{k_2 A} + \dfrac{L_3}{k_3 A}} \tag{2.11}$$

The denominator appearing on the right side of Eq. (2.11) is the *overall thermal resistance* of the three-layer sandwich. It has as many terms as there are resistances in the string shown on the right side of Figure 2.2.

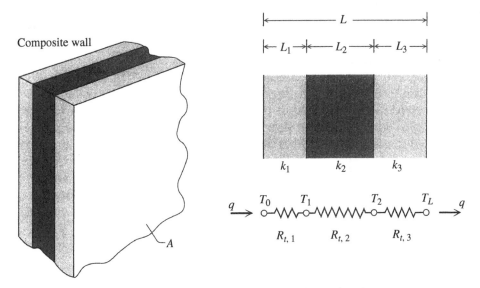

Figure 2.2 Composite wall and the structure of its thermal resistance.

2.1.3 Overall Heat Transfer Coefficient

In the preceding analysis it was assumed that the extreme temperatures (T_0, T_L) of the thin wall are known. That would be true in an application where both the inner body enveloped by the thin wall and the outer body are isothermal media (e.g. high-conductivity solids) at T_0 and T_L, respectively.

More frequent are the applications in which the two sides of the thin wall are exposed to fluids in motion. A common example is the metallic wall that separates two streams in a heat exchanger. Another example is the single-pane glass window. Both examples are represented by Figure 2.3.

The distinguishing feature of this new configuration is that the known temperatures are the extreme fluid temperatures (T_{hot} and T_{cold}), not the surface temperatures of the thin wall. Recall that these extreme fluid temperatures represent either bulk temperatures when the respective fluids flow through ducts or free-stream temperatures when the respective flows are of the "external" type (see Section 1.4). For each face of the thin wall, we can invoke the convective heat flux relation:

$$T_{hot} - T_0 = \frac{q''}{h_{hot}} \tag{2.12}$$

$$T_L - T_{cold} = \frac{q''}{h_{cold}} \tag{2.13}$$

where h_{hot} and h_{cold} are the respective heat transfer coefficients (these are assumed known from analyses of the convective heat transfer processes on the two fluid sides). The heat flux q'' is conserved as it passes from the hot fluid into the cold fluid. The same q'' overcomes the thermal resistance of the solid wall itself; therefore, according to Eq. (2.5),

$$T_0 - T_L = \frac{L}{k}q'' \tag{2.14}$$

Adding Eqs. (2.12)–(2.14) side by side, we obtain a compact proportionality between the heat flux and the overall (fluid-to-fluid) temperature difference:

$$T_{hot} - T_{cold} = \left(\frac{1}{h_{hot}} + \frac{L}{k} + \frac{1}{h_{cold}} \right) q'' \tag{2.15}$$

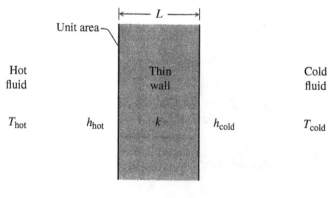

Figure 2.3 Thin wall sandwiched between two flows: the definition of overall heat transfer coefficient.

The form of this equation is similar to that of Eqs. (2.12) and (2.13), in which the denominator of the right-hand side was occupied by a heat transfer coefficient. For this reason it is customary to rewrite Eq. (2.15) as

$$T_{\text{hot}} - T_{\text{cold}} = \frac{q''}{U} \tag{2.16}$$

or as $q'' = U(T_{\text{hot}} - T_{\text{cold}})$ where the *overall heat transfer coefficient U* is defined by

$$\frac{1}{U} = \frac{1}{h_{\text{hot}}} + \frac{L}{k} + \frac{1}{h_{\text{cold}}} \tag{2.17}$$

Equation (2.17) and the electrical resistance analog of this heat transfer configuration (Figure 2.3) show that the inverse of U is the overall thermal resistance. The right side of Eq. (2.17) stresses the conceptual difference between heat transfer coefficient and thermal conductivity: the units of h are W/m²·K, whereas the units of k are W/m·K.

Example 2.1 *Composite Wall: Overall Heat Transfer Coefficient*
The wall of a large incubator for eggs contains an 8-cm-thick layer of fiberglass sandwiched between two plywood sheets with a thickness of 1 cm. The outside temperature is $T_c = 10\,°C$, and the heat transfer coefficient at the outer plywood surface is $h_1 = 5\,W/m^2·K$. The corresponding conditions on the wall surface that faces the eggs are $T_h = 40\,°C$ and $h_3 = 20\,W/m^2·K$. The heat transfer coefficient is higher on the warm side of the wall because a fan recirculates the air that comes in contact with the eggs. Calculate the heat flux through the wall of the incubator.

Solution

According to Eq. (2.16), to calculate the heat flux through the wall, we must first determine the overall heat transfer coefficient U (Figure E2.1):

$$
\begin{aligned}
\frac{1}{U} &= \frac{1}{h_3} + \frac{L_3}{k_3} + \frac{L_2}{k_2} + \frac{L_1}{k_1} + \frac{1}{h_1} \\
&= \left(\frac{1}{20} + \frac{0.01}{0.11} + \frac{0.08}{0.035} + \frac{0.01}{0.11} + \frac{1}{5} \right) \frac{m^2 \cdot K}{W} \\
&= (0.05 + 0.09 + 2.29 + 0.09 + 0.2)\, m^2 \cdot K/W \\
&= 2.72\, m^2 \cdot K/W
\end{aligned} \tag{1}
$$

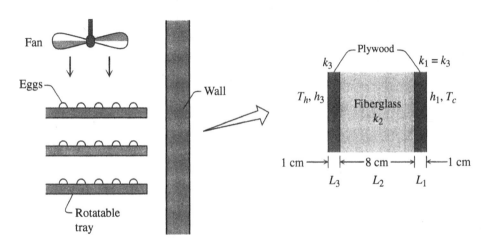

Figure E2.1

The conductivities of plywood (k_1, k_3) and fiberglass (k_2) are from Appendix B. The third line in the calculation of $1/U$ shows that the overall thermal resistance of the composite wall is dominated by the fiberglass layer. The heat flux is

$$q'' = U\left(T_h - T_c\right) = \frac{1}{2.72}\frac{\text{W}}{\text{m}^2 \cdot \text{K}}\,(40 - 10)\,^{\circ}\text{C}$$
$$= 11\,\text{W}/\text{m}^2 \tag{2}$$

2.2 Cylindrical Shells

A wall with non-negligible thickness must be analyzed by a method that takes the wall curvature into account. Figure 2.4 shows a cylindrical shell of length l, inner radius r_i, and outer radius r_o. The temperatures of the inner and outer cylindrical surfaces are T_i and T_o. Heat does not flow in the longitudinal direction (z). Think of the l-tall cylinder as a segment of a long pipe that extends above and below the portion shown in Figure 2.4.

The challenge is to predict the total heat transfer rate q driven through the shell by the temperature difference $T_i - T_o$. The calculation would be simple if we knew the radial heat flux through the inner surface, q_i'', because the total heat transfer rate is the area integral of the heat flux:

$$q = \left(2\pi r_i l\right) q_i'' \tag{2.18}$$

We focus therefore on q_i'':

$$q_i'' = -k\left(\frac{dT}{dr}\right)_{r=r_i} \tag{2.19}$$

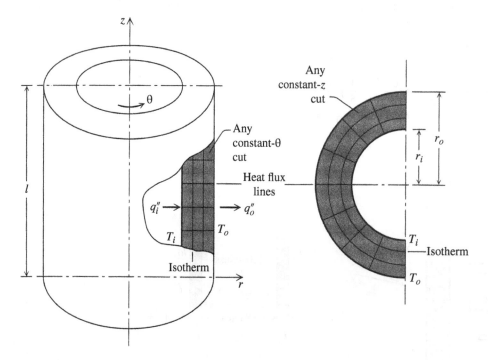

Figure 2.4 Radial conduction through a cylindrical shell.

which brings us to the problem of determining the temperature distribution $T(r)$, and consists of solving the conduction equation in a medium with constant thermal conductivity and without volumetric heat generation:

$$\frac{1}{r}\frac{d}{dr}\left(r\frac{dT}{dr}\right) = 0 \tag{2.20}$$

The temperature boundary conditions are

$$T - T_i \text{ at } r = r_i \tag{2.21}$$

$$T - T_o \text{ at } r = r_o \tag{2.22}$$

The solution is obtained by integrating Eq. (2.20) twice in r, in the following sequence:

$$\frac{d}{dr}\left(r\frac{dT}{dr}\right) = 0 \tag{2.23}$$

$$r\frac{dT}{dr} = C_1 \tag{2.24}$$

$$\frac{dT}{dr} = \frac{C_1}{r} \tag{2.25}$$

$$T = C_1 \ln r + C_2 \tag{2.26}$$

By subjecting the solution (2.26) to the boundary conditions, we obtain

$$T_i = C_1 \ln r_i + C_2 \tag{2.27}$$

$$T_o = C_1 \ln r_o + C_2 \tag{2.28}$$

and, after eliminating C_2,

$$C_1 = \frac{T_i - T_o}{\ln\left(r_i/r_o\right)} \tag{2.29}$$

Finally, we subtract Eq. (2.27) from Eq. (2.26):

$$T - T_i = C_1 \ln\frac{r}{r_i} \tag{2.30}$$

and use the C_1 value determined in Eq. (2.29):

$$T = T_i - \left(T_i - T_o\right)\frac{\ln\left(r/r_i\right)}{\ln\left(r_o/r_i\right)} \tag{2.31}$$

In combination with Eqs. (2.18) and (2.19), this solution yields

$$q = \frac{2\pi kl}{\ln\left(r_o/r_i\right)}\left(T_i - T_o\right) \tag{2.32}$$

In conclusion, the thermal resistance of the cylindrical shell increases as the logarithm of the radii ratio:

$$R_t = \frac{\ln\left(r_o/r_i\right)}{2\pi kl} \tag{2.33}$$

This thermal resistance becomes identical to that of the thin wall, Eq. (2.7), in the limit where r_i approaches r_o. The effect of the wall curvature is to increase the density of the heat flux lines toward smaller radii (Figure 2.4), as a sign that the heat flux q'' in that region is larger. The same effect can be demonstrated

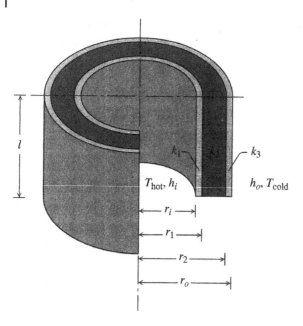

Figure 2.5 Composite cylindrical shell with convective heat transfer on both sides.

analytically, by noting that it is the total heat transfer rate q (and not the heat flux q'') that is conserved as it flows radially outward through the shell:

$$q = \left(2\pi r_i l\right) q_i'' = (2\pi r l)\, q'' \tag{2.34}$$

From this we learn that at any radial distance inside the shell, r, the heat flux varies as $1/r$. For the three-layer shell illustrated in Figure 2.5, the heat transfer rate is given by the familiar formula

$$q = \frac{T_{\text{hot}} - T_{\text{cold}}}{R_t} \tag{2.35}$$

in which the overall thermal resistance of the shell, R_t, is defined by

$$R_t = \frac{1}{h_i A_i} + \frac{\ln\left(r_1/r_i\right)}{2\pi k_1 l} + \frac{\ln\left(r_2/r_1\right)}{2\pi k_2 l} + \frac{\ln\left(r_o/r_2\right)}{2\pi k_3 l} + \frac{1}{h_o A_o} \tag{2.36}$$

In this expression, A_i and A_o are the areas of the innermost and outermost cylindrical surfaces, $A_i = 2\pi r_i l$ and $A_o = 2\pi r_o l$, and h_i and h_o are the respective heat transfer coefficients. Note the presence of three conduction-type terms on the right side of Eq. (2.36); the number of these terms matches the number of layers in the constitution of the composite shell.

2.3 Spherical Shells

With reference to the notation defined in Figure 2.6, the total heat current q (watts) that traverses the shell from T_i to T_o is

$$q = 4\pi k \frac{r_i r_o}{r_o - r_i}\left(T_i - T_o\right) \tag{2.37}$$

Figure 2.6 Radial conduction through a spherical shell.

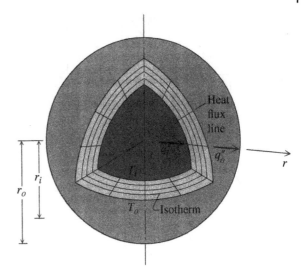

The thermal resistance of a spherical shell is

$$R_t = \frac{1}{4\pi k}\left(\frac{1}{r_i} - \frac{1}{r_o}\right) \tag{2.38}$$

which becomes identical to Eq. (2.7) in the thin-wall limit $r_i \rightarrow r_o$.

If the shell experiences convection on both sides (h_i, h_o) and if it is composed of two layers (k_1, k_2), then its overall thermal resistance is

$$R_t = \frac{1}{h_i A_i} + \frac{1}{4\pi k_1}\left(\frac{1}{r_i} - \frac{1}{r_1}\right) + \frac{1}{4\pi k_2}\left(\frac{1}{r_1} - \frac{1}{r_o}\right) + \frac{1}{h_o A_o} \tag{2.39}$$

The radius r_1 indicates the position of the interface between the k_1 and k_2 layers, while $A_i = 4\pi r_i^2$ and $A_o = 4\pi r_o^2$ are the inner and outer areas of the spherical shell.

2.4 Critical Insulation Radius

The function of the insulation positioned between radii r_i and r_o in Figure 2.7 is to reduce the total heat transfer rate between the inner body and the ambient fluid, T_∞. The total heat transfer rate q behaves as the inverse of the overall thermal resistance R_t, because $q = (T_i - T_\infty)/R_t$. Assuming that the heat transfer coefficient between the insulating layer and the ambient fluid is a known constant h, the overall thermal resistance is the sum of the "cylindrical shell" resistance of the insulation, plus the external convective heat transfer resistance, Eq. (2.36):

$$R_t = \frac{\ln\left(r_o/r_i\right)}{2\pi k l} + \frac{1}{\left(2\pi r_o l\right) h} \tag{2.40}$$

By solving the equation $\partial R_t/\partial r_o = 0$, we find that R_t is *minimum* (or q is maximum) at

$$R_{t,\min} = \frac{\ln\left(k/hr_i\right) + 1}{2\pi k l} \tag{2.41}$$

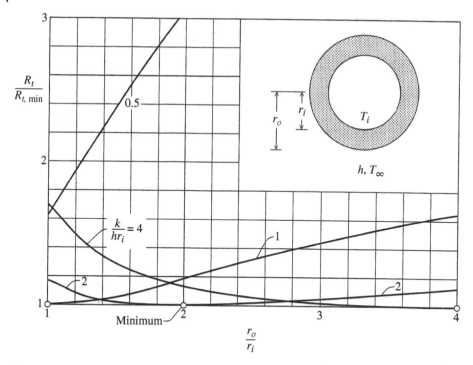

Figure 2.7 Effect of the outer radius on the overall thermal resistance of a cylindrical insulation layer.

and that this minimum occurs when the outer surface of the insulation reaches the *critical radius*:

$$r_{o,c} = \frac{k}{h} \quad \text{(cylinder)} \tag{2.42}$$

Figure 2.7 shows the behavior of the thermal resistance in terms of $R_t/R_{t,\text{min}}$ versus r_o/r_i. The R_t minimum shifts toward the lower r_o values as the dimensionless group k/hr_i decreases, that is, as r_i increases and the inner cylinder becomes "thick." The effect of building a thicker layer of insulation depends on how the radius of the bare cylinder, r_i, compares with the critical outer radius $r_{o,c}$.

In the case of "thick" bare cylinders, $r_i > r_{o,c}$ or $k/hr_i < 1$, the addition of insulation always translates into a higher R_t value, that is, into an "insulation effect." In the opposite extreme, where the bare cylinder is "thin," $r_i < r_{o,c}$ or $k/hr_i > 1$, the wrapping of the very first layer of insulation induces a decrease in the overall thermal resistance. This initial effect is one of heat transfer enhancement, not insulation. Only when enough material has been added so that r_o exceeds $r_{o,c}$, the thickening of the insulation layer increases the R_t value and reduces q.

The behavior of R_t in the thin bare cylinder limit may seem paradoxical, because an insulation effect is intuitively expected when a finite-thickness layer of finite-k material is wrapped on anything. The explanation for the initial drop in R_t as r_o increases lies in the fact that, when the bare cylinder is sufficiently thin, the thickening of the insulation layer leads to an *increase* in the external area ($2\pi r_o l$) that is bathed by fluid. For example, the thermal resistance between a thin electrical wire and the surrounding air can be reduced by coating the wire with a layer of dielectric material.

The same analysis can be performed for an insulation built around a spherical object of radius r_i. The critical outer radius of the insulating layer is in this case

$$r_{o,c} = 2\frac{k}{h} \quad \text{(sphere)} \tag{2.43}$$

where k and h are the thermal conductivity of the layer material and the heat transfer coefficient at the layer-ambient interface, respectively.

There is no "critical thickness" of insulation in an application where both the bare surface and the insulating layer are sufficiently plane, as in the thin-wall systems. The thickening of the insulation layer always leads to a higher R_t value. This conclusion can be drawn without performing an analysis, because the "plane" bare surface (or "thin" insulation layer) is the most extreme case of the class of thick bare cylinders.

Example 2.2 *Cylindrical Shell: Critical Radius*

An uninsulated wire suspended in air generates Joule heating at the rate $q' = 1\,\text{W/m}$. The wire is a bare cylinder of radius $r_i = 0.5\,\text{mm}$, and the temperature difference between it and the atmosphere is 30 °C. It is proposed to cover this wire with a plastic sleeve of electrical insulation, the outer radius of which will be $r_o = 1\,\text{mm}$. The thermal conductivity of the plastic material is $k = 0.35\,\text{W/m·K}$. Will the plastic sleeve improve the wire-ambient thermal contact, or will it provide a thermal insulation effect? To verify your answer, calculate the new wire-ambient temperature difference when the wire is encased in plastic.

Solution

The plastic sleeve has a heat transfer augmentation effect (i.e. it reduces the overall thermal resistance) if the radius of the bare wire is smaller than the critical radius of insulation. To calculate $r_{o,c}$, Eq. (2.42), we must first calculate the heat transfer coefficient (Figure E2.2):

$$h = \frac{q'}{2\pi r_i \left(T_i - T_\infty\right)} = 1\frac{\text{W}}{\text{m}}\frac{1}{2\pi}\frac{1}{0.5 \times 10^{-3}\,\text{m}}\frac{1}{30\,\text{K}}$$

$$= 10.6\,\text{W/m}^2 \cdot \text{K} \tag{1}$$

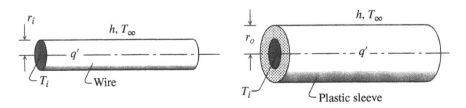

Figure E2.2

The critical radius

$$r_{o,c} = \frac{k}{h} = \frac{0.35\,\text{W}}{\text{m} \cdot \text{K}}\frac{\text{m}^2 \cdot \text{K}}{10.6\,\text{W}} = 3.3\,\text{cm} \tag{2}$$

is much greater than the radius of the bare wire and greater than the outer radius of the plastic sleeve. We can expect, then, a heat transfer augmentation effect (a decrease in $T_i - T_\infty$) from the presence of the plastic sleeve. The new wire-ambient temperature difference $T_i - T_\infty$ follows from the R_t definition $T_i - T_\infty = R_t q = R_t l q'$ where, according to Eq. (2.42), the $R_t l$ value of the wire encased in plastic is

$$R_t l = \frac{\ln\left(r_o/r_i\right)}{2\pi k} + \frac{1}{2\pi r_o h}$$

$$= \frac{\ln\left(1/0.5\right)}{2\pi} \frac{\mathrm{m \cdot K}}{0.35\,\mathrm{W}} + \frac{1}{2\pi \times 10^{-3}\,\mathrm{m}} \frac{\mathrm{m^2 \cdot K}}{10.6\,\mathrm{W}}$$

$$= 15.3\,\mathrm{m \cdot K/W} \tag{3}$$

The new temperature difference is half of what it was before installing the coating:

$$T_i - T_\infty = R_t l q' = 15.3 \frac{\mathrm{m \cdot K}}{\mathrm{W}} 1 \frac{\mathrm{W}}{\mathrm{m}} = 15.3\,\mathrm{K} \tag{4}$$

2.5 Variable Thermal Conductivity

In the problems treated so far, the thermal conductivity was assumed constant. In this section we learn that the preceding thermal resistance formulas are simple cases of a more general result that holds for an arbitrary thermal conductivity function $k(T)$.

The most basic feature of the thin-wall shell, cylindrical or spherical, is the unidirectionality of the heat flow. This is shown in Figure 2.8, in which $A(r)$ is the cross-sectional area penetrated by q at the position r along its path. The heat current q is conserved as it flows through its heat tube; however, the cross-sectional area A may vary with r. In a cylindrical shell, A is proportional to r, whereas in the thin-wall limit, A is a constant.

At any r, the material has a unique temperature $T(r)$ and, since $k = k(T)$, a unique thermal conductivity. The thermal resistance formula follows from

$$q = A(r)\,q''(r) = -A(r)\,k(T)\frac{dT}{dr} \tag{2.44}$$

where the T and r variables can be separated:

$$q\frac{dr}{A(r)} = -k(T)\,dT \tag{2.45}$$

This can be readily integrated because q is constant:

$$q\int_{r_i}^{r_o} \frac{dr}{A(r)} = -\int_{T_i}^{T_o} k(T)\,dT \tag{2.46}$$

The k integral can be rewritten in terms of the *thermal potential* function [1]

$$\theta(T) = \int_{T_{\mathrm{ref}}}^{T} k(T')\,dT' \tag{2.47}$$

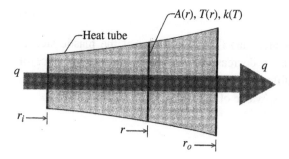

Figure 2.8 Unidirectional conduction through a solid with temperature-dependent thermal conductivity.

in which T_{ref} is a reference temperature, for example, the standard environmental temperature (25 °C, or 298.15 K) in applications near room temperature. Note that the entire $k(T)$ information of Figure 1.5 can be projected on a corresponding $\theta(T)$ chart. The heat transfer rate (2.46) assumes the new form [1]

$$q = \frac{\theta\left(T_i\right) - \theta\left(T_o\right)}{\int_{r_i}^{r_o} \frac{dr}{A(r)}} \tag{2.48}$$

and is equivalent to writing that the thermal resistance, $R_t = (T_i - T_o)/q$, is

$$R_t = \frac{1}{k_{\text{avg}}} \int_{r_i}^{r_o} \frac{dr}{A(r)} \tag{2.49}$$

where k_{avg} is the average thermal conductivity defined by

$$k_{\text{avg}} = \frac{\theta\left(T_i\right) - \theta\left(T_o\right)}{T_i - T_o} \tag{2.50}$$

2.6 Internal Heat Generation

Here we assume that the internal heating is steady and uniform. Consider the thin-wall system shown in Figure 2.9a. The conducting material experiences a uniform volumetric heating rate $\dot{q}$ (W/m^3): its temperature distribution $T(x)$ reaches a steady state because the wall is bathed by a fluid of fixed temperature T_∞, which plays the role of heat sink. The same fluid flow washes both sides of the wall; therefore, the heat transfer coefficient h (assumed constant) has the same value on both sides.

The key unknown here is not the total heat transfer rate, because this is already known as the product between $\dot{q}$ and the total volume of the conducting slab. In the steady state all the heat that is being generated inside the slab is transferred to the fluid reservoir. How much warmer the innards of the slab must become to be able to drive the heat current to the sides?

Since the unknown is $T(x)$, we start with Eq. (1.34) where the right-hand side is zero:

$$\frac{d^2T}{dx^2} + \frac{\dot{q}}{k} = 0 \tag{2.51}$$

The two boundary conditions required by this second-order ordinary differential equation are of the "convective" type:

$$-q'' = h\left(T - T_\infty\right) \quad \text{at} \quad x = -\frac{L}{2} \tag{2.52}$$

$$q'' = h\left(T - T_\infty\right) \quad \text{at} \quad x = \frac{L}{2} \tag{2.53}$$

The negative sign is needed on the left side of Eq. (2.52) because q'' is considered positive when pointing in the x direction, and in the h definition (1.50), q'' was assumed positive when pointing toward the fluid. After invoking the Fourier law on the left side of both Eqs. (2.52) and (2.53), the boundary conditions become

$$k\frac{dT}{dx} = h\left(T - T_\infty\right) \quad \text{at} \quad x = -\frac{L}{2} \tag{2.54}$$

$$-k\frac{dT}{dx} = h\left(T - T_\infty\right) \quad \text{at} \quad x = \frac{L}{2} \tag{2.55}$$

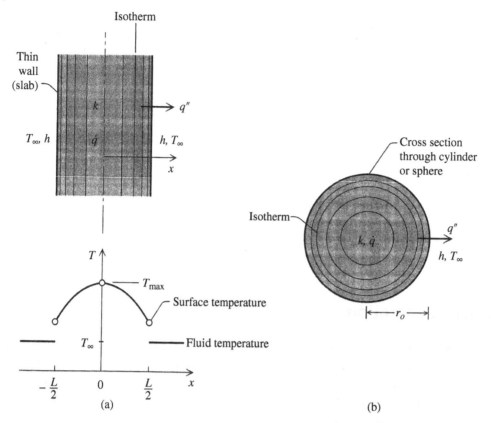

Figure 2.9 Steady temperature distribution due to uniform internal heat generation in a slab (a) and in a cylinder or sphere (b).

Note that the geometry and boundary conditions in Figure 2.9a are symmetric about the midplane. Consequently, we can replace one of these conditions by the statement that the temperature must also be symmetric: $dT/dx = 0$ at $x = 0$.

The solution to Eq. (2.51) has the general form $T = (-\dot{q}/k)(x^2/2) + c_1 x + c_2$. The integration constants are determined by invoking the boundary conditions:

$$T(x) - T_\infty = \frac{\dot{q}L^2}{8k}\left[1 - \left(\frac{x}{L/2}\right)^2\right] + \frac{\dot{q}L}{2h} \tag{2.56}$$

This parabolic temperature distribution is illustrated in the lower part of Figure 2.9a. The isotherms are closer together near the $x = -L/2$ and $x = L/2$ boundaries, indicating that the heat flux (or the slope $|dT/dx|$) is greatest on the boundaries. The maximum temperature occurs in the midplane of the slab, $T_{max} = T(0)$, and its value increases as $\dot{q}$ increases:

$$T_{max} - T_\infty = \frac{\dot{q}L^2}{8k}\left(1 + \frac{4}{Bi}\right) \tag{2.57}$$

Related results are the surface temperature $T(L/2) = T_\infty + (\dot{q}L/2h)$ and the temperature difference between the midplane and the slab surface, $T(0) - T(L/2) = \dot{q}L^2/8k$. The dimensionless factor Bi is the

Biot number[1] based on the overall thickness L:

$$\mathrm{Bi} = \frac{hL}{k} \tag{2.58}$$

The Biot number can be viewed as a "dimensionless heat transfer coefficient." The order of magnitude of Bi divides the applications of Figure 2.9a into two categories:

Bi ≫ 1. In this range the thermal contact between the solid boundary and fluid flow is "good," so that the boundary temperature $T(\pm L/2)$ approaches the temperature of the fluid T_{∞}.

Bi ≪ 1. In cases where the solid–fluid thermal contact is "poor," or when the solid is an extremely good thermal conductor, the boundary temperature is nearly the same as the maximum (midplane) temperature of the slab. The temperature profile illustrated in Figure 2.9a becomes flatter as Bi decreases.

Similar conclusions are reached in the study of a *cylindrical body* that is being heated volumetrically, Figure 2.9b. The radial distribution of temperature in the cylinder is

$$T(r) - T_{\infty} = \frac{\dot{q} r_o^2}{4k} \left[1 - \left(\frac{r}{r_0} \right)^2 \right] + \frac{\dot{q} r_o}{2h} \tag{2.59}$$

Similarly, the temperature distribution through a volumetrically heated *spherical body* is

$$T(r) - T_{\infty} = \frac{\dot{q} r_o^2}{6k} \left[1 - \left(\frac{r}{r_0} \right)^2 \right] + \frac{\dot{q} r_o}{3h} \tag{2.60}$$

where r_o is the outer radius of the sphere and h is the convective heat transfer coefficient.

2.7 Evolutionary Design: Extended Surfaces (Fins)

2.7.1 The Enhancement of Heat Transfer

In this section we see how a change in the geometry of the surface leads to a better thermal contact with the fluid. The design change consists of shaping certain portions of the surface so that they protrude into the fluid: the external surface of the solid protrusions constitutes the *extended surface* of the wall–fluid contact area, and the protrusions themselves are the *fins*. This is illustrated in Figure 2.10. In the absence of fins, the heat transfer rate between the solid wall and the external flow is

$$q_0 = hA_0 \left(T_b - T_{\infty} \right) \tag{2.61}$$

where A_0 is the area of the original (bare) surface and h is the wall–fluid heat transfer coefficient. The heat flux q_0/A_0 is distributed uniformly over the bare surface, because both h and the wall–fluid temperature difference $T_b - T_{\infty}$ are assumed constant.

The right side of Figure 2.10 shows what happens to the distribution of wall heat flux when fins are present. For clarity, it was assumed that the h value on the fin surfaces is the same as on the bare portions of the base surface. In practice, the heat transfer coefficient would be lower on the base surface than on the fin.

---

1 Jean Baptiste Biot (1774–1862) was a French physicist who worked for the first time on the problem of solid-body conduction with convection at the surface. He is best known for his work on electromagnetism (e.g. the Biot–Savart law), electrostatics, the properties of water at saturation, and the first balloon flight for scientific research.

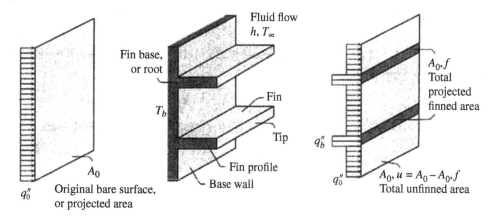

Figure 2.10 Increase in wall heat flux over the area covered by fins.

The use of fins leads to an increase in the total area of contact between solid surfaces and fluid, because the sum of the total external area of all the fins plus the original area without any fins is greater than the bare surface A_0. This does not guarantee an increase in the total heat transfer rate q between the solid body and the fluid. A net increase (i.e. $q > q_0$) is registered when the fin material has a sufficiently high-thermal conductivity so that the temperature of the fin surface is comparable with the temperature of the base surface, T_b. In such a case the heat flux that passes into the fluid through the lateral surface of the fin (i.e. vertically in Figure 2.10) is comparable in magnitude with the bare surface heat flux q_0/A_0. And since the total heat transfer rate that is discharged into the fluid through the wetted (exposed) surface of the fin must be equal to the heat transfer rate that – through its root – the fin "pulls" out of the wall, the fins augment the distribution of heat flux over the projected area A_0.

The new heat flux distribution q/A_0 has the original bare surface heat flux q_0/A_0 as background level for a series of heat flux "spikes." Each spike occurs over that spot of the projected area that is occupied by the base (root) of one fin. The degree to which the addition of fins augments the total heat transfer rate between solid wall and fluid is expressed by the overall projected *surface effectiveness* ratio:

$$\varepsilon_0 = \frac{q}{q_0} = \frac{q}{hA_0\left(T_b - T_\infty\right)} \tag{2.62}$$

Considering the difficulty of manufacturing a fin-covered surface, meaningful augmentation of heat transfer occurs when ε_0 is significantly greater than 1.

The design of finned surfaces reduces to being able to predict the heat transfer rate q, or the overall effectiveness ε_0, that is associated with a certain fin surface geometry, temperature difference $T_b - T_\infty$, convective heat transfer coefficient h, and thermal conductivity of fin material. Regard the original bare surface A_0 as the sum of the total projected area covered by the roots of fins $A_{0,f}$ and the remaining unfinned portion $A_{0,u}$:

$$A_0 = A_{0,f} + A_{0,u} \tag{2.63}$$

Then, by looking at the ragged shape of the heat flux distribution when fins are present, q/A_0, we see that the total heat transfer rate q is the sum of two contributions:

$$q = q_b'' A_{0,f} + hA_{0,u}\left(T_b - T_\infty\right) \tag{2.64}$$

In the first term on the right side, q_b'' is the average heat flux through the base of one fin. Therefore, the calculation of the fin-base heat flux is the focus of the analysis described next.

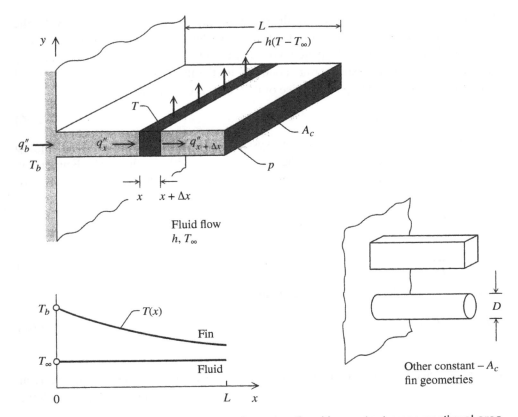

Figure 2.11 Longitudinal conduction through a fin with constant cross-sectional area.

2.7.2 Constant Cross-Sectional Area

The simplest fin geometry to consider goes back to the founding fathers of conduction heat transfer, Biot and Fourier [2]: the fin whose cross-sectional area A_c is independent of the longitudinal position x along the fin. Figure 2.11 features one example of such a fin, namely, the plate fin geometry that appeared earlier in Figure 2.10. Other examples are the cylindrical spine with constant circular cross section (the "pin fin") and the prismatic fin inserted in the lower-right portion of Figure 2.11.

2.7.2.1 The Longitudinal Conduction Model

The key assumption on which the analysis rests is that the temperature T inside the fin is only a function of x. Accordingly, heat is conducted through the fin longitudinally, even though in reality most of this heat current escapes into the fluid through the lateral (exposed) surfaces. The $T = T(x)$ assumption is a good approximation only under special conditions, which will be determined soon. The benefit derived from making this assumption is that the chief unknown (the fin-base heat flux) is the y-independent quantity:[2]

$$q_b'' = -k \left(\frac{dT}{dx} \right)_{x=0} \tag{2.65}$$

---

2 If the $T = T(x)$ assumption does not apply, then T is a function of both x and y, where y is the vertical coordinate in Figure 2.11. The base heat flux in this case is still given by Eq. (2.65); however, q_b'' is a function of y.

The next step is the derivation of the fin temperature distribution $T(x)$. For this, we cannot invoke the conduction equation for one-dimensional conduction in Cartesian coordinates, Eq. (1.21), because unlike the solid bar of Figure 1.3, the lateral surface of the plate fin of Figure 2.11 is not insulated. Indeed, the only reason the plate "works" as a fin is that it makes thermal contact with the fluid (h, T_∞) all along its lateral surface. The steady-state conservation of energy in the slice of thickness Δx in Figure 2.11 is

$$q_x'' A_c - q_{x+\Delta x}'' A_c - (p\Delta x)\, h\left(T - T_\infty\right) = 0 \tag{2.66}$$

The last term represents the heat transfer rate from the Δx system to the surrounding fluid. The length p is the *wetted perimeter* of the fin cross section, that is, the line of contact with the fluid. The Fourier law and the assumption that the thermal conductivity k is constant allow us to rewrite the first two terms of Eq. (2.66) as

$$A_c\left(q_x'' - q_{x+\Delta x}''\right) = -A_c \frac{dq_x''}{dx}\Delta x = -A_c \frac{d}{dx}\left(-k\frac{dT}{dx}\right)\Delta x$$

$$= kA_c \frac{d^2 T}{dx^2}\Delta x \tag{2.67}$$

Combining Eqs. (2.66) and (2.67) and dividing by Δx, we obtain

$$kA_c \underbrace{\frac{d^2 T}{dx^2}}_{\substack{\text{Longitudinal}\\\text{conduction}}} - \underbrace{hp\left(T - T_\infty\right)}_{\substack{\text{Lateral}\\\text{convection}}} = 0 \tag{2.68}$$

This equation expresses the balance between the net conduction heat current that arrives longitudinally at the location x and the convective heat transfer that leaves the fin through the lateral line of contact with the fluid.

2.7.2.2 Long Fin

Assume that the fin is so long that its tip region is in thermal equilibrium with the surrounding fluid:

$$T \to T_\infty \quad \text{as} \quad x \to \infty \tag{2.69}$$

The other boundary condition on the fin temperature $T(x)$ is that the temperature at the root of the fin is the same as that of the base wall:

$$T = T_b \quad \text{at} \quad x = 0 \tag{2.70}$$

The $T(x)$ problem consists of solving Eq. (2.68) subject to Eqs. (2.69) and (2.70); it is customary to restate the problem in terms of the *excess temperature* functions

$$\theta(x) = T(x) - T_\infty \tag{2.71}$$

$$\frac{d^2\theta}{dx^2} - m^2\theta = 0 \tag{2.72}$$

$$\theta = \theta_b \quad \text{at} \quad x = 0 \quad \left(\text{note: } \theta_b = T_b - T_\infty\right) \tag{2.73}$$

$$\theta \to 0 \text{ as } x \to \infty \tag{2.74}$$

where m is a crucial parameter of the fin–fluid arrangement:

$$m = \left(\frac{hp}{kA_c}\right)^{1/2} \tag{2.75}$$

In what follows, we assume that h has the same value along the fin surface; therefore, we treat m as a constant. The general solution to Eq. (2.72) is

$$\theta(x) = C_1 \exp(-mx) + C_2 \exp(mx) \tag{2.76}$$

for which the constants C_2 and C_1 are determined by invoking the boundary conditions (2.74) and (2.73):

$$C_2 = 0 \quad C_1 = \theta_b \tag{2.77}$$

In conclusion, the temperature of the fin material decays exponentially to the fluid temperature sufficiently far from the base:

$$\theta = \theta_b \exp(-mx) \tag{2.78}$$

that is, when the dimensionless product mx is significantly greater than 1. The lateral convective heat flux, $h(T - T_\infty) = h\theta$, decays to zero in the same manner as the temperature excess function $\theta(x)$. If L is the actual length of the fin, Figure 2.11, then the fin can be regarded as being "long enough" (as in the boundary condition (2.74)) when

$$mL \gg 1 \tag{2.79}$$

The heat flux through the base is calculated next by using Eq. (2.65), which yields $q_b'' = k\theta_b m$. Therefore, the total heat current that the fin pulls from the base wall is

$$q_b = q_b'' A_c = \theta_b \left(kA_c hp\right)^{1/2} \tag{2.80}$$

In this remarkably simple result, we see how all the physical parameters of the fin–fluid configuration affect the ultimate size of the important quantity q_b. The total heat transfer rate drawn from the base wall increases if any of the parameters θ_b, k, A_c, h, or p increases. In particular, q_b is equally sensitive to changes in the fin conductivity and the fin–fluid heat transfer coefficient.

2.7.2.3 Fin with Insulated Tip

Most fin designs do not meet the long-fin criterion (2.79); therefore, their "finite" length L must be taken into consideration. The physical consequence of the finite length is that the temperature of the tip $T(L)$ is higher than the ambient temperature T_∞. This temperature difference drives a heat current through the tip of the fin:

$$q_{tip} = hA_c \left[T(L) - T_\infty\right] \tag{2.81}$$

For simplicity, it is assumed that the h value on the tip is the same as that on the lateral surfaces of the fin. An intermediate step toward the more general case considered in Section 2.7.2.4 is the fin with insulated tip:

$$\frac{dT}{dx} = 0 \quad \text{or} \quad \frac{d\theta}{dx} = 0 \quad \text{at} \quad x = L \tag{2.82}$$

This is a good approximation when the heat current that passes through the tip is negligible relative to the total heat current drawn from the base wall:

$$q_b \gg q_{tip} \tag{2.83}$$

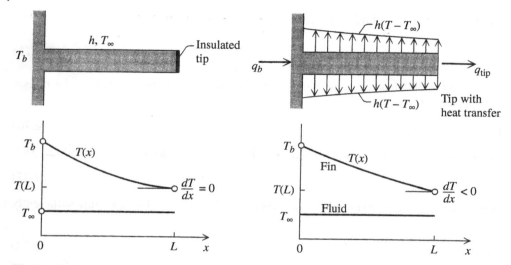

Figure 2.12 Fin with insulated tip versus fin with finite heat transfer rate through the tip.

The general solution to the fin conduction equation is in Eq. (2.76) or, alternatively,[3]

$$\theta\,(x) = C_1' \sinh\,(mx) + C_2' \cosh\,(mx) \tag{2.84}$$

The boundary conditions (2.73) and (2.82) produce, in order,

$$\theta_b = C_2' \tag{2.85}$$

$$0 = C_1' \cosh\,(mL) + C_2' \sinh\,(mL) \tag{2.86}$$

and, after substituting C_1' and C_2' into Eq. (2.84),

$$\theta = -\theta_b \tanh\,(mL)\sinh\,(mx) + \theta_b \cosh\,(mx)$$
$$= \theta_b \frac{\cosh\,[m\,(L - x)]}{\cosh\,(mL)} \tag{2.87}$$

This solution is illustrated in Figure 2.12a. The excess temperature that occurs at the tip can be deduced by setting $x = L$ in Eq. (2.87):

$$\theta\,(L) = \frac{\theta_b}{\cosh\,(mL)} \tag{2.88}$$

The total heat transfer rate flowing from the base wall into the root of the fin is

$$q_b = A_c\left(-k\frac{dT}{dx}\right)_{x=0}$$
$$= \theta_b\left(kA_c hp\right)^{1/2} \tanh\,(mL) \tag{2.89}$$

---

3 The relations between exponentials and hyperbolic sine and cosine are

$$\exp\,(u) = \cosh\,(u) + \sinh\,(u)$$
$$\exp\,(-u) = \cosh\,(u) - \sinh\,(u)$$

Other relations are listed in Appendix E. If we replace the exponentials, we obtain

$$\theta\,(x) = \left(C_2 - C_1\right)\sinh\,(mx) + \left(C_1 + C_2\right)\cosh\,(mx)$$

This shows that the new constants C_1' and C_2' in Eq. (2.85) are, respectively, $C_1 - C_2$ and $C_1 + C_2$.

The long-fin results of the Section 2.7.2.2 are the limit $mL \to \infty$ of the solution that was just completed. In that limit the tip excess temperature $\theta(L)$ of Eq. (2.88) approaches zero, as in Eq. (2.74), and the total heat transfer rate through the fin, Eq. (2.89), approaches Eq. (2.88).

The "insulated tip" assumption is an adequate approximation when the tip heat transfer condition (2.83) is satisfied. Combining Eqs. (2.81) and (2.88) into a formula for q_{tip}, we can use also Eq. (2.89) to rewrite the (2.83) inequality as

$$\frac{q_{tip}}{q_b} = \frac{1}{\sinh(mL)} \left(\frac{hA_c}{kp} \right)^{1/2} \ll 1 \tag{2.90}$$

In the case of finite-length fins, mL and $\sinh(mL)$ are finite quantities. This leaves hA_c/kp as the principal dimensionless group that governs the relative size of the heat transfer rate through the tip. A fin with a heat transfer coefficient and a cross-sectional area so small that the tip thermal conductance hA_c is much smaller than kp is represented adequately by the insulated tip model. Equation (2.90) shows that the same model is also applicable in the limit $mL \to \infty$ (recall that in the long fin case, the temperature gradient at the tip is zero, because the temperature near the tip is constant and equal to T_∞).

2.7.2.4 Heat Transfer Through the Tip

The insulated tip assumption can be relaxed by replacing the boundary condition (2.82) with

$$-kA_c \frac{d\theta}{dx} = hA_c\theta \quad \text{at} \quad x = L \tag{2.91}$$

in which we continue to assume that the h value on the tip surface is the same as on the lateral surfaces. In this boundary condition the left side represents the conduction heat transfer rate that arrives at the solid side of the tip surface, while the right side is the convective heat current removed by the fluid flow. Both sides of the equation are equal to q_{tip}, Eq. (2.81). The temperature distribution in this general case (Figure 2.12b) can be determined by combining the solution (2.84) with the boundary conditions (2.73) and (2.91):

$$\theta = \theta_b \frac{\cosh[m(L-x)] + (h/mk)\sinh[m(L-x)]}{\cosh(mL) + (h/mk)\sinh(mL)} \tag{2.92}$$

The total heat transfer rate could be calculated by using the top line of Eq. (2.89). However, a much more compact q_b formula is recommended by an approximate argument due to Harper and Brown [3]. The upper part of Figure 2.13 shows the actual fin with heat transfer through the tip. The fin length is L. The lower part of Figure 2.13 shows an alternative exit for the tip heat transfer rate, namely, a fin length $(L_c - L)$ that is attached to the end of the original fin. The right side of this additional chunk is insulated. The heat current q_{tip} enters it from the left (through the $x = L$ plane) and leaves it by crossing the top and bottom surfaces.

The total heat transfer rate of the original fin, q_b, is the same as that of a longer fin with insulated tip. According to Eq. (2.89), then,

$$q_b = \theta_b (kA_c hp)^{1/2} \tanh(mL_c) \tag{2.93}$$

in which L_c is the *corrected length* that defines the longer fin. The problem of calculating q_b reduces to that of estimating the proper L_c value for the original fin. This last step consists of invoking the conservation of energy in the chunk of length $L_c - L$:

$$hA_c \left[T(L) - T_\infty \right] = hp \left(L_c - L \right) \left[T(L) - T_\infty \right] \tag{2.94}$$

where the left side is the heat current through the tip of the actual fin and the right side is the convective heat transfer through the lateral area of the added segment, $p(L_c - L)$. We are assuming that the $L_c - L$ segment is

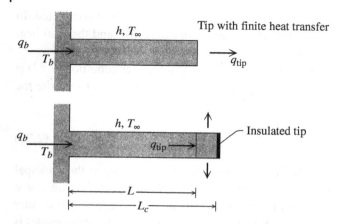

Figure 2.13 Geometrical reasoning behind the concept of corrected length, L_c.

short enough so that its temperature is approximately equal to the tip temperature of the original fin, $T(L)$. The corrected length formula dictated by Eq. (2.94) is

$$L_c = L + \frac{A_c}{p} \tag{2.95}$$

For example, in the case of a plate fin of thickness t and width W, we have $A_c = tW$ and $p = 2(W + t) \cong 2W$; therefore,

$$L_c = L + \frac{t}{2} \quad \text{(plate fin)} \tag{2.96}$$

Similarly, the geometry of a pin fin of constant diameter D is characterized by $A_c = \pi D^2/4$ and $p = \pi D$. Therefore,

$$L_c = L + \frac{D}{4} \quad \text{(pin fin, or cylindrical spine)} \tag{2.97}$$

The corrected length (2.93) is always an accurate substitute for the exact q_b formula based on Eq. (2.92) when the fin is "long" ($mL \gg 1$), because in this limit the heat transfer through the tip is negligible, Eq. (2.90). In the opposite extreme, $mL \ll 1$, the error associated with using the approximate (2.93) instead of the exact q_b expression based on Eq. (2.92) is approximately equal to $hA_c/3kp$, provided the Biot number based on fin thickness is small.

2.7.2.5 Fin Efficiency

A ratio that describes the performance of the fin is the *fin efficiency* η:

$$\eta = \frac{\text{actual heat transfer rate}}{\substack{\text{maximum heat transfer rate when} \\ \text{the entire fin is at } T_b}} = \frac{q_b}{hpL_c\theta_b} \tag{2.98}$$

The denominator in this definition is greater than q_b because in reality the fin temperature decreases along the fin. Consequently, the actual convective heat flux that crosses the wetted surface, $h\theta$, is consistently smaller than the ideal value $h\theta_b$ listed in the denominator (see the graphic definition of η in Figure 2.14). The range covered by the fin efficiency of any fin is therefore $0 < \eta < 1$.

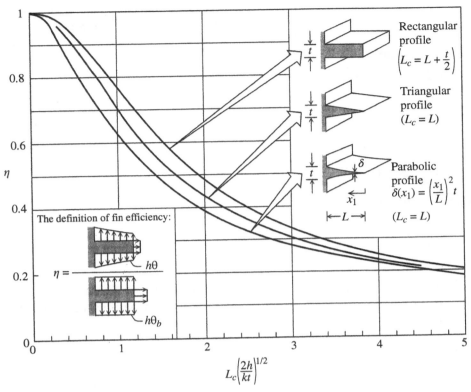

Figure 2.14 Efficiency of two-dimensional fins with rectangular, triangular, and parabolic profiles. Source: Gardner [4].

In view of Eq. (2.93), the efficiency of a fin with constant cross section is

$$\eta = \frac{\tanh\left(mL_c\right)}{mL_c} \tag{2.99}$$

for which m is defined in Eq. (2.75). This function is displayed in Figure 2.14, next to the efficiency curves of other fin geometries. Note that in the case of the wide plate fin, the abscissa parameter used in Figure 2.14, $L_c(2h/kt)^{1/2}$, is identical to the group mL_c. The efficiency drops significantly when the abscissa parameter is considerably greater than 1: long fins are "poor" because their tip temperature approaches the fluid temperature, and the convective heat fluxes $h\theta$ that cover the wetted surface of the fin are considerably smaller than their ceiling value $h\theta_b$ (see the lower left corner of Figure 2.14).

2.7.2.6 Fin Effectiveness

An alternative figure of merit is the *fin effectiveness* (ε_f):

$$\varepsilon_f = \frac{\text{total fin heat transfer}}{\substack{\text{the heat transfer that would have} \\ \text{occurred through the base area} \\ \text{in the absence of the fin}}} = \frac{q_b}{hA_c\theta_b} \tag{2.100}$$

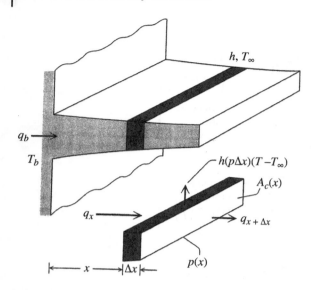

Figure 2.15 Longitudinal conduction through a fin with variable cross-sectional area and wetted perimeter.

If the fin is to perform its heat transfer augmentation function properly, then ε_f must be significantly greater than 1. It follows that in the case of a good fin the effectiveness value is greater than the efficiency value, the relation between the two being

$$\frac{\varepsilon_f}{\eta} = \frac{pL_c}{A_c} = \frac{\text{total fluid contact area}}{\text{cross-sectional area}} \tag{2.101}$$

The fin effectiveness ε_f is also greater than the overall projected-surface effectiveness ε_0, which was defined in Eq. (2.62). The relationship between ε_0 and ε_f is

$$\varepsilon_0 = \varepsilon_f \frac{A_{0,f}}{A_0} + \frac{A_{0,u}}{A_0} \tag{2.102}$$

To summarize, the total heat transfer rate q_b can be calculated based on formulas such as Eq. (2.93) or on fin efficiency charts of the type exhibited in Figure 2.14. Many more q_b results have been developed for other fin geometries [5].

2.7.3 Variable Cross-Sectional Area

Fins with variable cross-sectional area can be analyzed similarly. Central is the assumption that the conduction heat transfer is oriented longitudinally through the fin. Figure 2.15 shows a fin geometry in which the cross-sectional area A_c and wetted perimeter p are two functions of the longitudinal position x. The conduction equation for this configuration can be deduced by recognizing first the steady-state energy conservation requirement for the slice of thickness Δx:

$$q_x - q_{x+\Delta x} - (p\Delta x)h\left(T - T_\infty\right) = 0 \tag{2.103}$$

in which the longitudinal heat current is proportional to the local temperature gradient, $q_x = -kA_c(dT/dx)$. When A_c varies along the fin, Eq. (2.103) becomes

$$\frac{d}{dx}\left(kA_c\frac{dT}{dx}\right) - hp\left(T - T_\infty\right) = 0 \tag{2.104}$$

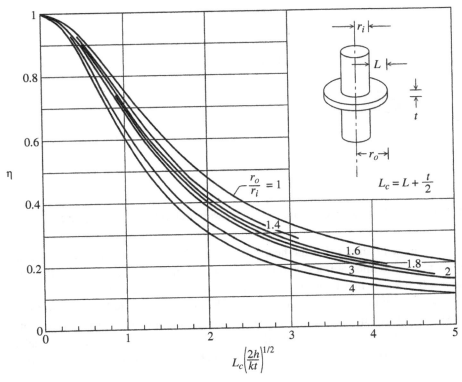

Figure 2.16 Efficiency of annular fins with constant thickness. Source: Gardner [4].

The objective in the analysis of a specified variable geometry, $A_c(x)$ and $p(x)$, is to calculate the total heat transfer rate between the fin and the surrounding fluid. This heat transfer rate is equal to the total heat current that passes through the base of the fin:

$$q_b = -\left(kA_c(x)\frac{dT}{dx}\right)_{x=0} \tag{2.105}$$

The end result is the fin efficiency η defined in Eq. (2.98):

$$\eta = \frac{q_b}{hA_{exp}(T_b - T_\infty)} \tag{2.106}$$

where A_{exp} is the exposed surface of the fin, that is, the area bathed by fluid. Three examples of fin efficiencies of variable-A_c fin geometries are presented in Figures 2.14 and 2.16. Note that in the two-dimensional fins with triangular and parabolic profiles shown in Figure 2.14, only the cross-sectional area varies with the longitudinal position (the wetted perimeter is twice the width of the fin, i.e. constant). In a disc-shaped fin, both A_c and p vary with the local radial position r, which here plays the role of longitudinal coordinate for the unidirectional heat current, as it did in Figure 2.8.

2.7.4 Scale Analysis: When the Unidirectional Conduction Model Is Valid

The fin analyses discussed until now are all based on the assumption that the conduction current is unidirectional. This is why "extended surfaces" belong in this chapter. In fins, however, the unidirectional conduction

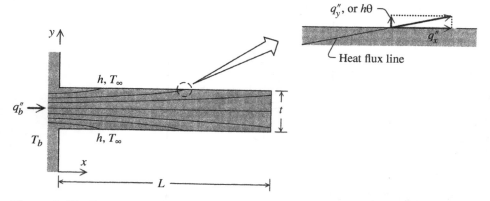

Figure 2.17 Pattern of heat flux lines through a two-dimensional fin with rectangular profile (plate fin).

model is valid approximately. Figure 2.17 shows that the heat flux lines inside the fin are not truly longitudinal and parallel, because most of these lines are destined to cross the lateral fin surface that is bathed by fluid. Even in the base plane $x = 0$, the heat flux lines are not truly equidistant and parallel.

The key question is: "Under what conditions is the longitudinal conduction model valid?" We answer this by the method of scale analysis [6]. Consider the two-dimensional fin with rectangular profile (the plate fin, Figure 2.17). A heat flux line that intersects the exposed surface of the fin is oriented longitudinally "sufficiently" if the longitudinal heat flux q_x'' is considerably greater than the transversal heat flux q_y'' that crosses the exposed area:

$$q_x'' \gg q_y'' \tag{2.107}$$

As the "scale," or representative order of magnitude of q_x'', we can use the heat flux through the base of the fin, $q_b'' = q_x/A_c$, which follows from Eq. (2.93). Next, the transversal heat flux q_y'' is the same as the convective heat flux on the fluid side of the exposed surface, $h\theta$; the order of magnitude of this heat flux is $h\theta_b$. Putting these conclusions together, in place of the inequality (2.107), we have

$$\frac{q_b}{A_c} \gg h\theta_b \tag{2.108}$$

In view of Eq. (2.100), this validity condition is the same as $\varepsilon_f \gg 1$.

In conclusion, the validity of the unidirectional conduction model improves as the single-fin effectiveness ε_f becomes considerably greater than 1. Conversely, the unidirectional conduction model is a poor description for a fin whose ε_f value is only marginally greater than 1. Such a low ε_f value is first and foremost a warning that the formulas that were employed in the calculation (e.g. Eq. (2.93)) *do not apply*. In other words, a calculated value of $\varepsilon_f \cong 1$ using a unidirectional conduction model does not necessarily mean that the fin is ineffective as a heat transfer augmentation device. It means that a two-dimensional conduction theory (e.g. Chapter 3) should be used for the purpose of calculating the true heat transfer rate through the base of the fin and, eventually, the correct ε_f value.

The validity criterion (2.108) acquires new meaning if we rewrite it using Eqs. (2.89) and (2.100):

$$\left(\frac{kp}{hA_c}\right)^{1/2} \tanh{(mL)} \gg 1 \tag{2.109}$$

The order of magnitude of $\tanh(mL)$ is 1, especially for long fins (Section 2.7.2.2). What remains is the factor, $(kp/hA_c)^{1/2}$, which assumes special forms for specific fin geometries. For example, for a plate fin of thickness t and width W, the order-of-magnitude inequality (2.109) reduces to

$$\left(\frac{ht}{k}\right)^{1/2} \ll 1 \quad \text{(plate fin)} \tag{2.110}$$

The validity criterion for unidirectional conduction in a pin fin of length L and diameter D is

$$\left(\frac{hD}{k}\right)^{1/2} \ll 1 \quad \text{(pin fin)} \tag{2.111}$$

Note that in accord with the method of scale analysis [6], numerical factors of the order of 1 (e.g. the factor 2) have been left out of the order-of-magnitude statements (2.110) and (2.111).

On the left-hand side of both Eqs. (2.110) and (2.111), we see the square root of the *Biot number* based on the transversal length scale of the fin. In conclusion, the use of the unidirectional conduction model is particularly appropriate in the analysis of fins where the thermal conductivity is so great (or the convective heat transfer coefficient is so small) that the thickness-based Biot number is smaller than 1.

2.7.5 Fin Shape Subject to Volume Constraint

The sizing procedure for a fin generally requires the selection of more than one physical dimension (length). In a plate fin, for example, there are two such dimensions, the length of the fin, L, and the thickness, t. One of these dimensions can be selected independently when the total volume of fin material is fixed. Such a constraint is often justified by the high cost of the high-thermal conductivity metals that are employed in the manufacture of finned surfaces (e.g. copper, aluminum) and by the cost associated with the weight of the fin (e.g. the finned cylinders of air-cooled motors for model airplanes and motorcycles).

To select one of the two dimensions of the fin means to select its profile shape such that the total heat transfer rate q_b is maximum when the fin volume V is fixed:

$$V = LtW \tag{2.112}$$

For simplicity, let us assume that the fin profile is sufficiently slender ($t \ll L$) so that $L_c \cong L$. This means that q_b can be calculated by using the insulated tip formula (2.89), in which $A_c = tW$ and $p = 2W$, where W is the width of the plate fin. Eliminating L using Eq. (2.112), we obtain

$$q_b = at^{1/2} \tanh\left(bt^{-3/2}\right) \tag{2.113}$$

in which a and b are two constants:

$$a = \theta_b W(2kh)^{1/2} \quad b = \frac{V}{W}\left(\frac{2h}{k}\right)^{1/2} \tag{2.114}$$

The maximum of the total heat current q_b with respect to the plate fin thickness t is found next by solving the equation $dq_b/dt = 0$, which is the same as the transcendental equation

$$\sinh\left(2bt^{-3/2}\right) = 6bt^{-3/2} \tag{2.115}$$

The solution is $bt^{-3/2} = 1.4192$, which yields the optimum plate fin thickness

$$t = 0.998\left(\frac{V}{W}\right)^{2/3}\left(\frac{h}{k}\right)^{1/3} \tag{2.116}$$

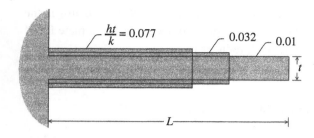

Figure 2.18 Scale drawing of the optimum profile of a plate fin of fixed volume.

In view of the volume constraint (2.112), the corresponding length of the plate fin is

$$L = 1.002 \left(\frac{V}{W}\right)^{1/3} \left(\frac{h}{k}\right)^{-1/3} \tag{2.117}$$

The corresponding slenderness ratio of the rectangular profile of the plate fin is

$$\frac{t}{L} = 0.996 \left(\frac{ht}{k}\right)^{1/2} \tag{2.118}$$

The conclusion drawn at the end of the Section 2.7.4 was that the thickness-based Biot number (ht/k) must be small if the longitudinal conduction model is to be valid. Consequently, the slenderness ratio recommended by this optimum design must also be smaller than 1. This discovery, incidentally, justifies the assumption adopted immediately under Eq. (2.112), at the start of this section.

Figure 2.18 shows a sequence of three shapes calculated with Eq. (2.118); the optimum rectangular profile becomes more slender as the Biot number (ht/k) decreases. Throughout this sequence the volume of the plate fin (in Figure 2.18, the profile area $t \times L$) remains constant. It can also be shown that the q_b extremum pinpointed by the solution to Eq. (2.115) is indeed a maximum [7, 8]:

$$q_{b,\mathrm{max}} = 1.258 k W \theta_b \left(\frac{ht}{k}\right)^{1/2} \tag{2.119}$$

2.7.6 Heat Tube Shape

In each of the conduction configurations treated in this chapter, the total heat current flows through a bundle of heat tubes that are defined by the lines (or surfaces) along which the heat flux is oriented. Heat tubes are illustrated in Figures 2.1, 2.4, 2.8, and 2.17. In most cases, the heat tubes have variable cross section. Which heat tube shape offers better access to the heat current that is guided by it?

This problem was formulated and solved in Ref. [8] with regard to Figure 2.19. The area normal to the heat current q is an unspecified function of longitudinal position, $A(x)$. There is one material, and its conductivity is $k(T)$. The heat current is driven by the temperature difference ($T_0 - T_L$), which is applied between the $x = 0$ and L ends of the heat tube. The length of the heat tube (L) and its total volume (V) are fixed:

$$\int_0^L A(x)\,\mathrm{d}x = V, \text{ constant} \tag{2.120}$$

The heat current q is not a function of x:

$$q = -k(T)A(x)\frac{\mathrm{d}T}{\mathrm{d}x} \tag{2.121}$$

Figure 2.19 One-dimensional conduction along a heat tube with fixed length and volume. Source: Jany and Bejan [8].

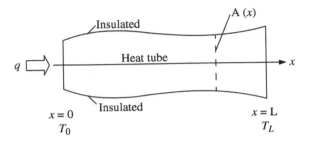

Integrating the above equation over the entire path traveled by q, we obtain

$$\int_0^L \frac{dx}{A(x)} = \frac{1}{q} \int_{T_L}^{T_0} k(T)\, dT \tag{2.122}$$

The thermal conductivity integral appearing on the right-hand side may be rewritten using the thermal potential function, Eq. (2.47):

$$\int_0^L \frac{dx}{A(x)} = \frac{\theta(T_0) - \theta(T_L)}{q} \tag{2.123}$$

The physical meaning of this result and Eq. (2.48) is that the heat current q is impeded by a thermal resistance proportional to $\int_0^L dx/A$.

The function $A(x)$ that minimizes the thermal resistance integral of Eq. (2.122) subject to volume integral (2.120) was determined by variational calculus [8]. The problem is equivalent to minimizing the aggregate integral

$$I = \int_0^L \left(\frac{1}{A} + \lambda A\right) dx \tag{2.124}$$

subject to no constraints. Note in the integrand of this new integral the linear combination of the integrands of the volume constraint (2.120) and thermal resistance integral (2.123). The variational calculus solution is

$$A_{opt} = \lambda^{-1/2}, \text{constant} \tag{2.125}$$

for which the constant $\lambda^{-1/2}$ is determined by substituting Eq. (2.125) into the volume constraint. The final form of the solution obtained in this manner is

$$A_{opt} = \frac{V}{L}, \text{constant} \tag{2.126}$$

Schmidt [7] invoked this design intuitively and discovered the fin profile with parabolic shape, as shown in Problem 2.19 for a fin with $k = $ constant. In that case the temperature varies linearly along the fin.

The design conclusions associated with the preceding analysis are more general, as the variational calculus solution (2.126) applies to a conducting heat tube of arbitrary conductivity, $k(T)$. Optimum fin profiles may be pursued in the same manner, by combining the differential energy equation of the extended surface with the conclusion

$$k(T) \frac{dT}{dx} = \text{constant} \tag{2.127}$$

where k is a known function of T. Equation (2.127) follows from (2.126) and (2.121). The optimum temperature distribution along the heat tube (and along the fins) is obtained by solving Eq. (2.127) for the unknown function $T(x)$.

The more recent literature on the designs and performance of tapered fins and of their counterparts that leak heat by convection is reviewed in Ref. [9].

2.7.7 Rewards from Freedom

The enhancement of heat transfer for cooling applications has generated a wide variety of surface configurations such as extended surfaces (fins) and "heat spreaders," which are conducting plates that make the wall temperature more uniform in the vicinity of the heat sources that reside on the wall. The heat spreader [10] performs as a fin, except that one of its lateral surfaces (bottom, Figure 2.20) is cooled by contact with a solid wall. The literature on such techniques is voluminous (for reviews, see Refs. [11, 12]).

Heat spreading and finning together are one technique (Figure 2.20) that illustrates the phenomenon of evolutionary design [10]. According to the constructal law, improvements in the overall performance of a

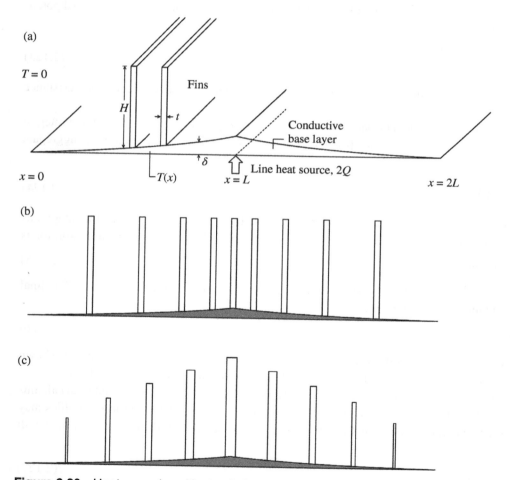

Figure 2.20 Heat spreader with steady line heat source and fins on the upper side [10]: (a) variable base thickness, (b) uniform fins spread nonuniformly, and (c) nonuniform fins spread equidistantly. Source: Matsushima et al. [10].

flow system follow as a result of endowing the flow system with more freedom to morph. This means two things: (i) the heat spreader (the base) that conducts heat away from the source, along the wall, is free to morph its thickness and (ii) the population of fins that are installed on the base is free to morph as well. The fins are not identical and are not spread equidistantly.

The freedom to change the configuration increases systematically, in three parts, from the simpler toward the more complicated. First, the thickness of the base layer is free to morph, and the population of fins is uniform (identical fins spread equidistantly, as in Figure 2.20a). Second, the base thickness and the fin-to-fin spacings are free to morph, while the fins are identical (Figure 2.20b). Third, the base thickness and the fin sizes are free to morph, while the fins are distributed equidistantly (Figure 2.20c).

The main thread of the evolutionary design is the constructal law – the idea that with more freedom to morph, the flow architecture improves progressively toward easier flowing. In the configurations explored in Figure 2.20a–c, the global flow performance is expressed by the peak temperature T, which must be decreased when the steady heat source Q is specified.

How well does the infusion of more freedom reward the performance of the flow system?

To answer, consider as reference the classical design where the degrees of freedom adopted in Figure 2.20a–c are absent. This design is shown in Figure 2.21a. The base layer has a constant thickness ($\bar{\delta}$) and a population of equidistant identical fins, which can be represented by a constant apparent heat transfer coefficient (h_b) on the finned side of the base layer. The classical configuration shown in Figure 2.21b has heat transfer (h_b) on both sides; therefore in order to comply with Figure 2.21a, its thickness is $2\bar{\delta}$ and its heat current entering from the right is $2Q$. According to an early manifestation of constructal design (Section 2.7.5), if the shape of the profile of the object of Figure 2.21b is free to morph, then the aspect ratio of its profile ($2\bar{\delta}/L$) is the same as in Eq. (2.118), and its thermal conductance or peak temperature T_* follows from Eq. (2.119). This is how Figure 2.21a is discovered and now serves as a reference (common denominator) for the T_*-performance of the evolving heat spreader with fins.

The effect of injecting more freedom in morphing the evolutionary design is evident as we compare the peak temperature ratios reported on the ordinate of Figure 2.21c. The peak temperature of the best classical design (Figure 2.21a) decreases progressively in the direction toward more freedom: from Figures 2.21a to 2.20a,b (or c). A bird's-eye view of this evolutionary phenomenon is presented in Figure 2.21c.

Shape and performance is the direction in which this field of research and application is accelerating. Advances were made on pin fin arrays [13–17], tree-shaped fin assemblies [18–21], fins with radiation and natural convection [22], animal hair and fur viewed as fin arrays [23], and tree-shaped cavities under the surface of a conducting body [24]. This direction is reviewed further in Ref. [25].

Example 2.3 *Long Fin: Tip Temperature*

During an investigation of the effect of temperature on the life of a battery, the battery is positioned on a plate heater of temperature $T_b = 50\,°C$. The battery is cylindrical, with a diameter $D = 3$ cm and a height $L = 6$ cm. The ambient temperature is $T_\infty = 20\,°C$, and the heat transfer coefficient at the outer cylindrical surface is $h = 5$ W/m²·K. The outer skin of the battery is made out of stainless-steel sheet with a thickness $t = 0.5$ mm. The interior of the battery is such a poor conductor that the heat transfer between it and the stainless-steel shell can be neglected.

a) Determine whether the stainless steel shell can be treated as a long fin.
b) Calculate the temperature at the top edge of the shell, that is, at $x = L$.

(a)

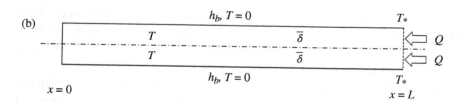

$h_b =$ constant, $T = 0$

T $\overline{\delta}$ T_* $\Leftarrow Q$

$\delta =$ constant

(b)

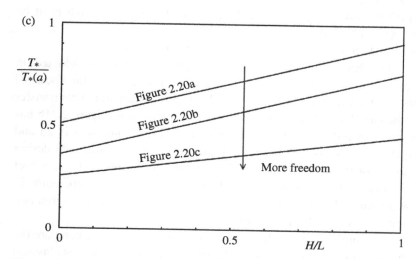

$h_b,\ T = 0$ T_*

T $\overline{\delta}$ $\Leftarrow Q$

T $\overline{\delta}$ $\Leftarrow Q$

$h_b,\ T = 0$ T_*

$x = 0$ $x = L$

(c)

Figure 2.20a

Figure 2.20b

Figure 2.20c

More freedom

$\dfrac{T_*}{T_*(a)}$

H/L

Figure 2.21 (a) The design of Figure 2.20a in the limit where the base thickness is constant, and the population of fins is uniform. (b) The classical fin with convection on both sides, cf. Figure 2.11. (c) Rewards from freedom: how the peak temperature T_* decreases as the freedom to morph the evolutionary design increases.

Solution

To determine whether the stainless-steel skin is a "long fin," Eq. (2.79), we first calculate the m parameter, Eq. (2.75):

$$m = \left(\frac{hp}{kA_c}\right)^{1/2} = \left(\frac{hp}{kpt}\right)^{1/2} = \left(\frac{h}{kt}\right)^{1/2}$$

$$= \left(\frac{5\,\mathrm{W}}{\mathrm{m}^2 \cdot \mathrm{K}}\frac{\mathrm{m} \cdot \mathrm{K}}{15\,\mathrm{W}}\frac{1}{0.5 \times 10^{-3}\,\mathrm{m}}\right)^{1/2} = 25.8\ \mathrm{m}^{-1} \tag{1}$$

The long-fin criterion (2.79) is not satisfied, because in the present case mL is not much greater than 1:

$$mL = \left(25.8\ \mathrm{m}^{-1}\right)(0.06\ \mathrm{m}) = 1.55 \tag{2}$$

The top-edge temperature can be estimated by modeling the edge as a tip with heat transfer (Figure E2.3). Substituting $x = L$ in Eq. (2.92), we write

$$\frac{\theta}{\theta_b} = \frac{1}{\cosh(mL) + \frac{h}{mk}\sinh(mL)}$$

in which $mL = 1.55$ and

$$\frac{h}{mk} = \frac{5\,W}{m^2 \cdot K}\,\frac{m}{25.8}\,\frac{m \cdot K}{15W} = 0.013$$

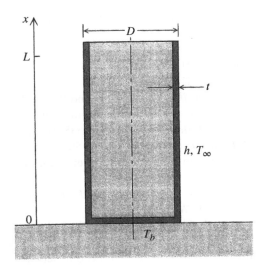

Figure E2.3

We obtain in this way

$$\frac{\theta}{\theta_b} = 0.402 = \frac{T - T_\infty}{T_b - T_\infty} \tag{3}$$

which means that the top-edge temperature is $T = 32.1\,°C$. The top edge could have also been modeled as an insulated tip (cf. Eq. (2.88)):

$$\frac{\theta}{\theta_b} = \frac{1}{\cosh(mL)} = 0.407 \tag{4}$$

and the top-edge temperature would have been $T \cong 32.2\,°C$. Since this estimate is practically equal to the one based on the more rigorous (2.92), we conclude that the effect of heat transfer through the top edge (as a fin tip) is negligible.

It is worth going back to the long-fin criterion (2) to see the error that would have been made if the steel skin were treated as a long fin. Under the long-fin assumption, the top-edge temperature is given by Eq. (2.78) with $x = L$:

$$\frac{\theta}{\theta_b} = \exp(-mL) = 0.212 \tag{5}$$

which means that $T = 26.4\,°C$. This estimate is lower than the correct value of $32.1\,°C$, because the long-fin assumption means that the fin continues beyond $x = L$ and that the cooling effect provided by this extension depresses the temperature at $x = L$.

References

1 Kirchhoff, G. (1894). *Vorlesungen über die Theorie der Wärme*, 13. Leipzig: Teubner.

2 Grattan-Guiness, I. and Ravetz, J.R. (1972). *Joseph Fourier 1768–1830*. Cambridge, MA: M.I.T. Press.

3 Harper, W.B. and Brown, D.R. (1922). Mathematical Equations for Heat Conduction in the Fins of Air-Cooled Engines. *NACA Report No. 158*. NACA.

4 Gardner, K.A. (1945). Efficiency of extended surfaces. *Trans. ASME* 67: 621–631.

5 Bejan, A. and Kraus, A.D. (2003). *Heat Transfer Handbook*. Hoboken, NJ: Wiley.

6 Bejan, A. (2013). *Convection Heat Transfer*, Section 1.5, 4e. Hoboken, NJ: Wiley.

7 Schmidt, E. (1926). Die Wärmeübertragung durch Rippen. *Z. Ver. Dtsch. Ing.* 70: 885–889 and 947–951.

8 Jany, P. and Bejan, A. (1988). Ernst Schmidt's approach to fin optimization: an extension to fins with variable conductivity and the design of ducts for fluid flow. *Int. J. Heat Mass Transfer* 31: 1635–1644.

9 Bejan, A. (2019). Heat tubes: conduction and convection. *Int. J. Heat Mass Transfer* 137: 1258–1262.

10 Matsushima, H., Almerbati, A., and Bejan, A. (2018). Evolutionary design of conducting layers with fins and freedom. *Int. J. Heat Mass Transfer* 126: 926–934.

11 Yovanovich, M.M. and Marotta, E.E. (2003). Thermal spreading and contact resistances. In: *Heat Transfer Handbook*, Chapter 4 (eds. A. Bejan and A.D. Kraus). Hoboken, NJ: Wiley.

12 Aziz, A. (2003). Conduction heat transfer. In: *Heat Transfer Handbook*, Chapter 3 (eds. A. Bejan and A.D. Kraus). Hoboken, NJ: Wiley.

13 Ledezma, G.A. and Bunker, R.S. (2015). The optimal distribution of pin fins for blade tip cap underside cooling. *J. Turbomach.* 137: 011002.

14 Almogbel, M. and Bejan, A. (2000). Cylindrical trees of pin fins. *Int. J. Heat Mass Transfer* 43: 4285–4297.

15 Zadhoush, M., Nadooshan, A.A., Afrand, M., and Ghafori, H. (2017). Constructal optimization of longitudinal and latitudinal rectangular fins used for cooling a plate under free convection by the intersection of asymptotes method. *Int. J. Heat Mass Transfer* 112: 441–453.

16 Hajmohammadi, M.R. (2018). Design and analysis of multi-scale annular fins attached to a pin fin. *Int. J. Refrig.* 88: 16–23.

17 Adewumi, O.O., Bello-Ochende, T., and Meyer, J.P. (2013). Constructal design of combined microchannel and micro pin fins for electronic cooling. *Int. J. Heat Mass Transfer* 66: 215–323.

18 Kephart, J. and Jones, G.F. (2016). Optimizing a functionally graded metal-matrix heat sink through growth of a constructal tree of convective fins. *J. Heat Transfer* 138: 082802.

19 Bejan, A. and Almogbel, M. (2000). Constructal T-shaped fins. *Int. J. Heat Mass Transfer* 43: 2101–2115.

20 Almogbel, M. and Bejan, A. (2001). Constructal optimization of nonuniformly distributed tree-shaped flow structures for conduction. *Int. J. Heat Mass Transfer* 44: 4185–4195.

21 Lorenzini, G., Biserni, C., Correa, R.L. et al. (2014). Constructal design of T-shaped assemblies of fins cooling a cylindrical solid body. *Int. J. Thermal Sci.* 83: 96–103.

22 Calamas, D. and Baker, J. (2013). Tree-like branching fins: performance and natural convective heat transfer behavior. *Int. J. Heat Mass Transfer* 62: 350–361.

23 Bejan, A. (1990). Theory of heat transfer from a surface covered with hair. *J. Heat Transfer* 112: 662–667.

24 Biserni, C., Rocha, L.A.O., and Bejan, A. (2004). Inverted fins: geometric optimization of the intrusion into a conducting wall. *Int. J. Heat Mass Transfer* 47: 2577–2586.

25 Miguel, A.F. and Rocha, L.A.O. (2018). *Tree-Shaped Fluid Flow and Heat Transfer*. Cham: Springer.

Problems

Steady Conduction Without Internal Heating

2.1 Consider the thin-wall limit of the thermal resistance of a cylindrical shell with isothermal inner and outer surfaces, Eq. (2.33). Writing L for the shell thickness and A for the size of the cylindrical surface in this limit, show that Eq. (2.33) reduces to the thin-wall formula (2.7).

2.2 Derive the expression for the thermal resistance of a spherical shell with constant thermal conductivity, Eq. (2.38). Show that when the shell is sufficiently thin, this expression becomes identical to Eq. (2.7). *Hint*: Follow the analysis exhibited for the cylindrical shell in Section 2.2.

2.3 An isothermal spherical object of radius r_i is coated with a constant-k material of outer radius r_o and exposed to an external flow. The heat transfer coefficient h is a known constant. Determine analytically the critical outer radius of this spherical shell insulation. Is the overall thermal resistance minimum or maximum when r_o is equal to the critical radius?

2.4 Derive the steady-state temperature distribution through a cylindrical body that is being heated internally such that the volumetric rate $\dot{q}$ is a constant (Figure 2.9b). The thermal conductivity of the solid material (k) and the cylinder–fluid heat transfer coefficient (h) are also constant.

2.5 The pipe shown in Figure P2.5 carries a water stream of temperature $T_w = 4\,°C$. It functions in a cold climate, the effect of which is to lower the pipe wall temperature to $T_s = -10\,°C$. A cylindrical sheet of ice builds over the pipe wall. The inner diameter of the pipe wall is 8 cm, and the convective heat transfer coefficient at the inner surface of the ice layer is $h = 1000\,W/m^2{\cdot}K$. Calculate the thickness of the ice layer (Figure P2.5).

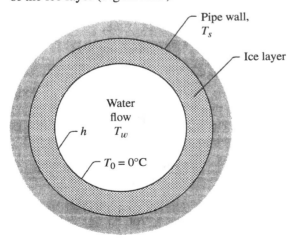

Figure P2.5

2.6 The spherical probe shown in Figure P2.6 is immersed in a bath of paraffin (*n*-octadecane) of temperature $T_\infty = 35\,°C$. The wall of the spherical probe is maintained at a lower temperature, $T_s = 10\,°C$. The solidification point of this paraffin falls between these two temperatures, $T_m = 27.5\,°C$, and this leads to the formation of a layer of solid paraffin all around the spherical object. Calculate the thickness of the solidified paraffin layer, considering that the radius of the spherical object is $r_i = 1$ cm, the conductivity of the solid paraffin is $k = 0.36$ W/m·K, and the heat transfer coefficient at the outer surface of the solidified layer is assumed constant, $h = 100$ W/m²·K.

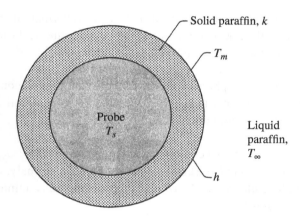

Figure P2.6

2.7 The insulated box shown in Figure P2.7 is designed to keep the air trapped in it at a high temperature, $T_h = 50\,°C$. The outside temperature is $T_c = 10\,°C$. To keep the temperature of the trapped air constant, the heat leak through the insulation, q, is made up by an electrical resistance heater placed in the center of the box. Calculate the electrical power dissipated in the heater.

The dimensions of the internal (air) space are $x = 1$ m, $y = 0.4$ m, and $z = 0.3$ m. The insulating wall consists of a 10-cm-thick layer of fiberglass sandwiched between two plywood sheets. The plywood sheet is 1 cm thick. The heat transfer coefficients on the internal and external surfaces of the wall are, respectively, $h_h = 5$ W/m²·K and $h_c = 15$ W/m²·K.

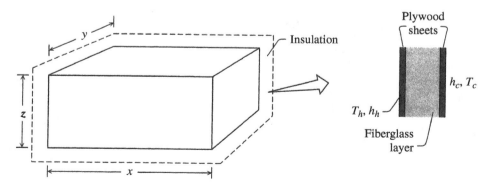

Figure P2.7

2.8 The homogeneous and isotropic porous medium shown on the left side of Figure P2.8 is saturated with a stagnant fluid of thermal conductivity k_f. The thermal conductivity of the solid material (the matrix) is k_s. The porosity (void fraction) of the medium, or the ratio of the fluid space divided by the total space, is equal to φ.

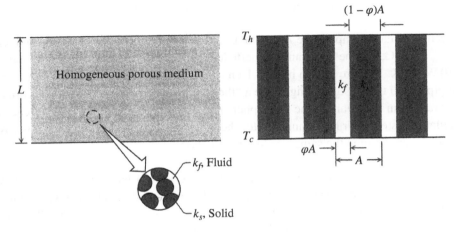

Figure P2.8

This L-thick layer of porous material is placed between two parallel walls at different temperatures, T_h and T_c. To determine the conduction heat flux from T_h to T_c across the layer, model the porous medium as a succession of parallel columns of solid and fluid arranged in such a way that the void fraction is equal to the same φ. Show that the heat flux is

$$q'' = k_{\text{eff}} \frac{T_h - T_c}{L}$$

in which k_{eff} is the "effective" thermal conductivity of the porous medium:

$$k_{\text{eff}} = (1 - \varphi) k_s + \varphi k_f$$

2.9 The temperature of the human body drops from the core (arterial blood) level $T_a = 36.5\,°\text{C}$ to the skin level T_w across a subskin layer of thickness $\delta \sim 1\,\text{cm}$ and conductivity $k \cong 0.42\ \text{W/m} \cdot \text{K}$ (Figure P2.9).

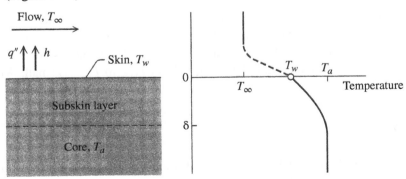

Figure P2.9

a) Calculate the skin temperature when the body is swept by the flow of 20 °C air with a heat transfer coefficient of 30 W/m^2·K.

b) Calculate the skin temperature when the external fluid is 10 °C water and the skin–water heat transfer coefficient is 500 W/m^2·K.

c) How much greater is the rate of body heat loss to water (b) than to air (a)?

2.10 The wall of a steam pipe has a temperature $T_i = 100$ °C and an outer radius $r_i = 4$ cm. The ambient air temperature is $T_\infty = 15$ °C, and the heat transfer coefficient between the cylindrical surface and ambient air is $h = 10$ W/m^2·K. You are told to build a 1-cm-thick layer of polystyrene insulation around this bare steam pipe. Will this design change have a "thermal insulation" effect? In other words, will the heat transfer rate from the steam to the atmosphere decrease if you install the polystyrene shell? To answer, calculate the steam–air heat transfer rate when the pipe wall is bare and when it is covered by the polystyrene layer. Assume the same h value in both cases.

2.11 The mechanical support for a low-temperature apparatus (a cryostat) has the geometry shown in Figure P2.11. It consists of two pieces joined end-to-end, their lengths and cross-sectional areas being L_1, A_1 and L_2, A_2, respectively. The two pieces are made out of the same material. The thermal conductivity of the material varies strongly with the temperature, $k(T)$.

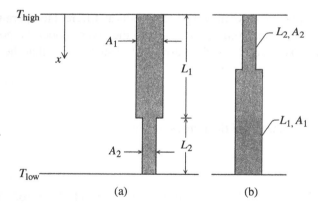

(a)   (b)

Figure P2.11

The lateral surface of the entire support is insulated perfectly. Both pieces are sufficiently slender, so that the conduction current is unidirectional through the entire support (i.e. in the x direction in Figure P2.11). You have a choice between design "a" and design "b." The good design is the one in which the conduction heat transfer rate through the support (from T_{high} to T_{low}) is small. Which of the two designs is better?

Internal Heat Generation

2.12 Consider a solid sphere immersed in a fluid flow, with which it communicates thermally through a constant heat transfer coefficient h (Figure 2.9b). The rate of internal heat generation in the sphere, $\dot{q}$, is constant. Determine the steady-state temperature distribution inside the spherical body, by assuming that its thermal conductivity k is a known constant.

2.13 The resistivity of a cylindrical electrical conductor increases linearly with the temperature. The outer surface (radius r_o) is exposed to a fluid of temperature T_∞ through a constant heat transfer coefficient h. The thermal conductivity of the conductor, k, is also constant. Using integral analysis, show that the conductor temperature is steady and finite if

$$J < \frac{2^{3/2}}{r_o} \left(\frac{k}{\rho'_e} \right)^{1/2} \left(\frac{\text{Bi}}{\text{Bi} + 4} \right)^{1/2}$$

where $\rho'_e = d\rho_e / dT$, $\text{Bi} = hr_o/k$, and J is the electric current density.

2.14 The coal powder shown in Figure P2.14 is stockpiled in a layer of constant thickness $H = 2$ m. This layer experiences volumetric heating at the rate $\dot{q} = 50$ W/m^3, which is due to the chemical reaction between coal particles and the interstitial air. The effective thermal conductivity of the coal layer is $k = 0.2$ W/m^2·K, and the heat transfer coefficient at the upper surface is $h = 2$ W/m^2·K. The temperature of the lower surface of the coal layer equals the temperature of the ground, $T_0 = 25$ °C. The atmospheric air temperature is also 25 °C.

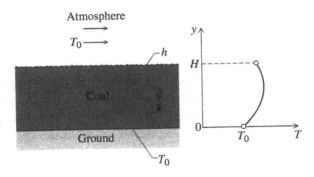

Figure P2.14

a) Show that the maximum temperature registered inside the coal layer is

$$T_{\max} = T_0 + \frac{\dot{q}H^2}{2k} \left(\frac{1 + (\text{Bi}/2)}{1 + \text{Bi}} \right)^2$$

where Bi is the Biot number hH/k. Show that the temperature maximum is located in a horizontal plane situated between $y = H/2$ and $y = H$.

b) Calculate numerically the maximum temperature and the position of this hottest plane relative to ground level.

2.15 Inserted in the middle of the conducting slab shown in Figure P2.15 is an infinitesimally thin blade that generates heat at the rate q'' per unit of blade area. The generated heat proceeds to the right and to the left, away from the midplane. The temperature of the two lateral surfaces is fixed at T_o.

Determine the steady temperature distribution in the slab of thickness L. Derive an expression for the blade temperature.

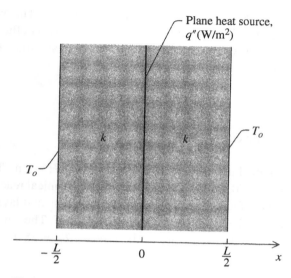

Figure P2.15

2.16 The long cylindrical rod shown in Figure P2.16 is being heated steadily by a line-shaped heat source situated along its axis. The source generates heat at the rate q' per unit axial length. The temperature of the external surface $r = r_o$ is fixed at T_o. Determine analytically the steady temperature distribution in the rod cross section. Show that the temperature at points close to the line heat source blows up as $\ln(r_o/r)$ when $r \to 0$.

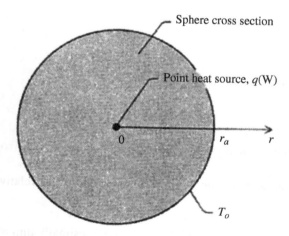

Figure P2.16

2.17 Embedded in the center of a conducting spherical body is a steady, point-size source of heat q. The temperature of the spherical surface is maintained at T_o by contact with an external fluid. Determine the temperature distribution in the spherical body, and show that the core temperature blows up when $r \to 0$.

Evolutionary Design

2.18 Show that the heat current through the base of a fin is equal to the integral of the heat flux over the total area bathed by fluid (A_{total}):

$$q_b = \int_{A_{total}} h\left(T - T_\infty\right) dA_{total}$$

Demonstrate this in two ways:
a) Integrate the fin conduction equation, Eq. (2.68), from $x = 0$ to $x = L$, while looking at Figure 2.11.
b) Apply the first law of thermodynamics to the entire fin as a control volume. You will find that the relation listed earlier holds for any fin geometry.

2.19 The temperature distribution along the two-dimensional fin with sharp tip shown in Figure P2.19 is *linear*:

$$\theta\left(x\right) = \frac{x}{L}\theta_b$$

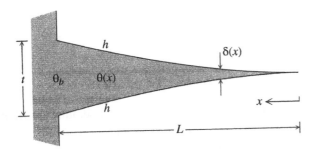

Figure P2.19

The width of this fin, W, is not shown (it is normal to the plane of Figure P2.19). Note also that x points toward the base of the fin and that the tip is in thermal equilibrium with the surrounding fluid, $\theta(0) = 0$. Recognizing that this fin must have a *variable* cross-sectional area (because a constant-A_c fin does not have a linear temperature distribution), show that the profile thickness δ is parabolic in x:

$$\delta = \frac{h}{k}x^2$$

Derive an expression for the total heat transfer rate through the base of this fin, q_b. Show that if the total volume of the parabolic profile fin, V, is the same as the volume of the fin shape optimized in Section 2.7.5, then q_b is 14.8% larger than the maximum heat transfer rate that can be accommodated by the plate fin, Eq. (2.119).

2.20 Consider the process of longitudinal conduction through the two-dimensional fin of triangular profile shown in Figure P2.20. The width of the fin (not shown) is W. Demonstrate that the temperature distribution along this fin, $\theta(x)$, must satisfy the differential equation

$$x\frac{d^2\theta}{dx^2} + \frac{d\theta}{dx} - a\theta = 0$$

where $a = 2hL/kt$. Solve this equation by assuming the series expression

$$\theta = c_0 + c_1 x + c_2 x^2 + \cdots + c_n x^n + \cdots$$

where $c_0, c_1, c_2 \ldots$ are constant coefficients. Show that the $\theta(x)$ solution is

$$\theta(x) = \theta_b \frac{F(ax)}{F(aL)} \quad \text{where} \quad F(ax) = \sum_{n=0}^{\infty} \frac{(ax)^n}{(n!)^2}$$

The $F(ax)$ series is better known as the modified Bessel function of the first kind, of order zero, $I_0[2(ax)^{1/2}]$.

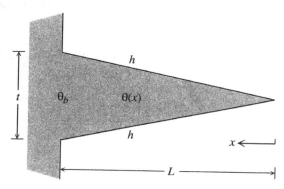

Figure P2.20

2.21 Derive an expression for the total heat transfer rate q_b corresponding to the temperature distribution (2.92). Compare this result with the approximate corrected-length (2.93), and show that:
a) When the fin is "long," $mL \gg 1$, the approximate formula is accurate.
b) When the fin is "short," $mL \ll 1$, the error associated with using Eq. (2.93) instead of Eq. (2.92) is approximately equal to $hA_c/(3kp)$. Show further that this error is small when the Biot number based on fin transversal dimension (e.g. thickness, if plate fin) is small when compared with 1.

2.22 The assumption of a constant fluid temperature (T_∞) around a fin is an approximation. In laminar flow parallel to the base wall, for example, the fluid temperature varies in the direction normal to the wall, $T_\infty(x)$. Consider the case where the variation of T_∞ mimics that of the fin temperature, $T(x)$, such that at every point along the fin, the fin–fluid temperature difference is a known constant:

$$T(x) - T_\infty(x) = \Delta T \quad \text{(constant)}$$

Assume that the remaining parameters of the fin (k, A_c, h, p) are known constants and that there is heat transfer through the tip. Determine the temperature distribution along the fin, $T(x)$, and the total heat transfer rate through the base, q_b.

2.23 The ring shown in Figure P2.23 is made out of a thin wire of length L, cross-sectional area A_c, cross-sectional perimeter p, and thermal conductivity k. The heat transfer coefficient between the wire and the ambient air is h. One point $(x=0)$ on the ring is heated by internal means so that its temperature T_b is maintained above the ambient temperature T_∞. Determine the following analytically:
a) The temperature distribution along the ring.
b) The temperature at the point that is diametrically opposed to $x=0$.
c) The total heat transfer rate q_b that must be applied at $x=0$.

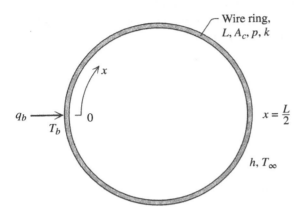

Figure P2.23

2.24 The Japanese teapot *kyusu* has a hollow handle that feels cool to the touch even when the pot contains boiling water. This handle can be modeled as a hollow-cylinder fin, as shown on the right side of Figure P2.24, where $L=7\,\text{cm}$ and

$$D_o = 2\ \text{cm} \quad h_o = 10\,\text{W/m}^2 \cdot \text{K} \quad T_b = 50°\text{C}$$
$$D_i = 1.5\ \text{cm} \quad h_o = 2\,\text{W/m}^2 \cdot \text{K} \quad T_\infty = 15°\text{C}$$

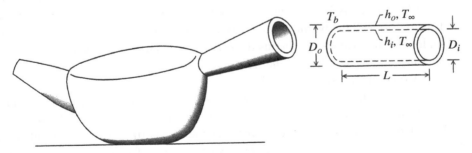

Figure P2.24

The handle is made out of porcelain. The heat transfer coefficient on the outer surface of the handle is considerably larger than on the inner surface, because the buoyancy-driven air flow is more intense around the handle.

a) Calculate the parameter m and show that the hollow-cylinder fin is a "long fin."

b) Calculate the distance from the base of the handle to the point where the handle temperature is 30 °C.

c) Calculate the instantaneous heat transfer rate through the base of the handle.

2.25 The copper fin (k) with the shape and dimensions shown in Figure P2.25 is soldered to a wall of temperature T_w. The solder film (k_s) has the thickness t, which is exaggerated in Figure P2.25. The heat transfer coefficient between the fin and the surrounding fluid has the same value h on all the exposed surfaces. Derive an expression for the total heat transfer rate through the fin, as a function of only the dimensions and properties identified on the illustration.

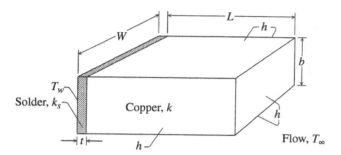

Figure P2.25

2.26 Like everything else that we see from the first moments of conscious life, hair and fur are things that we take for granted. Their purpose seems obvious not only with respect to our own needs but also for the survival of other species. The insulation effect of hair can be explained through the argument that when the hair strands are sufficiently dense, they trap a blanket of air in the tight spaces created between the strands. Well known for its low thermal conductivity, this air blanket insulates the warm skin against the colder atmosphere.

The hair-covered surface is considerably more interesting when approached from the point of view – the function – of the individual hair strand. The thermal conductivity of hair material (roughly that of human skin, 0.37 W/m·K) is 14 times greater than that of ambient air. It seems that one effect of the hair strands is to "extend" the warm surface into the cold atmosphere, in the same way that a population of cylindrical spines (pin fins) extends the heat transfer surface of a heat exchanger. This finning effect is an unwanted by-product of the true function of the hair strands, which is to slow down the breeze that would otherwise blow by [23].

The fur of a lynx, for example, has a density of 9000 strands/cm² and a cylindrical hair strand with a diameter of 27×10^{-6} m. The heat transfer coefficient between the hair strand and the surrounding air is 100 W/m²·K, and the heat transfer coefficient between the bare portions of the skin and the same air flow is 10 W/m²·K. The temperature difference between the skin surface and the air flow happens to be 10 K. The fur-covered area under consideration is a square with side 10 cm (Figure P2.26).

a) Calculate the heat transfer rate that leaves the skin through the roots of all the hair strands. Treat each hair strand as a long fin.

b) Calculate the heat transfer rate released by the bare portions of the skin. Compare this contribution with the heat transfer current calculated in part (a).

c) Estimate the order of magnitude of the distance of conduction penetration along the hair strand. In other words, for what length near its root is the hair strand significantly warmer than the surrounding air?

d) Verify that heat transfer through one hair strand can be modeled as unidirectional conduction.

Figure P2.26

2.27 The junction between two electrical leads suspended in air is defective (i.e. resistive) and generates heat at the known rate q. On both sides of the plane of the junction, the area and wetted perimeter of the lead cross section are A_c and p, respectively. The thermal conductivities and heat transfer coefficients are different, k_1, h_1 and k_2, h_2 (Figure P2.27).

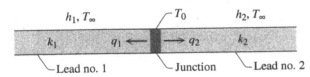

Figure P2.27

a) Develop the formula for calculating the rise of the junction temperature T_0 above the ambient temperature T_∞.

b) Show that the fraction of the junction heating rate that escapes to one side of the junction (q_1/q or q_2/q) is a function of only the parameter $R_{1,2} = [(k_1 h_1)/(k_2 h_2)]^{1/2}$.

2.28 A semi-infinite electrical cable with constant parameters (k, A_c, h, p) is attached to a base wall of temperature T_b. The electrical resistivity of the cable material is such that the Joule heating effect $\dot{q}$ increases linearly with the local temperature of the cable, $T(x)$:

$$\frac{\dot{q}}{k} = C_1 + C_2 \left(T - T_\infty \right)$$

In this expression, C_1 and C_2 are two known constants, whereas T_∞ (also constant) is the temperature of the surrounding fluid. Determine the temperature distribution $T(x)$ and the condition that must be met to avoid the thermal runaway (burnup) of the cable.

2.29 Determine the diameter D for which the total heat transfer rate q_b through a pin fin is maximum, when the total volume V of the fin is fixed. Assume that the pin fin is sufficiently slender, so that $L_c \cong L$, where L is the length of the fin. Determine the maximum heat transfer rate that corresponds to this optimum design, $q_{b,\max}$. Show that under the same conditions the pin slenderness ratio D/L is proportional to the square root of the Biot number based on diameter, hD/k.

2.30 Consider a long bar of heat-conducting material (thermal conductivity k, length L) that has two sections: one thin (length L_1, cross-section A_1) and the other thicker (length L_2, cross-section A_2) (Figure P2.30). The total volume of conducting material (V) is fixed. Heat is conducted in one direction (along the bar) in accordance with the Fourier law. The global thermal resistance of the bar is $\Delta T / q$, where ΔT is the temperature drop along the distance L, and q is the end-to-end heat current. Show analytically that $\Delta T / q$ is minimum when $A_1 = A_2$, which represents the uniform distribution of flow strangulation.

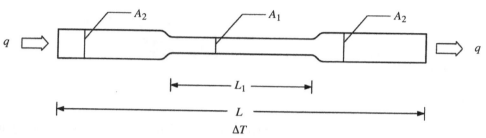

Figure P2.30

2.31 Curves may look the same, but their curvatures may be different. The cubic has a rounder (more blunt) nose than the parabola. This geometric feature is useful in heat transfer designs where the curve represents the temperature distribution through a conducting body.

Consider the heat conduction problem defined in the upper part of Figure P2.31. A slab of thickness L is heated volumetrically at the rate $q'''[\mathrm{W/m^3}]$, which in general is not uniform:

$$q'''(x) = C\,x^p$$

The left surface of the slab is adiabatic, $dT/dx = 0$ at $x = 0$. The right side is the heat sink, at $T = 0$. The total heat generation rate is the heat flux escaping to the left:

$$q'' = \int_0^L q''' dx = -k\left(\frac{dT}{dx}\right)_{x=L}$$

Determine $T(x)$ by solving the heat conduction equation

$$\frac{d^2 T}{dx^2} + \frac{q'''}{k} = 0$$

Derive the expressions for T at $x = 0$ and dT/dx at $x = L$, and show that

$$q'' = k\,(p+2)\,\frac{\Delta T}{L}$$

where $\Delta T = T(0) - T(L)$. This q'' result looks as if the heat flux is driven by a bigger temperature difference, $(p+2)\,\Delta T$, across a slab without internal heat generation.

The q'' result can be written by inspection, without solving the conduction problem. The observation is that when q''' is distributed as x^p, $T(x)$ is distributed as x^{p+2}. When q''' is uniform ($p = 0$), T is a parabola (as x^2). When q''' is linear ($p = 1$), T is a cubic (as x^3). The graphic difference between the parabola and the cubic is in the height of the tent made by the tangent to the curve at the bottom extremity of the curve. The cubic fits under a taller tent than the parabola.

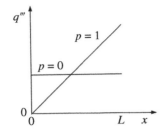

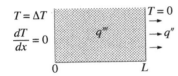

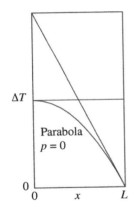

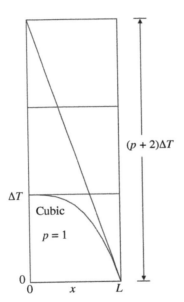

Figure P2.31

3

Multidirectional Steady Conduction

3.1 Analytical Solutions

3.1.1 Two-Dimensional Conduction in Cartesian Coordinates

In this chapter we focus on a more complicated class of conduction problems, the distinguishing feature of which is the multidirectionality of the heat flow. A situation of this kind was identified already in the discussion of when the unidirectional conduction model is no longer appropriate for description of the heat transfer through a fin (Section 2.7.4). We learned then that the heat flux lines flare out in two or more directions when the order of magnitude of the Biot number based on thickness (ht/k) is larger than 1.

As a first example of steady conduction in two dimensions, consider the rectangular-profile plate fin precisely in the limit where the unidirectional conduction assumption breaks down, $Bi \to \infty$. An infinite Biot number ht/k means that the heat transfer coefficient h is infinite. Furthermore, if the heat flux through the fluid-bathed surface is finite, the "infinite h" limit is one where the temperature difference between surface and fluid is zero. Figure 3.1 shows the temperature boundary conditions that surround the rectangular profile in this limit. The thickness of the profile is now labeled H instead of t, because soon t will be used as symbol for time. Note further the constant temperature T_b maintained along the base ($x = 0$), and the fact that the remaining three sides of the profile are in thermal equilibrium with the ambient fluid (T_∞).

3.1.1.1 Homogeneous Boundary Conditions

The problem consists of determining the temperature field inside the two-dimensional rectangular domain, $T(x, y)$, because only when $T(x, y)$ is known can we evaluate the total heat transfer rate through the base:

$$q'_b = \int_0^H q''_b \, dy = -k \int_0^H \left(\frac{\partial T}{\partial x} \right)_{x=0} dy \tag{3.1}$$

The total heat current q'_b (W/m) is expressed per unit length in the direction normal to the plane of Figure 3.1. The temperature field $T(x, y)$ must satisfy the equation for two-dimensional steady conduction in a medium with constant thermal conductivity and zero internal heat generation rate:

$$\frac{\partial^2 T}{\partial x^2} + \frac{\partial^2 T}{\partial y^2} = 0 \tag{3.2}$$

Heat Transfer: Evolution, Design and Performance, First Edition. Adrian Bejan.
© 2022 John Wiley & Sons, Inc. Published 2022 by John Wiley & Sons, Inc.
Companion website: www.wiley.com/go/bejan/heattransfer

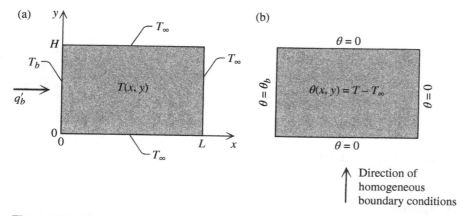

Figure 3.1 Two-dimensional conducting medium with isothermal boundaries (a), and the emergence of a direction of homogeneous boundary conditions (b).

This equation is known as *Laplace's equation*. The boundary conditions indicated in Figure 3.1 are written finally as

$$T = T_b \quad \text{at } x = 0 \tag{3.3a}$$

$$T = T_\infty \quad \text{at } x = L \tag{3.3b}$$

$$T = T_\infty \quad \text{at } y = 0 \tag{3.3c}$$

$$T = T_\infty \quad \text{at } y = H \tag{3.3d}$$

The most important aspect of the following analysis is the *structure* or the sequence of steps that must be recognized en route to the final answer. The same structure holds in the analysis of any other multidirectional steady conduction problem. The first step is the adoption of the excess temperature function as the unknown:

$$\theta = T - T_\infty \tag{3.4}$$

In terms of θ, the two-dimensional conduction equation is now written as

$$\frac{\partial^2 \theta}{\partial x^2} + \frac{\partial^2 \theta}{\partial y^2} = 0 \tag{3.5}$$

This equation must be solved subject to four conditions:

$$\theta = \theta_b \quad \text{at } x = 0 \tag{3.6a}$$

$$\theta = 0 \quad \text{at } x = L \tag{3.6b}$$

$$\theta = 0 \quad \text{at } y = 0 \tag{3.6c}$$

$$\theta = 0 \quad \text{at } y = H \tag{3.6d}$$

The effect of using θ (instead of T) as the unknown of the problem is that the last three boundary conditions are now *homogeneous*. Note the zeros appearing on the right side of Eqs. (3.6b)–(3.6d); in the original formulation, Eqs. (3.3b)–(3.3d), the right side of each equation was occupied by a constant.

Special among the homogeneous conditions mentioned earlier are the last two, because both refer to special positions marked along the same direction (y). In the x direction, only the $x = L$ condition is homogeneous. We conclude that in the present problem the y direction is *the direction of homogeneous boundary conditions*. The identification of such a direction is necessary; this is why the original problem statement, Eqs. (3.2)–(3.3), which contained no homogeneous boundary conditions, had to be restated in terms of the excess temperature θ.

3.1.1.2 Separation of Variables

The second step is the search for a product-type solution:

$$\theta(x, y) = X(x)Y(y) \tag{3.7}$$

where $X(x)$ and $Y(y)$ are two unknown functions. Substituting this expression into Eq. (3.5) and dividing the result by the product XY yields

$$\underbrace{\frac{X''}{X}}_{\substack{\text{Not a} \\ \text{function} \\ \text{of } y}} + \underbrace{\frac{Y''}{Y}}_{\substack{\text{Not a} \\ \text{function} \\ \text{of } x}} = 0 \tag{3.8}$$

The product expression (3.7) effects the separation of the x and y variables in the conduction equation. The first term in Eq. (3.8) is, at best, a function of x only. Similarly, the second term can be either a constant or a function of y. The only way in which both terms can coexist in the same equation for *any* values of x and y is if both terms are constant.

Let the positive constant λ^2 be the value of the first term, X''/X. Equation (3.8) generates two equations, for $X(x)$ and $Y(y)$:

$$X'' - \lambda^2 X = 0 \tag{3.9}$$

$$Y'' + \lambda^2 X = 0 \tag{3.10}$$

The general solutions to these linear ordinary differential equations are

$$X = C_1 \sinh(\lambda x) + C_2 \cosh(\lambda x) \tag{3.11}$$

$$Y = C_3 \sin(\lambda y) + C_4 \cosh(\lambda y) \tag{3.12}$$

Substituting them into Eq. (3.7) yields the general solution for θ:

$$\theta = K[\sinh(\lambda x) + A \cosh(\lambda x)][\sin(\lambda y) + B \cos(\lambda y)] \tag{3.13}$$

where the new constants K, A, and B are shorthand for $C_1 C_3, C_2/C_1$, and C_4/C_3, respectively. Note that in the special case when $\lambda = 0$, the general solution (3.11) and (3.12) degenerates into the linear forms $X = C_1 x + C_2$ and $Y = C_3 y + C_4$. In the same case, Eq. (3.13) becomes $\theta_0 = K(x + A)(y + B)$. Soon we shall find that the $\lambda = 0$ case (i.e. θ_0) contributes zero to the ultimate solution that we seek. For this reason we continue the analysis by writing the general solution as in Eq. (3.13).

As an aside, note that in the statement above, Eq. (3.9), we had the choice to assign the positive constant λ^2 to the second term of Eq. (3.8), writing instead of Eqs. (3.9)–(3.10)

$$X'' + \lambda^2 X = 0 \tag{3.9'}$$

$$Y'' - \lambda^2 Y = 0 \tag{3.10'}$$

The end result of this alternative route is the general solution:

$$\theta = K' \left[\sin(\lambda x) + A' \cos(\lambda x)\right] \left[\sinh(\lambda y) + B' \cosh(\lambda y)\right] \tag{3.13'}$$

in which K', A', and B' are three new constants. The general solution (3.13′) is equivalent to the earlier form (3.13), which can be seen by replacing λ by $i\lambda$ in Eq. (3.13).

To choose between a θ solution written as in Eq. (3.13) and one arranged as in Eq. (3.13′) is to choose between y and x, respectively, as the direction in which the temperature variation is described by *multivalued functions* (sine, cosine). This choice is dictated by the boundary conditions of the problem. In Section 3.1.1.3 we will learn that in the present problem the correct choice is to orient the multivalued functions (called *characteristic functions*) in the y direction, Eq. (3.13), because in Figure 3.1, the y direction is the direction of homogeneous boundary conditions.

3.1.1.3 Orthogonality

Continuing then with the θ expression listed in Eq. (3.13), we note that the general solution depends on four unknowns, namely, K, A, B, and λ. These unknowns will be pinpointed one by one by the four boundary conditions (3.6a)–(3.6d), of which the lone nonhomogeneous condition (3.6a) will be invoked last.

We begin with the tip condition (3.6b), which, combined with Eq. (3.13), requires that

$$\sinh(\lambda L) + A\,\cosh(\lambda L) = 0 \tag{3.14}$$

This equation pinpoints the value of the A constant; by substituting this value in the first pair of brackets of the θ expression (3.13), we obtain

$$\sinh(\lambda x) + A\,\cosh(\lambda x) = \frac{\sinh[\lambda(x-L)]}{\cosh(\lambda L)} \tag{3.15}$$

According to the observation made under Eq. (3.13), in the special case of $\lambda = 0$, we obtain $A = -L$ in place of Eq. (3.14), so that the θ solution in this case reduces to $\theta_0 = K(x-L)(y+B)$, which was mentioned under Eq. (3.13).

Next, the statement that the lower side of the rectangular profile is isothermal, Eq. (3.6c), is obeyed by the θ expression (3.13) if

$$B = 0 \tag{3.16}$$

In summary, the two boundary conditions that have been invoked thus far have the effect of trimming the θ expression down to the new form:

$$\theta = K\frac{\sinh[\lambda(x-L)]}{\cosh(\lambda L)}\sin(\lambda y) \tag{3.17}$$

Note again that in the case of $\lambda = 0$, the boundary condition (3.6c) requires $B = 0$, and the solution that replaces Eq. (3.17) is $\theta_0 = K(x-L)y$.

The remaining two constants, λ and K, require special care. The isothermal surface condition at $y = H$, Eq. (3.6d), is satisfied by Eq. (3.17) only if

$$\sin(\lambda H) = 0 \tag{3.18}$$

This equation has an infinity of solutions[1] λH, which are called *characteristic values*, *proper values*, or *eigenvalues*[2]:

$$\lambda H = n\pi \quad (n = 0, \pm 1, \pm 2, \dots) \tag{3.19}$$

---

1 Had we continued the analysis with Eq. (3.13′) instead of Eq. (3.13), we would have been unable to satisfy the boundary condition at $y = H$, that is, unable to determine λ. In conclusion, the choice between Eqs. (3.13) and (3.13′) must be made in such a way that the multivalued functions (sine, cosine) are aligned in the same direction as the pair of homogeneous boundary conditions.

2 The term "eigenvalue" originates from the German words *eigen*, which means "proper," and *Eigenwerte*, which refers to a hidden intrinsic feature or capacity.

Among these, $n = 0$ represents the $\lambda = 0$ case that we have been commenting on. In that special case, the (3.6d) boundary condition is satisfied by the θ_0 solution only if $K = 0$. In conclusion, the solution that accounts for the case $\lambda = 0$ is the banal solution, $\theta_0 = 0$. This contributes zero to the complete θ solution that we seek, and, for this reason, we drop the case $n = 0$ from the summation that will be made in Eq. (3.21).

If we substitute $\lambda = n\pi/H$ into the θ expression (3.17), we find that we can write a θ_n solution (led by its own constant K_n) for each value of the integer n:

$$\theta_n = K_n \frac{\sinh[n\pi(x - L)/H]}{\cosh(n\pi L/H)} \sin\left(\frac{n\pi y}{H}\right) \tag{3.20}$$

All these θ_n solutions are part of the θ solution. We have no right to discard any of the θ_ns other than θ_0; therefore, we retain all of them in the sum

$$\theta = \sum_{n=1}^{\infty} \theta_n = \sum_{n=1}^{\infty} K_n \frac{\sinh[n\pi(x - L)/H]}{\cosh(n\pi L/H)} \sin\left(\frac{n\pi y}{H}\right) \tag{3.21}$$

Since each θ_n satisfies the conduction equation (3.5) and the boundary conditions (3.6a)–(3.6d), their sum θ satisfies the same problem statement. Note further that it is sufficient to extend the sum over the nonnegative values of the integer n, because the θ_n contribution to the sum accounts for the contribution that would have been made by the θ_{-n} term (both θ_n and θ_{-n} depend in the same manner on y and x).

The values of the leading coefficients K_n are determined by invoking the remaining boundary condition, Eq. (3.6a). In view of Eq. (3.21), this condition requires

$$\theta_b = -\sum_{n=1}^{\infty} K_n \tanh\left(\frac{n\pi L}{H}\right) \sin\left(\frac{n\pi y}{H}\right) \tag{3.22}$$

Key is the *orthogonality property* of the sine function, which is expressed by

$$\int_0^H \sin\left(n\frac{\pi y}{H}\right) \sin\left(m\frac{\pi y}{H}\right) dy = \begin{cases} 0 & \text{if } m \neq n \\ H/2 & \text{if } m = n \end{cases} \tag{3.23}$$

This integral has a finite value only when the new integer m is equal to n. Therefore, if we multiply both sides of Eq. (3.22) by $\sin(m\pi y/H)$ and then integrate both sides from $y = 0$ to $y = H$, the only term that yields a non-zero value on the right side of Eq. (3.22) is the $n = m$ term. This operation consists of writing, in order,

$$\int_0^H \theta_b \sin\left(m\frac{\pi y}{H}\right) dy = -\int_0^H \sum_{n=1}^{\infty} K_n \tanh\left(n\frac{\pi L}{H}\right) \sin\left(n\frac{\pi y}{H}\right) \sin\left(m\frac{\pi y}{H}\right) dy$$

$$= -K_m \tanh\left(m\frac{\pi L}{H}\right) \frac{H}{2} \tag{3.24}$$

The integral on the left side reduces to

$$\int_0^H \theta_b \sin\left(m\frac{\pi y}{H}\right) dy = \theta_b \frac{H}{m\pi} [1 - \cos(m\pi)] \tag{3.25}$$

where

$$1 - \cos(m\pi) = \begin{cases} 0 \text{ if } m = \text{even} \\ 2 \text{ if } m = \text{odd} \end{cases} \tag{3.26}$$

Combining Eqs. (3.24)–(3.26), we conclude that the only nonzero K_m coefficients correspond to the odd values of m:

$$K_m = -\frac{4\theta_b}{m\pi \, \tanh(m\pi L/H)} \quad (m = 1, 3, 5, \ldots) \tag{3.27}$$

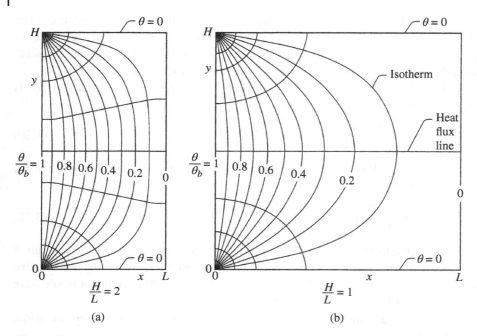

Figure 3.2 Isotherms and heat flux lines in the two-dimensional rectangular domain defined in Figure 3.1.

Looking back at the analysis started with Eq. (3.23), we see that the orthogonality property has the power to sift out of the infinite series (3.22) the value of a particular coefficient K_n, namely, the coefficient that corresponds to $n = m$. The complete θ solution is now easy to write, by substituting Eq. (3.27) into the sum (3.21) and by choosing the group $2n + 1$ as the new counter of odd integers (note that n starts now from zero, $n = 0, 1, 2, \ldots$):

$$\theta = \frac{4\theta_b}{\pi} \sum_{n=0}^{\infty} \frac{\sinh[(2n+1)\pi(L-x)/H]}{\sinh[(2n+1)(\pi L/H)]} \frac{\sin[(2n+1)(\pi y/H)]}{2n+1} \tag{3.28}$$

The network of isotherms and flux lines (or heat tubes) that corresponds to this solution is presented in Figure 3.2. Each heat tube carries one tenth of the total heat transfer rate q_b. The heat tubes are denser near the lower-left and upper-left corners because in the corner regions two isothermal boundaries intersect (note the finite temperature difference θ_b between the two intersecting boundaries). When the rectangular domain becomes taller (i.e. when H/L increases), the isotherms become vertical and equidistant. The $H/L \to \infty$ limit of this trend was recognized already in the thin wall (or plane wall) illustrated in Figure 2.1.

The total heat transfer rate through the $x = 0$ base is obtained by integrating the heat flux from $y = 0$ to $y = H$:

$$q_b' = -k \int_0^H \left(\frac{\partial \theta}{\partial x} \right)_{x=0} dy$$

$$= \frac{8}{\pi} k\theta_b \sum_{n=0}^{\infty} \frac{1}{(2n+1) \tanh[(2n+1)(\pi L/H)]} \tag{3.29}$$

This quantity is expressed in watts per meter, that is, per unit length in the direction normal to the plane of Figure 3.1.

At the end of any stretch of analysis, especially after a lengthy one, it is useful to check whether the end result agrees with much simpler solutions that are already known. In the limit $L/H \to 0$, for example, the rectangular profile of Figure 3.1 reduces to that of a "thin wall" (Figure 2.1) through which the conduction is oriented in the x direction. Therefore, the value of q_b' in this limit should be simply $kH\theta_b/L$. The q_b' formula (3.29) agrees with the expected result:

$$\lim_{L/H \to 0} q_b' = \frac{8}{\pi^2} k\theta_b \frac{H}{L} \left(1 + \frac{1}{3^2} + \frac{1}{5^2} + \cdots\right) = \frac{kH\theta_b}{L} \tag{3.30}$$

because the infinite sum listed in parentheses has the finite value of $\pi^2/8$. In the opposite limit, $L/H \to \infty$ the total heat transfer rate is infinite:

$$\lim_{L/H \to \infty} q_b' = \frac{8}{\pi} k\theta_b \left(1 + \frac{1}{3} + \frac{1}{5} + \cdots\right) = \infty \tag{3.31}$$

because this time the series inside the parentheses does not converge.

The practical result of the analysis is the formula for the total heat transfer rate driven by the temperature difference between the base and the remaining three sides of the rectangular profile ($\theta_b = T_b - T_\infty$). If W is the width measured in the direction normal to the plane of Figure 3.1 (note that $W \gg H$ if the present problem is to be two-dimensional), then the heat current that passes through the $W \times H$ base is

$$q_b = q_b' W \tag{3.32}$$

In view of Eq. (3.29), this total heat transfer rate can be expressed as

$$q_b = Sk\theta_b \tag{3.33}$$

where S is the *shape factor* defined as

$$S = \frac{8}{\pi} W \sum_{n=0}^{\infty} \frac{1}{(2n+1)\tanh[(2n+1)(\pi L/H)]} \tag{3.34}$$

The units of S are (m). The magnitude of S depends only on the geometry (shape, dimensions) of the body.

In this alternative formulation, the challenge of calculating the total heat transfer rate q_b reduces to figuring out the shape factor: it is the formula for S, Eq. (3.34), that the *Fourier analysis* of Sections 3.1.2–3.1.4 makes possible. The solutions to steady conduction problems in a diversity of multidimensional configurations have been cataloged in terms of the respective S formulas, as will be seen in Table 3.3.

Example 3.1 *Temperature Series Solution*

To see how the temperature series solution (3.28) converges, evaluate the temperature in the center of the rectangular domain of Figure 3.1 at $x = L/2$ and $y = H/2$. Consider three cases: $L/H = 1$, 2, and 3. Calculate the dimensionless center temperature θ_c/θ_b by retaining the first meaningful terms of the series expression. Show that the series converges faster as L/H increases.

Solution

Substituting $x = L/2$ and $y = H/2$ on the right side of Eq. (3.28), we obtain the center temperature θ_c:

$$\frac{\theta_c}{\theta_b} = \frac{4}{\pi} \sum_{n=0}^{\infty} \frac{\sinh[(2n+1)(\pi/2)(L/H)]}{\sinh[(2n+1)\pi L/H]} \frac{\sin[(2n+1)\pi/2]}{2n+1}$$

$$= \frac{2}{\pi} \sum_{n=0}^{\infty} \frac{\sin[(2n+1)\pi/2]}{(2n+1)\cosh[(2n+1)(\pi/2)(L/H)]}$$

where we had to invoke the identity $\sinh(2u) = 2 \sinh u \cosh u$.

The dimensionless center temperature θ_c/θ_b is a function of the slenderness ratio L/H and decreases monotonically as L/H increases. To illustrate the effect of the slenderness ratio on the convergence of the infinite series solution, we evaluate the first meaningful terms of the series, for specific values of L/H:

$$\frac{L}{H} = 1: \frac{\theta_c}{\theta_b} = 0.2537 - 0.0038 + 0.0001 - \cdots = 0.250$$

$$\frac{L}{H} = 2: \frac{\theta_c}{\theta_b} = 0.054\,92 - 0.000\,03 + \cdots = 0.0549$$

$$\frac{L}{H} = 3: \frac{\theta_c}{\theta_b} = 0.011\,44 - 3.08 \times 10^{-7} + \cdots = 0.011\,44$$

The convergence of the series improves as the ratio L/H increases.

3.1.2 Heat Flux Boundary Conditions

The analytical solutions for steady conduction in two or more dimensions are as diverse as the shapes and the boundary conditions of the conduction domains. A very large volume of such solutions has been generated in the 200 years since Fourier [1], that is, since the discovery of the analytical method. A small representative sample of this volume is found in the advanced treatises on conduction heat transfer and on mass diffusion, for example, in Refs. [2–4]. What unites these analytical solutions is the method, which – by reviewing the headings of the preceding analysis – consists of three important steps:

1. The identification of the direction of homogeneous boundary conditions.
2. The separation of variables.
3. The use of the orthogonality property of the characteristic functions, to determine the coefficients of the series solution.

To illustrate the generality of the method and the diversity of conduction problems, consider the phenomenon of steady conduction in the two-dimensional domain shown in Figure 3.3. The conduction pattern is two-dimensional (in the plane of Figure 3.3) because the solid body has a width W much larger than the thickness H, the width W being measured in the direction normal to the plane of Figure 3.3.

This new geometry is related to the one already treated in Figure 3.1. There are, however, two important differences: first, the *infinite* length of the rectangular cross section in Figure 3.3 and, second, the thermal contact with a fluid flow across a *finite* heat transfer coefficient h, or a finite Biot number hH/k. This new configuration is the same as an infinitely long plate fin, Figure 2.11, except that this time we treat the temperature field as two-dimensional (i.e. as illustrated in Figure 2.17).

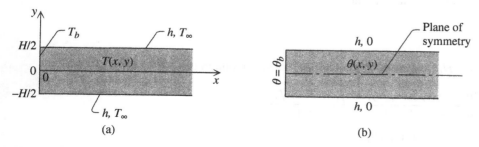

Figure 3.3 Infinitely long plate fin with finite heat transfer coefficient on both sides.

The problem statement for $T(x, y)$ consists of the conduction equation (3.2) and four boundary conditions:

$$T = T_b \quad \text{at } x = 0 \tag{3.35a}$$

$$T \to T_\infty \quad \text{as } x \to \infty \tag{3.35b}$$

$$k\frac{\partial T}{\partial y} = h(T - T_\infty) \quad \text{at } y = -\frac{H}{2} \tag{3.35c}$$

$$-k\frac{\partial T}{\partial y} = h(T - T_\infty) \quad \text{at } y = \frac{H}{2} \tag{3.35d}$$

The last boundary condition states that the conduction heat flux that arrives (from below) at the $y = H/2$ surface, $-k(\partial T/\partial y)$, is equal to the convective heat flux that enters the fluid realm, $h(T - T_\infty)$. The similar boundary condition along the lower surface, Eq. (3.35c), does not have a minus sign in front of the $k(\partial T/\partial y)$ term because on the lower surface the heat flux $h(T - T_\infty)$ points in the negative y direction.

Before plunging into any kind of laborious problem solving, it pays to examine the configuration, to see whether some of its features can be exploited for the purpose of simplifying the analysis. In this case, Figure 3.3 shows that the y boundary conditions are *symmetric* about the horizontal midplane of the rectangular profile; this is why the x axis was positioned already in that plane. The symmetry about the $y = 0$ plane allows us to replace either one of Eqs. (3.35c) and (3.35d) with the simpler adiabatic (zero heat flux) midplane condition:

$$\frac{\partial T}{\partial y} = 0 \quad \text{at } y = 0 \tag{3.36}$$

The identification of a zero heat flux condition is useful because Eq. (3.36) is a *homogeneous* boundary condition. The search for a direction of homogeneous boundary conditions is not over, however, because the remaining condition in y (say, Eq. (3.35d)) is nonhomogeneous (note the presence of the $-hT_\infty$ constant on the right side). The y direction emerges as the direction of homogeneous boundary conditions after restating the problem in terms of the excess temperature function $\theta = T - T_\infty$. The new problem statement consists of the conduction equation (3.5) and the θ counterparts of the boundary conditions:

$$\theta = \theta_b \quad \text{at } x = 0 \tag{3.37a}$$

$$\theta \to 0 \quad \text{as } x \to \infty \tag{3.37b}$$

$$\frac{\partial \theta}{\partial y} = 0 \quad \text{at } y = 0 \tag{3.37c}$$

$$\frac{h}{k}\theta + \frac{\partial \theta}{\partial y} = 0 \quad \text{at } y = \frac{H}{2} \tag{3.37d}$$

The separation of variables and the selection of y as the direction of characteristic functions leads to the θ expression (3.13). That expression cannot be used directly in the present analysis, because at first glance it does not satisfy the boundary condition (3.37b). More useful is an equivalent of Eq. (3.13), which is arrived at by noting that

$$C_1 \sinh(\lambda x) + C_2 \cosh(\lambda x) = C_3 e^{\lambda x} + C_4 e^{-\lambda x} \tag{3.38}$$

where $C_3 = (C_1 + C_2)/2$ and $C_4 = (C_2 - C_4)/2$. It is permissible to replace Eq. (3.13) with

$$\theta = (A_1 e^{\lambda x} + A_2 e^{-\lambda x})[B_1 \sin(\lambda y) + B_2 \cos(\lambda y)] \tag{3.39}$$

Table 3.1 The first six roots of Eq. (3.42).

hH/2k	a_1	a_2	a_3	a_4	a_5	a_6
0	0	π	2π	3π	4π	5π
0.01	0.0998	3.1448	6.2848	9.4258	12.5672	15.7086
0.1	0.3111	3.1731	6.2991	9.4354	12.5743	15.7143
1	0.8603	3.4256	6.4373	9.5293	12.6453	15.7713
3	1.1925	3.8088	6.7040	9.7240	12.7966	15.8945
10	1.4289	4.3058	7.2281	10.2003	13.2142	16.2594
30	1.5202	4.5615	7.6057	10.6543	13.7085	16.7691
100	1.5552	4.6658	7.7764	10.8871	13.9981	17.1093
∞	$\dfrac{\pi}{2}$	$\dfrac{3\pi}{2}$	$\dfrac{5\pi}{2}$	$\dfrac{7\pi}{2}$	$\dfrac{9\pi}{2}$	$\dfrac{11\pi}{2}$

Source: Carslaw and Jaeger [2].

in which A_1, A_2, B_1, and B_2 are all constant coefficients. Two of these coefficients can be determined by simply inspecting the boundary conditions (3.37b) and (3.37c):

$$A_1 = 0 \quad B_1 = 0 \tag{3.40}$$

In this way the θ expression reduces to

$$\theta = K e^{-\lambda x} \cos(\lambda y) \tag{3.41}$$

where K replaces now the constant product $A_2 B_2$. Note that in the special case when $\lambda = 0$, Eq. (3.13) reduces to $\theta_0 = K(x+A)(y+B)$. This expression satisfies the boundary condition (3.37b) if $K = 0$. In conclusion, θ_0 contributes zero to the ultimate solution, and for this reason we continue the analysis with Eq. (3.41).

The expression (3.41) contains only two constants, K and λ, because there are only two boundary conditions that have not been invoked yet, namely, Eqs. (3.37a) and (3.37d). Substituting the θ expression (3.41) first into the boundary condition (3.37d) leads to the transcendental equation

$$a \tan(a) = \frac{hH}{2k} \tag{3.42}$$

where a is the dimensionless group:

$$a = \lambda \frac{H}{2} \tag{3.43}$$

The constant appearing on the right side of Eq. (3.42) is the Biot number based on the half-thickness of the plate. Since $\tan(a)$ is periodic in a, Eq. (3.42) has an infinity of solutions a_n ($n = 1, 2, 3, \ldots$) for any fixed value of $hH/2k$. Furthermore, if a_n is a solution of Eq. (3.42), then $-a_n$ is also a solution.

The first six of the positive characteristic values a_n are listed in Table 3.1. A look at Table 3.1 teaches us that the form of the characteristic values of a certain conduction configuration is rarely as simple as in Eq. (3.19). More often, the characteristic values must be determined numerically by solving multiple-solutions equations such as Eq. (3.42). Each characteristic value is a function of a dimensionless parameter, namely, the Biot number based on the solid–fluid heat transfer coefficient and a linear dimension of the conducting body.

The series expression for θ is constructed in the same manner as in Eq. (3.21):

$$\theta = \sum_{n=1}^{\infty} K_n e^{-\lambda_n x} \cos(\lambda_n y) \tag{3.44}$$

where $\lambda_n = 2a_n/H$. Only the positive λ_n values are counted in this series, because the corresponding negative values do not contribute new terms to the series (note that $\cos(-\lambda_n y) = \cos(\lambda_n y)$).

The final step of the analytical method – the use of the orthogonality property of $\cos(\lambda_n y)$ – delivers the values of the coefficients K_n. This step consists of multiplying both sides of Eq. (3.44) by $\cos(\lambda_m y)$ and then integrating both sides from $y = 0$ to $y = H/2$. The final result can be written as

$$K_n = 2\theta_b \frac{\sin(a_n)}{a_n + \sin(a_n)\cos(a_n)} \tag{3.45}$$

In combination with the series (3.44), this result completes the solution to the steady two-dimensional conduction problem. This series solution is noted for its rapid convergence. Even though today such series are evaluated routinely on the computer, it is worth trying out a few numerical examples manually, to develop a feel for how the series behaves.

The corresponding solution for the total heat transfer rate (watts) through the $x = 0$ surface of the solid body is

$$q_b = W \int_{-H/2}^{H/2} -k\left(\frac{\partial \theta}{\partial x}\right)_{x=0} dy$$

$$= 4k\theta_b W \sum_{n=1}^{\infty} \frac{\sin^2(a_n)}{a_n + \sin(a_n)\cos(a_n)} \tag{3.46}$$

It is instructive to verify numerically that when the Biot number $hH/2k$ is smaller than 1, the q_b series (3.46) converges on values that agree within a few percentage points with the one-dimensional conduction estimate made possible by Eq. (2.97). The one-dimensional model overestimates the q_b value produced by the more exact, two-dimensional model (3.46). The comparison of Eq. (3.46) with Eq. (2.97) forms the subject of Problem 3.1. With regard to plate fins, even the two-dimensional model of Figure 3.3 is approximate, because the actual fin root temperature is depressed below T_b by the heat current q_b that is pulled out of the base wall. To the rest of the base wall (as a semi-infinite solid of temperature T_b), the root of the fin looks like a heat sink; therefore, the fin temperature at $x = 0$ must be lower than T_b.

3.1.3 Superposition of Solutions

Table 3.2 summarizes the three types of boundary conditions we have encountered thus far. Their homogeneous counterparts are listed in the right-hand column. Worth noting is that the last condition – specified heat transfer coefficient and fluid temperature (h, T_∞) – is somewhat more general than the first two. For example, if the heat transfer coefficient is infinite, then the boundary temperature becomes specified, because in this limit it is equal to the specified fluid temperature T_∞. On the other hand, if the boundary temperature T_w is specified *in addition* to h and T_∞, then the boundary heat flux is also specified, because $q'' = h(T_w - T_\infty)$.

Table 3.2 Three types of boundary conditions and the corresponding homogeneous forms.

Type of boundary condition[a]		Nonhomogeneous form	Homogeneous form
1. Specified boundary temperature (T_w)	T_w	$T = T_w$	$T = 0$
2. Specified boundary heat flux (q'')	$\vec{n}$ $\vec{q}''$	$-k\dfrac{\partial T}{\partial n} = q''$	$\dfrac{\partial T}{\partial n} = 0$
3. Specified heat transfer coefficient and fluid temperature (h, T_∞). Note that the sign in front of $\partial T/\partial n$ depends on the direction of n	$\vec{n}$ h, T_∞ T	$-k\dfrac{\partial T}{\partial n} = h(T - T_\infty)$	$\dfrac{h}{k}T + \dfrac{\partial T}{\partial n} = 0$

a) These three boundary condition types are recognized (respectively) by the alternative names:
  1. Boundary condition of the *first kind*, or of *Dirichlet* type.
  2. Boundary condition of the *second kind*, or of *Neumann* type.
  3. Boundary condition of the *third kind*, or of *Robin* type.
The conditions specified at infinity (e.g. in infinite or semi-infinite solids. Chapter 4) are said to be of the *zeroth kind*.

A step-by-step comparison of the two analyses of the problems posed in Figures 3.1 and 3.3 shows that the boundary conditions can involve the boundary temperature, the heat flux, or a relationship between the boundary temperature and heat flux. Figures 3.1 and 3.3 represent two of the simplest cases, because the specified boundary temperature (e.g. Eq. (3.3a)), boundary heat flux (e.g. Eq. (3.37c)), heat transfer coefficient h, and fluid temperature T_∞ are all constant.

In these simplest cases it was possible to identify the direction of homogeneous boundary conditions by just using the transformation $\theta = T - T_\infty$. There are other heat transfer configurations that require additional ingenuity. In Figure 3.4, for example, the use of the excess temperature function $\theta = T - T_\infty$ is not sufficient, because the temperatures of *two* boundaries (T_a, T_b) differ from the background temperature (T_∞). Note that with a finite θ_b on the left surface and a finite θ_a on the bottom surface, neither x nor y is a direction of homogeneous boundary conditions.

The $\theta(x, y)$ problem can still be solved analytically by noting the relationship

$$\theta(x, y) = \theta_1(x, y) + \theta_2(x, y) \tag{3.47}$$

where θ_1 and θ_2 are the solutions of two simpler conduction problems (Figure 3.4):

Conduction Equations

$$\frac{\partial^2 \theta_1}{\partial x^2} + \frac{\partial^2 \theta_1}{\partial y^2} = 0 \qquad \frac{\partial^2 \theta_2}{\partial x^2} + \frac{\partial^2 \theta_2}{\partial y^2} = 0 \tag{3.48a}$$

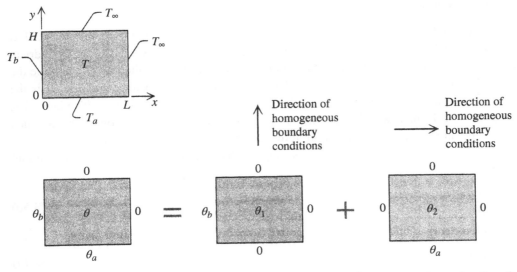

Figure 3.4 The superposition of two known solutions $(\theta_1 + \theta_2)$ as a way of constructing the solution for a problem without a direction of homogeneous boundary conditions (θ).

Boundary Conditions

$$x = 0: \quad \theta_1 = \theta_b \quad \theta_2 = 0 \tag{3.48b}$$

$$x = L: \quad \theta_1 = 0 \quad \theta_2 = 0 \tag{3.48c}$$

$$y = 0: \quad \theta_1 = 0 \quad \theta_2 = \theta_a \tag{3.48d}$$

$$y = H: \quad \theta_1 = 0 \quad \theta_2 = 0 \tag{3.48e}$$

Both $\theta_1(x, y)$ and $\theta_2(x, y)$ have analytical solutions because they both have one direction of homogeneous boundary conditions (y and x, respectively). The θ_1 solution is the same as in Eq. (3.28), whereas the θ_2 solution is obtained from Eq. (3.28) by replacing θ_b with θ_a, (x, L) with (y, H), and (y, H) with (x, L). It is easy to verify that superimposing θ_1 and θ_2, the conduction equations *and* their respective boundary conditions (i.e. by writing $\theta_1 + \theta_2$, Eq. (3.47)) lead back to the original problem statement for the $\theta(x, y)$ problem (see Figure 3.4, left side):

$$\frac{\partial^2 \theta}{\partial x^2} + \frac{\partial^2 \theta}{\partial y^2} = 0 \tag{3.49a}$$

$$\theta = \theta_b \quad \text{at } x = 0 \tag{3.49b}$$

$$\theta = 0 \quad \text{at } x = L \tag{3.49c}$$

$$\theta = \theta_a \quad \text{at } y = 0 \tag{3.49d}$$

$$\theta = 0 \quad \text{at } y = H \tag{3.49e}$$

Heat transfer configurations with more complicated boundary conditions than in Figure 3.4 may require superposition rules more complicated than Eq. (3.47). For example, it may be necessary to use more than two solutions $(\theta_1, \theta_2, \theta_3, \ldots)$ to construct the needed temperature field (θ). The efficient use of the method depends on the problem-solver's understanding of the configuration of the problem and, certainly, on experience with similar problems.

3.1.4 Cylindrical Coordinates

The analytical method outlined so far consists of the three basic steps reviewed at the start of Section 3.1.2. This section and Section 3.1.5 will show that the analytical method is general. Consider the cylindrical object shown in Figure 3.5. The lateral surface and the left end of this object are isothermal (T_∞), while the right end is maintained at a different temperature (T_b). This conduction heat transfer configuration is the "cylindrical" counterpart of the two-dimensional Cartesian problem solved in connection with Figure 3.1. The present temperature field is also two-dimensional, $T(r, z)$, because the boundary conditions are such that there can be no temperature variation in the circumferential direction (θ in Figure 1.8).

The problem statement begins with the equation for steady conduction in a constant-k medium without internal heat generation:

$$\frac{\partial^2 T}{\partial r^2} + \frac{1}{r}\frac{\partial T}{\partial r} + \frac{\partial^2 T}{\partial z^2} = 0 \tag{3.50}$$

and four boundary conditions:

$$T = T_\infty \quad \text{at } z = 0 \tag{3.51a}$$

$$T = T_b \quad \text{at } z = L \tag{3.51b}$$

$$\frac{\partial T}{\partial r} = 0 \quad \text{at } r = 0 \tag{3.51c}$$

$$T = T_\infty \quad \text{at } r = r_o \tag{3.51d}$$

Novel among these conditions is the third, the zero-heat flux condition (3.51c), which states that the temperature distribution $T(r, z)$ is symmetric about the centerline.

After introducing $\theta = T - T_\infty$ as the new unknown, the radial direction r emerges as the direction of homogeneous boundary conditions:

$$\frac{\partial^2 \theta}{\partial r^2} + \frac{1}{r}\frac{\partial \theta}{\partial r} + \frac{\partial^2 \theta}{\partial z^2} = 0 \tag{3.52}$$

$$\theta = 0 \quad \text{at } z = 0 \tag{3.53a}$$

$$\theta = \theta_b \quad \text{at } z = L \tag{3.53b}$$

$$\frac{\partial \theta}{\partial r} = 0 \quad \text{at } r = 0 \tag{3.53c}$$

$$\theta = 0 \quad \text{at } r = r_o \tag{3.53d}$$

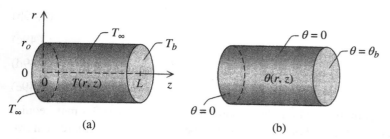

Figure 3.5 Cylindrical object with a temperature difference between one end face and the rest of its surface.

Next, the separation of variables is achieved by setting $\theta(r, z) = R(r)Z(z)$ in Eq. (3.52):

$$\frac{R''}{R} + \frac{1}{r}\frac{R'}{R} + \frac{Z''}{Z} = 0 \tag{3.54}$$

This result generates two equations for $R(r)$ and $Z(z)$:

$$\frac{R''}{R} + \frac{1}{r}\frac{R'}{R} = -\lambda^2 \tag{3.55}$$

$$\frac{Z''}{Z} = \lambda^2 \tag{3.56}$$

where λ^2 is a positive constant. The solution to Eq. (3.56) is the same as in Eq. (3.9):

$$Z = C_3 \sinh(\lambda z) + C_4 \cosh(\lambda z) \tag{3.57}$$

The $R(r)$ equation (3.55) can be solved by expanding R in a power series in r, by following the method of solution described in Problem 2.20. The general solution assumes the form

$$R = C_1 J_0(\lambda r) + C_2 Y_0(\lambda r) \tag{3.58}$$

where J_0 is the zeroth-order Bessel function of the first kind and Y_0 is the zeroth-order Bessel function of the second kind. The respective infinite series expressions for J_0 and Y_0 can be found in more advanced treatises [2–5]. Figure 3.6 illustrates the behavior of these two functions, which play the role of characteristic functions in the present problem.

Figure 3.6 shows that Y_0 becomes infinite on the cylinder centerline and that $dJ_0/dr = 0$ at $r = 0$. Consequently, the centerline boundary condition (3.53c) requires that $C_2 = 0$ in Eq. (3.58). The left-end boundary condition (3.53a), on the other hand, requires that $C_4 = 0$ in Eq. (3.57). Combining the resulting (shorter) versions of Eqs. (3.57) and (3.58) into the product expression $\theta = RZ$ yields

$$\theta = K \sinh(\lambda z)J_0(\lambda r) \tag{3.59}$$

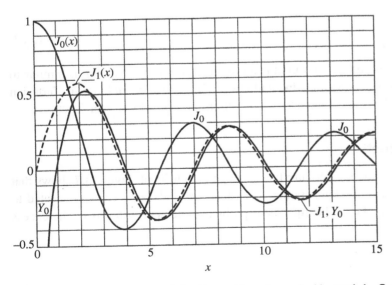

Figure 3.6 The behavior of the Bessel functions J_0, Y_0, and J_1. Source: Abramowitz and Stegun [6].

In the special case of $\lambda = 0$, the general solution $\theta = RZ$ degenerates into $\theta_0 = K(\ln r + A)(z + B)$, where K, A, and B are constants. The centerline boundary condition (3.53c) requires $K = 0$, meaning that $\theta_0 = 0$. In conclusion, θ_0 contributes zero to the final θ solution, and it is correct to continue the analysis with Eq. (3.59), which holds for $\lambda \neq 0$.

The $r = r_o$ boundary condition, Eq. (3.53d), pinpoints the characteristic values:

$$J_0(\lambda r_o) = 0 \tag{3.60}$$

According to Figure 3.6, there is an infinity of dimensionless characteristic values $\lambda_n r_o$ ($n = 1, 2, \ldots$) that satisfy Eq. (3.60); the first of these values is $\lambda_1, r_o = 2.405$. The θ solution can be constructed as an infinite series of J_0 characteristic functions:

$$\theta = \sum_{n=1}^{\infty} K_n \, \sinh(\lambda_n z) J_0(\lambda_n r) \tag{3.61}$$

The final step amounts to combining the last of the boundary conditions, Eq. (3.53b),

$$\theta_b = \sum_{n=1}^{\infty} K_n \, \sinh(\lambda_n L) J_0(\lambda_n r) \tag{3.61'}$$

with the orthogonality property of the J_0 function, which in cylindrical coordinates is expressed by the integral

$$\int_0^{r_o} r J_0(\lambda_n r) J_0(\lambda_m r) dr = 0 \quad \text{if } m \neq n \tag{3.62}$$

The K_n coefficients of the θ series can be identified by multiplying both sides of Eq. (3.61) by $r J_0(\lambda_m r)$ and integrating from $r = 0$ to $r = r_o$. This final step is omitted, because it requires the use of certain identities that exist between Bessel functions (see, for example, Ref. [3], pp. [132–143]). The end result for the steady-state temperature distribution $\theta(r, z)$ is

$$\theta = 2\theta_b \sum_{n=1}^{\infty} \frac{J_0(\lambda_n r) \sinh(\lambda_n z)}{\lambda_n r_o J_1(\lambda_n r_o) \sinh(\lambda_n L)} \tag{3.63}$$

where J_1 is the first-order Bessel function of the first kind (Figure 3.6). This series solution is similar to Eq. (3.28): the main difference is that instead of the sine and cosine functions encountered in Cartesian coordinates, in cylindrical coordinates the role of characteristic functions is played by Bessel functions.

3.1.5 Three-Dimensional Conduction

Consider the more general case in which the temperature field is three-dimensional. Figure 3.7 shows that five faces of a parallelepiped are cooled to the temperature T_∞, while the sixth face ($x = 0$) is heated to a different temperature (T_b). Figure 3.7b shows the equivalent problem statement in terms of the excess temperature function $\theta = T - T_\infty$:

Conduction Equation

$$\frac{\partial^2 \theta}{\partial x^2} + \frac{\partial^2 \theta}{\partial y^2} + \frac{\partial^2 \theta}{\partial z^2} = 0 \tag{3.64}$$

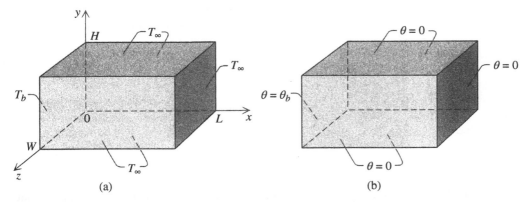

Figure 3.7 Three-dimensional (finite width) generalization of the heat transfer configuration treated in Figure 3.1.

Boundary Conditions

$$\theta = \theta_b \quad \text{at } x = 0 \tag{3.65a}$$

$$\theta = 0 \quad \text{at } x = L \tag{3.65b}$$

$$\theta = 0 \quad \text{at } y = 0 \tag{3.65c}$$

$$\theta = 0 \quad \text{at } y = H \tag{3.65d}$$

$$\theta = 0 \quad \text{at } z = 0 \tag{3.65e}$$

$$\theta = 0 \quad \text{at } z = W \tag{3.65f}$$

The variables in Eq. (3.64) can be separated by adopting the product solution $\theta = X(x)Y(y)Z(z)$,

$$\underbrace{\frac{X''}{X}}_{\lambda^2+\mu^2} + \underbrace{\frac{Y''}{Y}}_{-\lambda^2} + \underbrace{\frac{Z''}{Z}}_{-\mu^2} = 0 \tag{3.66}$$

where λ^2 and μ^2 are *two* positive numbers. Equation (3.66) generates three ordinary differential equations, the general solutions to which contribute as factors in the assumed product solution for θ:

$$\theta = \left\{ C_1 \sinh\left[(\lambda^2 + \mu^2)^{1/2}x\right] + C_2 \cosh\left[(\lambda^2 + \mu^2)^{1/2}x\right] \right\} \left[C_3 \sin(\lambda y) + C_4 \cos(\lambda y)\right]$$
$$\times \left[C_5 \sin(\mu z) + C_6 \cos(\mu z)\right] \tag{3.67}$$

Beyond this point the solution proceeds along the same path as in Sections 3.1.1–3.1.4. One important difference is that in the present problem *two* directions of homogeneous boundary conditions are needed (y and z) to determine two infinite series of characteristic functions, λ_n and μ_n.

3.2 Integral Method

The physical configuration and boundary conditions are in many cases more complicated than in Sections 3.1.1–3.1.5. When an exact analytical solution does not exist, or even when one exists but it is

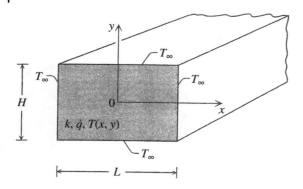

Figure 3.8 Bar with rectangular cross section and uniform rate of internal heat generation.

hampered by a lengthy analysis and numerical work, progress can still be made based on approximate methods.

As a first example, consider the integral method [7]. We focus on Figure 3.8. The rectangular area emphasized in Figure 3.8 represents a cross section through an electrical conductor with uniform rate of internal heat generation $\dot{q}$.

When the external heat transfer coefficient (or the Biot number hL/k, or hH/k) is high enough, the temperature of the outer edge of the cross section is equal to the ambient temperature T_∞. The temperature everywhere inside the conductor, $T(x, y)$, must be greater than T_∞ to drive the locally generated q toward the periphery of the cross section. The maximum temperature occurs, therefore, in the center of the cross section, because the point $(x = 0, y = 0)$ is located the farthest from all the boundaries. In problems such as this (finite-size body with internal heat generation), the key challenge is to predict the maximum temperature that occurs inside the body, no matter where it occurs (cf. Ref. [8], which was published on 1 November 1996 as shown in Fig. 3.3 of Ref. [9]).

Assuming that the thermal conductivity of the solid is constant, the equation for steady conduction in x–y coordinates is

$$\frac{\partial^2 T}{\partial x^2} + \frac{\partial^2 T}{\partial y^2} = -\frac{\dot{q}}{k} \tag{3.68}$$

In this equation $\dot{q}$ is a constant, that is, it has the same value throughout the conducting medium. The isothermal boundary conditions specified around the periphery,

$$T = T_\infty \quad \text{at } x = \pm\frac{L}{2} \tag{3.69a, b}$$

$$T = T_\infty \quad \text{at } y = \pm\frac{H}{2} \tag{3.69c, d}$$

can all be made homogeneous by adopting the excess temperature $\theta = T - T_\infty$ as the new unknown. Unfortunately, the next step of the exact analysis (the separation of variables) is not nearly as straightforward, because it requires the imaginative use of the superposition of simpler solutions (see Problem 3.2). The source of this difficulty is the additional term $\dot{q}/k$ that appears now on the right side of Eq. (3.68); because of this constant, Eq. (3.68) is nonhomogeneous, in contrast to the equation for steady conduction without internal heat generation, Eq. (3.2), which is homogeneous. Equations (3.68) and (3.2) are, respectively, known as Poisson's and Laplace's equations.

A much simpler approach to estimating T_{max} is based on the integral method, which consists of two steps. The first step is the assumption of a reasonable shape for the $T(x, y)$ surface, for example,

$$T(x, y) = T_\infty + \theta_{max} \left[1 - \left(\frac{x}{L/2} \right)^2 \right] \left[1 - \left(\frac{y}{H/2} \right)^2 \right] \tag{3.70}$$

This expression satisfies the boundary conditions (3.69a–d). The constant θ_{max} is unknown and equal to the maximum temperature difference, $\theta_{max} = T(0, 0) - T_\infty$.

The second step consists of substituting the assumed $T(x, y)$ expression into Eq. (3.68) and then integrating both sides of Eq. (3.68) over the entire rectangular domain:

$$\int_{-L/2}^{L/2} \int_{-H/2}^{H/2} \left(\frac{\partial^2 T}{\partial x^2} + \frac{\partial^2 T}{\partial y^2} \right) dy\, dx = -\frac{\dot{q}}{k} HL \tag{3.71}$$

This equation delivers the wanted result:

$$\theta_{max} = \frac{3\dot{q}}{16k} \frac{H^2 L^2}{H^2 + L^2} \tag{3.72}$$

The maximum temperature difference increases proportionally with the ratio $\dot{q}/k$ and with the square of the smaller of the two sides of the rectangular cross section. The formula (3.72) matches the exact solution when the cross section is flat ($H \gg L$, or $H \ll L$). It is the least accurate in the case of a square cross section, where it overestimates the exact maximum temperature difference by only 27%.

3.3 Scale Analysis

By focusing again on Figure 3.8, we have the opportunity to learn an even simpler and more direct method, namely, *scale analysis* or *order-of-magnitude analysis* [10]. The unknown is still the maximum temperature difference between the center of the cross section and the periphery, θ_{max}. This time, however, we estimate only the order of magnitude of each of the three terms of the conduction, Eq. (3.68). From the approximate relationship that emerges, we deduce the value of θ_{max}.

The first term, $\partial^2 T/\partial x^2$, represents the curvature of the temperature distribution in the x direction. The curvature is the change in the slope $\partial T/\partial x$; therefore,

$$\frac{\partial^2 T}{\partial x^2} \sim \frac{\left(\frac{\partial T}{\partial x} \right)_{x=L/2} - \left(\frac{\partial T}{\partial x} \right)_{x=0}}{(L/2) - 0} \tag{3.73}$$

in which the sign $\sim$ means *is of the same order of magnitude as*. Symmetry dictates that $(\partial T/\partial x)_{x=0} = 0$. The most we can say about the temperature gradient at the surface is that it is proportional to the maximum temperature difference and inversely proportional to the distance between the point of temperature maximum and the edge of the cross section:

$$\left(\frac{\partial T}{\partial x} \right)_{x=L/2} \sim -\frac{\theta_{max}}{L/2} \tag{3.74}$$

Combining Eqs. (3.74) and (3.73) leads to the x curvature, which must be negative:

$$\frac{\partial^2 T}{\partial x^2} \sim -\frac{\theta_{max}}{(L/2)^2} \tag{3.75}$$

The same argument can be used for the second term of Eq. (3.68):

$$\frac{\partial^2 T}{\partial y^2} \sim -\frac{\theta_{max}}{(H/2)^2} \tag{3.76}$$

The order of magnitude on the right side of Eq. (3.68) is indicated already, $\dot{q}/k$. Substituting these three scales in place of the respective terms in Eq. (3.68) yields

$$\frac{\theta_{max}}{(L/2)^2} + \frac{\theta_{max}}{(H/2)^2} \sim \frac{\dot{q}}{k} \tag{3.77}$$

In other words,

$$\theta_{max} \sim \frac{\dot{q}}{4k}\frac{H^2 L^2}{H^2 + L^2} \tag{3.78}$$

This scale analysis result is only 33% larger than the integral analysis estimate (3.72). Its deviation from the exact result (Problem 3.2) is as much as 70% when the cross section is square. In conclusion, the scale analysis produces a compact and extremely inexpensive result that matches within a factor of order 1 the exact solution to the same problem. For this reason the method of scale analysis is used extensively in heat transfer [7].

3.4 Evolutionary Design

3.4.1 Shape Factors

The heat transfer literature contains a large volume of shape factor results for the more common geometrical configurations that are encountered in applications. Table 3.3 shows a representative sample of what is presently available. Complementary tabulations can be found in Refs. [11–13], which served as sources for the information compiled in Table 3.3. Most of these results have been determined based on advanced analytical methods (e.g. conformal mapping, the superposition of heat sources and sinks) and experiments that exploit the analogy between the conduction of heat and electricity. An approximate formula for calculating the conduction shape factor for a layer of uniform thickness surrounding a body of arbitrary shape has been developed by Hassani and Hollands [14]. The heat transfer concept of "shape factor" was introduced in 1913 by Langmuir et al. [15].

Example 3.2 *Shape Factors*
The cubical experimental chamber shown on the left side of Figure E3.2 is surrounded by a layer of fiberglass insulation of thickness $L = 0.15$ m. The side of the cubical chamber is $H = 1$ m, and the chamber wall temperature is $T_h = 50\,°C$. The temperature of the external surface of the insulating layer is $T_c = 20\,°C$. Calculate the total heat transfer rate that leaks through the insulating layer.

Table 3.3 Shape factors (S) for several configurations with isothermal surfaces, $q_{1 \to 2} = Sk(T_1 - T_2)$.

Plane walls

Plane wall with isothermal surfaces (T_1, T_2)	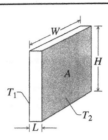	$S = \dfrac{A}{L} = \dfrac{WH}{L}$ $\left(H > \dfrac{L}{5}, \ W > \dfrac{L}{5} \right)$
Edge prism, joining two plane walls with isothermal surfaces	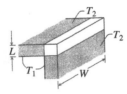	$S = 0.54W$ $\left(W > \dfrac{L}{5} \right)$

T_1 = temperature of internal surfaces of plane walls

T_2 = temperature of external surfaces of plane walls

T_2 = temperature of two exposed faces of the prism

Corner cube, joining three plane walls. The temperature difference $T_1 - T_2$ is maintained between the inner and outer surfaces of the three-wall structure	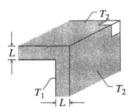 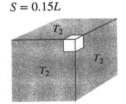	$S = 0.15L$

Cylindrical surfaces

Bar with square cross section, isothermal outer surface, and cylindrical hole through the center		$S = \dfrac{2\pi L}{\ln(1.08H/D)}$ (L = length normal to the plane of the figure)
Infinite slab with isothermal surfaces (T_2) and thickness $2H$, with an isothermal cylindrical hole (T_1) positioned midway between the surfaces	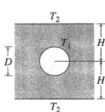	$S = \dfrac{2\pi L}{\ln(8H/\pi D)}$ ($L \gg D$, $H > D/2$, L = length of cylindrical hole)

Table 3.3 (Continued)

Cylinder of length L and diameter D, with cylindrical hole positioned eccentrically	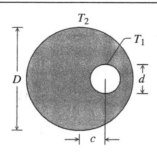	$$S = \frac{2\pi L}{\cosh^{-1}\left(\dfrac{D^2 + d^2 - 4c^2}{2Dd}\right)}$$ $(L \gg D)$
Semi-infinite medium with isothermal surface (T_2) and isothermal cylindrical hole (T_1) of length L, parallel to the surface (i.e. normal to the plane of the figure)	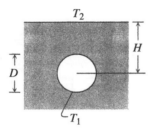	$$S = \frac{2\pi L}{\cosh^{-1}(2H/D)} \quad (L \gg D)$$ $$S \cong \frac{2\pi L}{\ln(4H/D)} \quad \text{if } H/D > 3/2$$
Infinite medium with two parallel isothermal cylindrical holes of length L	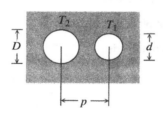	$$S = \frac{2\pi L}{\cosh^{-1}\left(\dfrac{4p^2 - D^2 - d^2}{2Dd}\right)}$$ $(L \gg D, d, p)$
Semi-infinite medium with isothermal surface (T_2) and with a cylindrical hole (T_1) drilled to a depth H normal to the surface. The hole is open. The medium temperature far from the hole is T_2	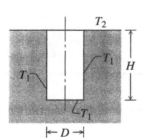	$$S = \frac{2\pi H}{\ln(4H/D)}$$ $(H \gg D)$

Spherical surfaces

Semi-infinite medium with isothermal surface (T_2) and isothermal spherical cavity	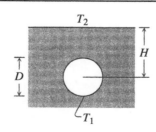	$$S = \frac{2\pi D}{1 - (D/4H)}$$ $(H > D/2)$ Spherical cavity in an infinite medium: $S \cong 2\pi D \quad \text{if } H/D > 3$

Table 3.3 (Continued)

Semi-infinite medium with insulated surface and far-field temperature T_2 containing an isothermal spherical cavity (T_1)	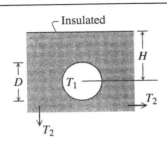	$$S = \frac{2\pi D}{1 + (D/4H)}$$ $$(H > D/2)$$
Hemispherical isothermal dimple (T_1) into the insulated surface of a semi-infinite medium with far-field temperature T_2	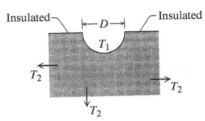	$S = \pi D$

Thin discs and plates

Semi-infinite medium with isothermal surface (T_2) and isothermal disc (T_1) parallel to the surface	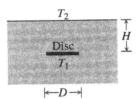	$$S = \frac{2\pi D}{(\pi/D) - \tan^{-1}(D/4H)}$$ $$(H/D > 1)$$ Disc attached to the surface: $S = 2D \quad (H = 0)$
Semi-infinite medium with insulated surface and far-field temperature T_2, containing an isothermal disc (T_1) parallel to the surface	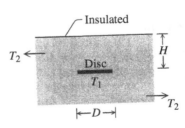	$$S = \frac{2\pi D}{(\pi/2) + \tan^{-1}(D/4H)}$$ $$(H/D > 1)$$
Infinite medium with two parallel, coaxial, and isothermal discs	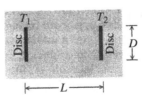	$$S = \frac{2\pi D}{(\pi/2) - \tan^{-1}(D/2L)}$$ $$(L/D > 2)$$
Semi-infinite medium with isothermal surface (T_2) and isothermal plate (T_1) parallel to the surface (W is the length of the plate, in the direction normal to the plane of the figure)		$$S = \frac{2\pi L}{\ln(4L/W)} \quad (H \gg L)$$ Plate attached to the surface: $$S = \frac{\pi L}{\ln(4L/W)} \quad (H = 0)$$

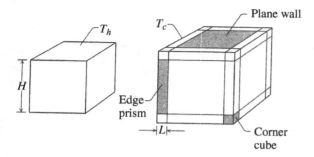

Figure E3.2

Solution

The total heat current through the insulation is due to several currents:

$$q = 6q_{\substack{\text{plane} \\ \text{wall}}} + 12q_{\substack{\text{edge} \\ \text{prism}}} + 8q_{\substack{\text{corner} \\ \text{cube}}}$$

$$= 6S_{\substack{\text{plane} \\ \text{wall}}} k\Delta T + 12S_{\substack{\text{edge} \\ \text{prism}}} k\Delta T + 8S_{\substack{\text{corner} \\ \text{cube}}} k\Delta T$$

$$= Sk\Delta T \tag{1}$$

where $\Delta T = T_h - T_c$. The overall shape factor S follows from Table 3.3:

$$S = 6S_{\substack{\text{plane} \\ \text{wall}}} + 12S_{\substack{\text{edge} \\ \text{prism}}} + 8S_{\substack{\text{corner} \\ \text{cube}}}$$

$$= \left[6\frac{1}{0.15} + (12 \times 0.54 \times 1) + (8 \times 0.15 \times 0.15) \right] \text{m}$$

$$= (40 + 6.48 + 0.18) \text{ m} = 46.66 \text{ m} \tag{2}$$

The total heat transfer rate is therefore equal to

$$q = Sk\Delta T = 46.66 \text{ m} \times 0.035 \text{ W/m·K} \times (50 - 20)°\text{C}$$

$$= 49 \text{ W}$$

The decomposition of the overall shape factor S into three contributions, Eq. (2), shows that the six "plane walls" of the insulating wrapping account for $40/46.66$ or 86% of the entire heat transfer rate q.

3.4.2 Trees: Volume–Point Flow

The emergence of novel flow architectures in the direction of greater performance is illustrated most vividly by flows that connect volumes (or areas) with discrete points. The direction of the flow, volume-to-point versus point-to-volume, is not the phenomenon. The phenomenon is the emergence and evolution of the freely morphing and seemingly stable architecture. The evolutionary design is tree shaped (arborescent, dendritic), and it is alive because it flows and morphs with freedom in a discernible direction in time. This "movie of evolution" is played every rainy season on the river plain and the river delta (note the area–point and the point–area directions in the same tree-shaped drawing), and it is still playing in the human lung (exhaling and inhaling, in the same arborescent architecture).

The volume–point evolutionary design was proposed in 1996, along with the law of physics that governs all such phenomena: the constructal law. The first three papers unveiled the phenomenon in the context of urban flow [16], heat conduction [8, 9], and fluid flow [17]. This has become a significant new field in heat transfer science and physics, which, for the students, is reviewed regularly in textbooks [10, 18–20]. In this section, I outline the 1996 heat conduction version of the volume–point flow evolution.

In the cooling of electronic components and packages, the objective is to install more electronics (heat generation) in a fixed volume in such a way that the peak temperature does not exceed a certain level. Most of the cooling techniques that are in use today rely on convection or conjugate convection and conduction, where the coolant is either a single phase fluid or one that boils. The frontier is being pushed in the direction of smaller package dimensions. There comes a point where miniaturization makes convection cooling impractical, because the ducts through which the coolant must flow take too much space. The only way to channel the generated heat out of the electronic package is by conduction, and this conduction path will have to be very effective (with high thermal conductivity, k_p).

Conduction paths, too, take space: Designs with fewer and smaller paths are better suited for the miniaturization evolution. The fundamental point volume heat flow problem formulated in 1996 is as follows:

> Consider a finite-size volume in which heat is being generated at every point, and which is cooled through a small patch (heat sink) located on its boundary. A finite amount of high conductivity (k_p) material is available. Determine the distribution of k_p material through the given volume such that the peak temperature is minimal.

The function of the electronic package or assembly of packages is the generation of heat at a certain rate (q) in a fixed volume (V), so that the temperature difference between the hot spot (the heart of the package) and the heat sink (on the side of the package) will not exceed a certain value. The heat generation rate per unit volume is $q''' = q/V$; for simplicity, we assume that q''' is constant (i.e. q is distributed uniformly), although this assumption was easily abandoned in follow-up studies of the fundamental problem.

The fraction of the volume V that is occupied by all the high-conductivity paths is V_P. This fraction too is fixed, although as noted already, a smaller ratio V_P/V is better for miniaturization. For simplicity, we assume that $V_P \ll V$. The thermal conductivity of the conducting paths is constant (k_p) and much larger than the thermal conductivity of the material (k_0) that occupies the rest of the volume.

The easiest way to present the emergence of configuration is in a plane. For this we make the assumption that the heat generation and temperature fields are two-dimensional.

The function of any portion of the conducting path is to be in touch with the material that generates heat volumetrically. In this way we arrive at a problem of allocating a conducting path length to a volume of heat-generating material. It is important to note that the allocation cannot be made at infinitesimally small scales throughout V, because the conducting paths must be of finite length so that they can be interconnected to channel the total q to one side of V, where the heat sink is located.

We illustrate the allocation of length to volume at the smallest volume scale, which is the elemental volume shown in Figure 3.9. The elemental volume is finite and fixed ($H_0 L_0 W$), where W is the dimension in the direction perpendicular to the plane (x, y). It is a small part of the volume V of the actual device, and its size is dictated by manufacturing constraints. The dimensions H_0 and L_0 may vary, while their product $A_0 = H_0 L_0$ remains fixed. The chief unknown in this generic problem is the shape of the elemental

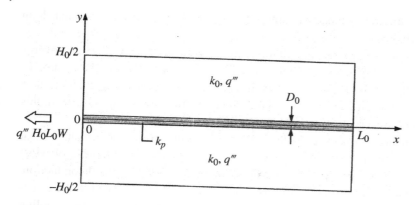

Figure 3.9 Slender elemental volume with uniform volumetric heat generation and one high-conductivity path along its axis of symmetry.

volume (or the geometric aspect ratio H_0/L_0) that facilitates the access of the heat generated ($q'''H_0L_0W$) to the lone conducting path that services the elemental volume. Symmetry and the requirement to reduce the peak temperatures are why the conducting path is placed on one of the axes of symmetry of the elemental volume. The entire heat generation rate is channeled to the exterior through *one end* of its allocated conducting path. With the exception of the origin $(0, 0)$, the boundary of the rectangular area $H_0 \times L_0$ is insulated.

Assume that the shape of the rectangular area is sufficiently slender, $H_0 \ll L_0$, and that the conduction through the heat-generating material (k_0) is oriented in the y direction. Integrating the equation for steady conduction with uniform heat generation in the k_0 material,

$$\frac{\partial^2 T}{\partial y^2} + \frac{q'''}{k_0} = 0 \tag{3.79}$$

subject to the boundary conditions $\partial T/\partial y = 0$ at $y = H_0/2$ and $T = T_0(x)$ at $y = 0$, we obtain

$$T(x, y) = \frac{q'''}{2k_0}(H_0 y - y^2) + T_0(x) \tag{3.80}$$

Conduction along the x axis is governed by the fin-type equation:

$$k_p D_0 \frac{d^2 T_0}{dx^2} + q''' H_0 = 0 \tag{3.81}$$

where $q'''H_0$ accounts for the rate at which the generated heat is being collected by the high-conductivity path. Integrating Eq. (3.81) subject to $dT_0/dx = 0$ at $x = L_0$ and $T_0 = T(0, 0)$ at $x = 0$ and substituting $T_0(x)$ into Eq. (3.80), we obtain

$$T(x, y) - T(0, 0) = \frac{q'''}{2k_0}(H_0 y - y^2) + \frac{q'''H_0}{k_p D_0}\left(L_0 x - \frac{x^2}{2}\right) \tag{3.82}$$

This expression is strictly valid for $y > 0$. The corresponding solution for $y < 0$ is obtained by replacing H_0 with $-H_0$ in the first term on the right-hand side.

In conclusion, Eq. (3.82) shows that the peak temperature in Figure 3.9 occurs at the corners situated the farthest from the origin: at $x = L_0$ and $y = \pm H_0/2$. We call this peak temperature difference ΔT_0, and we nondimensionalize it by noting that the area ($A_0 = H_0 L_0$) and the volumetric heat generation rate q''' are fixed:

$$\frac{\Delta T_0}{q''' H_0 L_0 / k_0} = \frac{1}{8}\left(\frac{H_0}{L_0}\right) + \frac{k_0 H_0}{2 k_p D_0}\frac{L_0}{H_0} \tag{3.83}$$

On the right side, the ratio D_0/H_0 is a manufacturing constant accounting for the proportion in which high-conductivity material is sandwiched with blades of the original material at the elemental level. The shape H_0/L_0 varies freely. Equation (3.83) shows that ΔT_0 can be minimized with respect to the *shape* of the elemental system (H_0/L_0). The results are

$$\frac{H_0}{L_0} = 2\left(\frac{k_0 H_0}{k_p D_0}\right)^{1/2} \quad \text{and} \quad \frac{\Delta T_0}{q''' H_0 L_0 / k_0} = \frac{1}{2}\left(\frac{k_0 H_0}{k_p D_0}\right)^{1/2} \tag{3.84}$$

They are valid when the elemental system is slender, $H_0 \ll L_0$. This restriction means that the conductivity ratio k_p/k_0 should be much greater than 1, such that

$$\frac{k_p}{k_0} \gg \frac{H_0}{D_0} \gg 1 \tag{3.85}$$

Two additional features of this design are worth stressing. First, the temperature difference along the x axis (from $x = L_0$ to $x = 0$) is the same $\frac{1}{2}\Delta T_0$ as the temperature drop from the hot spot ($x = L_0$, $y = H_0/2$) to its projection on the x axis. The second feature can be seen by combining Eq. (3.84):

$$\Delta T = \frac{q''' H_0^2}{4 k_0} \tag{3.86}$$

This means that at the elemental level the excess temperature decreases as H_0^2, from which the miniaturization incentive discussed in greater detail in Sections 5.6.2 and 7.5.2.

3.4.3 Rewards from Freedom

The evolution of configuration at scales larger than elemental has been pursued in several ways. In the first and simplest approach, elemental building blocks such as Figure 3.9 were assembled into progressively larger constructs. In 1997, a fully numerical approach [21] replaced the construction sequence with numerical simulations of conduction in the composite domain (k_0, k_p), with freedom to vary all the geometric features of the emerging tree structure.

For example, at the elemental level Ref. [21] abandoned the assumption that the high-conductivity insert is a blade of constant thickness. As shown in Figure 3.10, the profile of the k_p insert should be such that D_0 increases as $x^{1/2}$, where x is measured away from the tip. Relative to Eq. (3.84), the decrease in the global thermal resistance of the elemental volume is 6%. Even greater reductions in global thermal resistance were obtained when the assumption that the element is rectangular was abandoned. Figure 3.10 shows the emergence of a leaflike elemental shape and a tapered shape for the k_p fiber.

Increasing the freedom to morph the structure leads to higher performance levels and to designs that look *natural* [22]. Angled branches increase the global performance and make the tree architecture look even more natural, cf. Figure 3.11. Tree-shaped flow architectures that have evolved only partially are *robust*.

Resistance

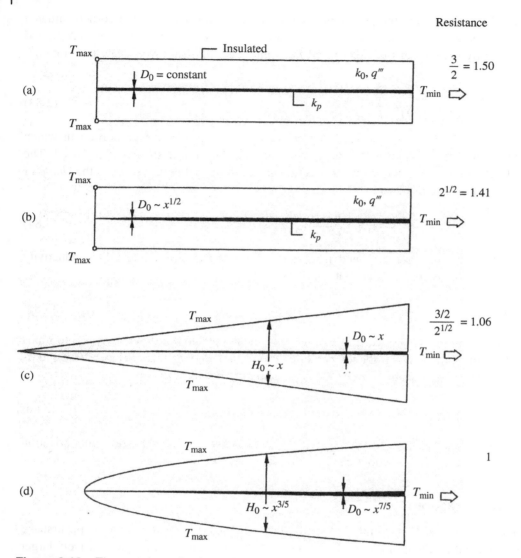

(a) T_{max} Insulated k_0, q''' $\frac{3}{2} = 1.50$
      D_0 = constant
      T_{max} k_p T_{min}

(b) T_{max} k_0, q''' $2^{1/2} = 1.41$
      $D_0 \sim x^{1/2}$
      T_{max} k_p T_{min}

(c) T_{max} $D_0 \sim x$ $\frac{3/2}{2^{1/2}} = 1.06$
      $H_0 \sim x$ T_{min}
      T_{max}

(d) T_{max} 1
      $H_0 \sim x^{3/5}$ $D_0 \sim x^{7/5}$ T_{min}
      T_{max}

Figure 3.10 Elemental conduction volume with progressively greater freedom to morph and with progressively higher performance.

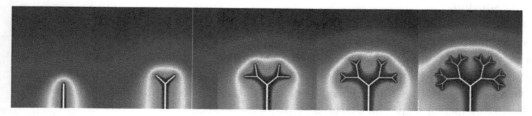

Figure 3.11 Evolutionary invasion of a conducting tree into a conducting body.

Table 3.4 How, not what: the balancing of high resistivity flow with low resistivity flow in a wide diversity of flow systems.

Flow system	What	How	
		Interstices: high resistance at the smallest scale	Channels: low resistance at larger scales
Electronics packages	Heat	Low-conductivity substrate	High-conductivity inserts (blades, needles)
River basins	Water	Darcy flow through porous media	Rivulets and rivers
Lungs	Air	Diffusion in alveoli, tissues	Bronchial passages
Circulatory systems	Blood	Diffusion in capillaries, tissues	Vessels, capillaries, arteries, veins
Turbulent flow	Momentum	Laminar, viscous diffusion	Streams, eddies
Urban traffic	People	Walking	Street traffic
Economics	Goods	Hand delivery and hand collection	Freight, rail, truck, air, ship

Source: Private communication from S. Périn.

The literature on tree-shaped paths for conduction is growing rapidly [23–29] and shows the merits of increasing the number of degrees of freedom in the evolution of the flow design.

In drawings such as Figures 3.9–3.11, we discover tree patterns that facilitate flow access. Every detail of the tree geometry is the result of invoking the constructal law. The discovery of the tree as the flow architecture that facilitates access between one point and an infinity of points (area, volume) is general – it is not restricted to trees of streets, rivers, and high-conductivity inserts.

The generality of the tree discovery is stressed by Table 3.4, which shows that the "how" unites and the "what" divides. How the tree is generated (through a balance between high resistivity and low resistivity) is the same in many classes of flow systems, regardless of the diversity of the currents that flow through them. The unique principle that *unites* is a lot harder to see than the obvious observation that *diversity* characterizes the flow systems of nature and engineering. The common architecture generated by the constructal law is subtle, while the observed diversity is banal.

References

1 Fourier, J. (1878). *Analytical Theory of Heat* (trans., with notes, by A. Freeman). New York: G. E. Stechert & Co. (The French edition appeared in 1822, as *Théorie Analytique de la Chaleur*; this theory was described first in a paper submitted by Fourier to the *Institut de France* in 1807.).

2 Carslaw, H.S. and Jaeger, J.C. (1959). *Conduction of Heat in Solids*, 2e. Oxford: Oxford University Press.

3 Arpaci, V.S. (1966). *Conduction Heat Transfer*. Reading, MA: Addison Wesley.

4 Kakac, S. and Yener, Y. (1985). *Heat Conduction*. Washington, DC: Hemisphere.

5 Watson, G.N. (1966). *Theory of Bessel Functions*. Cambridge: Cambridge University Press.

6 Abramowitz, M. and Stegun, I. (eds.) (1964). *Handbook of Mathematical Functions*. Washington, DC: NBS AMS 55.

7 Bejan, A. (2013). *Convection Heat Transfer*, Section 2.4, 4e. Hoboken, NJ: Wiley.

8 Bejan, A. (1997). Constructal-theory network of conducting paths for cooling a heat generating volume. *Int. J. Heat Mass Transfer* 40: 799–816, published on 1 November 1996 as shown in Ref. [9].

9 Bejan, A. (2013). Technology evolution, from the constructal law. *Adv. Heat Transfer* 45: 183–207.

10 Bejan, A. (2013). *Convection Heat Transfer*, Section 1.5, 4e. Hoboken, NJ: Wiley.

11 Andrews, R.V. (1955). Solving conductive heat transfer problems with electrical-analogue shape factors. *Chem. Eng. Prog.* 51 (2): 67–71.

12 Sunderland, J.E. and Johnson, K.R. (1964). Shape factors for heat conduction through bodies with isothermal or convective boundary conditions. *Trans. ASHRAE* 70: 237–241.

13 Hahne, E. and Grigull, U. (1975). Shape factor and shape resistance for steady multidimensional heat conduction (in German). *Int. J. Heat Mass Transfer* 18: 751–767.

14 Hassani, A.V. and Hollands, K.G.T. (1990). Conduction shape factor for a region of uniform thickness surrounding a three-dimensional body of arbitrary shape. *J. Heat Transfer* 112: 492–495.

15 Langmuir, I., Adams, E.Q., and Meikle, G.S. (1913). Flow of heat through furnace walls: the shape factor. *Trans. Am. Electrochem. Soc.* 24: 53–81.

16 Bejan, A. (1996). Street network theory of organization in nature. *J. Adv. Transp.* 30: 85–107.

17 Bejan, A. (1997). Constructal tree network for fluid flow between a finite-size volume and one source or sink. *Rev. Gén. Therm.* 36: 592–604.

18 Bejan, A. (2016). *Advanced Engineering Thermodynamics*, Chapter 13, 4e. Hoboken, NJ: Wiley.

19 Bejan, A. and Lorente, S. (2008). *Design with Constructal Theory*. Hoboken, NJ: Wiley.

20 Miguel, A.F. and Rocha, L.A.O. (1997). *Tree-Shaped Fluid Flow and Heat Transfer*. Cham: Springer.

21 Ledezma, G.A., Bejan, A., and Errera, M.R. (1997). Constructal tree networks for heat transfer. *J. Appl. Phys.* 82: 89–100.

22 Bejan, A. (2015). Constructal law: optimization as design evolution. *J. Heat Transfer* 137: 061003.

23 Manuel, M.C.E. and Lin, P.T. (2017). Design exploration of heat conductive pathways. *Int. J. Heat Mass Transfer* 104: 835–851.

24 Errera, M.R., Frigo, A.L., and Segundo, E.H.V. (2014). The emergence of the constructal element in tree-shaped flow organization. *Int. J. Heat Mass Transfer* 78: 181–188.

25 Kobayashi, H., Maeno, T., Lorente, S., and Bejan, A. (2014). Double tree structure in a conducting body. *Int. J. Heat Mass Transfer* 77: 140–146.

26 Li, B., Hong, J., and Ge, L. (2017). Constructal design of internal cooling geometries in heat conduction system using the optimality of natural branching structures. *Int. J. Therm. Sci.* 115: 16–28.

27 Li, B., Hong, J., Ge, L., and Xuan, C. (2017). Designing biologically inspired heat conduction paths for 'volume-to-point' problems. *Mater. Des.* 130: 317–326.

28 Biserni, C., Rocha, L.A.O., and Bejan, A. (2004). Inverted fins: geometric optimization of the intrusion into a conducting wall. *Int. J. Heat Mass Transfer* 47: 2577–2586.

29 Hajmohammadi, M.R. (2017). Introducing ψ-shaped cavity for cooling a heat generating medium. *Int. J. Therm. Sci.* 121: 204–212.

Problems

Analytical Methods

3.1 Compare the two-dimensional conduction estimate (3.46) with the simpler one-dimensional formula (2.97), and show that the ratio $q_{b, (3.46)}/q_{b, (2.97)}$ depends only on the Biot number $hH/2k$. Using the first six characteristic values listed in Table 3.1, show that the q_b ratio decreases steadily as the Biot number increases. In other words, show that the one-dimensional model (2.97) overestimates the value of q_b and that it is accurate only when the Biot number is small.

3.2 The exact analytical solution for the temperature distribution inside the cross section with uniform heat generation shown in Figure 3.8 in the text can be constructed by adding together the solutions to two simpler problems (see Figure P3.2):

$$\theta(x, y) = \theta_1(y) + \theta_2(x, y)$$

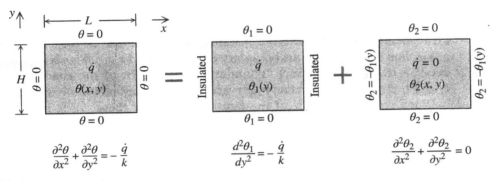

Figure P3.2

The volumetric rate of internal heat generation is uniform, $\dot{q} = $ constant. The first contribution (θ_1) is the solution to a unidirectional conduction problem with internal heat generation, whereas the second contribution (θ_2) is the solution to a two-dimensional conduction problem without internal heat generation.

a) Determine the complete analytical solution for $\theta(x, y)$. You may wish to position the origin of the (x, y) system in the center of the cross section.

b) Show that when the cross section is square, the maximum temperature at the center of the cross-section is

$$\theta_{max} = 0.07367 \frac{\dot{q} H^2}{k}$$

Compare this result with the approximate result based on integral analysis, Eq. (3.72).

3.3 Reconsider the two-dimensional plate fin problem of Figure 3.1 by focusing on the case when L may be regarded as infinitely greater than H. Step by step, perform the equivalent of the analysis contained between Eqs. (3.2) and (3.28), and show that the temperature distribution is

$$\theta = \frac{4\theta}{\pi} \sum_{n=0}^{\infty} \frac{1}{2n+1} \exp[-(2n+1)(\pi x / H)] \sin[(2n+1)(\pi y / H)]$$

3.4 The semi-infinite plate of thickness H shown in Figure P3.4 has a linear temperature distribution imposed on the left boundary:

$$\theta = by \quad \text{at } x = 0$$

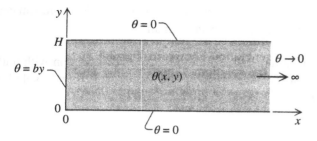

Figure P3.4

The remaining boundaries and the distant end to the right ($x \to \infty$) are all at the same temperature ($\theta = 0$). By following the steps outlined in Sections 3.1.2–3.1.4 in the text, show that the temperature distribution inside the plate is

$$\theta = bH\frac{2}{\pi}\sum_{n=0}^{\infty}\frac{(-1)^{n+1}}{n}\exp\left(-n\pi\frac{x}{H}\right)\sin\left(n\pi\frac{y}{H}\right)$$

where the constant b is the temperature gradient along the left boundary.

3.5 A temperature difference θ_a is maintained between the top and bottom surfaces of the semi-infinite plate shown on the left side of Figure P3.5. The temperature distribution inside the plate, $\theta(x, y)$, cannot be determined directly; note that y is not a direction of homogeneous boundary conditions and x is not such a direction either because there is no distinct boundary to the right of the conducting medium. Instead, $\theta(x, y)$ can be obtained by adding the solutions to the simpler problems outlined on the right side of Figure P3.5. The first of these, $\theta_1(y)$, is the solution to one-dimensional conduction in the y direction. The $\theta_2(x, y)$ problem is the same as Problem 3.4. Determine the temperature distribution $\theta(x, y)$.

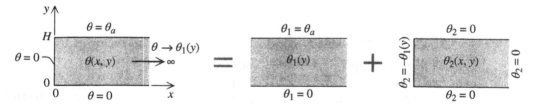

Figure P3.5

3.6 Figure P3.6 shows the actual distribution of temperature in the cross section through the semi-infinite plate shown on the left of Figure P3.5 in Problem 3.5. This plot was made using the series solution derived in Problem 3.5. Note that the temperature field $\theta(x, y)$ has two distinct regions – a distant region (far to the right) in which θ is practically independent of x and an "end" region of approximate length δ, in which θ depends on both x and y.

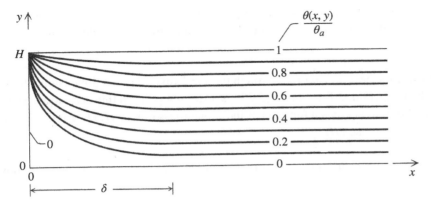

Figure P3.6

Write down the conduction equation satisfied by θ in the end region, and replace each term in this equation with its respective scale. From the approximate algebraic equation that results, deduce the proper scale of the end-region length δ. Compare your conclusion with the length scale revealed by the exact drawing.

3.7 Figure P3.7 shows the triangular cross section through a long bar. A finite temperature difference (θ_b) is maintained between the two sides that are mutually perpendicular. The hypotenuse is perfectly insulated. Determine analytically the temperature distribution for steady conduction in the triangular area, $\theta(x, y)$. (*Hint*: Exploit the geometrical relationship that might exist between the given triangle and a square cross section of side L.)

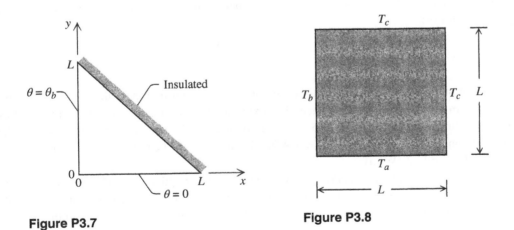

Figure P3.7

Figure P3.8

3.8 Figure P3.8 shows the square cross section of a long bar. Two adjoining sides are at the same temperature, T_c, while the remaining two sides are at different temperatures, T_a and T_b. Determine the steady-state temperature distribution inside the square cross section by superimposing two solutions of the type developed in the text for the problem of Figure 3.1; in other words, rely on Eq. (3.28) and the principle of superposition in order to deduce the series solution for the present problem.

3.9 Derive an expression for the heat transfer rate through the $x = L$ tip of the two-dimensional fin shown in Figure 3.1. Use the temperature distribution (3.28) as a starting point in this derivation. Show that the heat transfer rate through the tip is a function of the slenderness ratio L/H. Determine the tip heat transfer rate formula for the case $L/H = 2$.

3.10 a) Determine the temperature distribution $T(x, y)$ inside semi-infinite slab of thickness H shown in Figure P3.10.

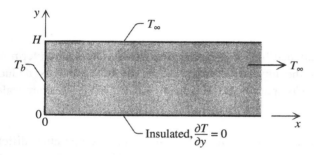

Figure P3.10

b) Note the geometrical relationship between the present problem and the problem solved in the text based on Figure 3.3. Write down the solution to the present problem by appropriately interpreting the solution listed in Eqs. (3.44) and (3.45). Compare your solution with the one developed step by step in part (a).

Shape Factor

3.11 The pipe that supplies drinking water to a community in a cold region of the globe is buried at a depth $H = 3$ m below the ground surface. The external surface of the pipe can be modeled as an isothermal cylinder of diameter $D = 0.5$ m and temperature $4\,°C$. The ground surface is at $0\,°C$, and the thermal conductivity of the soil is $k = 1$ W/m·K. Calculate the heat transfer rate from the pipe to the surrounding soil, q' (W/m).

During a very harsh winter, the top layer of the soil freezes to a depth of 1 m. This new position of the freezing front can be modeled as an isothermal plane of temperature $0\,°C$ situated at $H = 2$ m above the pipe centerline. Calculate again the per-unit-pipe-length heat transfer rate q', and the increase in q' that is due to the 1-m advancement of the freezing front (Figure P3.11).

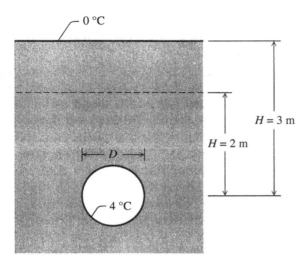

Figure P3.11

3.12 The sphere of radius r_i is surrounded by a constant thickness shell of conductivity k and outer radius r_o. This assembly is buried in an infinite stationary medium of conductivity k_∞.

Derive an expression for the overall thermal resistance R_t between the r_i sphere and the infinite medium. Show that R_t varies monotonically with the shell outer radius; in other words, show that the "critical radius" concept of Section 2.4 does not apply. Under what circumstances does the spherical shell provide a thermal insulation effect for the r_i sphere? Conversely, what relationship must exist between k and k_∞ if the addition of the shell is to reduce the thermal resistance between the r_i sphere and the infinite medium? (Figure P3.12)

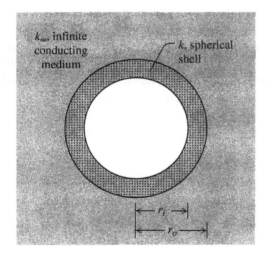

Figure P3.12

3.13 Figure P3.13 shows a cross-sectional view through a horizontal pipe buried at a depth $H = 2\,\text{m}$ beneath the ground surface. The pipe of outer diameter $D_i = 26\,\text{cm}$ carries pressurized steam of temperature $T_s = 120\,°\text{C}$. A 7-cm-thick layer of insulation of conductivity $k_{\text{ins}} = 0.2\,\text{W/m·K}$ is wrapped around the pipe wall. The ground may be modeled as dry soil with the thermal conductivity $k_\infty = 1\,\text{W/m·K}$. The ground surface is isothermal and at the average seasonal temperature $T_\infty = 10\,°\text{C}$.

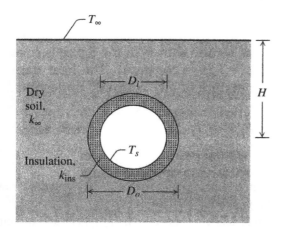

Figure P3.13

Calculate the heat transfer rate from the steam pipe, expressed per unit pipe length. Model the conduction heat transfer through the insulation as purely radial, and use Eq. (2.40) for the overall thermal resistance between T_s and T_∞. The apparent heat transfer coefficient h at the outer surface of the insulation can be evaluated based on the appropriate shape factor S listed in Table 3.3.

3.14 The working space of an experimental apparatus is shaped as a parallelepiped with the dimensions $x = 1\,\text{m}$, $y = 0.5\,\text{m}$, and $z = 0.3\,\text{m}$. This space is covered all around by a layer of fiberglass insulation, with a uniform thickness $L = 0.1\,\text{m}$. The temperature difference everywhere across this insulating layer is $\Delta T = 10\,°\text{C}$. Calculate the total heat transfer rate through the insulation.

4

Time-Dependent Conduction

4.1 Immersion Cooling or Heating

The most basic problem of time-dependent conduction is to predict the temperature history inside a conducting body that is immersed suddenly in a bath of fluid at a different temperature. This problem finds applications in many areas, for example, in the heat treating (e.g. quenching) of special alloys. It is represented by the model shown in Figure 4.1, in which a body of initial temperature T_i, volume V, density ρ, specific heat c, thermal conductivity k, and external (wetted) area A is surrounded by a fluid of temperature T_∞. The coefficient h for convective heat transfer across the exposed area of the body is assumed constant, for simplicity.

The first solid layers that feel the effect of thermal contact with the surrounding fluid are the "skin" under the wetted area A. This region assumes temperatures that bridge the gap between the core temperature (still at T_i) and the surface temperature T_0. The latter assumes an intermediate value between T_i and T_∞ and, as the time increases, approaches the bath temperature T_∞.

Think of a metal-quenching application, where $T_i > T_\infty$. This case is evident on the right side of Figure 4.1, which is an enlargement of the temperature distribution across the skin layer. The surviving hot core is surrounded by a colder region (the skin layer), and the size of the hot core shrinks one layer at a time as the thickness of the skin layer, δ, increases. There comes a time t_c when δ has grown all the way to the center of the body, that is, a time when the solid no longer has a distinct core and a distinct skin layer. When the time greatly exceeds this time of complete thermal penetration, the body temperature becomes essentially equal to the surface temperature T_0. In other words, at sufficiently long times, the temperature distribution through the body is for all practical purposes a function of time, $T(t)$.

The preceding description is illustrated in the three-frame scenario of Figure 4.2. Early in the life of the process, the body temperature is a function of both time and spatial position. The long-time behavior is quite different (i.e. much simpler) in that the entire body is represented by a single instantaneous temperature, $T(t)$. Important is the special time scale t_c that marks the transition between the two regimes. This time scale is determined based on the following scale analysis.

Examine the enlarged view of the temperature distribution across the skin layer, Figure 4.1. If the skin thickness δ is much smaller than the distance to the center of the body (labeled L or r_o in the charts of Figures 4.7–4.16), then the conduction through the skin region may be treated as *one-dimensional*, in the same manner as the curved thin wall in Figure 2.1. According to Eq. (1.21), the equation for time-dependent

Heat Transfer: Evolution, Design and Performance, First Edition. Adrian Bejan.
© 2022 John Wiley & Sons, Inc. Published 2022 by John Wiley & Sons, Inc.
Companion website: www.wiley.com/go/bejan/heattransfer

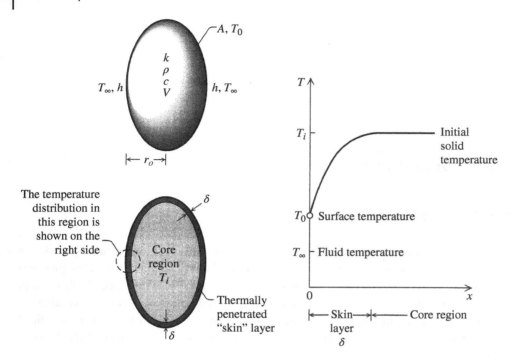

Figure 4.1 Thermally penetrated layer in a body immersed suddenly in a fluid.

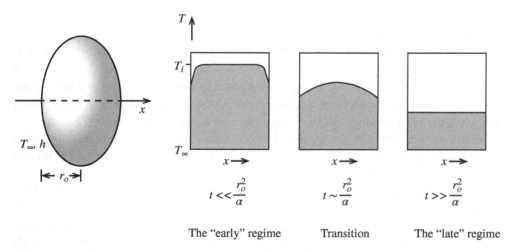

Figure 4.2 The transition from the early regime to the late regime.

conduction in the direction x, without internal heat generation, is

$$\frac{\partial^2 T}{\partial x^2} = \frac{1}{\alpha} \frac{\partial T}{\partial t} \tag{4.1}$$

Inside the skin region of thickness δ, the order of magnitude of the curvature of the temperature profile is the same as the change in the slope $\partial T/\partial x$ across the distance δ:

$$\frac{\partial^2 T}{\partial x^2} \sim \frac{\left(\frac{\partial T}{\partial x}\right)_{x\sim\delta} - \left(\frac{\partial T}{\partial x}\right)_{x=0}}{\delta - 0} \tag{4.2}$$

Figure 4.1 reveals these temperature gradient scales:

$$\left(\frac{\partial T}{\partial x}\right)_{x\sim\delta} \sim 0 \qquad \left(\frac{\partial T}{\partial x}\right)_{x=0} \sim \frac{T_i - T_0}{\delta} \tag{4.3}$$

Therefore, after substituting these into Eq. (4.2), the scale of the x-curvature becomes

$$\frac{\partial^2 T}{\partial x^2} \sim -\frac{T_i - T_0}{\delta^2} \tag{4.4}$$

Next, we focus on the right side of Eq. (4.1), which is the thermal inertia of the conducting material. The order of magnitude of that term follows from the observation that the average temperature of the δ-thick region drops from the initial level T_i to a value comparable[1] with T_0 during the time interval t:

$$\frac{\partial T}{\partial t} \sim \frac{T_0 - T_i}{t - 0} \tag{4.5}$$

Combining this result with Eq. (4.4) into the equality of scales required by Eq. (4.1),

$$-\frac{T_i - T_0}{\delta^2} \sim \frac{1}{\alpha}\frac{T_0 - T_i}{t} \tag{4.6}$$

we conclude that the skin thickness δ increases as the square root of the elapsed time:

$$\delta \sim (\alpha t)^{1/2} \tag{4.7}$$

The transition time t_c is reached when δ has grown such that it is comparable with the transversal dimension of the whole body:

$$\delta \sim r_o \quad \text{at} \quad t \sim t_c \tag{4.8}$$

In view of Eq. (4.7), the transition time scale is

$$t_c \sim \frac{r_o^2}{\alpha} \tag{4.9}$$

This time scale distinguishes between the *early regime*, when the skin layer and the untouched core are distinct:

$$t \ll \frac{r_o^2}{\alpha} \qquad T = T(x, t) \tag{4.10}$$

and the *late regime*, when the instantaneous temperature is uniform:

$$t \gg \frac{r_o^2}{\alpha} \qquad T \cong T(t) \tag{4.11}$$

The time criterion and the two regimes that have just been identified are the skeleton of an *approximate* description of the temperature history at a point inside a conducting body of arbitrary shape. In Sections 4.2 and 4.3, we analyze these regimes separately.

---

1 The lower the surface temperature T_0, the lower is the average temperature of the skin layer.

4.2 Lumped Capacitance Model (The "Late" Regime)

Consider the late regime when the temperature gradients inside the body have decayed and the body temperature is well approximated by a single value, $T(t)$. We treat this regime first because it is analytically the simplest. The analysis consists of writing the first law of thermodynamics for the body of Figure 4.1, by treating it as a closed system that experiences zero work transfer, Eq. (1.2):

$$q = \frac{dE}{dt} \tag{4.12}$$

The net heat transfer rate *into* the body is proportional to the fluid–body temperature difference and the wetted area:

$$q = hA(T_\infty - T) \tag{4.13}$$

The heat transfer coefficient h is assumed constant, although this is an approximation of the real situation (in natural convection, for example, h is larger at the bottom of the hot object, Chapter 7). The time rate of change in the energy inventory of the system can be rewritten by invoking the incompressible substance model (1.17):

$$\frac{dE}{dt} = \rho c V \frac{dT}{dt} \tag{4.14}$$

Substituting Eqs. (4.13) and (4.14) in Eq. (4.12) yields

$$-\frac{hA}{\rho c V}(T - T_\infty) = \frac{dT}{dt} \tag{4.15}$$

This equation for $T(t)$ can be integrated by noting the starting condition:

$$T = T_1 \quad \text{at} \quad t = t_c \tag{4.16}$$

The solution shows that the body–fluid temperature difference decays exponentially:

$$\frac{T - T_\infty}{T_1 - T_\infty} = \exp\left[-\frac{hA}{\rho c V}(t - t_c)\right] \tag{4.17}$$

and that the time scale of the decay is $\rho c V/hA$. It takes longer for the body to reach equilibrium with the surrounding fluid when its lumped capacitance $\rho c V$ is large and/or its product hA is small.

Under what circumstances does the exponential decay (4.17) *alone* characterize the temperature history inside the body of Figure 4.1? This happens when the starting temperature T_1 assumed in Eq. (4.16) is nearly the same as the initial temperature T_i specified in the original problem statement. At the start of the exponential decay, $t = t_c$, the skin layer had just reached the center of the body. The temperature gradient across the body at this moment is of order $(T_i - T_0)/r_o$; therefore, the conduction heat flux that escapes through the exposed area A is

$$q'' \sim k\frac{T_i - T_0}{r_o} \tag{4.18}$$

This flux equals the convective flux through the wetted side of the surface:

$$k\frac{T_i - T_0}{r_o} \sim h(T_0 - T_\infty) \tag{4.19}$$

which can be rearranged as

$$T_i - T_0 \sim \frac{Bi}{1 + Bi}(T_i - T_\infty) \quad \text{where } Bi = \frac{hr_o}{k} \tag{4.20}$$

to show that the temperature variation across the body, $T_i - T_0$, is negligible relative to the overall difference $T_i - T_\infty$ only when the Biot number (Bi) is small:

$$\frac{hr_o}{k} \ll 1 \tag{4.21}$$

Therefore, it is in the small-Bi limit that the starting temperature, T_1, used in Eq. (4.16) is nearly the same as the true initial temperature of the body, T_i. The exponential decay (4.17) is an adequate description of the body temperature history when the time exceeds the critical time t_c, Eq. (4.9). The lumped capacitance regime of Eq. (4.17) covers most of the life of the time-dependent conduction process when the decay time scale $\rho c V/(hA)$ greatly exceeds the duration of the early regime, t_c.

4.3 Semi-infinite Solid Model (The "Early" Regime)

4.3.1 Constant Surface Temperature

The analysis of the early regime is more complicated because when the skin layer is distinct, the temperature is a function of both x and t. Relative to the thin layer of thickness δ, which steadily expands according to Eq. (4.7), the isothermal core looks like a *semi-infinite solid* of temperature T_i. The unsteady conduction in the skin layer can be studied in the one-dimensional system of Figure 4.3, in which the x axis points toward the core.

Consider first the case in which the heat transfer coefficient is so large that, once exposed, the surface assumes the same temperature as the fluid, $T_0 = T_\infty$. The complete problem statement for determining $T(x, t)$ near the surface consists of Eq. (4.1) and the initial and boundary conditions shown in the figure:

Conduction Equation.

$$\frac{\partial^2 T}{\partial x^2} = \frac{1}{\alpha}\frac{\partial T}{\partial t} \tag{4.22}$$

Initial Condition.

$$T = T_i \quad \text{at} \quad t = 0 \tag{4.23}$$

Boundary Conditions.

$$T = T_\infty \quad \text{at} \quad x = 0 \tag{4.24}$$

$$T \to T_i \quad \text{as} \quad x \to \infty \tag{4.25}$$

From these conditions and the statement that δ increases as $(\alpha t)^{1/2}$, Eq.(4.7), we can sketch the general outlook of the family of curves $T(x, t)$ – see the left side of Figure 4.3. The temperature gradient along each $T - x$ curve decreases as the time increases, that is, as the effect of having dropped the surface temperature from T_i to T_∞ diffuses into the body.

The key to solving the problem (4.22)–(4.25) is the observation that all the $T - x$ curves are *similar*. Each curve starts out from $T = T_\infty$ and aims asymptotically for T_i as x becomes sufficiently large. Furthermore, each curve has only one knee. The scale analysis completed in Section 4.2 showed that the skin

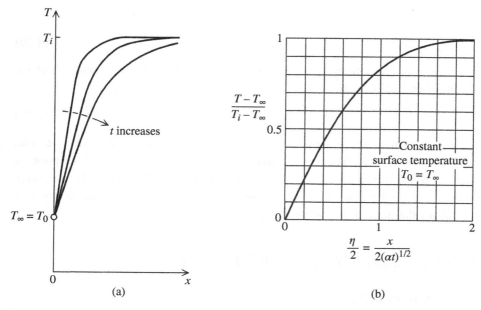

Figure 4.3 Penetration of heat conduction into a semi-infinite solid with isothermal surface (a) and the collapse of the *T–x* curves into a similarity profile (b).

layer (the curved portion of the $T - x$ curve) expands to the right as $(\alpha t)^{1/2}$; therefore, it is possible to replot the entire $T(x, t)$ family as a single curve $T(\eta)$, where the dimensionless *similarity variable* η is the ratio between the actual x and the scale of the skin thickness:

$$\eta = \frac{x}{(\alpha t)^{1/2}} \tag{4.26}$$

This graphic alternative is shown on the right side of Figure 4.3. What remains is to transform the $T(x, t)$ problem (4.22)–(4.25) into a new version for the *similarity* temperature profile $T(\eta)$. Equation (4.22) is transformed as follows:

$$\frac{\partial T}{\partial x} = \frac{dT}{d\eta}\frac{\partial \eta}{\partial x} = \frac{dT}{d\eta}\frac{1}{(\alpha t)^{1/2}} \tag{4.27}$$

$$\frac{\partial^2 T}{\partial x^2} = \frac{d}{d\eta}\left(\frac{\partial T}{\partial x}\right)\frac{\partial \eta}{\partial x} = \frac{d^2 T}{d\eta^2}\frac{1}{\alpha t} \tag{4.28}$$

$$\frac{\partial T}{\partial t} = \frac{dT}{d\eta}\frac{\partial \eta}{\partial t} = \frac{dT}{d\eta}\left(-\frac{x}{2\alpha^{1/2}t^{3/2}}\right) \tag{4.29}$$

Substituting Eqs. (4.28) and (4.29) in Eq. (4.22), we obtain an equation for $T(\eta)$:

$$\frac{d^2 T}{d\eta^2} + \frac{\eta}{2}\frac{dT}{d\eta} = 0 \tag{4.30}$$

The initial and boundary conditions (4.23)–(4.25) are represented by

$$T = T_\infty \quad \text{at} \quad \eta = 0 \tag{4.31}$$

$$T \to T_i \quad \text{as} \quad \eta \to \infty \tag{4.32}$$

in which Eq. (4.32) accounts for Eqs. (4.25) and (4.23). The $T(\eta)$ problem (4.30)–(4.32) is solved by separating the variables in Eq. (4.30):

$$\frac{d(T')}{T'} = -\frac{\eta}{2}d\eta \quad \text{where } T' = \frac{dT}{d\eta} \tag{4.33}$$

Integrated twice in η, this equation yields sequentially

$$\ln T' = -\frac{\eta^2}{4} + \ln C_1 \tag{4.34}$$

$$\frac{dT}{d\eta} = C_1 \exp\left(-\frac{\eta^2}{4}\right) \tag{4.35}$$

$$T = C_1 \int_0^\eta \exp\left(-\frac{\beta^2}{4}\right) d\beta + C_2 \tag{4.36}$$

where β is a dummy variable and, according to Eq. (4.31), $C_2 = T_\infty$:

$$T - T_\infty = C_1 \int_0^\eta \exp\left[-\left(\frac{\beta}{2}\right)^2\right] d\beta \tag{4.37}$$

On the right side, we see the emergence of the *error function* described in Appendix E:

$$\text{erf}(x) = \frac{2}{\pi^{1/2}} \int_0^x \exp\left(-m^2\right) dm \tag{4.38}$$

which has these noteworthy features:

$$\text{erf}(0) = 0 \tag{4.39a}$$

$$\text{erf}(\infty) = 1 \tag{4.39b}$$

$$\frac{d}{dx}[\text{erf}(x)]_{x=0} = \frac{2}{\pi^{1/2}} = 1.1284 \tag{4.40}$$

The right side of Eq. (4.37) can be reshaped by introducing the notation $m = \beta/2$:

$$
\begin{aligned}
T - T_\infty &= 2C_1 \int_0^\eta \exp\left[-\left(\frac{\beta}{2}\right)^2\right] d\left(\frac{\beta}{2}\right) \\
&= 2C_1 \int_0^{\eta/2} \exp(-m^2) \, dm \\
&= 2C_1 \underbrace{\frac{\pi^{1/2}}{2}}_{C_3} \frac{2}{\pi^{1/2}} \int_0^{\eta/2} \exp(-m^2) \, dm \\
&= C_3 \, \text{erf}\left(\frac{\eta}{2}\right)
\end{aligned}
\tag{4.41}
$$

The constant C_3 is finally pinpointed by invoking the remaining boundary condition, Eq. (4.32), while keeping in mind Eq. (4.39b). This operation yields $C_3 = T_i - T_\infty$. Replacing η with $x/(\alpha t)^{1/2}$ in Eq. (4.41), the analytical solution for $T(x, t)$ becomes

$$\frac{T - T_\infty}{T_i - T_\infty} = \text{erf}\left[\frac{x}{2(\alpha t)^{1/2}}\right] \tag{4.42}$$

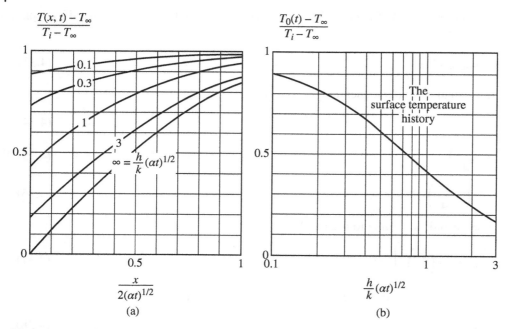

Figure 4.4 Temperature distribution in an isothermal semi-infinite solid (T_i) placed suddenly in contact with fluid flow (T_∞).

This solution is displayed in two forms, $T(x, t)$ and $T(\eta)$, in Figure 4.3. The corresponding instantaneous surface heat flux formula is

$$q''(t) = -k\left(\frac{\partial T}{\partial x}\right)_{x=0} = -k\frac{T_i - T_\infty}{(\pi\alpha t)^{1/2}} \tag{4.43}$$

where the sign of q'' is positive when q'' points in the positive x direction.

This solution is valid in the early regime, Eq. (4.10), *and* when the surface temperature can indeed be modeled as fixed and equal to the fluid temperature, $T_0 = T_\infty$. (Review the large-h assumption made before adopting the isothermal surface condition, Eq. (4.24)). We will learn that the $T_0 = T_\infty$ assumption holds true when the Biot number based on the thermal penetration depth δ, namely $(\alpha t)^{1/2}h/k$, is considerably greater than 1 (see the right side of Figure 4.4).

4.3.2 Constant Heat Flux Surface

There are several other analytical solutions that, although unrelated to the metal quenching problem of Figure 4.1, are quite useful in the calculation of transient temperatures and heat transfer in the superficial layers of objects heated by thermal radiation. One is temperature near the surface of a semi-infinite solid that is exposed at $t = 0$ to a constant (i.e. time-independent) heat flux q'':

$$T(x, t) - T_i = 2\frac{q''}{k}\left(\frac{\alpha t}{\pi}\right)^{1/2}\exp\left(-\frac{x^2}{4\alpha t}\right) - \frac{q''}{k}x\,\text{erfc}\left[\frac{x}{2(\alpha t)^{1/2}}\right] \tag{4.44}$$

where T_i is the initial temperature of the solid. The new function erfc( ) is the complementary error function that, in association with Eq. (4.38), is defined by

$$\text{erfc}(x) = 1 - \text{erf}(x) \tag{4.45}$$

The surface temperature history $T_0(t)$ that corresponds to the constant flux solution is

$$T_0(t) - T_i = T(0, t) - T_i = 2\frac{q''}{k}\left(\frac{\alpha t}{\pi}\right)^{1/2} \quad (q'' = \text{constant}) \tag{4.46}$$

The temperature of the surface with q'' increases as $t^{1/2}$. Note the similarity between this formula and isothermal surface (4.43), which can be rearranged as

$$T_\infty - T_i = \frac{q''(t)}{k}(\pi\alpha t)^{1/2} \quad (T_0 = T_\infty) \tag{4.47}$$

In Eqs. (4.47) and (4.46), the ratio $(T_0 - T_i)/q''$ is proportional to the group $(\alpha t)^{1/2}/k$, which is the same as δ/k. This conclusion can also be drawn based on scale analysis by stating that the conduction heat flux (on the solid side of the exposed surface) is proportional to the local temperature gradient, $q'' \sim k(T_0 - T_i)/\delta$.

4.3.3 Surface in Contact with Fluid Flow

The more general version of the isothermal surface problem treated in Section 4.3.1 is the problem of time-dependent conduction in a semi-infinite solid in contact with a fluid of a different temperature. This problem is the same as Eqs. (4.22)–(4.25), except that the surface condition (4.24) is replaced by

$$\underbrace{h(T_\infty - T)}_{\substack{\text{Convection heat} \\ \text{flux arriving from} \\ \text{the fluid}}} = \underbrace{-k\left(\frac{\partial T}{\partial x}\right)}_{\substack{\text{Conduction heat} \\ \text{flux entering} \\ \text{the solid}}} \quad \text{at} \quad x = 0 \tag{4.48}$$

The solution for the instantaneous temperature distribution through the solid is [1]

$$\frac{T(x, t) - T_\infty}{T_i - T_\infty} = \text{erf}\left[\frac{x}{2(\alpha t)^{1/2}}\right] + \exp\left(\frac{hx}{k} + \frac{h^2\alpha t}{k^2}\right)\text{erfc}\left[\frac{x}{2(\alpha t)^{1/2}} + \frac{h}{k}(\alpha t)^{1/2}\right] \tag{4.49}$$

This temperature distribution depends on two dimensionless groups, the dimensionless distance from the surface, x/δ, and the Biot number based on the thermal penetration depth, $h\delta/k$, or, as plotted on Figure 4.4, $x/2(\alpha t)^{1/2}$ and $h(\alpha t)^{1/2}/k$. The left side of Figure 4.4 shows the effect of the Biot number $h(\alpha t)^{1/2}/k$ on the temperature profile. The lowest curve shown on this graph is the same as the curve plotted on the right side of Figure 4.3, because the constant surface temperature model of Figure 4.3 is valid when the δ-based Biot number is much greater than 1.

The right side of Figure 4.4 shows the evolution of the surface temperature as the time increases. If the heat transfer coefficient is finite, the earliest temperatures of the surface resemble the initial temperature of the solid, T_i, no matter how large the value of h. The surface temperature $T_0(t)$ approaches the temperature of the fluid bath, T_∞, only after a time interval that is long enough:

$$\frac{h}{k}(\alpha t)^{1/2} \gg 1 \tag{4.50}$$

This time criterion can be rewritten in terms of the dimensionless time group brought to light by Eqs. (4.10) and (4.11):

$$\frac{hr_o}{k}\left(\frac{\alpha t}{r_o^2}\right)^{1/2} \gg 1 \tag{4.51}$$

where hr_o/k is the Biot number based on the transversal dimension of the body.

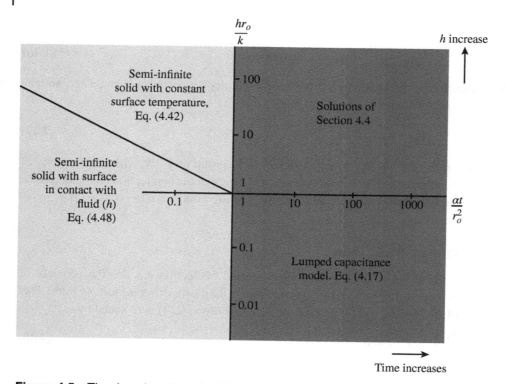

Figure 4.5 The domains of applicability of the lumped capacitance and semi-infinite solid models.

The combined message of the "transitions" identified by the time inequalities (4.10), (4.11), and (4.51) is presented in Figure 4.5. The time increases from left to right. The right half of the plane represents the late regime, when the lumped system model of Section 4.2 is valid. The left half of the plane accounts for the shorter times $\alpha t/r_o^2$, that is, for the early regime. The line of slope $-\frac{1}{2}$ (note the log–log grid) represents Eq. (4.51); only above this line, namely, when the Biot number hr_o/k is sufficiently large, does the constant surface temperature solution of Section 4.3.1 hold true. Below the inclined line, the appropriate solution to use is Eq. (4.48).

The behavior of the transient conduction phenomenon inside a body immersed suddenly in a fluid is now clear. Regardless of the geometric shape of the body (slab, cylinder, sphere), the conduction phenomenon is described adequately by an appropriate sequence of much simpler regimes for which the temperature distribution and heat flux solutions are analytically simple. In particular, for applications in which the Biot number hr_o/k is smaller than 1, it is permissible to use the finite-h semi-infinite solid result (4.48) when $\alpha t/r_o^2 \ll 1$, followed by the lumped capacitance model (4.17) when $\alpha t/r_o^2 \gg 1$.

Figure 4.5 also identifies the only segment of the domain in which the simple solutions developed until now are not adequate. That segment is the upper-right quadrant (large hr_o/k and large $\alpha t/r_o^2$), that is, applications in which (i) the surface temperature has already been decreased to a level comparable with the temperature of the surrounding fluid and (ii) the conduction effect has penetrated to the center of the body. For this reason we must consider the exact solutions that are described next, one exact solution for each body shape that is specified. These exact solutions are valid for all values of hr_o/k and $\alpha t/r_o^2$, not just for the upper-right quadrant of Figure 4.5. They constitute, therefore, the means by which one can assess the accuracy in using the lumped and semi-infinite solid models.

Example 4.1 *Semi-infinite Medium: Fixed Surface Temperature*

Hot tea is poured into a porcelain cup whose wall is initially at the temperature $T_i = 25\,°C$. Assume that the surface of the porcelain assumes instantly the tea temperature $T_\infty = 70\,°C$. The thickness of the porcelain wall is 6 mm. Estimate the time until the wall temperature rises to $30\,°C$ at a point situated at 2 mm under the wetted surface. Show that during this short time the conduction process may be modeled according to the semi-infinite medium with fixed side temperature treated in Figure 4.3.

Solution

The temperature $(T = 30\,°C)$ at the location $x = 2$ mm under the heated surface is nearly the same as the initial temperature at every point inside the porcelain wall. This means that most of the heating experienced by the wall is located to the left of $x = 2$ mm, near the $T_\infty = 70\,°C$ surface. The model we can use is the unidirectional conduction through a semi-infinite solid, Eq. (4.42):

$$\frac{T - T_\infty}{T_i - T_\infty} = \mathrm{erf}\left[\frac{x}{2(\alpha t)^{1/2}}\right]$$

$$\frac{30 - 70}{25 - 70} = 0.889 = \mathrm{erf}\left[\frac{x}{2(\alpha t)^{1/2}}\right]$$

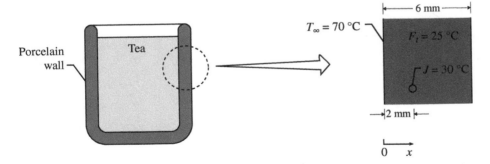

Figure E4.1

Interpolating the error function values listed in Appendix E, we find

$$\frac{x}{2(\alpha t)^{1/2}} \cong 1.14$$

$$t \cong \frac{x^2}{(2.28)^2 \alpha} = \frac{4\ \mathrm{mm}^2}{5.2}\frac{\mathrm{s}}{0.004\ \mathrm{cm}^2} = 1.92\ \mathrm{s}$$

Now that we have an estimate for the time needed to witness the temperature $30\,°C$ at the depth $x = 2$ mm, we can evaluate the goodness of the semi-infinite medium model adopted when we used Eq. (4.42). We invoke Eq. (4.7) and calculate the scale of thermal penetration:

$$\delta \sim (\alpha t)^{1/2} = \left(0.004\frac{\mathrm{cm}^2}{\mathrm{s}}1.92\ \mathrm{s}\right)^{1/2}$$

$$= 0.88\ \mathrm{mm}$$

This length is much smaller than the thickness of the wall; therefore, the time t is short enough so that, to the heating that proceeds from the left, the wall looks semi-infinite.

Example 4.2 *Lumped Capacitance: Varying Ambient Temperature*

Consider a conducting body of density ρ, specific heat c, volume V, and surface area A that can be treated as a lumped thermal capacitance of temperature $T(t)$. Beginning with the time $t = 0$, this body makes thermal contact with a fluid flow through a constant heat transfer coefficient h. The ambient fluid temperature increases linearly in time:

$$T_\infty = at \quad (a = \text{constant})$$

Initially, the body and the surrounding fluid are in thermal equilibrium, $T(0) = 0$. Determine the temperature history of the body, $T(t)$.

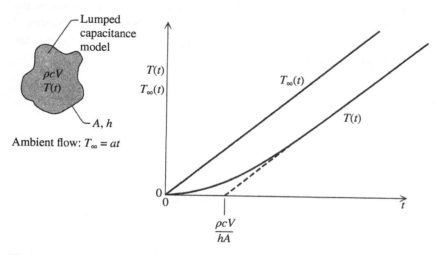

Figure E4.2

Solution

When the temperature of the surrounding fluid is a function of time, $T_\infty(t)$, the lumped capacitance energy conservation Eq. (4.15) becomes

$$\frac{dT}{dt} + \frac{hA}{\rho c V}T - \frac{hA}{\rho c V}T_\infty(t) = 0 \tag{1}$$

This is a linear differential equation of type

$$\frac{dy}{dx} + P(x)y + Q(x) = 0 \tag{2}$$

the general solution of which contains a constant of integration, C:

$$y(x) = \exp\left[-\int P(x)\,dx\right] \cdot \left\{C - \int Q(x)\exp\left[\int P(x)\,dx\right]dx\right\} \tag{3}$$

In this example the ambient temperature increases linearly in time:

$$T_\infty = at \quad (a = \text{constant}) \tag{4}$$

and the solution (3) reduces to

$$T(t) = C\exp\left(-\frac{hA}{\rho c V}t\right) + a\left(t - \frac{\rho c V}{hA}\right) \tag{5}$$

From the condition that at $t = 0$ the body and the ambient are in equilibrium,

$$T(0) = 0 \tag{6}$$

we learn that $C = a\rho c V/hA$, so that the $T(t)$ solution becomes

$$T(t) = a\frac{\rho c V}{hA}\left[\exp\left(-\frac{hA}{\rho c V}t\right) - 1\right] + at \tag{7}$$

The behavior of this temperature function is illustrated in the figure. The body temperature remains unchanged in the early stages of the heat transfer process:

$$\frac{dT}{dt} = 0 \quad \text{at} \quad t = 0 \tag{8}$$

while in the late stages it increases linearly in time:

$$T \cong a\left(t - \frac{\rho c V}{hA}\right) \tag{9}$$

The time represented by $\rho c V/hA$ is the interval by which the linear rise of the body temperature is delayed with respect to the linear rise of the ambient temperature.

4.4 Unidirectional Conduction

4.4.1 Plate

Consider first the plane wall geometry, or, more generally, the thin wall geometry discussed in Figure 2.1. Figure 4.6 shows a slab of thickness $2L$ and initial temperature T_i, in which both sides are exposed suddenly to a convective medium of distant temperature T_∞. The heat transfer coefficient is equal to the same constant h on both surfaces of the slab. In view of the large height and width of the slab, we expect the temperature distribution inside the solid to depend only on the transversal position x and time t. The only difference between the present problem and the semi-infinite solid placed in contact with fluid (Section 4.3.3) is that now the solid is finite in the x direction.

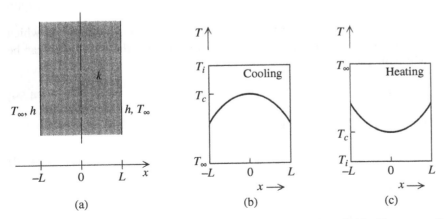

Figure 4.6 Plate of thickness $2L$ immersed suddenly in a fluid with convection.

In terms of the excess temperature function $\theta(x, t) = T(x, t) - T_\infty$, the mathematical statement of the problem of Figure 4.6 is the following:

Conduction Equation.

$$\frac{\partial^2 \theta}{\partial x^2} = \frac{1}{\alpha} \frac{\partial \theta}{\partial t} \tag{4.52}$$

Initial Condition.

$$\theta = \theta_i \quad \text{at} \quad t = 0 \tag{4.53}$$

Boundary Conditions.

$$\frac{\partial \theta}{\partial x} = 0 \quad \text{at} \quad x = 0 \tag{4.54}$$

$$-k\frac{\partial \theta}{\partial x} = h\theta \quad \text{at} \quad x = L \tag{4.55}$$

where $\theta_i = T_i - T_\infty$. The boundary conditions (4.54) and (4.55) are both homogeneous. This means that x is a direction of homogeneous boundary conditions (review Section 3.1.1), which is why we chose $\theta(x, t)$ as the unknown instead of $T(x, t)$. The separation of variables in Eq. (4.52) is achieved by assuming (i.e. by trying out) the product solution $\theta(x, t) = X(x)\tau(t)$,

$$\underbrace{\frac{X''}{X}}_{-\lambda^2} = \underbrace{\frac{1}{\alpha} \frac{\tau'}{\tau}}_{-\lambda^2} \tag{4.56}$$

Equation (4.56) generates two separate equations, for $X(x)$ and $\tau(t)$, in which λ^2 is an undetermined positive number. The general solutions to these two equations can be combined into the product solution:

$$\theta(x, t) = [C_1 \sin(\lambda x) + C_2 \cos(\lambda x)] \exp(-\alpha\lambda^2 t) \tag{4.57}$$

In the special case $\lambda = 0$, this solution degenerates into $\theta_0 = C_1 x + C_2$. The boundary conditions (4.54) and (4.55) require $C_1 = 0$ and $C_2 = 0$, so that $\theta_0 = 0$. In this way the $\lambda = 0$ solution contributes zero to the $\theta(x, t)$ solution, and this means that we can proceed with the general solution (4.57), which holds for $\lambda \neq 0$.

Invoking first the symmetry condition (4.54), we learn that C_1 must be zero. Next, the surface condition (4.55) and what remains of the solution (4.57) require

$$a_n \tan(a_n) = \frac{hL}{k} \tag{4.58}$$

where $a_n = \lambda_n L$ are the characteristic values. The first six roots of Eq. (4.58) are listed in Table 3.1, in which the left column represents now the values of the Biot number hL/k. In summary, the $\theta(x, t)$ solution can be constructed as an infinite series with unknown coefficients:

$$\theta(x, t) = \sum_{n=1}^{\infty} K_n \cos(\lambda_n x) \exp(-\alpha\lambda_n^2 t) \tag{4.59}$$

The last step consists of applying the initial condition (4.53):

$$\theta_i = \sum_{n=1}^{\infty} K_n \cos(\lambda_n x) \tag{4.60}$$

and using the orthogonality property of the characteristic functions $\cos(\lambda_n x)$,

$$\theta_i \int_0^L \cos(\lambda_n x)\, dx = K_n \int_0^L \cos^2(\lambda_n x)\, dx \tag{4.61}$$

Following the evaluation of the integrals, this step identifies the coefficients K_n, which substituted back into the series (4.59) complete the solution:

$$\frac{\theta(x,t)}{\theta_i} = \frac{T(x,t) - T_\infty}{T_i - T_\infty}$$

$$= 2\sum_{n=1}^{\infty} \frac{\sin(a_n)}{a_n + \sin(a_n)\cos(a_n)} \cos\left(a_n \frac{x}{L}\right) \exp\left(-a_n^2 \frac{\alpha t}{L^2}\right) \tag{4.62}$$

Expressed in dimensionless form as $(T - T_\infty)/(T_i - T_\infty)$, the temperature distribution determined earlier depends on three dimensionless groups:

$$\frac{x}{L}, \quad \frac{\alpha t}{L^2} = Fo, \quad \frac{hL}{k} = Bi \tag{4.63}$$

The Biot number hL/k influences the solution (4.62) through the characteristic values a_n, as shown in Eq. (4.58). The characteristic values a_n have been listed in Table 3.1, in which the left most column contains the values of hL/k. The temperature in one plane (x/L = constant) depends only on the Biot number hL/k and the Fourier number, $Fo = \alpha t/L^2$. "Fourier number" is another way of saying "dimensionless time."

Figure 4.7 shows the history of the temperature in the midplane of the plate, $T_c(t) = T(0, t)$. The curves correspond to fixed values of the inverse Biot number, k/hL. These lines appear almost straight because of the time exponentials of the solution (4.62) and the semilogarithmic scale of Figure 4.7. The sharp breaks (the corners) in the lines are due to the changes in the size of the divisions marked on the abscissa: there are four such division sizes. This method of plotting the temperature history had been used earlier by Hottel. The great merit of the method is that it expands greatly the Fourier number range that is covered by the abscissa.

The temperature in a plane other than the midplane of the plate can be calculated by multiplying the readings furnished by Figures 4.7 and 4.8. The latter shows how the excess temperature in an arbitrary plane, $T(x, t) - T_\infty$, compares with the corresponding (simultaneous) value in the midplane, $T_c(t) - T_\infty$. The temperature in any plane is the product of two readings:

$$\left(\frac{T(x,t) - T_\infty}{T_i - T_\infty}\right)_{\text{plate}} = \left(\frac{T(x,t) - T_\infty}{T_c(t) - T_\infty}\right)_{\text{Figure 4.8}} \times \left(\frac{T_c(t) - T_\infty}{T_i - T_\infty}\right)_{\text{Figure 4.7}} \tag{4.64}$$

Figure 4.7 Temperature history in the midplane of a plate immersed suddenly in a fluid of a different temperature (L = plate half-thickness). Source: Drawn after Heisler [2].

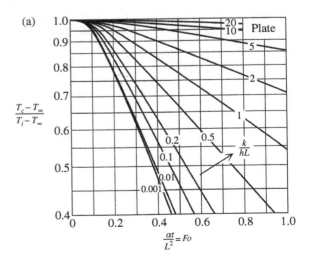

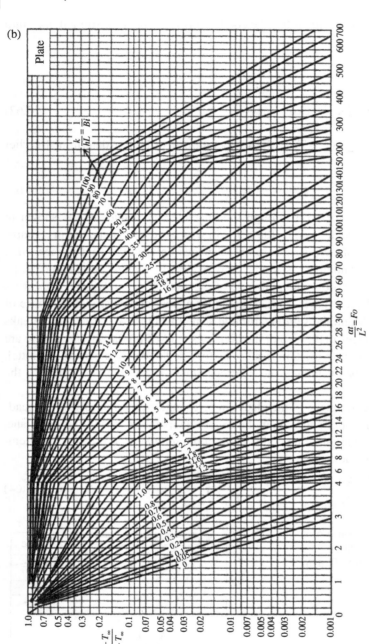

Figure 4.7 *(Continued)*

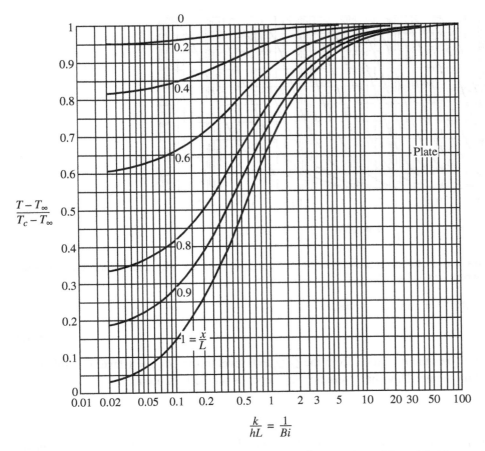

Figure 4.8 Relationship between the temperature in any plane (x) and the temperature in the midplane (x = 0, Figure 4.7) of a plate immersed suddenly in a fluid of a different temperature (L = plate half-thickness). Source: Drawn after Heisler [2].

Another quantity of interest is the total heat transfer between the solid and the ambient fluid during the finite time interval $0 - t$. Considering only the right half of the plate ($0 < x < L$, Figure 4.6), the maximum heat transfer that can take place is

$$Q_i = \rho WHLc(T_i - T_\infty) \tag{4.65}$$

where H and W are the large height and large width of the plate, respectively. The area $W \times H$ is normal to the x direction. The maximum heat transfer Q_i (J) is equal to the drop in the half-plate internal energy, from the initial state (T_i), to a state of thermal equilibrium with the ambient (T_∞). The actual heat transfer during the time interval $0 - t$ is smaller than Q_i: its value is given by the integral

$$Q(t) = WH \int_0^t q'' dt \tag{4.66}$$

where q'' is the conduction heat flux that leaves the half-plate through the $x = L$ surface.

Figure 4.9 shows Gröber's classical chart [3] for the ratio $Q(t)/Q_i$ obtained by dividing Eqs. (4.65) and (4.66). This ratio starts from zero and approaches 1. The dimensionless time on the abscissa is the same as

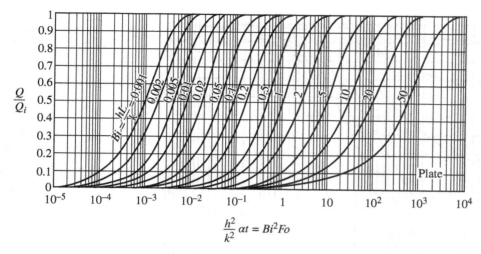

Figure 4.9 Total heat transfer between a plate and the surrounding fluid, as a function of the total time of exposure *t*. Source: Drawn after Gröber et al. [3].

the Fourier number times the Biot number squared:

$$\left(\frac{h}{k}\right)^2 \alpha t = \left(\frac{hL}{k}\right)^2 \frac{\alpha t}{L^2} \tag{4.67}$$

Each constant-hL/k curve makes the transition from the $Q/Q_i = 0$ level to the $Q/Q_i = 1$ level in a relatively narrow, "characteristic" time range. This characteristic time depends on the Biot number.

4.4.2 Cylinder

Consider next the time-dependent temperature field in a long cylindrical rod of radius r_o and initial temperature T_i, which is placed suddenly in contact with a fluid flow of temperature T_∞ through a constant heat transfer coefficient h. The only feature that differentiates between this problem and the slab of Figure 4.6 is the outer radius r_o of the cylinder, which now becomes the transversal dimension of the body (i.e. the length to be used in defining the Biot and Fourier numbers). Since the temperature distribution inside the rod depends only on radial position and time, by reading Eqs. (4.62) and (4.63), we expect a dimensionless temperature $[T(r, t) - T_\infty]/(T_i - T_\infty)$ that depends on three dimensionless groups:

$$\frac{r}{r_o}, \quad \frac{\alpha t}{r_o^2} = Fo, \quad \frac{hr_o}{k} = Bi \tag{4.68}$$

The exact solution for the temperature field is available as an infinite series [4]:

$$\frac{T(r, t) - T_\infty}{T_i - T_\infty} = \sum_{n=1}^{\infty} K_n J_0\left(b_n \frac{r}{r_o}\right) \exp\left(-b_n^2 Fo\right) \tag{4.69}$$

in which the K_n coefficients are given by

$$K_n = \frac{2Bi}{(b_n^2 + Bi^2)J_0(b_n)} \tag{4.70}$$

and where the characteristic values b_n are the roots of the equation

$$b_n J_1(b_n) - Bi\, J_0(b_n) = 0 \tag{4.71}$$

Table 4.1 Constants for the solution for temperature in a cylinder, Eq. (4.69).

$Bi = hr_o/k$	b_1	K_1	b_2	K_2
0.01	0.141 2	1.002 5	3.834 3	−0.003 38
0.03	0.243 9	1.007 5	3.839 5	−0.010 12
0.1	0.441 7	1.024 6	3.857 7	−0.033 4
0.3	0.746 5	1.071 2	3.909 1	−0.098 3
1	1.255 8	1.207 1	4.079 5	−0.290 4
3	1.788 7	1.419 1	4.463 4	−0.631 5
10	2.179 5	1.567 7	5.033 2	−0.958 1
30	2.326 1	1.597 3	5.341 0	−1.048 7
100	2.380 9	1.601 5	5.465 2	−1.060 5

The Bessel functions of the first kind, J_0 and J_1 are shown in Figure 3.6. The first six roots of Eq. (4.71) are reported as functions of the Biot number in Ref. [1]. Table 4.1 shows only the first two characteristic values and the values of the corresponding K_n coefficients. In many cases the series (4.69) converges so rapidly that it is approximated adequately by only the first term or by the first two terms. The approximation is particularly good in the late regime, that is, when the Fourier number at/r_o^2 is of order 1 or greater (Figure 4.5).

The same solution is displayed in Figure 4.10, which shows the temperature history on the centerline of the cylinder, $T_c(t) = T(0, t)$. The temperature at other radial positions can be calculated by multiplying the readings provided by Figures 4.10 and 4.11:

$$\left(\frac{T(r,t) - T_\infty}{T_i - T_\infty}\right)_{\text{cylinder}} = \left(\frac{T(r,t) - T_\infty}{T_c(t) - T_\infty}\right)_{\text{Figure 4.11}} \times \left(\frac{T_c(t) - T_\infty}{T_i - T_\infty}\right)_{\text{Figure 4.10}} \qquad (4.72)$$

Finally, Figure 4.12 shows the evolution of the ratio $Q(t)/Q_i$, where $Q(t)$ is the total heat transfer between solid and fluid during the time interval $0 - t$ and Q_i is the maximum heat transfer that occurs when the cylinder

Figure 4.10 Temperature history on the centerline of a cylinder immersed suddenly in a fluid of a different temperature (r_o = cylinder radius). Source: Drawn after Heisler [2].

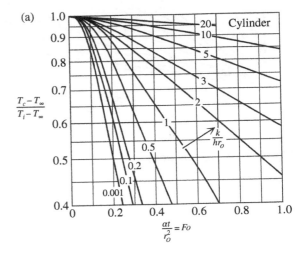

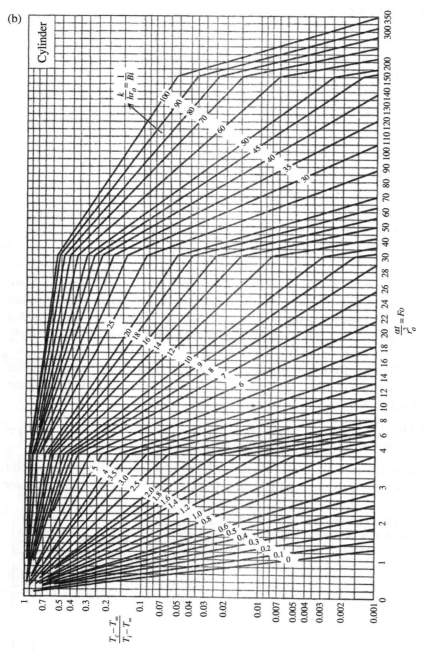

Figure 4.10 (*Continued*)

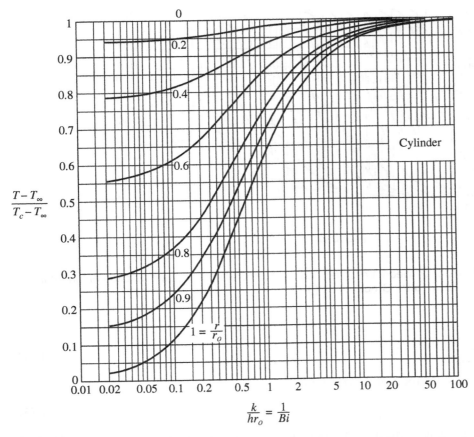

Figure 4.11 Relationship between the local temperature (r) and the centerline temperature ($r = 0$) of a cylinder immersed suddenly in a fluid of a different temperature. Source: Drawn after Heisler [2].

reaches equilibrium with the fluid, $Q_i = \rho Vc(T_t - T_\infty)$, where V is the cylinder volume. The behavior of the Q/Q_i ratio of a cylinder, Figure 4.12, is very similar to that of the Q/Q_i, ratio of a constant-thickness plate, Figure 4.9.

4.4.3 Sphere

Similar means exist for calculating the temperature distribution inside a sphere that is placed in contact with a fluid flow of a different temperature. The transversal dimension in this case is the sphere radius r_o. The dimensionless temperature difference $[T(r, t) - T_\infty]/(T_i - T_\infty)$ depends on the same dimensionless parameters as the solution for the cylinder, except that Fo and Bi are based on the radius of a sphere.

The series solution for the instantaneous temperature distribution in a sphere is [4]

$$\frac{T(r, t) - T_\infty}{T_i - T_\infty} = \sum_{n=1}^{\infty} K_n \frac{\sin(s_n r/r_0)}{s_n r/r_0} \exp(-s_n^2\, Fo) \qquad (4.73)$$

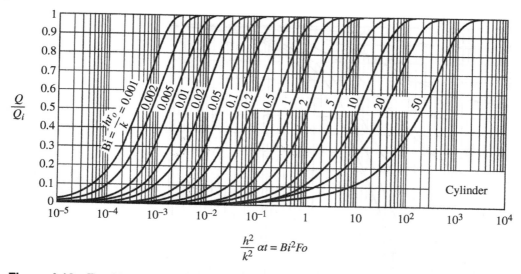

$$\frac{h^2}{k^2}\alpha t = Bi^2 Fo$$

Figure 4.12 Total heat transfer between a cylinder and the surrounding fluid, as a function of the total time of exposure t. Source: Drawn after Gröber et al. [3].

where

$$K_n = 2\frac{\sin(s_n) - s_n \cos(s_n)}{s_n - \sin(s_n) \cos(s_n)} \tag{4.74}$$

The characteristic values s_n are the roots of the equation

$$s_n \cot(s_n) = 1 - Bi \tag{4.75}$$

The first two roots (s_1, s_2) and the corresponding coefficients (K_1, K_2) are presented in Table 4.2. The first term of the series (4.73) is a very good approximation for the entire series when the Fourier number is not much smaller than 1. Note the presence of $-s_n^2$ in the argument of the exponential. When Fo is not negligibly small, the argument becomes very large (and negative), rendering the second and the subsequent terms in the series negligible relative to the first term.

The temperature history in the sphere is presented in Figures 4.13–4.15. The first of these graphs shows the history of the temperature right in the center of the sphere, $T(0, t) = T_c(t)$. Figure 4.14 provides the correction factor needed for calculating the simultaneous temperature at any other radius, $T(r, t)$:

$$\left(\frac{T(r,t) - T_\infty}{T_i - T_\infty}\right)_{\text{sphere}} = \left(\frac{T(r,t) - T_\infty}{T_c(t) - T_\infty}\right)_{\text{Figure 4.14}} \times \left(\frac{T_c(t) - T_\infty}{T_i - T_\infty}\right)_{\text{Figure 4.13}} \tag{4.76}$$

The total heat transfer ratio $Q(t)/Q_i$ is shown as a function of Fourier number and Biot number in Figure 4.15. The denominator Q_i is the ceiling value to which $Q(t)$ aspires as t increases, namely, $Q_i = \rho V c(T_i - T_\infty)$, where V is the sphere volume, $4\pi r_0^3/3$.

4.4.4 Plate, Cylinder, and Sphere with Fixed Surface Temperature

Figures 4.9, 4.12, and 4.15 show that, in general, the actual/maximum heat transfer ratio Q/Q_i is a function of the Biot number and the Fourier number. This relationship is considerably simpler in the $Bi \to \infty$ limit,

Table 4.2 Constants for the solution for temperature in a sphere, Eq. (4.73).

$Bi = hr_o/k$	s_1	K_1	s_2	K_2
0.01	0.173 0	1.003 0	4.495 6	−0.004 49
0.03	0.299 1	1.009 0	4.500 1	−0.013 69
0.1	0.542 3	1.029 8	4.515 7	−0.045 44
0.3	0.920 8	1.088 0	4.560 1	−0.1345
1	1.570 8	1.273 2	4.712 4	−0.4244
3	2.288 9	1.622 7	5.087 0	−1.0288
10	2.836 3	1.924 9	5.717 2	−1.7381
30	3.037 2	1.989 8	6.074 0	−1.9593
100	3.110 2	1.999 0	6.220 4	−1.9961

where the heat transfer coefficient is sufficiently large so that the surface of the immersed body assumes the temperature of the surrounding fluid, T_∞. In this limit the Q/Q_i ratio is a function of time, or Fourier number.

Figure 4.16 shows these limiting relations for the three geometries considered in this section – plate, cylinder, and sphere. Plotted on the ordinate is the group $1 - (Q/Q_i)$, whose value at $t = 0$ is 1. This group is the same as the temperature difference ratio $(\overline{T} - T_\infty)/(T_i - T_\infty)$, in which $\overline{T}$ is the instantaneous temperature averaged over the entire volume V of the body, $\overline{T} = V^{-1}\int_V T\, dV$. Written in terms of $\overline{T}$, the total heat transfer from the body to the ambient during the interval $0 - t$ is $Q = \rho Vc(T_i - \overline{T})$.

Example 4.3 *The Immersion Cooling of a Steel Plate*

The purpose of this exercise is to illustrate the use of the charts exhibited in this section and to show that alternative estimates can be made using the analytical solutions on which the charts are based. Consider a 1.6-cm-thick plate of carbon steel at the initial temperature $T_i = 600\,°C$. This plate is plunged at $t = 0$

Figure 4.13 Temperature history in the center of a sphere (r_o = sphere radius). Source: Drawn after Heisler [2].

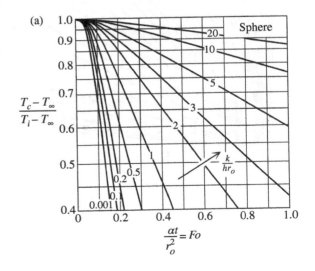

(b)

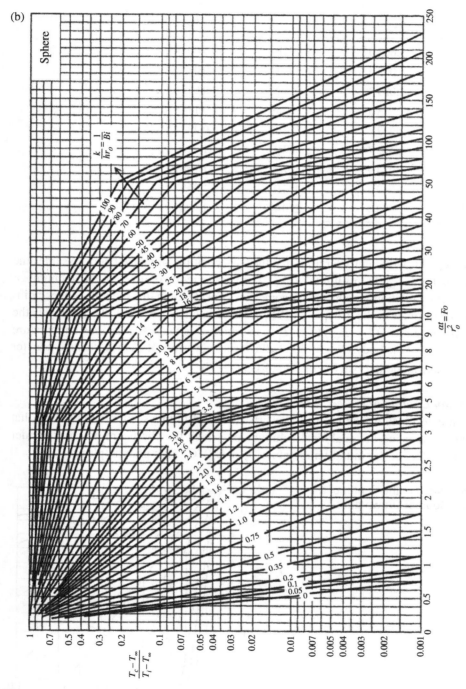

Figure 4.13 (*Continued*)

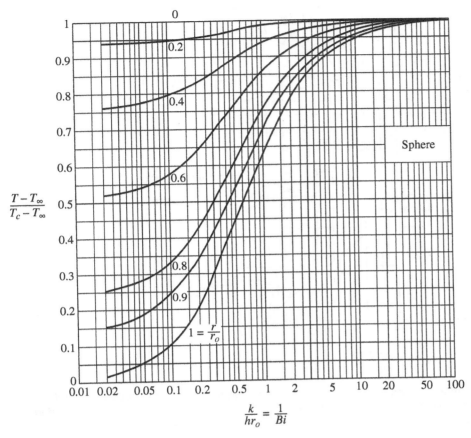

Figure 4.14 Relationship between the temperature at any radius (r) and the temperature in the center ($r = 0$, Figure 4.13) of a sphere immersed suddenly in a fluid flow. Source: Drawn after Heisler [2].

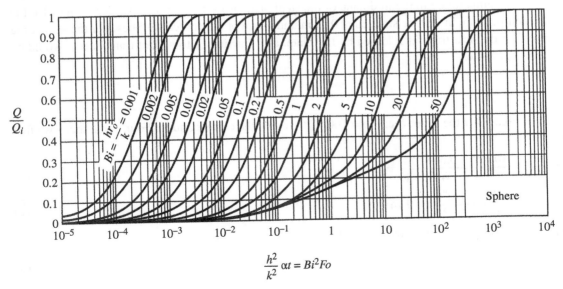

Figure 4.15 Total heat transfer between a sphere and the surrounding fluid, as a function of the total time of exposure t. Source: Drawn after Gröber et al. [3].

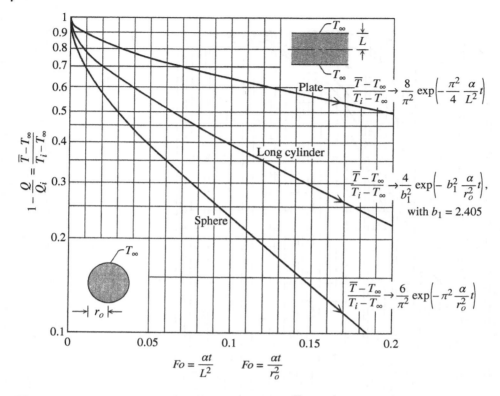

Figure 4.16 The volume-averaged temperature $\overline{T}$ and the total heat transfer from a body with fixed surface temperature. The three curves in this figure represent the $Bi = \infty$ limit of the charts presented in Figures 4.9, 4.12, and 4.15.

in a bath of water at the temperature $T_\infty = 15\,°C$. The heat transfer coefficient is assumed constant and equal to $h = 10^4\,W/m^2 \cdot K$. The properties of carbon steel are $k = 40\,W/m \cdot K$ and $\alpha = 0.1\,cm^2/s$. Calculate the time t when the temperature in the midplane of the steel plate drops to $T_c = 100\,°C$. Determine also the corresponding temperature in a plane situated at 0.2 cm under one of the cooled surfaces. For the same time interval t, calculate the heat transfer released by the plate, as a fraction of total heat transfer that would be released in the limit $t \to \infty$.

Solution

To deduce from Figure 4.7 the time when $T_c = 100\,°C$ (namely, the corresponding Fourier number), we first calculate

$$Bi = \frac{hL}{k} = \frac{10^4\,W}{m^2 \cdot K} \frac{0.008\,m}{40\,W/m \cdot K} = 2 \tag{1}$$

$$\frac{T_c - T_\infty}{T_i - T_\infty} = \frac{(100 - 15)°C}{(600 - 15)°C} = 0.145 \tag{2}$$

Next, we identify on Figure 4.7 the curve labeled $1/Bi = 0.5$, and on the abscissa we read

$$\frac{\alpha t}{L^2} \cong 1.8 \tag{3}$$

which corresponds to the time interval

$$t = 1.8 \frac{0.8^2 \text{ cm}^2}{0.1 \text{ cm}^2/\text{s}} = 11.5 \text{ s} \tag{4}$$

A more exact alternative is to rely on Eq. (4.62) instead of Figure 4.7 and to retain only the first term in the series

$$\frac{T_c - T_\infty}{T_i - T_\infty} \cong 2 \frac{\sin a_1}{a_1 + \sin a_1 \cos a_1} \exp(-a_1^2 \, Fo) \tag{5}$$

We know that the first term (i.e. a single exponential) is an adequate approximation because in the *Fo* range of interest the $(T_c - T_\infty)/(T_i - T_\infty)$ curve appears as a straight line on the semilogarithmic grid of Figure 4.7. The exact eigenvalue a_1, follows from the source for Table 3.1, in which the half-thickness Biot number listed in the leftmost column is equal to 2:

$$a_1 = 1.0769 \tag{6}$$

This value can be approximated by interpolating between the a_1 values for the Biot numbers 1 and 3. Linear interpolation yields $a_1 = 1.0264$. Interpolating linearly between the logarithms of the Biot numbers yields $a_1 = 1.07$, which is more accurate. After we substitute a_1, T_c, T_∞, and T_i in Eq. (5), we obtain the Fourier number

$$Fo = 1.807 \tag{7}$$

which corresponds to $t = 11.6$ seconds. This time estimate is nearly the same as the one obtained graphically, Eq. (4).

Consider now the temperature located at $x = L - 0.2 \text{ cm} = 0.6 \text{ cm}$, at the time $t = 11.6 \text{ s}$. For this we use Figure 4.8, in which we know $x/L = 6/8 = 0.75$ and $1/Bi = 0.5$. The correct point is located between the curves labeled $x/L = 0.6$ and $x/L = 0.8$, and closer to the $x/L = 0.8$ curve. Therefore, we conclude *approximately* that

$$\frac{T - T_\infty}{T_c - T_\infty} \cong 0.7 \tag{8}$$

which means that the sought temperature is

$$T \cong 15°C + 0.7(100-15)°C = 74.5°C \tag{9}$$

A more accurate alternative is to use the first term of the series solution (4.62):

$$\frac{T - T_\infty}{T_i - T_\infty} = 2 \frac{\sin a_1}{a_1 + \sin a_1 \cos a_1} \cos\left(a_1 \frac{x}{L}\right) \exp(-a_1^2 \, Fo) \tag{10}$$

for which we know a_1, x/L, and *Fo*. The numerical value of the entire right side of Eq. (10) turns out to be 0.1001; therefore, the temperature in the $x = 0.6 \text{ cm}$ plane is

$$T = 15°C + 0.1001(600 - 15)°C = 73.6°C \tag{11}$$

Finally, we read Figure 4.9 and estimate the total heat transfer that left the slab during the interval $0 - t$, as a fraction of the heat transfer total Q_i. For this reading we need two numbers, $Bi = 2$ and

$$Bi^2 \, Fo = 4 \times 1.807 = 7.23 \tag{12}$$

and these lead to the ordinate value

$$\frac{Q}{Q_i} \cong 0.84 \qquad (13)$$

4.5 Multidirectional Conduction

The preceding three sections outlined analytical and graphic methods for calculating temperature distributions and total heat transfer interactions in *unidirectional* time-dependent conduction configurations. The reason why so much space was dedicated to unidirectional problems is that their solutions can be combined in order to construct the solutions for time-dependent conduction in more complex geometries.

Consider the immersion cooling or heating of a long bar with rectangular cross section. This problem is stated analytically in the left column of Figure 4.17. It consists of the following:

The rectangular domain of half-length L and half-height H.
The equation for simultaneous conduction in the x and y directions, Eq. (a) in Figure 4.17.
The boundary conditions in the x direction, Eqs. (b) and (c) in Figure 4.17.
The boundary conditions in the y direction, Eqs. (d) and (e) in Figure 4.17.
The initial condition, Eq. (f) in Figure 4.17.

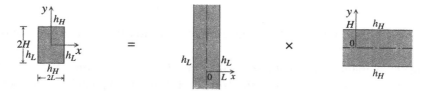

	Long bar with rectangular cross section, $2L \times 2H$		Plate, L-half-thickness		Plate, H-half-thickness	
Equation:	$\dfrac{\partial^2 \theta}{\partial x^2} + \dfrac{\partial^2 \theta}{\partial y^2} = \dfrac{1}{\alpha}\dfrac{\partial \theta}{\partial t}$	(a)	$\dfrac{\partial^2 \theta_L}{\partial x^2} = \dfrac{1}{\alpha}\dfrac{\partial \theta_L}{\partial t}$	$(a)_L$	$\dfrac{\partial^2 \theta_H}{\partial y^2} = \dfrac{1}{\alpha}\dfrac{\partial \theta_H}{\partial t}$	$(a)_H$
$x = 0$:	$\dfrac{\partial \theta}{\partial x} = 0$	(b)	$\dfrac{\partial \theta_L}{\partial x} = 0$	$(b)_L$	$\dfrac{\partial \theta_H}{\partial x} = 0$	$(b)_H$
$x = L$:	$\dfrac{h_L}{k}\theta + \dfrac{\partial \theta}{\partial x} = 0$	(c)	$\dfrac{h_L}{k}\theta_L + \dfrac{\partial \theta_L}{\partial x} = 0$	$(c)_L$	$\theta_H = \theta_H(y,t)$	$(c)_H$
$y = 0$:	$\dfrac{\partial \theta}{\partial y} = 0$	(d)	$\dfrac{\partial \theta_L}{\partial y} = 0$	$(d)_L$	$\dfrac{\partial \theta_H}{\partial y} = 0$	$(d)_H$
$y = H$:	$\dfrac{h_H}{k}\theta + \dfrac{\partial \theta}{\partial y} = 0$	(e)	$\theta_L = \theta_L(x,t)$	$(e)_L$	$\dfrac{h_H}{k}\theta_H + \dfrac{\partial \theta_H}{\partial y} = 0$	$(e)_H$
$t = 0$:	$\theta = \theta_i$	(f)	$\theta_L = \theta_{i,L}$	$(f)_L$	$\theta_H = \theta_{i,H}$	$(f)_H$

Figure 4.17 The time-dependent temperature in a bar immersed in fluid, as the product of the temperature distributions in two perpendicular plates.

This problem is formulated in terms of the excess temperature function $\theta(x, y, t) = T(x, y, t) - T_\infty$, in which T_∞ is the temperature of the fluid.

The second and third columns contain two simpler problems. The $\theta_L(x, t)$ problem shown in the second column is identical to that of Figure 4.6, namely, the unidirectional time-dependent distribution of temperature in a plate of half-thickness L. The solution to this problem is documented in Section 4.4.1, in which x represented the distance to the midplane of the plate. Geometrically, the bar with rectangular cross section is the "intersection" in space of the two plates sketched at the top of Figure 4.17.

The $\theta_H(y, t)$ problem listed in the right column of Figure 4.17 is the same as the unidirectional problem of Section 4.4.1, with the only difference that the distance to the midplane is called y and the half-thickness of the plate is H. Therefore, the $\theta_H(y, t)$ distribution can be calculated based on Figures 4.7 and 4.8, by replacing x with y, and L with H. It remains to show that the solution $\theta(x, y, t)$ is equal to the product of the one-dimensional solutions $\theta_L(x, t)$ and $\theta_H(y, t)$:

$$\theta(x, y, t) = \theta_L(x, t) \cdot \theta_H(y, t) \tag{4.77}$$

when θ_L and θ_H obey the equations aligned in the middle and right columns of Figure 4.17. To begin with, substituting the product formula (4.77) into the conduction equation of the original problem, Eq. (a) in Figure 4.17, yields

$$\underbrace{\left(\frac{\partial^2\theta_L}{\partial x^2} - \frac{1}{\alpha}\frac{\partial\theta_L}{\partial t}\right)}_{\substack{=0\\ \text{Eq. (a)}_L,\ \text{Figure 4.17}}}\theta_H + \underbrace{\left(\frac{\partial^2\theta_H}{\partial y^2} - \frac{1}{\alpha}\frac{\partial\theta_H}{\partial t}\right)}_{\substack{=0\\ \text{Eq. (a)}_H,\ \text{Figure 4.17}}}\theta_L = 0 \tag{4.78}$$

This result shows that the original Eq. (a) in Figure 4.17 is satisfied if the unidirectional temperature distributions satisfy their conduction equations, Eqs. (a)$_L$ and (a)$_H$.

Next, any of the four original boundary conditions is satisfied when the corresponding boundary conditions are satisfied by θ_L and θ_H. For example, by combining the boundary conditions (b)$_L$ and (b)$_H$ through the product formula (4.77), we obtain the original boundary condition for the vertical midplane, Eq. (b). Note that the h value need not be the same all around the rectangular cross section; rather, the h value must be the same only on a pair of parallel faces.

The last observation in this argument concerns the initial condition of the original problem. According to the product solution (4.77), this condition is respected only if the initial temperatures of the long plates ($\theta_{i,L}$, $\theta_{i,H}$) are such that

$$\theta_i = \theta_{i,L} \cdot \theta_{i,H} \tag{4.79}$$

Dividing Eqs. (4.77) and (4.79) side by side, we find that the dimensionless temperature distribution of the two-dimensional problem, θ/θ_i, is the same as the product of the dimensionless temperature distributions in the two "plate" configurations:

$$\left[\frac{\theta(x, y, t)}{\theta_i}\right]_{\substack{\text{bar,}\\ 2L \times 2H}} = \left[\frac{\theta(x, t)}{\theta_i}\right]_{\substack{\text{plate,}\\ L = \text{half-thickness}}} \times \left[\frac{\theta(y, t)}{\theta_i}\right]_{\substack{\text{plate,}\\ H = \text{half-thickness}}} \tag{4.80}$$

In conclusion, the value of $\theta(x, y, t)/\theta_i$, for the long bar of rectangular cross section can be calculated by reading Figures 4.7 and 4.8 twice, for the plate of half-thickness L and later for the plate of half-thickness H.

As a more recent extension to the theorem outlined in Figure 4.17, it has been shown [5] that a special formula can be used to calculate the total heat transfer Q (J) that crosses the boundary of the rectangular domain during the time interval $0 - t$:

$$\frac{Q(t)}{Q_i} = \left(\frac{Q}{Q_i}\right)_L + \left(\frac{Q}{Q_i}\right)_H - \left(\frac{Q}{Q_i}\right)_L \left(\frac{Q}{Q_i}\right)_H \tag{4.81}$$

In this expression the subscripts L and H denote the plates of half-thickness L and H, respectively. Each of the Q_i denominators represents the maximum value of the respective $Q(t)$ function, that is, $Q_i = Q(t \to \infty)$.

Solutions analogous to Eqs. (4.80) and (4.81) hold in the case of other multidirectional configurations. The rule in constructing the appropriate factors that participate in the product is to recognize the two simpler bodies, which if intersected form is the body of interest. Figure 4.18 shows that a *short cylinder* of length $2L$ and radius r_o is the intersection of a long cylinder and a perpendicular plate, each with its own h value at the exposed surfaces. Therefore, the dimensionless temperature distribution inside a short cylinder immersed suddenly in a fluid can be calculated with the formula

$$\left[\frac{\theta(r,y,t)}{\theta_i}\right]_{\substack{\text{short} \\ \text{cylinder,} \\ L = \text{half-length} \\ r_o = \text{radius}}} = \left[\frac{\theta(r,t)}{\theta_i}\right]_{\substack{\text{long} \\ \text{cylinder,} \\ r_o = \text{radius} \\ \text{Figures 4.10} \\ \text{and 4.11}}} \times \left[\frac{\theta(y,t)}{\theta_i}\right]_{\substack{\text{plate,} \\ L = \text{half-thickness} \\ \text{Figures 4.7} \\ \text{and 4.8}}} \tag{4.82}$$

Additional examples of two-body intersections that can be used to reconstruct solutions for multidirectional time-dependent conduction are the semi-infinite counterparts of the rectangular cross section shown on the left side of Figure 4.17 and the cylinder drawn on the left side of Figure 4.18. The temperature distribution in a semi-infinite plate of half-thickness L, immersed suddenly in a fluid at a different temperature, is therefore (Figure 4.19, top)

$$\left[\frac{\theta(x,y,t)}{\theta_i}\right]_{\substack{\text{semi-infinite} \\ \text{plate,} \\ L = \text{half-thickness}}} = \left[\frac{\theta(x,t)}{\theta_i}\right]_{\substack{\text{infinite} \\ \text{plate,} \\ L = \text{half-thickness} \\ \text{Figs.4.7} \\ \text{and 4.8}}} \times \left[\frac{\theta(y,t)}{\theta_i}\right]_{\substack{\text{semi-infinite} \\ \text{medium} \\ y = \text{normal to} \\ \text{the surface} \\ \text{eq.(4.48)}}} \tag{4.83}$$

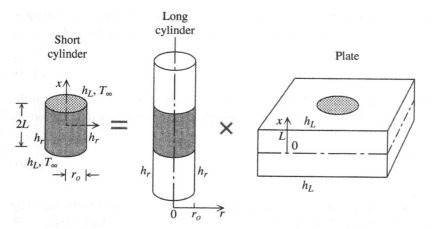

Figure 4.18 The time-dependent temperature of a short cylinder immersed in fluid, as the product of the temperatures in a long cylinder and in a perpendicular plate.

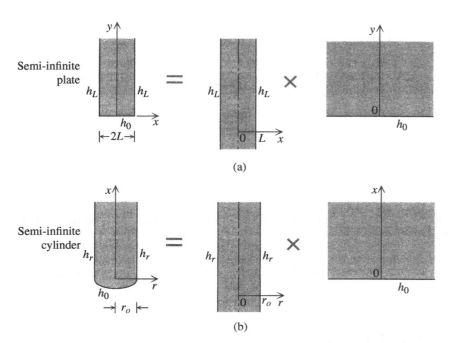

Semi-infinite plate

$$h_L \quad h_L$$
$$h_0$$
$$|\leftarrow 2L \rightarrow|$$

$$= \quad h_L \quad h_L \quad 0 \ L \ x$$

$$\times$$

$$0 \quad h_0$$

(a)

Semi-infinite cylinder

$$h_r \quad h_r$$
$$h_0$$
$$\rightarrow| r_o |\leftarrow$$

$$= \quad h_r \quad h_r \quad 0 \ r_o \ r$$

$$\times$$

$$0 \quad h_0$$

(b)

Figure 4.19 Multiplication rules for the temperature distribution in a semi-infinite plate (a) and the temperature distribution in a semi-infinite cylinder (b).

Likewise, the temperature distribution in a semi-infinite cylinder exposed to a fluid along the base and the cylindrical surface is (Figure 4.19, bottom)

$$\left[\frac{\theta(r,x,t)}{\theta_i}\right]_{\substack{\text{semi-infinite} \\ \text{cylinder,} \\ r_o = \text{radius}}} = \left[\frac{\theta(r,t)}{\theta_i}\right]_{\substack{\text{infinite} \\ \text{cylinder,} \\ r_o = \text{radius} \\ \text{Figs.4.10} \\ \text{and 4.11}}} \times \left[\frac{\theta(x,t)}{\theta_i}\right]_{\substack{\text{semi-infinite} \\ \text{medium} \\ x = \text{normal to} \\ \text{the surface} \\ \text{eq.(4.48)}}} \tag{4.84}$$

In these examples as well as in the case of the short cylinder (Figure 4.18), the total heat transfer interaction can be calculated using Eq. (4.81), in which a new set of subscripts may be used to indicate the two simpler bodies identified on the right side of each of Eqs. (4.82)–(4.84).

A more complicated body shape that can be handled in the same manner is the parallelepiped shown on the left side of Figure 4.20. This body can be viewed as the intersection of three plates that are mutually perpendicular. Therefore,

$$\left[\frac{\theta(x,y,z,t)}{\theta_i}\right]_{\substack{\text{parallelepiped} \\ 2L \times 2H \times 2W}} = \left[\frac{\theta(x,t)}{\theta_i}\right]_{\substack{\text{plate} \\ L = \text{half-thickness}}} \times \left[\frac{\theta(y,t)}{\theta_i}\right]_{\substack{\text{plate,} \\ H = \text{half-} \\ \text{thickness}}} \times \left[\frac{\theta(z,t)}{\theta_i}\right]_{\substack{\text{plate,} \\ W = \text{half-} \\ \text{thickness}}} \tag{4.85}$$

The figure shows that the values x, y, and z mark the position relative to a Cartesian system with the origin in the center of the body. This means that on the right side of Eq. (4.85), the x, y, and z values represent the distances to the midplanes of the respective plates. The total heat transfer between the three-dimensional body and the surrounding fluid during the interval $0 - t$ is [5]

$$\frac{Q(t)}{Q_i} = \left(\frac{Q}{Q_i}\right)_L + \left(\frac{Q}{Q_i}\right)_H \left[1 - \left(\frac{Q}{Q_i}\right)_L\right] + \left(\frac{Q}{Q_i}\right)_W \left[1 - \left(\frac{Q}{Q_i}\right)_L\right] \left[1 - \left(\frac{Q}{Q_i}\right)_H\right] \tag{4.86}$$

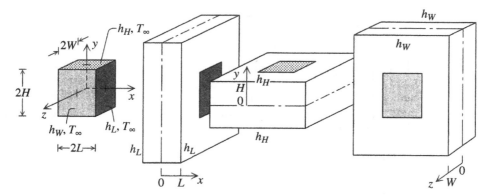

Figure 4.20 The time-dependent temperature of a parallelepiped immersed in fluid, as the product of the temperatures in three mutually perpendicular plates.

The subscripts L, H, and W are shorthand for the three plates identified in Figure 4.20 and on the right side of Eq. (4.85).

4.6 Concentrated Sources and Sinks

4.6.1 Instantaneous (One-Shot) Sources and Sinks

In this section and Section 4.6.2, we consider the class of time-dependent conduction problems in which the key feature is the generation (or absorption) of heat transfer in a very small region (source, sink) of the conducting medium. The small region will be modeled by means of a plane, a straight line, and a point.

When heat transfer is released into the medium from this small region, the process will be one of time-dependent conduction in the vicinity of a heat source. Common phenomena that can be described in terms of concentrated heat sources are underground fissures filled with geothermal steam, underground explosions, canisters of nuclear and chemical waste, and electrical cables buried underground.

When the small region receives heat transfer from the surrounding (infinite) medium, the region functions as a concentrated heat sink. An example of a heat sink is the buried heat exchanger (duct) through which a heat pump receives heat transfer from the ambient (the soil) in order to augment it and deposit it into a building.

Consider first the direction x through an infinite medium with constant properties (k, α, ρ, c), Figure 4.21. The equation for time-dependent conduction in the x direction is

$$\frac{\partial^2 \theta}{\partial x^2} = \frac{1}{\alpha} \frac{\partial \theta}{\partial t} \tag{4.87}$$

where $\theta(x, t) = T(x, t) - T_\infty$ is the excess temperature relative to the background temperature (T_∞). Note that Eq. (4.87) is satisfied at all times by a function of the form

$$\theta(x, t) = \frac{K}{(\alpha t)^{1/2}} \exp\left(-\frac{x^2}{4\alpha t}\right) \tag{4.88}$$

where K is a constant. The excess temperature distribution exhibits the behavior illustrated in Figure 4.21. At times close to $t = 0$, the $\theta - x$ curve has a very sharp spike in the plane $x = 0$. The height and

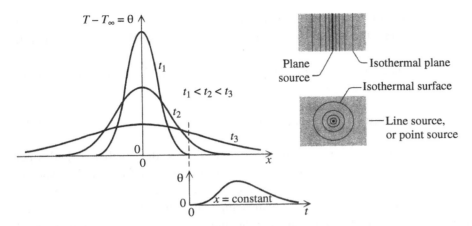

Figure 4.21 The temperature distribution in the vicinity of an instantaneous (one-shot) heat source.

sharpness of this spike diminish as the time increases; in fact, as $t \to \infty$, the excess temperature θ drops to zero.

Another interesting feature of the $\theta - x$ curve is that the area trapped under it is independent of time. The area can be evaluated by recognizing the error function definition (4.38) and Eq. (4.39b):

$$\int_{-\infty}^{\infty} \theta \, dx = 2\pi^{1/2}K \quad \text{(constant)} \tag{4.89}$$

The integral listed on the left side of Eq. (4.89) is proportional to the internal energy inventory of the entire medium:

$$\int_{-\infty}^{\infty} \rho(u - u_\infty)A \, dx = \int_{-\infty}^{\infty} \rho c(T - T_\infty)A \, dx$$
$$= \rho c A \int_{-\infty}^{\infty} \theta \, dx \tag{4.90}$$

where A is the large area of the plane normal to the x direction. The specific internal energy u is measured relative to the reference state provided by the distant temperature of the medium, $u_\infty = u(T_\infty)$. Together, Eqs. (4.89) and (4.90) indicate that the internal energy inventory of the medium is conserved in time. This inventory could have been deposited in the very beginning of the process (at $t = 0$) as a one-shot deposit of heat transfer Q (J) in the $x = 0$ plane:

$$\int_{-\infty}^{\infty} \rho(u - u_\infty)A \, dx = Q \tag{4.91}$$

Combining Eqs. (4.89)–(4.91), we find that the constant K is

$$K = \frac{Q''}{2\pi^{1/2}\rho c} \tag{4.92}$$

where $Q'' = Q/A$ is the strength of the instantaneous plane heat source. In summary, the excess temperature in the vicinity of the $x = 0$ plane in which Q'' (J/m²) is released once at $t = 0$ is [cf. Eq. (4.88)]

$$\theta(x, t) = \frac{Q''}{2\rho c(\pi \alpha t)^{1/2}} \exp\left(-\frac{x^2}{4\alpha t}\right) \quad \text{(instantaneous plane source)} \tag{4.93}$$

From this, we learn that the temperature of the heat source (the plane $x = 0$) decreases as $t^{-1/2}$. Figure 4.21 shows that in any other constant-x plane, the temperature reaches a maximum level at a particular time following the one-shot release of Q''.

Formulas similar to Eq. (4.93) have been derived for the temperature variation in the vicinity of sources of other shapes. For example, the excess temperature near a one-shot line heat source is

$$\theta(r, t) = \frac{Q'}{4\rho c\pi\alpha t} \exp\left(-\frac{r^2}{4\alpha t}\right) \quad \text{(instantaneous line source)} \tag{4.94}$$

In this expression, Q' is the strength of the instantaneous line source, that is, the number of joules per meter that are released only once at $t = 0$. The radial distance r is measured away from the line source. Equation (4.94) shows that the temperature along the line in which Q' was deposited ($r = 0$) decreases as t^{-1} as the time increases. This temperature decrease is steeper than the temperature in the plane of the Q'' source, Eq. (4.93), because the line source is exposed to the conducting medium in all directions (over an angle of 2π, or 360°) around the $r = 0$ axis.

The temperature history near an instantaneous point source is described by

$$\theta(r, t) = \frac{Q}{8\rho c(\pi\alpha t)^{3/2}} \exp\left(-\frac{r^2}{4\alpha t}\right) \quad \text{(instantaneous point source)} \tag{4.95}$$

The strength Q (J) represents the heat transfer released into the medium at the time $t = 0$, and the radial distance r is measured away from the point source ($r = 0$). The temperature at $r = 0$ decreases as $t^{-3/2}$ (i.e. faster than in the plane of the source (4.93) and in the line of the source (4.94)) because now the source is touched by a cold medium from all the directions of the sphere circumscribed to the point source.

4.6.2 Persistent (Continuous) Sources and Sinks

All the concentrated sources that have been covered in the preceding section (plane, line, point) are of the instantaneous (one-shot) kind. The time-dependent temperature distributions and conduction processes that are induced by sources that persist in time can be determined analytically by superposing the effects of a large number of instantaneous sources.

This superposition procedure can be illustrated by considering again the plane source geometry, Figure 4.21. Assume that at the time $t = 0$, the plane $x = 0$ receives the per-unit-area heat input Q_0''; the temperature distribution that results is indicated in Eq. (4.93), namely,

$$\theta_0(x, t) = \frac{Q_0''}{2\rho c(\pi\alpha t)^{1/2}} \exp\left(-\frac{x^2}{4\alpha t}\right) \tag{4.96}$$

Assume that at a subsequent moment, $t = t_1$, the plane $x = 0$ receives a new deposit of heat transfer per unit of plane area, Q_1''. If this instantaneous heat source were to occur alone, that is, in the absence of the original shot Q_0'', then the temperature variation (decaying hump) triggered by Q_1'' could also be written by reading Eq. (4.93):

$$\theta_1(x, t) = \frac{Q_1''}{2\rho c[\pi\alpha(t - t_1)]^{1/2}} \exp\left[-\frac{x^2}{4\alpha(t - t_1)}\right] \tag{4.97}$$

The value of $(t - t_1)$ counts the time that elapses after the release of Q_1''. If the Q_1'' source takes place in the presence (on the background) of the temperature hump created by Q_0'' at $t = 0$, then the temperature distribution after the time $t = t_1$, is simply the sum of $\theta_0(x, t)$ and $\theta_1(x, t)$. To summarize, the temperature distribution $\theta(x, t)$ during the entire time interval $t > 0$ is described by

$$\theta(x, t) = \begin{cases} \theta_0(x, t) & 0 < t < t_1 \\ \theta_0(x, t) + \theta_1(x, t) & t_1 < t \end{cases} \tag{4.98}$$

where θ_0 and θ_1 are the expressions listed in Eqs. (4.96) and (4.97). It can be shown that the sum $\theta = \theta_0 + \theta_1$ satisfies the conduction Eq. (4.87), the far-field conditions ($\theta \to 0$ as $x \to \pm \infty$), and the energy conservation statement that the internal energy inventory (per unit area of the source plane $x = 0$) in the entire medium is equal to $Q_0'' + Q_1''$ when $t > t_1$.

More entries (lines) can be added to the two-line sequence started in Eq. (4.98) if additional shots of size Q_i'' are deposited at the times t_i, in the source plane $x = 0$. For example, after the time $t = t_n$ (i.e. after $n + 1$ shots), the temperature distribution is given by

$$\theta(x, t) = \theta_0 + \theta_1 + \theta_2 + \cdots + \theta_n \tag{4.99}$$

A *continuous* heat source in the plane $x = 0$ has the same effect as a sequence of a very large number of small instantaneous plane sources of equal size:

$$\Delta Q'' = q'' \, \Delta t \tag{4.100}$$

where $q''(\text{W/m}^2)$ is the heat transfer deposited per unit area and unit time and Δt is the short duration of each shot. As Δt becomes infinitesimally small, the number of terms in the finite-time interval sum, Eq. (4.99), becomes infinite, and the sum is replaced by

$$\theta(x, t) = \int_0^t \frac{q''}{2\rho c[\pi \alpha(t - \tau)]^{1/2}} \exp\left[-\frac{x^2}{4\alpha(t - \tau)}\right] d\tau \tag{4.101}$$

Under the integral sign, the dummy variable τ marks the time when each additional instantaneous source $q'' \, d\tau$ springs into action. The integral (4.101) can be evaluated [1], and the resulting expression is the temperature distribution near the $x = 0$ plane in which the continuous source q'' is turned on at $t = 0$:

$$\theta(x, t) = \frac{q''}{\rho c}\left(\frac{t}{\pi \alpha}\right)^{1/2} \exp\left(-\frac{x^2}{4\alpha t}\right) - \frac{q''|x|}{2k} \text{erfc}\left[\frac{|x|}{2(\alpha t)^{1/2}}\right]$$
$$\text{(continuous plane source)} \tag{4.102}$$

Placing $x = 0$ in this expression, we learn that the temperature in the plane of the source increases as $t^{1/2}$ forever:

$$\theta(0, t) = \frac{q''}{\rho c}\left(\frac{t}{\pi \alpha}\right)^{1/2} \tag{4.103}$$

In other words, even though the plane source persists at the constant level q'', the temperature in the source plane (and, for that matter, everywhere else in the medium) increases as the elapsed time t increases.

The temperature distribution near a continuous line source can be determined similarly by superposing the effects of a large number of small instantaneous line sources of equal size. The result is

$$\theta(r, t) = \frac{q'}{4\pi k} \int_{r^2/4\alpha t}^{\infty} \frac{e^{-u}}{u} du \quad \text{(continuous line source)} \tag{4.104}$$

where q' (W/m) is the source strength. The integral on the right side of Eq. (4.104) is the *exponential integral* function, for which the values are tabulated in Appendix E. At sufficiently long times and/or small radial distances, where the group $r^2/4\alpha t$ is smaller than 1, the temperature distribution approaches [1]

$$\theta(r, t) \cong \frac{q'}{4\pi k} \left[\ln\left(\frac{4\alpha t}{r^2}\right) - 0.5772 \right] \qquad \left(\frac{r^2}{4\alpha t}\right) \ll 1 \tag{4.105}$$

The effect of a continuous point source $q(W)$ can also be determined by superposing the effects of a large number of instantaneous point sources of equal strength:

$$\theta(r, t) = \frac{q}{4\pi kr} \text{erfc} \left[\frac{r}{2(\alpha t)^{1/2}} \right] \qquad \text{(continuous point source)} \tag{4.106}$$

Noting that erfc(0) = 1, we conclude that as the time increases and the argument $r/[2(\alpha t)^{1/2}]$ becomes considerably smaller than 1, the temperature distribution stabilizes at the level

$$\theta(r, \infty) = \frac{q}{4\pi kr} \tag{4.107}$$

In this steady state, the temperature decreases as r^{-1} away from the point heat source. The existence of a steady temperature distribution at large times distinguishes the continuous point source from the continuous line and plane sources, for which steady-state solutions do not exist. The steady state can exist around a continuous point source because the spherical–radial geometry allows the released heat to escape in more directions than in the cylindrical–radial and one-dimensional Cartesian geometries. In other words, the infinite conducting medium provides a better (more effective) cooling effect to a point source than to a line source or a plane source.

Throughout the Sections 4.6.1 and 4.6.2, we referred to the concentrated heat transfer effect as that of a heat source. The same discussion and formulas apply to concentrated heat sinks: in these cases the numerical values of the source strengths (Q'', Q', Q, q'', q', q) are negative, and so are the corresponding excess temperatures θ.

4.6.3 Moving Heat Sources

One common feature of the concentrated sources and sinks treated until now is the symmetrical shape of the isotherms that develop around the location of the source. In this subsection we consider briefly the deformation of this temperature field when the source moves relative to the conductive medium in which it is imbedded. Specifically, we consider the long-time behavior of steady (continuous) sources that move with an unchanging velocity through the medium. We shall see that the temperature field that develops is a slender "thermal wake" that trails the source.

In Figure 4.22 a continuous *line source* of strength q' (W/m) moves with the constant speed U through the conductive medium (k, α). The two-dimensional system (x, y) is perpendicular to the line source; in the (x, y) plane, the medium moves in the positive x direction. The thermal wake left behind the source is two-dimensional and is represented by the temperature distribution $T(x, y)$, which is the unknown in this heat transfer problem. The temperature distribution $T(x, y)$ is steady in the eyes of an observer riding on the line source.

The model constructed on the left side of Figure 4.22 is a simplified representation of what happens in the vicinity of the line of contact between two semi-infinite plates that are being welded together. Seen from above, the electric arc plays the role of the line heat source $q' = q/\Delta z$, where $q(W)$ is the steady rate of heating provided by the arc to the plate-to-plate contact, and Δz is the thickness of the plates in the direction normal to the (x, y) plane. When the top and bottom surfaces of the plates are insulated enough (e.g. bathed in slowly

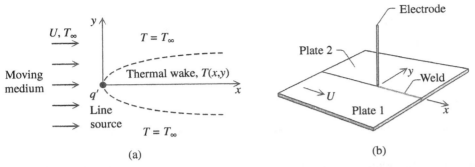

Figure 4.22 The thermal wake behind a continuous line source moving through a conductive medium (a) and the edge-to-edge arc welding of two semi-infinite plates (b).

moving air), the temperature distribution in the wake is nearly two-dimensional, $T(x, y)$. This thermal wake has the freshly formed weld as its line of symmetry.

The temperature distribution $T(x, y)$ is determined by solving the problem

$$U\frac{\partial T}{\partial x} = \alpha\frac{\partial^2 T}{\partial y^2} \qquad (4.108)$$

$$T = T_\infty \quad \text{at} \quad y = \pm\infty \qquad (4.109)$$

$$q' = \int_{-\infty}^{\infty} \rho c U(T - T_\infty)dy \qquad (4.110)$$

in which the energy Eq. (4.108) acquires one "convective" term on the left side, $U(\partial T/\partial x)$. The origin of this term will become clear in Section 5.2.3. The integral condition (4.110) states that the energy that flows through any constant-x plane is conserved and equal to the energy rate deposited by the line source in the moving medium. The solution to Eqs. (4.108)–(4.110) forms the subject of Problem 4.23 and is the same as the solution for the thermal wake behind a line source immersed in a turbulent stream with grid-generated (uniform) turbulence [6]:

$$T(x, y) - T_\infty = \frac{q'/\rho c}{(4\pi U\alpha x)^{1/2}} \exp\left(-\frac{Uy^2}{4\alpha x}\right) \qquad (4.111)$$

This solution shows that the excess temperature in the centerplane of the wake, $T(x, 0) - T_\infty$, decreases as $x^{-1/2}$ in the downstream direction. Note further that the argument of the exponential is constant when y^2/x is constant; this means that the width of the thermal wake increases as $x^{1/2}$ in the downstream direction. This growth is shown on the left side of Figure 4.22.

The thermal wake created behind a continuous point source $q(W)$ in a moving medium can be analyzed based on a similar model (Figure 4.23). In the cylindrical coordinate system (x, r) attached to the point source, the solution reads [6]

$$T(x, r) - T_\infty = \frac{q/\rho c}{4\pi\alpha x} \exp\left(-\frac{Ur^2}{4\alpha x}\right) \qquad (4.112)$$

The thermal wake is shaped of a body of revolution, the radius of which increases as $x^{1/2}$ in the downstream direction. The distribution of excess temperature in each constant-x plane has the same shape (Gaussian) as in the two-dimensional wake, Eq. (4.111). The centerline excess temperature, $T(x, 0) - T_\infty$, decreases as x^{-1} in the downstream direction, that is, faster than in the two-dimensional case.

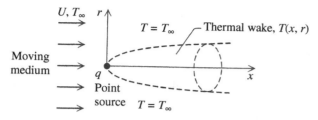

Figure 4.23 The thermal wake left behind a continuous point source in a moving medium.

An important aspect of any moving source problem is that the relative speed U must be sufficiently large for thermal wake solutions of types (4.111) and (4.112) to be valid. For example, in the extreme where the source moves so slowly that it is practically stationary, the temperature field is described by the persistent source solutions (4.104) and (4.106). Note that in the $U \to 0$ limit, the isotherms around the source are almost circular (cylindrical or spherical).

The thermal wake solutions in this subsection are valid when each wake is slender (elongated in the U direction), not round. According to the argument of the exponentials in Eqs. (4.111) and (4.112), the width of each thermal wake scales as $(\alpha x/U)^{1/2}$. Writing the wake slenderness condition as $(\alpha x/U)^{1/2} \ll x$, we conclude that the source speed is "large enough" when $Ux/\alpha \gg 1$. This condition also means that solutions of types (4.111) and (4.112) do not hold right next to the moving source, that is, at x values so small that the dimensionless number Ux/α is of order 1 or smaller. This dimensionless group is the Péclet number, which is used more frequently for convection, Eq. (5.70).

4.7 Melting and Solidification

In this section we revisit the time-dependent conduction in a semi-infinite solid exposed suddenly to a surface temperature T_0 that differs from the initial temperature T_i of the material (Section 4.3). Consider first the heating effect due to raising the surface temperature above the temperature of the solid, $T_0 > T_i$.

If the solid is a crystalline substance (e.g. ice, high-purity metal), a known property of the solid is also the melting point T_m. When the temperature of the heated surface is lower than the melting point, the semi-infinite medium remains solid and the formulas developed in Section 4.3 apply. If T_0 is greater than the melting point, a near-surface layer of finite thickness will melt. The conduction process continues to be time dependent; however, sandwiched between the semi-infinite solid and the heated boundary (T_0) is a liquid layer of finite thickness. This thickness and the temperatures of the still solid regions continue to increase in time.

The melting phenomenon is illustrated in Figure 4.24. Unlike in the all-solid case of Figure 4.3, the figure is oriented so that the conduction process proceeds in the vertical direction and the liquid layer floats on top of the solid. It is assumed that the conducting medium is a substance that expands upon melting and that its liquid phase expands upon heating. The downward orientation of conduction in Figure 4.24(a) is therefore a reminder that the entire medium (liquid + solid) is motionless in a gravitational field that points in the downward direction. The liquid layer is stably stratified because its uppermost layers are the lightest (for more on this, see Section 7.4.3 on Bénard convection).

Figure 4.24(b) shows the analogous problem of unidirectional solidification in a semi-infinite liquid pool that is cooled suddenly along the bottom surface, $T_0 < T_m$. If, as in most substances, the solid phase is denser

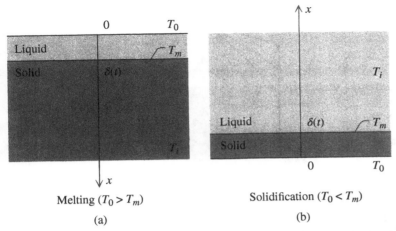

Melting ($T_0 > T_m$)

(a)

Solidification ($T_0 < T_m$)

(b)

Figure 4.24 Melting (a) and solidification (b) into a semi-infinite isothermal medium (T_i) with sudden temperature change at the boundary (T_0).

than the liquid phase, and if the liquid contracts upon cooling, the solid layer and the pool of liquid remain motionless.

The pure conduction analysis that is presented below applies to a perfectly motionless medium. In the case of water near the ice point (0 °C) – that is, in an anomalous case where the solid is lighter than the liquid and the liquid expands upon cooling – both sides of Figure 4.24a,b, would have to be rotated by 180°, so that motion is ruled out in a gravitational field that points downward.

The key unknown in melting and solidification processes is the movement of the melting front $x = \delta(t)$ or the melting/solidification rate $d\delta/dt$. Consider first the melting process (Figure 4.24a). Assume that the initial solid temperature is equal to the melting point, $T_i = T_m$. Since the interface between the solid and liquid regions (the melting front) is always at the melting point, the $T_i = T_m$ assumption means that the solid remains isothermal (at T_m, i.e. as a saturated solid) as the time increases. This feature is illustrated in Figure 4.25, which also shows that the temperature difference between the heated boundary (T_0) and the remaining solid (T_m) is bridged by a δ-thick layer of liquid.

The movement of the melting front is governed by the conservation of energy in the plane $x = \delta(t)$. Consider a thin control volume that contains the melting front and moves downward with the speed of the melting front. The right side of Figure 4.25 shows that a stream of solid flowrate $\rho A(d\delta/dt)$ enters the control volume from below, while another $\rho A(d\delta/dt)$ stream of liquid exits through the upper surface. The density ρ refers to the liquid,[2] that is, to the region whose thickness is δ.

The first law of thermodynamics states that the net enthalpy flow out of the control volume must be balanced by the net heat transfer rate received by the same control volume:

$$\left(\rho A \frac{d\delta}{dt}\right) h_f - \left(\rho A \frac{d\delta}{dt}\right) h_s = -kA \left(\frac{\partial T}{\partial x}\right)_{\substack{x = \delta,\,\text{liquid} \\ \text{side}}} \tag{4.113}$$

In this equation, A, h_f, and h_s are the frontal area of the control volume, the specific enthalpy of the liquid, and the specific enthalpy of the solid, respectively. The right side of Eq. (4.113) represents the heat transfer

---

2 Note that in this analysis the density of the solid (ρ_s) does not have to be exactly equal to the density of the liquid (ρ). The reason is the principle of mass conservation, according to which $\rho(d\delta/dt) = \rho_s v_s$, where v_s is the vertical velocity of the solid entering the control volume shown on the right side of Figure 4.25.

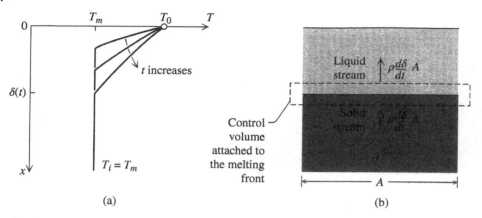

(a)　　　　　　　　　　　　　(b)

Figure 4.25 Melting of a semi-infinite solid at the melting point ($T_i = T_m$) and the flow through the infinitesimally thin control volume attached to the solid–liquid interface.

rate that arrives from above, that is, on the liquid side of the melting front. There is no heat transfer term for the solid side of the melting front, because the solid region is isothermal. The coefficient k is therefore the thermal conductivity of the liquid.

The exact analysis of this melting phenomenon would continue with the derivation of the $T(x, t)$ distribution inside the liquid layer, because that result would permit the evaluation of the right side of Eq. (4.113). The end result of the exact analysis is reported later, in Eq. (4.117). A much simpler approach is based on the observation that sufficiently early in the life of the melting process, when the melt layer is thin enough, the temperature distribution is linear:

$$\frac{T(x, t) - T_m}{T_0 - T_m} \cong 1 - \frac{x}{\delta(t)} \tag{4.114}$$

Combining this with Eq. (4.113) and the latent heat of melting $h_{sf} = h_f - h_s$, we obtain

$$\delta \frac{d\delta}{dt} \cong \frac{k}{\rho h_{sf}}(T_0 - T_m) \tag{4.115}$$

Equation (4.115) can be integrated from the starting condition of all solid, $\delta(0) = 0$, to show that the melt thickness increases as $t^{1/2}$:

$$\delta(t) \cong \left[2\frac{kt}{\rho h_{sf}}(T_0 - T_m)\right]^{1/2} \tag{4.116}$$

The exact solution for the melting of a solid at the melting point is [7]

$$\pi^{1/2}\lambda \exp(\lambda^2)\,\mathrm{erf}(\lambda) = \frac{c(T_0 - T_m)}{h_{sf}} \tag{4.117}$$

where c is the specific heat of the liquid and the dimensionless number λ is

$$\lambda = \frac{\delta}{2(\alpha t)^{1/2}} \tag{4.118}$$

The group on the right side of Eq. (4.117) is the Stefan number:

$$Ste = \frac{c(T_0 - T_m)}{h_{sf}} \tag{4.119}$$

The number λ is small relative to 1 in the case of water freezing by conduction to an environment a few degrees below $0\,°C$. In the case of materials with high melting points, such as metals and rocks cooled by contact with room temperature surfaces, the value of λ is of the order of 1.

The Stefan number is a measure of the degree of superheating $(T_0 - T_m)$ experienced by the liquid. The exact solution (4.117) is presented in Figure 4.26 next to the approximate solution (4.116). This figure shows that the approximate solution is quite accurate when the Stefan number is smaller than 1. In general, the approximate solution (4.116) overestimates the melt thickness $\delta(t)$.

When the solid is initially at a temperature lower than the melting point, $T_i < T_0$, the melting phenomenon analyzed earlier is accompanied by a process of time-dependent conduction into the solid, that is, on the solid side of the melting front. The analytical solution for $\delta(t)$ in this more general case can be found in Ref. [1]; it was reported first by Franz Neumann in 1860.

Consider now briefly the *solidification* process illustrated on the right side of Figure 4.24. In this case the bottom surface is colder than the melting front, $T_0 < T_m$. If the liquid is saturated $(T_i = T_m)$, the thickness of the solidified layer $\delta(t)$ can be calculated with Eqs. (4.116) and (4.117), in which the temperature difference $T_0 - T_m$ is replaced by $T_m - T_0$. In the resulting formulas, k, c, and α are now properties of the *solid* layer. The first analysis of solidification by time-dependent conduction was published in 1831 by Lamé and Clapeyron [8]. Other solutions for solidification and melting with plane or cylindrical liquid–solid interfaces can be found in Ref. [1].

The present treatment of melting and solidification was based on the assumption that the conduction process is unidirectional. Review the plane liquid–solid interfaces assumed in Figures 4.24 and 4.25. Real melting and solidification processes are often accompanied by additional effects that can deform severely the liquid–solid interface. In such cases the conduction process is no longer unidirectional.

One example is the effect of natural convection (buoyancy-driven flow) in the liquid if the melting front is rotated away from the stable (no flow) position chosen in Figure 4.24. A classical case of melting in the presence of natural convection is outlined in Figure 4.27, which results from rotating Figure 4.24a by 90°. The left side is heated to the temperature T_0, which is above the melting point of the solid, T_m. The entire

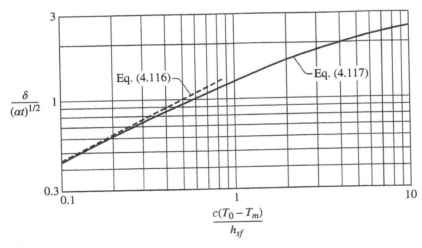

Figure 4.26 History of the melting (or solidification) front position in a semi-infinite medium of temperature equal to the melting point.

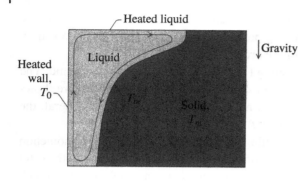

Figure 4.27 The deformed shape of the melting front when heating is from the side and the effect of natural convection is strong.

solid is at the melting point. The top and bottom sides of the two-dimensional system are impermeable and insulated. The liquid is driven clockwise by the heating imposed from the left side. The liquid rises along the heated wall because it is warmer and lighter than the liquid near T_m. The latter descends along the melting front, which is colder.

As time passes, the melting front of Figure 4.27 deviates considerably from the original (plane, vertical) shape. This deformation is due to the natural circulation, which brings the stream of T_0-heated liquid in direct contact with the upper end of the liquid–solid interface. This is why the melting process is considerably faster near the top of the system [9].

4.8 Evolutionary Design

4.8.1 Spacings Between Buried Heat Sources

Performance depends on configuration. When the configuration has freedom to change, the performance can be increased by evolving the design in the direction recognized as constructal law. In heat transfer, the method of evolutionary design covers the field, from fin shapes (Chapter 2) to spacings between plates with convection (Sections 6.5.3 and 7.5.1).

In this section we consider the fundamental problem of how to position in a conducting body several parallel lines that serve as time-dependent heat sources or heat sinks [10, 11]. The volume of the conducting body is finite. The objective is to distribute the lines in such a way that the overall thermal resistance between the volume and the sources is minimal. This means that when the heat generation rate is specified, the peak temperature difference between sources and volume should be minimal. Conversely, when the peak temperature difference is specified, the heat transfer rate between the lines and the finite volume should be maximal, i.e. the design represents the highest density of heat deposition in the volume, or heat extraction from the volume, or the best packing.

This fundamental problem has several important applications most notably in the design of heat pumps with heat extraction or release through pipes buried in the ground, and the positioning of decaying nuclear waste underground. In such applications, the heat transfer interaction between the line sources (or sinks) and the ground is time dependent for two reasons: transient heat conduction in the surrounding medium and the finite lifetime t_c of the line sources. For example, during heat pump operation, the lifetime is dictated by the daily cycle and the seasonal weather variations. In nuclear waste storage, the lifetime is controlled by the decay of the heat generation rate.

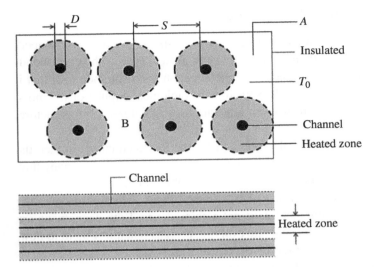

Figure 4.28 Conducting finite-size volume with several embedded line heat sources or sinks.

The method is illustrated with the model proposed in Ref. [10]. Heat sources are positioned as several parallel thin cylinders, which are viewed in the cross-sectional area A shown in Figure 4.28. The third dimension is perpendicular to A, and is aligned with the horizontal direction shown in the lower part of the figure. The volume is fixed. Each heat source generates heat in time-dependent fashion. The cylinders have the diameter D and are indicated by black discs. Because the available volume is finite, the boundary of A is modeled as insulated. In general, the number of cylinders and their placement on A are free to vary. The initial temperature of the surrounding medium is uniform, T_0. The effects of convection, phase change, and heterogeneity are assumed negligible.

The time scale of the heat interaction is t_c, and every cylinder generates at a decaying rate, $q(t) = q_0 \exp(-t/t_c)$, where t is the time and q_0 is the heat generation rate at $t = 0$. An alternative model is the step change (top hat q function) from constant q from $t = 0$ until $t = t_c$, to $q = 0$ after $t = t_c$. The release of heat creates a time-dependent temperature field over the domain A. All the thermophysical properties of the surrounding medium are assumed to be constant.

The temperature field in the conduction medium depends on time (t), source diameter (D), and spacing (S). The temperature field was simulated numerically for transient heat conduction in a two-dimensional domain of finite size where the heat sources were arranged in a square pattern. The spacing S was free to vary. Of chief interest was the peak temperature in the surrounding medium – its value, location, and time of occurrence.

When the spacing is large, the peak temperature is below the highest allowable level, indicating that more space is available for additional heat sources to be buried. When the spacing is too small, the peak temperature rises monotonically and overshoots the allowable ceiling. There exists an optimal spacing, and it occurs at the intersection of between the two asymptotic behaviors identified earlier. The numerical results for the optimal spacings for exponentially decaying and step-change $q(t)$ functions line up according to approximately $S \sim 3(\alpha t_c)^{1/2}$, where α is the thermal diffusivity of the medium. The same correlation was anticipated based on scale analysis. The method was extended to optimizing the packing of heat-generating channels of two sizes (diameters and heat source strengths).

4.8.2 The S-Curve Growth of Spreading and Collecting

When something spreads on a territory (area, or volume), the curve of territory size versus time is S-shaped; slow initial growth is followed by much faster growth and finally by slow growth again. The corresponding curve of the rate of spreading versus time is bell shaped. This phenomenon is so common [12] that it has generated entire fields of research that seem unrelated: the spreading of biological populations, tumors, chemical reactions, contaminants, languages, news, information, innovations, technologies, infrastructure, and economic activity.

The natural phenomenon is not a particular mathematical expression for the S-shaped curve; in fact, the S-shaped curves are not unique. The natural phenomenon is the very frequent observation that in many flow systems the covered territory increases in time according to a curve that resembles an S.

The prevalence of S-curve phenomena in nature rivals that of tree-shaped flows, which also cover the animate, inanimate, and human realms. Here we show that this is not a coincidence. Both phenomena are manifestations of the natural tendency of flow systems to generate evolving designs that allow them to flow, spread, and collect more easily. This tendency is summarized by the constructal law.

S-curve observations are often called "diffusion" of change, innovation, technology, infrastructure, etc. References [13, 14] showed that diffusion alone cannot account for the S shape. For example, if matter or energy were to diffuse from one boundary point into an area, the radius of the patch swept by diffusion would increase as $(\alpha t)^{1/2}$, where α (m²/s) is the diffusivity (mass, heat) relative to the medium that occupies the area and t is the time. The diffused area would grow as the radius squared (i.e. as t), and the curve would be a straight line, not an S.

From turbulence and river basins to human lungs, natural flow structures do not flow by diffusion alone. They combine two flow mechanisms (convection and diffusion) in order to facilitate the flow over the entire available space. Figure 4.29 shows how the flow architecture spreads from point to area in two successive phases. First, it *invades* the available area $(2L_1)^2$ with the speed V to distance $L = Vt$ during the time $t_{in} = 2L_1/V$, while diffusion spreads transversally, on both sides of L.

The following analysis is presented in terms of convective and conductive heat transfer, although similar two-mechanism analyses can be formulated for each of the flow architectures that exhibit the $S(t)$ curve while spreading or collecting.

The invading channel has a constant temperature $(T_0 + \Delta T)$, which is different than the initial temperature of the invaded medium (T_0). In the vicinity of the invading line, the thickness scale (δ) of the diffused area is such that the transversal diffusion time (δ^2/α) is the same as the time of longitudinal advance (L/V). From this follows $\delta \sim (\alpha L/V)^{1/2}$ and the finger-shaped area covered by diffusion $\delta L \sim \alpha^{1/2} V t^{3/2}$. At the end of invasion (t_{in}), the diffused area has grown to $A_{in} \sim \alpha^{1/2} V^{-1/2}(2L_1)^{3/2}$:

$$\frac{A}{A_{in}} \sim \left(\frac{t}{t_{in}}\right)^{3/2}$$

(4.120)

Equation (4.120) accounts for the early part of the S curve, because the rate dA/dt increases monotonically.

After t_{in}, the available area is filled by diffusion above and below the finger of length $2L_1$. This is the *consolidation* phase, and it lasts until $t_{co} \sim L_1^2/\alpha$. The thickness of the area covered by diffusion is $(\alpha t)^{1/2}$. The area swept by diffusion is $A_{co} \sim 2 \times (2L_1) \times (\alpha t)^{1/2}$, and its history is

$$\frac{A}{4L_1^2} \sim \left(\frac{t}{t_{co}}\right)^{1/2}$$

(4.121)

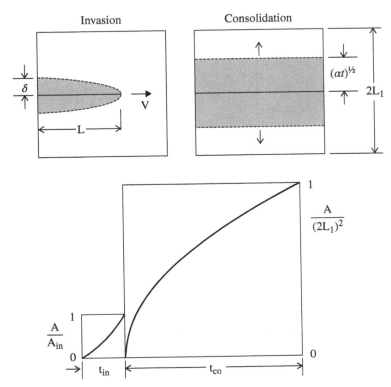

Figure 4.29 Line-shaped invasion, followed by consolidation by traversal diffusion. The predicted history of the area covered by diffusion reveals the S-shape curve.

This accounts for the late part of the S curve, because dA/dt decreases monotonically. The scales of the invasion and consolidation periods depend on the invasion speed and diffusivity:

$$\frac{t_{co}}{t_{in}} \sim \frac{L_1 V}{2\alpha} > 1 \tag{4.122}$$

$$\frac{A_{in}}{4L_1^2} \sim \left(\frac{\alpha}{2L_1 V}\right)^{1/2} < 1 \tag{4.123}$$

The group $L_1 V/\alpha$ must be greater than 1, because the invasion finger must be slender.

In summary, the invasion–consolidation analysis accounts for the S curve entirely, as shown in Figure 4.29. It also predicts that the S shape is not unique, because it depends on the group $L_1 V/\alpha$. Because the point–area spreading is natural, it has the tendency and freedom to generate configurations that allow it to cover the available area during a time shorter than $t_{co} \sim L_1^2/\alpha$.

The configuration that offers greater access than the line invasion is a tree-shaped invasion [13, 15]. Figure 4.30 shows the tree of mass or energy supply that emerges from the side and grows at constant speed V. We continue to model this invading tree in heat transfer terms, as lines with a temperature that differs from the ambient temperature. The first line is the trunk of the growing tree. When it reaches the center, it bifurcates and continues to invade at constant speed. The emergence of the snowflake is an example of this. The snowflake is a tree-shaped architecture that grows such that its ice volume has an S-shaped history. What flows in the snowflake is the latent heat of solidification, from the ice to the cold air [16].

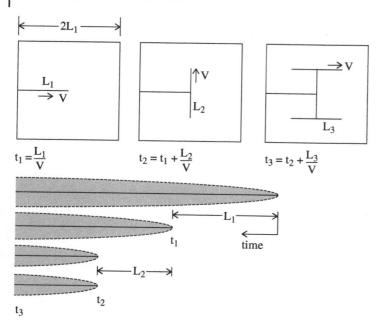

Figure 4.30 Tree-shaped invasion, showing the narrow regions covered by diffusion in the immediate vicinity of the invasion lines.

References

1 Carslaw, H.S. and Jaeger, J.C. (1959). *Conduction of Heat in Solids*, 71. Oxford: Oxford University Press.

2 Heisler, M.P. (1947). Temperature charts for induction and constant-temperature heating. *Trans. ASME* 69: 227–236.

3 Gröber, H., Erk, S., and Grigull, U. (1961). *Fundamentals of Heat Transfer*. New York: McGraw-Hill.

4 Arpaci, V.S. (1966). *Conduction Heat Transfer*, 284, 288. Reading, MA: Addison-Wesley.

5 Langston, L.S. (1982). Heat transfer from multidimensional objects using one-dimensional solutions for heat loss. *Int. J. Heat Mass Transfer* 25: 149–150.

6 Bejan, A. (2013). *Convection Heat Transfer*, 4e, 422–424. Hoboken, NJ: Wiley.

7 Stefan, J. (1891). Über die Theorie der Eisbildung, insbesondere über die Eisbildung im Polarmeere. *Ann. Phys. Chem.* 42: 269–286.

8 Lamé, G. and Clapeyron, E. (1831). Memoire sur la solidification par refroidissement d'un globe liquide. *Ann. Chim. Phys.* 47: 250–256.

9 Zhang, Z. and Bejan, A. (1989). Melting in an enclosure heated at constant rate. *Int. J. Heat Mass Transfer* 32: 1063–1076.

10 Jung, J., Lorente, S., Anderson, R., and Bejan, A. (2011). Configuration of heat sources or sinks in a finite volume. *J. Appl. Phys.* 110: 023502.

11 Baik, Y.-J., Ngo, I.-L., Park, J.M., and Byon, C. (2016). Generalized correlations for predicting optimal spacing of decaying heat sources in a conducting medium. *J. Heat Transfer* 138: 091301.

12 Bejan, A. (2016). *The Physics of Life: The Evolution of Everything*. New York: St. Martin's Press.

13 Bejan, A. and Lorente, S. (2011). The constructal law origin of the logistics S curve. *J. Appl. Phys.* 110: 024901.

14 Bejan, A. and Lorente, S. (2012). The physics of spreading ideas. *Int. J. Heat Mass Transfer* 55: 802–807.

15 Cetkin, E., Lorente, S., and Bejan, A. (2012). The steepest S curve of spreading and collecting flows: discovering the invading trees, not assuming it. *J. Appl. Phys.* 111: 114903.

16 Bejan, A., Lorente, S., Yilbas, B.S., and Sahin, A.Z. (2013). Why solidification has an S-shaped history. *Sci. Rep.* 3: 1711.

17 Fowler, A.J. and Bejan, A. (1991). The effect of shrinkage on the cooking of meat. *Int. J. Heat Fluid Flow* 12: 375–383.

Problems

Lumped Capacitance Model

4.1 A sphere of carbon steel (0.5%C) with a diameter of 1 cm and an initial temperature of $100\,°C$ is exposed for two minutes to the atmosphere of temperature $10\,°C$. The heat transfer coefficient between the surface and the air flow is $20\ \text{W/m}^2 \cdot \text{K}$.

    (a) Calculate the Biot number hr_o/k and the Fourier number $\alpha t/r_o^2$, and using Figure 4.5 demonstrate that the cooling of the steel ball is described adequately by the lumped capacitance model.

    (b) Calculate the temperature of the steel at the end of two minutes of exposure to the atmosphere.

    (c) For the same point in time, estimate the order of magnitude of the temperature difference between the center and the surface of the steel ball. (*Hint:* Write that the scale of the convection heat flux at the surface is the same as the scale of the radial conduction heat flux through the ball.)

4.2 A solid body of initial temperature $T_{1,0}$ is immersed suddenly in an amount of incompressible liquid of initial temperature $T_{2,0}$. The respective heat capacities of the immersed body and the liquid are $(mc)_1$ and $(mc)_2$, where m and c denote the respective masses and specific heats. The external (wetted) area of the immersed body is A, and the heat transfer coefficient between the body and the liquid is h (constant).

Treating the immersed body and the surrounding liquid as two lumped capacitance systems, show that their respective temperatures vary according to the following relations:

$$T_1(t) = T_{1,0} - \frac{T_{1,0} - T_{2,0}}{1 + [(mc)_1/(mc)_2]}(1 - e^{-nt})$$

$$T_2(t) = T_{2,0} + \frac{T_{1,0} + T_{2,0}}{1 + [(mc)_2/(mc)_1]}(1 - e^{-nt})$$

where

$$n = hA\,\frac{(mc)_1 + (mc)_2}{(mc)_1(mc)_2}$$

Note that what distinguishes this problem from the system treated in Section 4.2 is the finiteness of the liquid pool. Because the amount of liquid is finite, the lumped liquid temperature T_2 varies under the influence of the heat transfer across the wetted surface A.

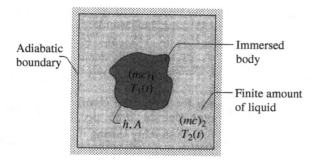

Adiabatic boundary

Immersed body

$(mc)_1$
$T_1(t)$

Finite amount of liquid

h, A

$(mc)_2$
$T_2(t)$

Figure P4.2

4.3 A small bottle containing milk of 60 °C temperature is cooled by sudden immersion in a vessel containing 200 cm³ tap water at 10 °C. The milk bottle and the amount of tap water can be modeled as two lumped capacitances, in accordance with the temperature history solutions reported in the preceding problem. The milk bottle is a cylinder of diameter $D = 4$ cm and height $H = 6$ cm. Its total mass and average specific heat are $m_1 = 0.125$ kg and $c_1 = 4$ kJ/kg K. The heat transfer coefficient between the bottle and the cold water is constant, $h = 350$ W/m² · K. The heat transfer between this ensemble and the atmosphere is negligible.

(a) Using the $T_1(t)$ solution listed in the preceding problem, calculate the temperature of the milk bottle after one minute of continuous immersion in cold water. Calculate the corresponding water temperature T_2.

(b) As a simpler alternative, calculate the bottle temperature after one minute by making the (incorrect) assumption that the temperature of the surrounding water is fixed at 10 °C. (c) Compare this new estimate with the T_1 value calculated in part (a).

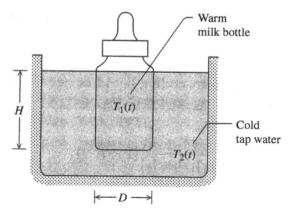

Warm milk bottle

$T_1(t)$

Cold tap water

$T_2(t)$

H

D

Figure P4.3

4.4 A body of volume V, surface area A, density ρ, and specific heat c is initially at the temperature T_0. The body is immersed at $t = 0$ in a fluid reservoir of a higher temperature, $T_0 + \Delta T$. At the time $t = t_1$, it is removed from the hot fluid and plunged into a cold bath of temperature $T_0 - \Delta T$. Derive an expression for the time $t = t_2$ when the body temperature returns to its initial level T_0. Show that the time spent in the cold bath can never be greater than the time spent in the hot fluid.

4.5 During the daily cycle, the temperature of ambient air T_∞ varies approximately as $T_\infty = T_0 + a \sin bt$. The average temperature T_0 and the oscillation amplitude a are known constants of the geographic region and annual season. The constant b is shorthand for 2π per day.

Consider the temperature history $T(t)$ of a body that is permanently exposed to $T_\infty(t)$. The body (e.g. a rock) is small enough to behave as a thermal capacitance (density ρ, specific heat c, volume V, surface area A). The heat transfer coefficient between A and the T_∞ air is the constant h.

Determine analytically the body temperature $T(t)$. Show that the body temperature oscillates sinusoidally and that this oscillation lags behind the atmospheric temperature oscillation. Determine the time lag between the two sinusoids, and the amplitude of the body temperature oscillation.

Unidirectional Time-Dependent Conduction

4.6 The surface temperature history curve plotted on the right side of Figure 4.4 can be approximated by using the simpler expression

$$\frac{T_0(t) - T_\infty}{T_i - T_\infty} = \frac{1}{1 + C(h/k)(\alpha t)^{1/2}}$$

where C is a constant approximately equal to 1. Show that this expression can be derived from the fluid contact boundary condition (4.48), by restating this condition as an equality between the respective scales of the two sides of the equal sign.

4.7 Consider the porcelain cup illustrated in Example 4.1 in the text, where the initial temperature of the porcelain wall is $T_i = 25\,°C$ and the liquid temperature is $T_i = 70\,°C$. The time when the hot liquid is poured into the cup is $t = 0$, and the location of the wetted surface of the porcelain wall is $x = 0$. Calculate the porcelain temperature at the time $t = 3$ seconds and at the distance $x = 2$ mm away from the wetted surface, using each of these models:

(a) The porcelain wall is a semi-infinite conducting medium where the $x = 0$ surface assumes the temperature of the liquid, $T_0 = T_\infty = 70\,°C$.

(b) The wall is a semi-infinite medium whose $x = 0$ surface is in contact with the T_∞ liquid across a heat transfer coefficient $h = 1200$ W/m$^2 \cdot$ K, with the solution presented graphically in Figure 4.4.

(c) The same model as in part (b), with the analytical solution given in Eq. (4.49).

4.8 The surface of a brick wall of uniform initial temperature $T_i = 10\,°C$ absorbs solar thermal radiation at the rate $q'' = 100$ W/m^2, beginning with the time $t = 0$. The thermal conductivity and diffusivity of brickwork are $k = 0.5$ W/m $\cdot$ K and $\alpha = 0.005$ cm^2/s. Calculate the following:

(a) The surface temperature after one hour of exposure to the heat flux q''.

(b) The temperature after one hour at an interior point in the brick wall situated at $x = 10$ cm beneath the exposed surface.

4.9 Hot coffee of temperature $T_\infty = 90\,°C$ is poured into a cup. The porcelain wall is initially at $T_i = 10\,°C$. The wall is 0.5 cm thick. The heat transfer coefficient between the coffee and the wetted surface is

$h = 700 \text{ W/m}^2 \cdot \text{K}$. The external surface of the cup may be modeled as perfectly insulated. Calculate the time that passes until the external surface of the cup becomes too hot to hold, for example, when it reaches $50\,°\text{C}$.

4.10 An isothermal icicle at the initial temperature $T_i = 0\,°\text{C}$ is placed inside a domestic freezer in which the air temperature is maintained at $T_\infty = -5\,°\text{C}$. The heat transfer coefficient between the ice surface and the surrounding air flow is $h = 10 \text{ W/m}^2 \cdot \text{K}$. The icicle can be modeled as a long cylinder with a diameter of $2\,\text{cm}$.

Calculate the temperature on the icicle centerline after 10 minutes in the freezer. Perform this calculation in two ways: first, by reading Figure 4.10 and later, by using the series solution for the time-dependent temperature in a long cylinder. Numerically, show that the first term of the series is a very good substitute for the sum of the entire series.

4.11 It is proposed to manufacture glass beads of diameter $0.5\,\text{mm}$ by spraying them in $20\,°\text{C}$ air and allowing them to harden as they fall to the ground. Assume that the initial temperature of the glass bead is $T_i = 500\,°\text{C}$, the bead-air heat transfer coefficient is $h = 324 \text{ W/m}^2 \cdot \text{K}$, and the terminal velocity is $U = 3.8 \text{ m/s}$. The glass properties are similar to those listed for window glass in Appendix B.

The design calls for the production of beads such that the center temperature does not exceed $40\,°\text{C}$. From what height should the beads be allowed to fall? Calculate the temperature of the bead surface when its center temperature reaches $40\,°\text{C}$. Compare these two temperatures, and comment on whether the lumped capacitance model is applicable in the late stages of this cooling process.

4.12 The shape in which a potato is cut has an important effect on how fast each piece becomes cooked. Study this effect by considering three different shapes, each containing $5\,\text{g}$ of potato matter:
(a) Sphere
(b) Cylinder with a length of $6\,\text{cm}$
(c) Thin disc with a diameter of $4\,\text{cm}$

Each piece is initially at the temperature $T_i = 30\,°\text{C}$. At the time $t = 0$, each piece is placed in a pot with water boiling at the temperature $T_\infty = 100\,°\text{C}$. The heat transfer coefficient is constant, $h = 2 \times 10^4 \text{ W/m}^2 \cdot \text{K}$, and the properties of potato matter are approximately $\rho = 0.9 \text{ g/cm}^3$, $k = 0.6 \text{ W/m} \cdot \text{K}$, and $\alpha = 0.0017 \text{ cm}^2/\text{s}$.

For each shape, calculate the time t until the volume-averaged temperature of the potato rises to $65\,°\text{C}$. Compare the three times calculated in this manner, and comment on why only rarely potato is shaped as balls before cooking. (*Hint:* Show that the Biot number is generally much greater than 1, and then rely exclusively on Figure 4.16 to calculate the respective *Fo* values.)

4.13 The cooking of a beef steak can be modeled as time-dependent conduction in a slab. The thickness of the steak is $4\,\text{cm}$, its initial temperature is $25\,°\text{C}$, and the properties of beef are $k = 0.4 \text{ W/m} \cdot \text{K}$ and $\alpha = 1.25 \times 10^{-7} \text{ m}^2/\text{s}$. The steak is cooked (with both sides exposed) in a convection oven of temperature $150\,°\text{C}$. The heat transfer coefficient between meat and oven air is $60 \text{ W/m}^2 \cdot \text{K}$.
(a) The steak is cooked "well done" when its centerplane temperature reaches $80\,°\text{C}$. Calculate the required cooking time by noting that the right side of Eq. (4.62) is approximated well by the first term of the series.

(b) Meat shrinks during cooking [17]. For example, the thickness of a steak cooked well done drops to about 75% of the original thickness. To see the extent to which the shrinking of meat shortens the cooking time, repeat part (a) by assuming that the steak thickness is equal to 3 cm throughout the cooking process.

(c) Is the real cooking time longer or shorter than the estimate obtained in part (b)?

4.14 Two solid blocks with different properties (see figure) and with the same initial temperature T_∞ are suddenly placed in contact and rubbed against each other. The friction between the two blocks generates the heat flux q'' in the interface. A fraction of this flux, q_1'', is absorbed by the upper block, while the remainder, q_2'', flows into the lower block. Determine the manner in which the properties of the two materials affect the ratio q_2''/q_1''. Derive the formulas needed for calculating q_1'', q_2'' and the interface temperature T_0.

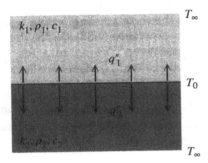

Figure P4.14

4.15 A slab of ice 1-cm-thick is pulled out of the freezer of a domestic refrigerator and exposed to room temperature air. Its initial temperature is $-10\,°C$. The immediate heating felt by the ice slab causes melting at the surface. The superficial melting effect can be modeled by setting the ice surface temperature equal to $0\,°C$, which is equivalent to assuming an infinite Biot number in Eq. (4.62). Calculate the time that passes until the temperature in the centerplane of the slab rises from -10 to $-0.1\,°C$. For this, it is sufficient to use only the first term in the series of Eq. (4.62).

4.16 A hot dog with a diameter of roughly 2 cm and initial temperature of $20\,°C$ is dropped into a pool of $95\,°C$ water. The heat transfer coefficient is assumed constant and equal to $200\ \mathrm{W/m^2 \cdot K}$. The meat thermal conductivity and thermal diffusivity are $0.4\ \mathrm{W/m \cdot K}$ and $0.0014\ \mathrm{cm^2/s}$.

(a) How long do we have to wait until the center of the hot dog warms up to $65\,°C$?

(b) Estimate the hot dog surface temperature at the time calculated in part (a).

4.17 The regenerative heat exchanger of a Stirling cycle heat engine consists of a thick stack of stainless steel wire screens (gauzes). During each half-cycle, hot gas at $600\,°C$ is forced to flow through the wire screens. The duration of this flow, or the exposure of the individual wire to hot gas, is 0.01 seconds. The heat transfer coefficient h between the wire surface and the hot gas varies inversely with the wire diameter D, such that $hD = 1\ \mathrm{W/m \cdot K}$.

The purpose of the wire screens is to store a large fraction of the energy carried by the hot gas, so that during the second half of the cycle, that energy can be used to preheat a cold stream. The design is considered successful when during each blow the wire stores 90% or more of the maximum energy that it can store. How thin must the wire be to meet this requirement?

Concentrated Sources and Sinks

4.18 The purpose of this exercise is to demonstrate that a function of the exponential type shown in Eq. (4.88) is a solution of the conduction Eq. (4.87).
  (a) Assume that the unspecified function $f(x, t)$ satisfies the conduction Eq. (4.87). Show that if this is true, then the new function $g = \partial f / \partial x$ also satisfies the conduction equation.
  (b) Recall the solution for time-dependent conduction in a semi-infinite solid, Eq. (4.41). Regard that solution as one example of an $f(x, t)$ function. Use the conclusion drawn in part (a), and report the form of the corresponding function $g(x, t)$. In this way you will arrive at the exponential-type expression (4.88), that is, you will expect that expression to be a solution of your Eq. (4.87).

4.19 Consider the temperature history in a constant-x plane parallel to the $x = 0$ plane of the instantaneous plane heat source Q''. Determine the time t when the temperature reaches a maximum in the constant-x plane. Derive an expression for the maximum temperature at that location.

4.20 Consider the temperature history at a fixed distance r away from an instantaneous line heat source Q'. Determine the time t when the temperature becomes maximum at this location. Show that the maximum temperature at the radius r decreases as r^{-2} as r increases.

4.21 Examine the temperature history at a point P situated at a distance r away from the instantaneous point heat source Q. Determine the time t when the temperature reaches a maximum at point P. Report the maximum temperature that is registered at this special moment.

4.22 At sufficiently long times, the temperature distribution around a continuous point source reaches the steady state described by Eq. (4.107). Use this result to determine the shape factor S for steady conduction between an isothermal spherical object (radius r_1 temperature T_1) and the infinite conducting medium (k, T_∞) in which the object is buried. Compare your result with the appropriate S value deduced from Table 3.3.

4.23 The two-dimensional wake problem of Figure 4.22 and Eqs. (4.108)–(4.110) can be solved by replacing $T(x, y)$ with the unknown $\theta(\eta)$, where

$$T(x, y) - T_\infty = \frac{q'/\rho c}{(U\alpha x)^{1/2}} \theta(\eta)$$

$$\eta = y\left(\frac{U}{\alpha x}\right)^{1/2}$$

Show that in this notation the problem statement (4.108)–(4.110) reads

$$-\frac{1}{2}(\eta\theta)' = \theta''$$

$$\theta = 0 \quad \text{at} \quad \eta = \pm\infty$$

$$\int_{-\infty}^{\infty} \theta \, d\eta = 1$$

Solve this problem for $\theta(\eta)$, and in this way prove the validity of Eq. (4.111).

4.24 One of the proposed methods for the extraction of geothermal energy is the "hot dry rock" scheme illustrated in the figure. The rock is fractured hydraulically to generate a crack through which cold high-pressure water will later be circulated. The crack space is approximated fairly well by a disc of diameter D, which has a crack thickness much smaller than D. The two surfaces of this crack become the heat transfer area A between the hot rock and the cold water that circulates through the crack. The cold stream is pumped down through one well and returned to ground level through a parallel well. Assume that the crack surface always has the same temperature as the cold water, namely, $T_c = 25\,^\circ$C. The temperature of the rock sufficiently far from the crack is $T_\infty = 200\,^\circ$C. The rock properties are approximately the same as those of granite, $k = 2.9$ W/m·K and $\alpha = 0.012$ cm²/s. The cold water circulates through the crack beginning with the time $t = 0$. The crack diameter is $D = 10$ m.

(a) It is 10 days later. Calculate in an order-of-magnitude sense the thickness of the rock layer that has been affected (cooled) by the water flow. Verify that this thickness is considerably smaller than D. Calculate the total heat transfer rate from the rock medium to the water-cooled crack.

(b) After how many years will the cooled rock layer acquire a thickness of the same order as D? Assume that enough years have passed so that the cooled region can be modeled as a sphere of diameter D and temperature 25 °C. Note that at long times the rock temperature distribution around this sphere becomes steady (time independent) [cf. Eq. (4.107)]. Calculate the total heat transfer rate that "sinks" into the cooled region and is removed steadily by the water stream.

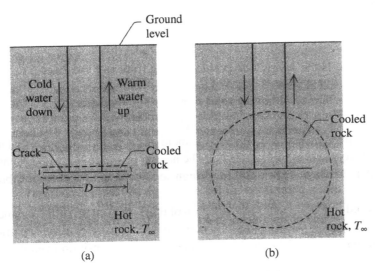

Figure P4.24 (a) After 10 days and (b) much later.

Melting and Solidification

4.25 The solid layer shown in the figure is initially at the melting point T_m. At a certain point in time ($t = 0$), the upper surface is heated to the temperature T_0, while the lower surface is cooled to the temperature T_L; in other words, $T_0 > T_m > T_L$. As the time increases, liquid is generated in a sublayer of thickness δ near the heated boundary.

Determine the relationship between δ and time t by assuming that the liquid and solid sublayers are sufficiently thin and the melting rate $d\delta/dt$ is sufficiently small so that temperature distributions in the two sublayers are nearly the same as in the case of steady unidirectional conduction. The liquid and solid phases have different thermal conductivities (k_f and k_s) and nearly the same density (ρ). Carry out this analysis in terms of the dimensionless front position and time:

$$X = \frac{\delta}{L} \qquad \tau = \frac{t}{\rho h_{sf} L^2 / [k_f (T_0 - T_m)]}$$

and the dimensionless degree of solid subcooling:

$$B = \frac{k_s (T_m - T_L)}{k_f (T_0 - T_m)}$$

What is the asymptotic position of the melting front as t increases (i.e. what is the true steady-state value of X)?

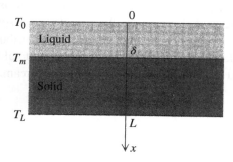

Figure P4.25

4.26 The semi-infinite solid shown in the figure is isothermal at the melting point, T_m. The $x = 0$ surface is placed in contact with a warmer convective fluid, $T_\infty > T_m$. The heat transfer coefficient h is constant and known. The liquid layer that develops in time is thin enough so that the temperature distribution across it is always linear.

Derive an expression for the liquid film thickness δ as a function of the time t and the other parameters noted in the figure. Show that in the very beginning δ increases as t and that later δ is proportional to $t^{1/2}$. Determine the order of magnitude of the thickness δ that marks the transition from the early regime to the late regime.

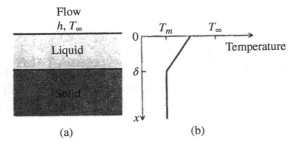

Figure P4.26

4.27 The chief unknown in the making of popsicles in the home freezer is the freezing time. The finished popsicle is a bar of flavored ice frozen around a stick. The liquid is poured into a cavity that gives the frozen bar the dimensions shown in the figure. The bar is roughly two-dimensional, that is, sufficiently wide in the direction perpendicular to the figure. It has a slight taper to allow for the expansion during freezing and to be pulled out of the cavity by means of the flat stick (tongue depressor) frozen in the middle.

Calculate the freezing time by assuming that all the liquid is originally at the freezing point. The properties of the frozen material are approximately equal to those of ice. The temperature of the cavity wall is maintained at $-5\,°C$.

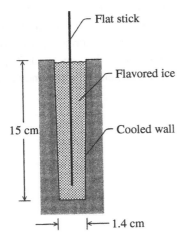

Figure P4.27

4.28 Fish is being frozen by exposure to a strong flow of $-20\,°C$ air. Each fish rests on a 30-m-long perforated conveyor belt, which exposes the fish to cold air and then dumps it frozen for packing and storage. Calculate the upper limit of the speed of the conveyor belt, that is, the speed that ensures the complete freezing of each fish. Model the fish as a 2-cm-thick slab with both sides maintained at $-20\,°C$. The original temperature of the fish is equal to the freezing point, which is $-2.2\,°C$. The approximate properties of frozen fish are $h_{sf} = 235$ kJ/kg, $c = 1.7$ kJ/kg $\cdot$ K, $k = 1$ W/m $\cdot$ K, and $\rho = 1000$ kg/m^3.

Evolutionary Design

4.29 Here, we learn that "shape" matters when one pursues performance. Meat stores and cookbooks usually recommend the cooking time in accord with the size of the piece of meat, so many hours for a piece that weighs so much. This kind of recommendation (cooking time as a function of one dimension, the size) is based on the assumption that the piece of meat is "round," like a sphere, with one dimension (the diameter).

In reality, a cut of meat has shape. The simplest way to account for shape is to view the body as a cylinder with the length L and diameter D, where $L > D$. The cooking time is the time when heating penetrates by thermal diffusion to a depth $\delta \sim (\alpha t)^{1/2}$ that matches the shorter dimension of the body of meat. That dimension is D in the two shapes recognized earlier, spherical and cylindrical.

Calculate the scales of the two cooking times, t_{sphere} and t_{cylinder}, and report the ratio $t_{\text{sphere}}/t_{\text{cylinder}}$ as a function of the shape parameter L/D. Which model would you recommend to the meat store for labeling the cooking time on pieces of meat? How grave is the cooking time error (the overestimate) if $L/D = 2$ and the cooking time is based on the spherical shape model?

4.30 According to the lumped capacitance model and Eq. (4.17), the difference between the heat source temperature (T_∞) and the temperature of the heated body (T) decreases exponentially as the time (t) increases. The initial temperature of the body is T_i. Imagine that the body is a quantity (modeled as mc and hA) that you are heating, for cooking. The desired temperature is T, and it is fixed. You would like to decrease the cooking time t to $t(1 - \varepsilon)$, where the dimensionless $\varepsilon \ll 1$ is a small fraction of the time t. According to the model (4.17), this can be accomplished by increasing the overall temperature difference $(T_\infty - T_i)$ by a small dimensionless fraction $\theta \ll 1$, to the new level $(T_\infty - T_i)$ $(1 + \theta)$. Determine the relation between the desired ε and the necessary θ, and report dimensionally the increase in $(T_\infty - T_i)$ that corresponds to the decrease in t.

5

External Forced Convection

5.1 Classification of Convection Configurations

Beginning with this chapter we turn our attention to the heat transfer mechanism called "convection," that is, the transport mechanism made possible by the motion of a fluid. What distinguishes a convection configuration from the conduction configurations is the fact that the "convective" heat transfer medium (e.g. fluid) is in motion.[1] Alternatively, conduction is the heat transfer mechanism that survives when the motion ceases throughout what once was a convective medium.

The basic problem in convection heat transfer is to determine the relationship between the heat flux through the solid wall (q'') and the temperature difference ($T_w - T_\infty$) between the wall and the fluid flow that bathes it. This problem amounts to determining the convective heat transfer coefficient h (W/m$^2 \cdot$K):

$$h = \frac{q''}{T_w - T_\infty} \tag{5.1}$$

In the infinitesimally thin fluid layer that sticks to the solid wall, the transversal heat flux q'' is due to pure conduction because the fluid in that layer is motionless. In that layer we invoke the Fourier law, Eq. (1.52), in which y is the transversal distance measured away from the solid surface (toward the fluid) and k is the thermal conductivity of the fluid. Another definition of h that follows from Eqs. (5.1) and (1.52) is Eq. (1.57):

$$h = -\frac{k}{T_w - T_\infty} \left(\frac{\partial T}{\partial y} \right)_{y=0+} \tag{5.2}$$

This expression shows that to determine h, we must first determine the temperature distribution in the thin fluid layer that coats the wall.

Equation (5.2) serves as an introduction and objective for the convection heat transfer analyses that we develop in this chapter. Recognize that the flow and heat transfer configurations that can be categorized as "convective" are extremely diverse. Figure 5.1 outlines one classification that distinguishes first between *external-flow* and *internal-flow* convection problems.

Figure 5.1 differentiates further between *forced convection* and *free convection* (*natural convection*) configurations, that is, between the origins of the motion experienced by the convective medium. The flow is "forced" when another (a different) entity pushes the fluid past the solid body against which the fluid rubs.

---

1 This important feature is stressed by the term "convection," which originates from the Latin verbs *convecto-are* and *convěho-věhěre* (to bring together, to carry into one place).

Heat Transfer: Evolution, Design and Performance, First Edition. Adrian Bejan.
© 2022 John Wiley & Sons, Inc. Published 2022 by John Wiley & Sons, Inc.
Companion website: www.wiley.com/go/bejan/heattransfer

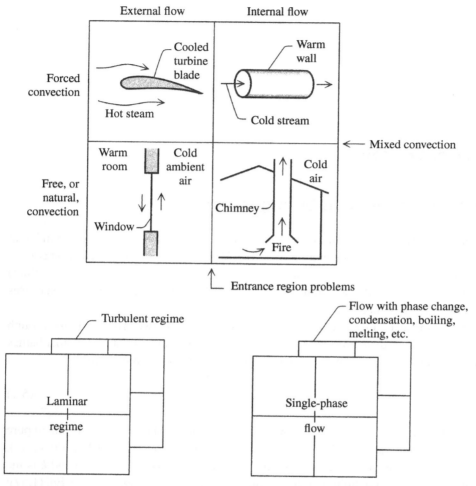

Figure 5.1 The field of convection heat transfer and the main configurations.

In internal forced convection (e.g. pipe flow), the external device is the pump or compressor. In an external forced convection configuration such as the flow parallel to a plate fin in a compact heat exchanger, the external device is the pump or compressor that pushes the entire fluid stream through the heat exchanger.

In "free" or natural convection, the fluid motion occurs "by itself," without the assistance of an external mechanism. In Chapter 7, for example, all the flows are due to the effect of relative buoyancy (lightness or heaviness) of the various regions of the flow field. It turns out that this motion-driving mechanism is intimately coupled to the heating or cooling effect experienced by the fluid because most fluids expand upon heating at constant pressure.

The ordering of convection heat transfer configurations is even more complicated than what is listed in the four-option menu shown in the upper part of Figure 5.1. At the interface between forced convection and natural convection we encounter a diversity of *mixed-convection* configurations. In a mixed-convection problem the flow is driven simultaneously by an external device (e.g. pump) and by a built-in effect such as that of buoyancy.

Furthermore, at the interface between external-flow and internal-flow configurations, there is also a group of in-between problems, which configurations are neither external flow nor internal flow. One example of this kind is in the entrance (the mouth) region of the flow into a duct (Section 6.1.1).

The convection classification scheme of Figure 5.1 continues on at least two additional planes. One of these is represented by the transition from the laminar flow regime to the turbulent flow regime, that is, from a smooth flow to one with complicated fluctuations that are dominated by eddies of several sizes (Section 5.4). The heat transfer characteristics of the flow change dramatically from the laminar regime to the turbulent regime, and vice versa.

Another direction of diversification is illustrated in the lower right-hand corner of Figure 5.1. The eight flow categories identified until now (external versus internal, forced versus free, laminar versus turbulent) were discussed in the context of a single-phase fluid. For example, a stream of nitrogen gas in a counterflow heat exchanger and the buoyancy-driven flow of air on the room side of a single-pane window remain single phase (i.e. gaseous). The same flow categories are found also when the convective medium undergoes a change of phase. In the film condensation of steam on a vertical cold surface, for example, the flow consists of two distinct regions: liquid water film attached to the wall and buoyancy-driven steam on the outer side of the water layer (Section 8.1.1).

With this diversity of configurations in mind, we will first review the three principles that govern the flow of fluid and the transport of energy through a convective medium. The purpose of the principles is to allow us to predict sequentially the following:

(a) The flow field.
(b) The temperature field – in particular, the temperature distribution near the solid wall.
(c) The heat transfer coefficient, or the relationship between wall heat flux and wall – fluid temperature difference.

In *forced convection* problems, an auxiliary result is the *friction force* between the flow and the solid wall. This result is needed to estimate the total power required by the external mechanism (e.g. pump) that forces the fluid to flow.

5.2 Basic Principles of Convection

5.2.1 Mass Conservation Equation

Consider the two-dimensional flow illustrated in Figure 5.2, where the fluid velocity at any point in the flow field is described by the velocity components u and v. These components are functions of location (x, y) and time (t). An important relation between the spatial changes in u and v is dictated by the principle of mass conservation. From thermodynamics, we recall that the conservation of mass in a control volume of instantaneous mass inventory m requires (see, for example, Ref. [1]) that

$$\frac{\partial m}{\partial t} = \sum_{in} \dot{m} - \sum_{out} \dot{m} \tag{5.3}$$

The mass conservation statement for the (u, v) flow of Figure 5.2 can be derived from the infinitesimal control volume $\Delta x \Delta y$ enlarged to the right of the figure. The left and bottom faces of the control volume are inlet ports in the sense of Eq. (5.3), whereas the top and right faces are outlet ports. For the $\Delta x \Delta y$ system,

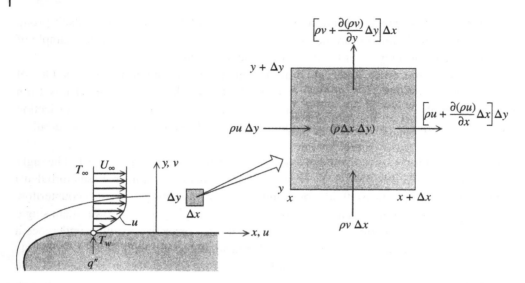

Figure 5.2 The conservation of mass in an infinitesimal control volume in a two-dimensional flow field.

Eq. (5.3) states that

$$\frac{\partial}{\partial t}(\rho \Delta x\,\Delta y) = \rho u\,\Delta y + \rho u\,\Delta x - \left[\rho u + \frac{\partial(\rho u)}{\partial x}\Delta x\right]\Delta y - \left[\rho v + \frac{\partial(\rho v)}{\partial y}\Delta y\right]\Delta x \qquad (5.4)$$

where ρ is the local density of the fluid. Dividing Eq. (5.4) by size $\Delta x \Delta y$, we obtain

$$\frac{\partial \rho}{\partial t} + \frac{\partial(\rho u)}{\partial x} + \frac{\partial(\rho v)}{\partial y} = 0 \qquad (5.5)$$

Of particular interest are the convection problems in which the fluid density may be *approximated* by a constant (ρ). According to the *nearly constant-density model*, the spatial and temporal variations in density are neglected relative to those of u and v, and Eq. (5.5) becomes

$$\frac{\partial u}{\partial x} + \frac{\partial v}{\partial y} = 0 \qquad (5.6)$$

This is the *mass conservation* equation for two-dimensional constant-density flow in Cartesian coordinates. The same result is also known as the *continuity equation*, because of the continuity of mass during its flow through the control volume $\Delta x \Delta y$. The three-dimensional counterpart of this equation is listed in Table 5.1. The corresponding equations for constant-density flow in cylindrical and spherical coordinates are listed in Tables 5.2 and 5.3.

5.2.2 Momentum Equations

The next principle is the *momentum principle*, or the *momentum theorem* [1]:

$$\frac{\partial}{\partial t}(m v_n) = \sum F_n + \sum_{\text{in}} \dot{m} v_n - \sum_{\text{out}} \dot{m} v_n \qquad (5.7)$$

where n is the direction for which Eq. (5.7) is written and v_n and F_n are the projections of the fluid velocities and forces on the n direction. The product $m v_n$ is the instantaneous inventory of n-direction momentum in the

Table 5.1 The governing equations for constant-property flow in Cartesian coordinates.

Mass conservation:

$$(m) \quad \frac{\partial u}{\partial x} + \frac{\partial v}{\partial y} + \frac{\partial w}{\partial z} = 0$$

Momentum equations:

$$(M_x) \quad \rho\left(\frac{\partial u}{\partial t} + u\frac{\partial u}{\partial x} + v\frac{\partial u}{\partial y} + w\frac{\partial u}{\partial z}\right) = -\frac{\partial P}{\partial x} + \mu\left(\frac{\partial^2 u}{\partial x^2} + \frac{\partial^2 u}{\partial y^2} + \frac{\partial^2 u}{\partial z^2}\right) + X$$

$$(M_y) \quad \rho\left(\frac{\partial v}{\partial t} + u\frac{\partial v}{\partial x} + v\frac{\partial v}{\partial y} + w\frac{\partial v}{\partial z}\right) = -\frac{\partial P}{\partial y} + \mu\left(\frac{\partial^2 v}{\partial x^2} + \frac{\partial^2 v}{\partial y^2} + \frac{\partial^2 v}{\partial z^2}\right) + Y$$

$$(M_z) \quad \rho\left(\frac{\partial w}{\partial t} + u\frac{\partial w}{\partial x} + v\frac{\partial w}{\partial y} + w\frac{\partial w}{\partial z}\right) = -\frac{\partial P}{\partial z} + \mu\left(\frac{\partial^2 w}{\partial x^2} + \frac{\partial^2 w}{\partial y^2} + \frac{\partial^2 w}{\partial z^2}\right) + Z$$

Energy equation:

$$(E) \quad \rho c_P\left(\frac{\partial T}{\partial t} + u\frac{\partial T}{\partial x} + v\frac{\partial T}{\partial y} + w\frac{\partial T}{\partial z}\right) = k\left(\frac{\partial^2 T}{\partial x^2} + \frac{\partial^2 T}{\partial y^2} + \frac{\partial^2 T}{\partial z^2}\right) + \dot{q} + \mu\,\Phi$$

where

$$\Phi = 2\left[\left(\frac{\partial u}{\partial x}\right)^2 + \left(\frac{\partial v}{\partial y}\right)^2 + \left(\frac{\partial w}{\partial z}\right)^2\right] + \left(\frac{\partial u}{\partial y} + \frac{\partial v}{\partial x}\right)^2 + \left(\frac{\partial v}{\partial z} + \frac{\partial w}{\partial y}\right)^2 + \left(\frac{\partial w}{\partial x} + \frac{\partial u}{\partial z}\right)^2$$

---

control volume. The terms F_n account for the forces that act on the control volume. The terms $\dot{m}v_n$ represent the flow of n-direction momentum into and out of the control volume. In summary, Eq. (5.7) is the control volume representation of Newton's second law of motion.

Consider the implications of Eq. (5.7) with respect to the flow through the control volume $\Delta x \Delta y$ enlarged in Figure 5.3. There are two directions, x and y, in which we can invoke Eq. (5.7). We begin with the x direction.

The upper drawing shows the impact and reaction due to the flow of momentum into and out of the control volume. The x momentum that enters through the left side of the control volume is $\rho u^2 \Delta y$, where $\rho u \Delta y$ is the stream flowrate and u is the velocity of that stream in the x direction. Entering through the bottom side of the control volume is a stream of flowrate $\rho v \Delta x$, and the x velocity of the fluid carried by that stream is u. Therefore, the x-momentum flowrate that enters through the bottom side is $(\rho v \Delta x)u$.

The bottom right of Figure 5.3 shows the forces that act on the control volume, namely, the forces due to the normal stress σ_{xx}, the tangential stress τ_{xy}, and the per-unit-volume body force in the x direction, X.

Table 5.2 The governing equations for constant-property flow in cylindrical coordinates.

Mass conservation:

$$(m) \quad \frac{\partial v_r}{\partial r} + \frac{v_r}{r} + \frac{1}{r}\frac{\partial v_\theta}{\partial \theta} + \frac{\partial v_z}{\partial z} = 0$$

Momentum equations:

$$(M_r) \quad \rho\left(\frac{\partial v_r}{\partial t} + v_r\frac{\partial v_r}{\partial r} + \frac{v_\theta}{r}\frac{\partial v_r}{\partial \theta} - \frac{v_\theta^2}{r} + v_z\frac{\partial v_r}{\partial z}\right) = -\frac{\partial P}{\partial r} + \mu\left(\frac{\partial^2 v_r}{\partial r^2} + \frac{1}{r}\frac{\partial v_r}{\partial r} - \frac{v_r}{r^2} + \frac{1}{r^2}\frac{\partial^2 v_r}{\partial \theta^2} - \frac{2}{r^2}\frac{\partial v_\theta}{\partial \theta} + \frac{\partial^2 v_r}{\partial z^2}\right) + F_r$$

$$(M_\theta) \quad \rho\left(\frac{\partial v_\theta}{\partial t} + v_r\frac{\partial v_\theta}{\partial r} + \frac{v_\theta}{r}\frac{\partial v_\theta}{\partial \theta} + \frac{v_r v_\theta}{r} + v_z\frac{\partial v_\theta}{\partial z}\right) = -\frac{1}{r}\frac{\partial P}{\partial \theta} + \mu\left(\frac{\partial^2 v_\theta}{\partial r^2} + \frac{1}{r}\frac{\partial v_\theta}{\partial r} - \frac{v_\theta}{r^2} + \frac{1}{r^2}\frac{\partial^2 v_\theta}{\partial \theta^2} + \frac{2}{r^2}\frac{\partial v_r}{\partial \theta} + \frac{\partial^2 v_\theta}{\partial z^2}\right)$$
$$+ F_\theta$$

$$(M_z) \quad \rho\left(\frac{\partial v_z}{\partial t} + v_r\frac{\partial v_z}{\partial r} + \frac{v_\theta}{r}\frac{\partial v_z}{\partial \theta} + v_z\frac{\partial v_z}{\partial z}\right) = -\frac{\partial P}{\partial z} + \mu\left(\frac{\partial^2 v_z}{\partial r^2} + \frac{1}{r}\frac{\partial v_z}{\partial r} + \frac{1}{r^2}\frac{\partial^2 v_z}{\partial \theta^2} + \frac{\partial^2 v_z}{\partial z^2}\right) + F_z$$

Energy equation:

$$(E) \quad \rho c_P\left(\frac{\partial T}{\partial t} + v_r\frac{\partial T}{\partial r} + \frac{v_\theta}{r}\frac{\partial T}{\partial \theta} + v_z\frac{\partial T}{\partial z}\right) = k\left[\frac{1}{r}\frac{\partial}{\partial r}\left(r\frac{\partial T}{\partial r}\right) + \frac{1}{r^2}\frac{\partial^2 T}{\partial \theta^2} + \frac{\partial^2 T}{\partial z^2}\right] + \dot{q} + \mu\,\Phi$$

where

$$\Phi = 2\left[\left(\frac{\partial v_r}{\partial r}\right)^2 + \left(\frac{1}{r}\frac{\partial v_\theta}{\partial \theta} + \frac{v_r}{r}\right)^2 + \left(\frac{\partial v_z}{\partial z}\right)^2\right] + \left(\frac{\partial v_\theta}{\partial r} - \frac{v_\theta}{r} + \frac{1}{r}\frac{\partial v_r}{\partial \theta}\right)^2 + \left(\frac{1}{r}\frac{\partial v_z}{\partial \theta} + \frac{\partial v_\theta}{\partial z}\right)^2 + \left(\frac{\partial v_r}{\partial z} + \frac{\partial v_z}{\partial r}\right)^2$$

Note that the forces due to the normal and tangential stresses act on, and are proportional to, the respective surfaces of the control volume. The body force associated with X is proportional to the volume of the control volume. An example of body force is the buoyancy effect that serves as driving mechanism in all the natural convection configurations of Chapter 7.

Adding the "arrows" indicated on the two frames of Figure 5.3, i.e. substituting the indicated terms in their appropriate places in the momentum theorem, Eq. (5.7), taking into account Eq. (5.6), and dividing the resulting equation by $\Delta x \Delta y$, we obtain [2]

$$\rho\left(\frac{\partial u}{\partial t} + u\frac{\partial u}{\partial x} + v\frac{\partial u}{\partial y}\right) = -\frac{\partial \sigma_{xx}}{\partial x} + \frac{\partial \tau_{xy}}{\partial y} + X \tag{5.8}$$

Table 5.3 The governing equations for constant-property flow in spherical coordinates.

Mass conservation:

$$(m) \quad \frac{1}{r}\frac{\partial}{\partial r}(r^2 v_r) + \frac{1}{\sin\phi}\frac{\partial}{\partial\phi}(v_\phi \sin\phi) + \frac{1}{\sin\phi}\frac{\partial v_\theta}{\partial\theta} = 0$$

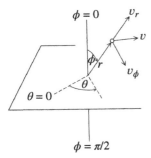

Momentum equations:

$$(M_r) \quad \rho\left(\frac{Dv_r}{Dt} - \frac{v_\phi^2 + v_\theta^2}{r}\right) = -\frac{\partial P}{\partial r} + \mu\left(\nabla^2 v_r - \frac{2v_r}{r^2} - \frac{2}{r^2}\frac{\partial v_\phi}{\partial\phi} - \frac{2v_\phi \cot\phi}{r^2} - \frac{2}{r^2\sin\phi}\frac{\partial v_\theta}{\partial\theta}\right) + F_r$$

$$(M_\phi) \quad \rho\left(\frac{Dv_\phi}{Dt} + \frac{v_r v_\phi}{r} - \frac{v_\theta^2 \cot\phi}{r}\right) = -\frac{1}{r}\frac{\partial P}{\partial\phi} + \mu\left(\nabla^2 v_\phi - \frac{2}{r^2}\frac{\partial v_r}{\partial\phi} - \frac{v_\phi}{r^2\sin^2\phi} - \frac{2\cos\phi}{r^2\sin^2\phi}\frac{\partial v_\theta}{\partial\theta}\right) + F_\phi$$

$$(M_\theta) \quad \rho\left(\frac{Dv_\theta}{Dt} + \frac{v_\theta v_r}{r} + \frac{v_\phi v_\theta \cot\phi}{r}\right) = -\frac{1}{r\sin\phi}\frac{\partial P}{\partial\theta} + \mu\left(\nabla^2 v_\theta - \frac{v_\theta}{r^2\sin^2\phi} + \frac{2}{r^2\sin\phi}\frac{\partial v_r}{\partial\theta} + \frac{2\cos\phi}{r^2\sin^2\phi}\frac{\partial v_\phi}{\partial\theta}\right) + F_\theta$$

where

$$\frac{D}{Dt} = \frac{\partial}{\partial t} + v_r\frac{\partial}{\partial r} + \frac{v_\phi}{r}\frac{\partial}{\partial\phi} + \frac{v_\theta}{r\sin\phi}\frac{\partial}{\partial\theta}$$

$$\nabla^2 = \frac{1}{r^2}\frac{\partial}{\partial r}\left(r^2\frac{\partial}{\partial r}\right) + \frac{1}{r^2\sin\phi}\frac{\partial}{\partial\phi}\left(\sin\phi\frac{\partial}{\partial\phi}\right) + \frac{1}{r^2\sin^2\phi}\frac{\partial^2}{\partial\theta^2}$$

Energy equation:

$$(E) \quad \rho c_P\left(\frac{\partial T}{\partial t} + v_r\frac{\partial T}{\partial r} + \frac{v_\phi}{r}\frac{\partial T}{\partial\phi} + \frac{v_\theta}{r\sin\phi}\frac{\partial T}{\partial\theta}\right) = k\left[\frac{1}{r^2}\frac{\partial}{\partial r}\left(r^2\frac{\partial T}{\partial r}\right) + \frac{1}{r^2\sin\phi}\frac{\partial}{\partial\phi}\left(\sin\phi\frac{\partial T}{\partial\phi}\right) + \frac{1}{r^2\sin^2\phi}\frac{\partial^2 T}{\partial\theta^2}\right]$$

$$+ \dot{q} + \mu\,\Phi$$

where

$$\Phi = 2\left[\left(\frac{\partial v_r}{\partial r}\right)^2 + \left(\frac{1}{r}\frac{\partial v_\phi}{\partial\phi} + \frac{v_r}{r}\right)^2 + \left(\frac{1}{r\sin\phi}\frac{\partial v_\theta}{\partial\theta} + \frac{v_r}{r} + \frac{v_\phi \cot\phi}{r}\right)^2\right] + \left[r\frac{\partial}{\partial r}\left(\frac{v_\phi}{r}\right) + \frac{1}{r}\frac{\partial v_r}{\partial\phi}\right]^2$$

$$+ \left[\frac{\sin\phi}{r}\frac{\partial}{\partial\phi}\left(\frac{v_\theta}{r\sin\phi}\right) + \frac{1}{r\sin\phi}\frac{\partial v_\theta}{\partial\theta}\right]^2 + \left[\frac{1}{r\sin\phi}\frac{\partial v_r}{\partial\theta} + r\frac{\partial}{\partial r}\left(\frac{v_\theta}{r}\right)\right]^2$$

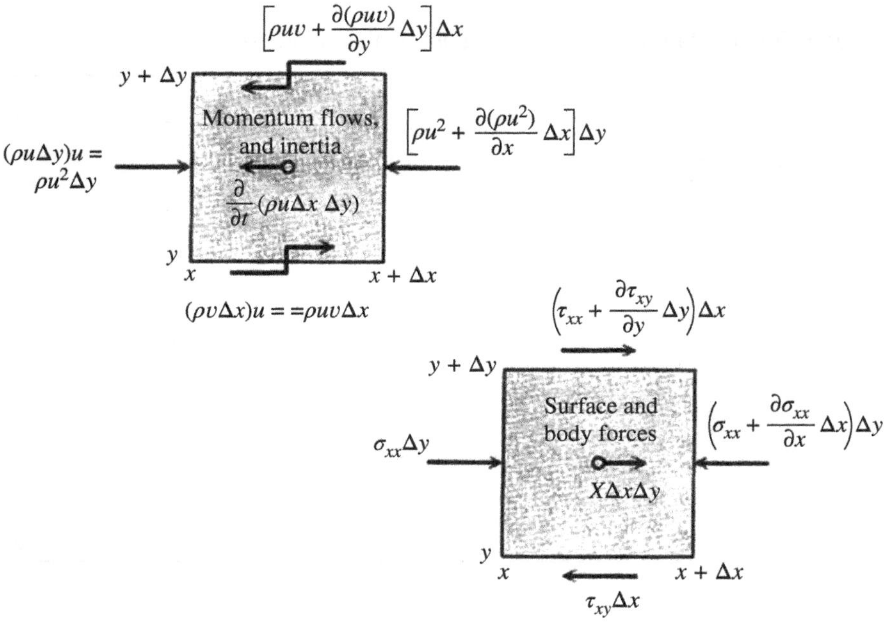

Figure 5.3 The development of the momentum equation for the *x* direction: the effects of momentum flows and inertia (top) and the surface and body forces (bottom).

This equation holds at every point (x, y) in the flow field. The three terms on the left side represent the effects of *x*-momentum accumulation and flow through the point (x, y). The three terms on the right side represent the normal, shear, and body forces.

The next step is the observation that a fluid packet can be deformed without resistance if, regardless of size, the deformation occurs infinitely slowly. The fluid packet poses an increasing resistance as it is being deformed faster. It is said that a fluid packet offers no resistance to a finite-size (infinitely slow) change of shape and that, instead, it resists the *time rate* of the given change of shape. This observation is the basis for a set of constitutive relations that we borrow from the field of fluid mechanics [2]:

$$\sigma_{xx} - P = -2\mu\frac{\partial u}{\partial x} + \frac{2}{3}\mu\left(\frac{\partial u}{\partial x} + \frac{\partial v}{\partial y}\right) \tag{5.9}$$

$$\tau_{xy} = \mu\left(\frac{\partial u}{\partial y} + \frac{\partial v}{\partial x}\right) \tag{5.10}$$

These equations relate the stresses to the local velocity gradients. On the left side of Eq. (5.9), the group $\sigma_{xx} - P$ is the "excess" normal stress, that is, the normal stress above the level of the local pressure P.

The coefficient μ is the *viscosity* of the fluid. The value of this physical property can be measured using Eq. (5.10); its units are $N \cdot s/m^2$ or $kg/(m \cdot s)$. Equation (5.10) states that the shear stress is proportional to the rate of angular deformation of the fluid packet: fluids that obey this proportionality are called *Newtonian*. All the analyses reported in the convection chapters of this course contain Eq. (5.10) as a basic assumption; consequently, all the fluids to which these analyses refer are Newtonian fluids.

The viscosity μ varies in general with the local temperature of the fluid. Several examples of the $\mu(T)$ dependence are exhibited in Appendixes C and D. When the temperature range spanned by the flow (e.g.

the difference $T_w - T_\infty$) is sufficiently small, the viscosity can be treated as a constant. In this limit the substitution of Eqs. (5.9) and (5.10) into Eq. (5.8) and the use of Eq. (5.6) lead to

$$\rho\left(\frac{\partial u}{\partial t} + u\frac{\partial u}{\partial x} + v\frac{\partial u}{\partial y}\right) = -\frac{\partial P}{\partial x} + \mu\left(\frac{\partial^2 u}{\partial x^2} + \frac{\partial^2 u}{\partial y^2}\right) + X \tag{5.11}$$

This is the *momentum equation* for the flow of a constant property fluid in the x direction. The momentum equations are known also as the *Navier–Stokes equations*, first presented in 1822 (published in 1827 [3]) by the French engineer Claude-Louis-Marie-Henri Navier (1785–1836). The properties modeled as constant are the density and the viscosity. The three-dimensional counterpart of Eq. (5.11) is listed in Table 5.1. The same table also shows the corresponding forms of the constant property momentum equations in the y and z directions of the Cartesian system.

The momentum equations for flows in cylindrical coordinates are presented in Table 5.2, where v_r, v_θ, and v_z are the velocity components and F_r, F_θ, and F_z are the components of the body force per unit volume. The equations for spherical coordinates are presented in Table 5.3, where the components of velocity and body force per unit volume are v_r, v_ϕ, and v_θ and F_r, F_ϕ, and F_θ, respectively. The D/Dt and ∇^2 notations are explained in Table 5.3.

5.2.3 Energy Equation

In forced convection configurations, the mass and momentum equations are sufficient for determining the flow field, that is, the velocity distribution through the fluid. The fluid temperature distribution can be determined as an afterthought, by using the just-determined velocity distribution and a new equation (conservation principle): the first law of thermodynamics.

The derivation of the differential equation that accounts for the first law of thermodynamics is a lengthy analysis that can be found in a more advanced treatment of convection (e.g. Ref. [2]). In this course we review only the structure of this derivation (Figure 5.4) and the most useful form of the resulting equation.

With reference to a control volume with inlet and outlet ports, heat and work transfer, and time-dependent energy inventory, the first law requires [1]

$$\frac{\partial}{\partial t}(me) = \sum_{in}\dot{m}e - \sum_{out}\dot{m}e + \sum_i q_i - \sum_j w_j \tag{5.12}$$

The left side accounts for the accumulation of energy inside the control volume, in which e is the local specific *internal* energy. The common thermodynamics notation for this property is u. In convection the symbol u is reserved for the velocity component in the x direction. On the right side of Eq. (5.12), the first two sums account for the inflows and outflows of energy that are associated with the mass flows across the control surface. The third sum represents the net rate of heat transfer into the control volume, in such a way that i indicates the boundary point (or portion) that is crossed by the individual heat current q_i. Finally, the fourth sum accounts for the net rate of work transfer delivered by the control volume to its environment. Note that in writing Eq. (5.12) we are neglecting the contributions of kinetic energy and gravitational potential energy to the local specific energy of the fluid. This assumption is appropriate in the convection heat transfer problems addressed in this course.

It remains to apply Eq. (5.12) to the infinitesimal control volume $\Delta x \Delta y$ of Figure 5.2. This amounts to identifying the appropriate expressions that must be substituted in place of the terms of Eq. (5.12). The four sums listed on the right side of Eq. (5.12) represent, in order, the effects of energy inflow (via fluid flow), energy outflow, heat transfer rate into the control volume, and work transfer rate out of the control volume.

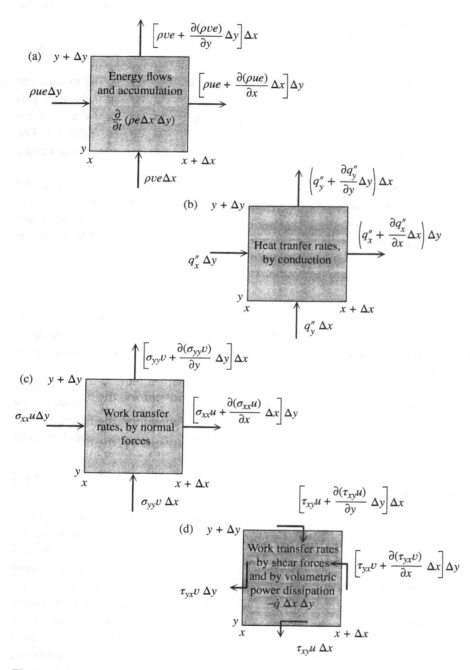

Figure 5.4 The four contributions to the statement of energy conservation in the control volume $\Delta x \Delta y$ of Figure 5.2.

The appropriate substitutions for Eq. (5.12) are identified in four stages in Figure 5.4. The top illustration shows the effects due to energy accumulation, energy inflow, and energy outflow. This first drawing takes care of the left side of Eq. (5.12) and the first two summations listed on the right side. The second drawing identifies the heat transfer rate interactions between the surrounding fluid and the system. For example, q_x'' is the conduction heat flux oriented in the x direction at the location (x, y) in the fluid. The corresponding heat flux in the next plane in the x direction can be approximated using a Taylor expansion:

$$q_{x+\Delta x}'' = q_x'' + \frac{\partial q_x''}{\partial x} \Delta x \tag{5.13}$$

The third drawing shows the work transfer rate interactions associated with the *normal* stresses (σ_{xx} and σ_{yy}) and the piston-like movement of fluid across the faces of the system. For example, entering through the left side of the control volume is the work transfer rate associated with the normal stress σ_{xx} and the volume swept per unit time, $u\Delta y$. This contribution represents work done by the surroundings on the $\Delta x\Delta y$ system; this is why the $\sigma_{xx} u\Delta y$ arrow points toward the system.

The bottom drawing of Figure 5.4 presents the work transfer rate contributed by the shear stresses that "work" along the four faces and the contribution made by an internal (volumetric) mechanism of power dissipation (e.g. Joule heating). The latter is listed with the minus sign $-\dot{q}\Delta x \Delta y$ because it represents power deposited by an external system into the control volume. The factor $\dot{q}$ is the volumetric rate of heat generation.

When we substitute the quantities of Figure 5.4 into the first law, Eq. (5.12), divide the resulting equation by $\Delta x\Delta y$, and invoke the constant-density model, we obtain [2]

$$\rho\left(\frac{\partial e}{\partial t} + u\frac{\partial e}{\partial x} + v\frac{\partial e}{\partial y}\right) = -\frac{\partial q_x''}{\partial x} - \frac{\partial q_y''}{\partial y} + \dot{q} + \mu\Phi \tag{5.14}$$

where Φ is shorthand for the viscous dissipation function for a constant-ρ fluid:

$$\Phi = \left(\frac{\partial u}{\partial y} + \frac{\partial v}{\partial x}\right)^2 + 2\left[\left(\frac{\partial u}{\partial x}\right)^2 + \left(\frac{\partial v}{\partial y}\right)^2\right] \tag{5.15}$$

The left side of Eq. (5.14) can be restated in terms of the local temperature T and its gradients by executing two steps. First, the specific internal energy e can be eliminated based on the definition of specific enthalpy (i):[2]

$$e = i - \frac{1}{\rho}P \tag{5.16}$$

The specific enthalpy derivatives that result from this operation can be eliminated using the general differential form for a single-phase fluid (see, for example, Ref. [1]):

$$di = c_P \, dT + (1 - \beta T)\frac{1}{\rho}dP \tag{5.17}$$

where c_P is not necessarily constant and where β is the coefficient of volumetric thermal expansion, also called volume expansivity:

$$\beta = -\frac{1}{\rho}\left(\frac{\partial P}{\partial T}\right)_P \tag{5.18}$$

---

2 The thermodynamics notation for specific enthalpy is h. In the field of heat transfer, the h symbol is reserved for the convective heat transfer coefficient.

In the end, if ρ continues to be treated as nearly constant, Eq. (5.14) becomes

$$\rho \left(\frac{\partial e}{\partial t} + u \frac{\partial e}{\partial x} + v \frac{\partial e}{\partial y} \right) = \rho c_P \left(\frac{\partial T}{\partial t} + u \frac{\partial T}{\partial x} + v \frac{\partial T}{\partial y} \right) - \beta T \left(\frac{\partial P}{\partial t} + u \frac{\partial P}{\partial x} + v \frac{\partial P}{\partial y} \right) \tag{5.19}$$

The contribution made by the pressure terms is negligible relative to that of the temperature terms; therefore, we abandon the P terms at this stage. This decision leaves the temperature-derivative terms on the left side of Eq. (5.14). On the right side of the equation, the heat fluxes q_x'' and q_y'' can be replaced with the respective Fourier laws, Eqs. (1.32), in which k is the thermal conductivity of the isotropic fluid.

In summary, the first law, Eq. (5.14), assumes the differential equation form

$$\rho c_P \left(\frac{\partial T}{\partial t} + u \frac{\partial T}{\partial x} + v \frac{\partial T}{\partial y} \right) = \frac{\partial}{\partial x} \left(k \frac{\partial T}{\partial x} \right) + \frac{\partial}{\partial y} \left(k \frac{\partial T}{\partial y} \right) + \dot{q} + \mu \, \Phi \tag{5.20}$$

This result constitutes the *energy equation* for the two-dimensional flow of a fluid with nearly constant properties (ρ, μ), variable conductivity (k), and prescribed internal heat generation ($\dot{q}$). The three-dimensional counterpart of Eq. (5.20) in Cartesian coordinates is listed as Eq. (E) in Table 5.1. Compare this equation with the conduction equation for the same coordinate system, Eq. (1.33); when the fluid comes to rest ($u = v = 0$, $\Phi = 0$), the energy equation reduces to the equation for conduction (thermal diffusion).

Tables 5.2 and 5.3 also contain the corresponding energy equations for constant property (ρ, μ, k) convection in cylindrical and spherical coordinates. In almost all the convection problems considered in this course, the $\dot{q}$ term is equal to zero, and the viscous dissipation effect ($\mu\Phi$) is negligible.

The constant-ρ approximation that led to Eq. (5.6) differs conceptually from the "incompressible substance model" of thermodynamics. The latter is considerably more restrictive than the "nearly constant" density model, Eq. (5.6). For example, a compressible substance such as air can flow in such a way that Eq. (5.6) is a very good approximation of Eq. (5.5).

For the restrictive class of fluids that are incompressible, the specific heat at constant pressure c_P can be replaced by the lone specific heat of the fluid, c, on the left side of Eq. (5.20). Water, liquid mercury, and engine oil are examples of fluids for which this substitution is justified. There are even convection problems in which the moving media are solid (e.g. a roller and its substrate, in the zone of elastic contact). In such cases the $c_P = c$ substitution is permissible; after all, this is why in the limit of pure conduction in a solid, the energy Eq. (1.33) has c as a factor in the energy–inertia term.

The specific heat that does not belong on the left side of Eq. (5.20) is the specific heat at constant volume, c_v. This observation is important because Fourier [4, 5], and later Poisson [6], who were the first to follow Navier [3] to derive the energy equation for a convective flow, wrote c on the left side of Eq. (5.20). They made this choice because their analyses were aimed specifically at *incompressible* fluids (liquids), for which c happens to have nearly the same value as c_v. Because of this choice, they did not have to account for the PdV-type work done by the fluid packet as it expands or contracts in the flow field. In the modern era, however, the use of c_v instead of c_P is an error (a misreading of the pioneering work) that continues to propagate through the convection literature.

The caloric origins of the science of convection heat transfer are also responsible for the "thermal energy equation" label that some prefer to attach to Eq. (5.20). This terminology is sometimes used to stress (incorrectly) the conservation of "thermal" energy in Eq. (5.20) as something distinct from "mechanical and thermal" energy. In thermodynamics, however, this distinction disappeared as soon as the first law of thermodynamics was enunciated, that is, as soon as the thermodynamic property "energy" was defined [1].

Equation (5.20) represents the first law of thermodynamics. This law proclaims the conservation of the sum of energy change (the property) and energy interactions (heat transfer and work transfer), subject to the simplifying assumptions that have been made en route to Eq. (5.20). The impression that mechanical effects (e.g. work transfer) are absent from Eq. (5.20) is wrong. The presence of c_P on the left side of the equation is a sign that each fluid packet expands or contracts (i.e. it does PdV-type work) as it rides on the flow. The terms $\dot{q}$ and $\mu\Phi$ are work transfer rate terms also.

5.3 Laminar Boundary Layer

5.3.1 Velocity Boundary Layer

Consider the flow and heat transfer near a flat wall, Figure 5.5. Sufficiently far away from the $y = 0$ plane, the fluid is isothermal (T_∞) and isobaric (P_∞) and flows to the right with a uniform velocity (U_∞). We assume that the leading edge of the flat plate is sharp enough so that the collision between the U_∞ flow and the leading edge does not disturb the parallel linear motion of every fluid packet in the vicinity of the $y = 0$ plane.

The flow field is steady, two-dimensional, and *laminar*, that is, a sandwich of smooth fluid blades of infinitesimal thickness, slipping past one another with different speeds. The word "laminar" originates from the Latin noun *lamina*, which means a thin piece of metal (blade), or a thin sheet of wood. In Eq. (5.85) we shall discover that the flow pattern is indeed laminar at a given location x if the free-stream velocity U_∞ is sufficiently small, or, when U_∞ is given, if the downstream length x does not exceed a critical value.

According to Eqs. (m), (M_x), and (M_y) of *Table 5.1*, the steady two-dimensional flow in Figure 5.5 is described by the velocity components $u(x, y)$ and $v(x, y)$ that, along with the pressure $P(x, y)$, must satisfy the system:

$$(m) \quad \frac{\partial u}{\partial x} + \frac{\partial v}{\partial y} = 0 \tag{5.21}$$

$$(M_x) \quad u\frac{\partial u}{\partial x} + v\frac{\partial u}{\partial y} = -\frac{1}{\rho}\frac{\partial P}{\partial x} + v\left(\frac{\partial^2 u}{\partial x^2} + \frac{\partial^2 y}{\partial y^2}\right) \tag{5.22}$$

$$(M_y) \quad u\frac{\partial v}{\partial x} + v\frac{\partial v}{\partial y} = -\frac{1}{\rho}\frac{\partial P}{\partial y} + v\left(\frac{\partial^2 v}{\partial x^2} + \frac{\partial^2 v}{\partial y^2}\right) \tag{5.23}$$

The factor v (m²/s) is the kinematic viscosity of the fluid:

$$v = \frac{\mu}{\rho} \tag{5.24}$$

This coefficient is a constant, because both ρ and μ are modeled as constant.

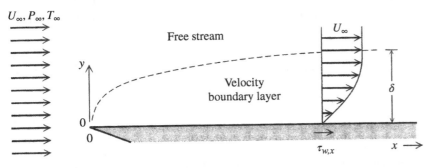

Figure 5.5 Velocity boundary layer in laminar flow near a plane wall.

The momentum equations that apply to the present flow are considerably simpler than Eqs. (5.22) and (5.23). Let δ be the characteristic size (length scale) of the transversal distance over which the horizontal velocity $u(x, y)$ changes from the free-stream value U_∞ to $u = 0$ at the solid wall. The order-of-magnitude definition of δ is

$$\frac{\partial u}{\partial y} \sim \frac{U_\infty}{\delta} \tag{5.25}$$

The velocity profile "thickness" δ is a function of the downstream position x. Let us assume that δ is negligible relative to x:

$$\delta \ll x \tag{5.26}$$

in other words, the flow region of length x and thickness δ (the region between the wall and the undisturbed free stream) is *slender*. Following Prandtl [7],[3] in convection heat transfer, a slender region of this kind is called a *boundary layer*.

The slenderness inequality (5.26) allows us to simplify the x-momentum Eq. (5.22) in two ways. First, the term $\partial^2 u/\partial x^2$ is neglected in favor of $\partial^2 u/\partial y^2$, which is much larger. The scales of these terms are (review the method of scale analysis, Section 3.2.2)

$$\frac{\partial^2 u}{\partial x^2} \sim \frac{\left(\frac{\partial u}{\partial x}\right)_{x=x} - \left(\frac{\partial u}{\partial x}\right)_{x=0}}{x - 0} \sim \frac{\frac{-U_\infty}{x} - 0}{x - 0} \sim \frac{-U_\infty}{x^2} \tag{5.27}$$

$$\frac{\partial^2 u}{\partial y^2} \sim \frac{\left(\frac{\partial u}{\partial y}\right)_{y=\delta} - \left(\frac{\partial u}{\partial y}\right)_{y=0}}{\delta - 0} \sim \frac{0 - \frac{U_\infty}{\delta}}{\delta - 0} \sim \frac{-U_\infty}{\delta^2} \tag{5.28}$$

Dividing the absolute values of these scales, we find that

$$\frac{|\partial^2 u/\partial x^2|}{|\partial^2 u/\partial y^2|} \sim \left(\frac{\delta}{x}\right)^2 \ll 1 \tag{5.29}$$

In conclusion, the slenderness inequality (5.26) implies that the $\partial^2 u/\partial x^2$ term can be neglected inside the boundary layer.

The second simplification of the x-momentum equation stems from the fact that the pressure P does not vary appreciably across the thin boundary layer:

$$P(x, y) \cong P(x) \tag{5.30}$$

The validity of Eq. (5.30) can be demonstrated based on the y-momentum equation, Eq. (5.23), and the slenderness (5.26) [2]. In other words, inside the boundary layer region, the y-momentum equation reduces to Eq. (5.30).

In the present problem the pressure inside the boundary layer, $P(x)$, must be the same as the pressure at the outer edge of the boundary layer, P_∞, which is constant. This means that the pressure gradient term in Eq. (5.22) is zero:

$$\frac{\partial P}{\partial x} \cong \frac{dP}{dx} = \frac{dP_\infty}{dx} = 0 \tag{5.31}$$

---

3 Ludwig Prandtl (1875–1953) was a mechanical engineer and a professor of applied mechanics at Göttingen University, Germany. Arguably the founder of modern fluid mechanics, he is best known for his boundary layer theory, wing theory, development of the wind tunnel, and research on supersonic flow and turbulence.

Equations (5.29) and (5.31) indicate the simplifications that result from the slenderness of the boundary layer region. Combining Eqs. (5.22), (5.29), and (5.31), we obtain the lone momentum equation of the flat-plate boundary layer flow:

$$\underbrace{u\frac{\partial u}{\partial x} + v\frac{\partial u}{\partial y}}_{\text{Inertia}} = \underbrace{\nu\frac{\partial^2 u}{\partial y^2}}_{\text{Friction}} \tag{5.32}$$

Equation (5.32) is the only momentum equation because its derivation is based on both (M_x) and (M_y). In qualitative terms the momentum Eq. (5.32) expresses a balance between the inertia (deceleration) of a fluid packet and the friction effect (restraining force) transmitted by the wall to the fluid packet via viscous diffusion in the y direction. Inertia equals friction.

The flow distribution (u, v) can be determined by solving the mass and momentum Eqs. (5.21) and (5.32). Term by term, the order-of-magnitude equations are

$$(m) \quad \frac{U_\infty}{x} \sim \frac{v}{\delta} \tag{5.33}$$

$$(M) \quad U_\infty\frac{U_\infty}{x}, v\frac{U_\infty}{\delta} \sim \nu\frac{U_\infty}{\delta^2} \tag{5.34}$$

Note first that the two inertia scales on the left side of Eq. (5.34) are the scale (cf. Eq. (5.33)); therefore, the momentum balance between inertia and friction is simply

$$\underbrace{U_\infty\frac{U_\infty}{x}}_{\text{Inertia}} \sim \underbrace{\nu\frac{U_\infty}{\delta^2}}_{\text{Friction}} \tag{5.35}$$

Equations (5.33) and (5.35) deliver the two unknown scales of the boundary layer flow:

$$\delta \sim \left(\frac{vx}{U_\infty}\right)^{1/2} \tag{5.36}$$

$$v \sim U_\infty\left(\frac{U_\infty x}{\nu}\right)^{-1/2} \tag{5.37}$$

This approximate solution contains an order-of-magnitude answer to the question about the wall shear stress at the location x:

$$\tau_{w,x} = \mu\left(\frac{\partial u}{\partial y}\right)_{y=0} \sim \mu\frac{U_\infty}{\delta} \sim \rho U_\infty^2\left(\frac{U_\infty x}{\nu}\right)^{-1/2} \tag{5.38}$$

The dimensionless version of the local wall shear stress is the local skin friction coefficient $C_{f,x}$,

$$C_{f,x} = \frac{\tau_{w,x}}{\frac{1}{2}\rho U_\infty^2} \tag{5.39}$$

Combining Eqs. (5.38) and (5.39) and neglecting[4] the numerical factor $\frac{1}{2}$, we conclude that the local skin friction coefficient is approximately

$$C_{f,x} \sim Re_x^{-1/2} \tag{5.40}$$

---

4 Factors of order 1 are neglected in an order-of-magnitude analysis.

where Re_x is the *Reynolds number*[5] based on the downstream distance x:

$$Re_x = \frac{U_\infty x}{\nu} \tag{5.41}$$

These results hold when the boundary layer is slender. Combining Eq. (5.26) with the δ scale (5.36), we find that the boundary layer is slender when

$$Re_x^{1/2} \gg 1 \tag{5.42}$$

This inequality teaches that the boundary layer theory results hold for positions x sufficiently far downstream from the leading edge of the wall, so that $Re_x^{1/2} \gg 1$. The theory breaks down at small enough x values (called the "tip region") where $Re_x^{1/2} \lesssim 1$.

The scale-analysis solution for wall friction, Eq. (5.40), can be improved. One alternative is the integral method outlined in Section 3.2. The step-by-step integral analysis of the δ-thin flow region forms the subject of Problem 5.11; the more refined $C_{f,x}$ formulas that are recommended by that analysis are listed in the problem statement.

Students who study integral analysis may find it inconsistent that the starting condition $\delta = 0$ at $x = 0$ is necessary in the development of the solution for the function $\delta(x)$. The stipulation of a condition (any condition) at $x = 0$ appears to contradict the observation that the boundary layer theory does not hold at $x = 0$. In fact, there is no inconsistency. The boundary layer theory holds sufficiently far downstream, over that long stretch where $Re_x^{1/2} \gg 1$ and the boundary layer is slender. The theory states (predicts) that over that downstream region the thickness varies in a certain way, $\delta(x)$, which turns out to be supported very well by experiments. That "certain" variation of $\delta(x)$ is such that when extrapolated to $x = 0$ it yields $\delta = 0$. In other words, the invocation of the tip condition $\delta(0) = 0$ is an important component of the theory, as the theory attempts to predict what happens sufficiently *far* from $x = 0$. The fact that the theoretical thickness varies as $x^{1/2}$ and the wall shear stress varies as $x^{-1/2}$ does not mean that the real thickness is zero and the real shear stress is infinite at $x = 0$. They are not.

The exact solution to Eqs. (5.21) and (5.32) proceeds from the idea that, regardless of x, the u–y profiles are similar. They all start with the no-slip condition at the wall ($u = 0$) and approach $u = U_\infty$ as y takes values of order δ. The $u(x, y)$ distribution is summarized by the universal profile

$$u(x, y) = U_\infty \text{ function } \left[\frac{y}{\delta(x)} \right] \tag{5.43}$$

where $\delta(x)$ is the scale determined in Eq. (5.36). The transversal coordinate of the universal profile is the similarity variable η:

$$\eta = \frac{y}{x} Re_x^{1/2} \tag{5.44}$$

The similarity solution for the laminar boundary layer flow was carried out by Blasius [8], who replaced the unknowns $u(x, y)$ and $v(x, y)$ with a single unknown – the stream function $\psi(x, y)$ defined by

$$u = \frac{\partial \psi}{\partial y} \qquad v = -\frac{\partial \psi}{\partial x} \tag{5.45}$$

---

5 Osborne Reynolds (1842–1912) was a professor of engineering at Owens College in Manchester, England. Among his many contributions are the discovery of the critical velocity for the laminar–turbulent transition in pipe flow (the critical Reynolds number), and the thin-film theory of lubrication.

The continuity, Eq. (5.21), is satisfied identically by ψ, and the momentum, Eq. (5.32), becomes

$$\frac{\partial \psi}{\partial y}\frac{\partial^2 \psi}{\partial x \partial y} - \frac{\partial \psi}{\partial x}\frac{\partial^2 \psi}{\partial y^2} = v\frac{\partial^3 \psi}{\partial y^3} \tag{5.46}$$

The appropriate boundary conditions to be satisfied by ψ are

$$\frac{\partial \psi}{\partial y} = 0 \quad \text{at} \quad y = 0 \quad (\text{no-slip}, u = 0) \tag{5.47a}$$

$$\psi = 0 \quad \text{at} \quad y = 0 \quad (\text{impermeable wall}, v = 0) \tag{5.47b}$$

$$\frac{\partial \psi}{\partial y} \to U_\infty \quad \text{as} \quad y \to \infty \quad (\text{free stream}, u \to U_\infty) \tag{5.47c}$$

The problem (5.46)–(5.47) can be restated in terms of the similarity variable η; however, this transformation [2] is beyond the level of this course, and what follows is only an outline of the method. Combining the similarity u profile (5.43) with the first of Eq. (5.45), we see that we must choose a stream function of the form

$$\psi(x, y) = (U_\infty vx)^{1/2} f(\eta) \tag{5.48}$$

The unknown function $f(\eta)$ is the similarity stream function profile; and, in view of Eqs. (5.43) and (5.45), the derivative $df/d\eta$ is the similarity longitudinal velocity profile,

$$u = U_\infty \frac{df}{d\eta} \tag{5.49}$$

The $f(\eta)$ formulation of the flow problem (5.46)–(5.47) is

$$2f''' + ff'' = 0 \tag{5.50}$$

$$f' = f = 0 \quad \text{at } \eta = 0 \tag{5.51a, b}$$

$$f' \to 1 \quad \text{as } \eta \to \infty \tag{5.51c}$$

The numerical solution to this problem [2] is presented in Figure 5.6 in terms of the velocity profile $df/d\eta$. The slope of the similarity velocity profile at the wall is

$$\left(\frac{d^2f}{d\eta^2}\right)_{\eta=0} = 0.332 \tag{5.52}$$

This slope is needed for calculating the exact value of the wall shear stress or the local skin friction coefficient:

$$C_{f,x} = \frac{\mu(\partial u/\partial y)_{y=0}}{\frac{1}{2}\rho U_\infty^2} = 2\left(\frac{d^2f}{d\eta^2}\right)_{\eta=0} Re_x^{-1/2}$$

$$= 0.664 Re_x^{-1/2} \tag{5.53}$$

This exact result is not far off from the integral analysis estimate (Problem 5.11) and the scale-analysis calculation that gave us Eq. (5.40).

The *total* shear force (per unit length in the direction normal to the plane of Figure 5.5) experienced by the plane wall of longitudinal length x is

$$\int_0^x \tau_{w,x}\, dx = x\overline{\tau}_{w,x} \tag{5.54}$$

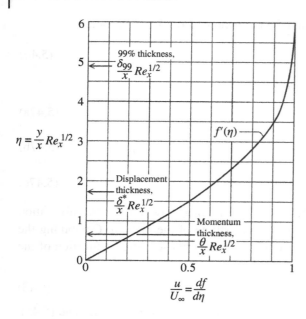

Figure 5.6 The similarity velocity profile for laminar boundary layer flow over a plane wall.

where $\bar{\tau}_{w,x}$ is the x-averaged value of the local shear stress $\tau_{w,x}$, which is recorded as a dimensionless average skin friction coefficient

$$\bar{C}_{f,x} = \frac{\bar{\tau}_{w,x}}{\frac{1}{2}\rho U_\infty^2} = 1.328 Re_x^{-1/2} \tag{5.55}$$

Figure 5.6 shows that u increases smoothly in the transversal direction η (or y) and approaches U_∞ asymptotically. For this reason, the actual "thickness" of the velocity boundary layer must be defined based on a precise definition. There are three such definitions. First, there is the δ_{99} thickness, which is defined (by convention) as the distance y where u reaches 99% of its maximum (free-stream) value:

$$u(x, \delta_{99}) = 0.99 U_\infty \tag{5.56}$$

The numerical solution of Figure 5.6 shows that $u = 0.99 U_\infty$ at $\eta = 4.92$,

$$\delta_{99} = 4.92 \, x \, Re_x^{-1/2} \tag{5.57}$$

Other definitions are the displacement thickness $\delta*$ and the momentum thickness θ:

$$\delta* = \int_0^\infty \left(1 - \frac{u}{U_\infty}\right) dy = 1.72 \, x \, Re_x^{-1/2} \tag{5.58}$$

$$\theta = \int_0^\infty \frac{u}{U_\infty}\left(1 - \frac{u}{U_\infty}\right) dy = 0.664 \, x \, Re_x^{-1/2} \tag{5.59}$$

Listed on the right side of these equations are the actual sizes revealed by the Blasius solution of Figure 5.6. The physical meaning and the explanation of the names of $\delta*$ and θ can be found in a more advanced treatment [2]. All three thicknesses (δ_{99}, $\delta*$, and θ) confirm the validity of the scale-analysis estimate, Eq. (5.36).

5.3.2 Thermal Boundary Layer

The temperature distribution $T(x, y)$ near the isothermal wall (T_w) swept by the parallel flow U_∞ is outlined in Figure 5.7. Let δ_T be the transversal length scale that represents the distance over which the fluid temperature makes the transition from the wall value T_w to the free-stream value T_∞. The thermal boundary layer thickness δ_T is defined therefore as the distance from the wall to the knee in the T-versus-y curve:

$$\left|\frac{\partial T}{\partial y}\right| \sim \frac{\Delta T}{\delta_T} \tag{5.60}$$

where ΔT is the imposed temperature difference $\Delta T = T_w - T_\infty$.

Assume next that the x-long and δ_T-thin region is slender,

$$\delta_T \ll x \tag{5.61}$$

namely, that it is a *thermal boundary layer region*. Based on a reasoning analogous to the one used in the development of Eq. (5.29), we conclude that in the steady two-dimensional energy equation (E) of Table 5.1 the $\partial^2 T/\partial x^2$ term is negligible:

$$\underbrace{u\frac{\partial T}{\partial x} + v\frac{\partial T}{\partial y}}_{\text{Convection}} = \underbrace{\alpha\frac{\partial^2 T}{\partial y^2}}_{\substack{\text{Transversal}\\\text{conduction}}} \tag{5.62}$$

This is the boundary layer-simplified form of the energy equation for the fluid that resides in the (x, δ_T) layer. The coefficient $\alpha(\text{m}^2/\text{s})$ is the thermal diffusivity of the fluid:

$$\alpha = \frac{k}{\rho c_P} \tag{5.63}$$

The unknown feature of the thermal boundary layer is its thickness δ_T. When δ_T is known, the wall heat flux can be evaluated (cf. Eq. (1.56)):

$$q''_{w,x} \sim k\frac{\Delta T}{\delta_T} \tag{5.64}$$

This is the scale-analysis estimate of the relationship between wall heat flux and imposed temperature difference. The δ_T scale needed for this estimate can be determined analytically in the following two limits.

5.3.2.1 Thick Thermal Boundary Layer

In this limit, the δ_T layer is "thick" relative to the velocity boundary layer thickness measured at the same x, Figure 5.7 (top):

$$\delta_T \gg \delta \tag{5.65}$$

Inside the δ_T layer, the respective scales of the three terms of the energy Eq. (5.62) are

$$u\frac{\Delta T}{x}, v\frac{\Delta T}{\delta_T} \sim \alpha\frac{\Delta T}{\delta_T^2} \tag{5.66}$$

According to Eq. (5.33), the v scale outside the thin velocity boundary layer (and inside the δ_T layer) is $v \sim U_\infty\delta/x$, therefore the second term on the left side of Eq. (5.66) is

$$v\frac{\Delta T}{\delta_T} \sim U_\infty\frac{\Delta T}{x}\frac{\delta}{\delta_T} \tag{5.67}$$

in which $\delta/\delta_T \ll 1$. The second term is therefore δ/δ_T times smaller than the first, and the left side of Eq. (5.66) is dominated by $u\Delta T/x$. The convection–conduction balance is

$$u\,\frac{\Delta T}{x} \sim \alpha\,\frac{\Delta T}{\delta_T^2} \tag{5.68}$$

What is the scale of the longitudinal velocity u inside the δ_T-thick layer? In the case of a thermal boundary layer that is thicker than the velocity boundary layer, Figure 5.7 (top) shows that $u \sim U_\infty$. Substituting U_∞ in Eq. (5.68) we obtain

$$\delta_T \sim x\,Pe_x^{-1/2} \tag{5.69}$$

where Pe_x is the *Péclet number*[6] based on the downstream position x:

$$Pe_x = \frac{U_\infty x}{\alpha} \tag{5.70}$$

The thermal boundary layer thickness increases as $x^{1/2}$ in the downstream direction, as illustrated in Figure 5.7. The Prandtl number (Pr) mentioned in Figure 5.7 will be defined in Eq. (5.74).

The local wall heat flux $q''_{w,x}$ can be deduced from Eq. (5.64). The result is reported as a dimensionless local Nusselt number[7] defined as

$$Nu_x = \frac{q''_{w,x}}{\Delta T}\frac{x}{k} = \frac{h_x\,x}{k} \tag{5.71}$$

where h_x is the local heat transfer coefficient, $h_x = q''_{w,x}/\Delta T$. Combining Eqs. (5.64) and (5.69), we conclude that the local Nusselt number is of order

$$Nu_x \sim Pe_x^{1/2} \quad (Pr \ll 1) \tag{5.72}$$

This conclusion holds as long as the assumption $\delta_T > \delta$ is correct. Substituting Eqs. (5.69) and (5.36) on the left and right sides of the inequality (5.65), we learn that the $\delta_T > \delta$ assumption is the same as writing

$$\alpha > \nu \tag{5.73}$$

The thermal boundary layer is thicker than the velocity layer when the Prandtl number

$$Pr = \frac{\nu}{\alpha} \tag{5.74}$$

is smaller than 1. This restriction has been added to the right of Eq. (5.72). Examples of low-Prandtl-number fluids are the liquid metals (mercury, lead, and sodium (Appendix C)).

5.3.2.2 Thermal Boundary Layer

What changes when the thermal layer is thinner than the superimposed velocity layer

$$\delta_T \ll \delta \tag{5.75}$$

---

6 Jean-Claude-Eugène Péclet (1793–1857) was a French physicist who wrote an influential 1829 treatise on heat transfer and its applications, which was translated into English in 1843.
7 Ernst Kraft Wilhelm Nusselt (1882–1957) was a professor of theoretical mechanics at the Technical University of Munich. He pioneered the dimensional analysis of convective processes, the dimensionless correlation of experimental data, and the analysis of laminar film condensation.

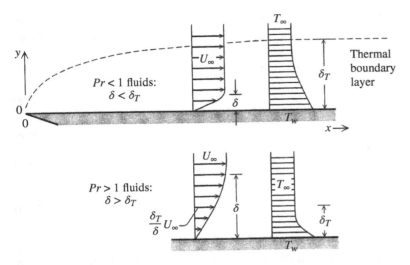

Figure 5.7 The thermal boundary layer in low-*Pr* fluids (a) and high-*Pr* fluids (b).

is the scale of the longitudinal velocity u in the δ_T-thin region. According to the configuration shown in the lower part of Figure 5.7, u scale is

$$u \sim \frac{\delta_T}{\delta} U_\infty \tag{5.76}$$

Combining this with the convection–conduction balance inside the δ_T-thin layer, Eq. (5.68), and using the velocity thickness δ, Eq. (5.36), we obtain

$$\delta_T \sim x \, Pr^{-1/3} \, Re_x^{-1/2} \tag{5.77}$$

The corresponding local heat flux and local Nusselt number follow from this conclusion and the definitions (5.64) and (5.71):

$$Nu_x \sim Pr^{1/3} \, Re_x^{1/2} \quad (Pr \gg 1) \tag{5.78}$$

These results (δ_T, Nu_x) are valid when δ_T is smaller than δ; combined, Eqs. (5.36), (5.75), and (5.77) show that the thermal layer is thinner than the velocity layer when the Prandtl number is greater than 1, as in nonmetallic liquids (e.g. water, oils).

The common gases (e.g. air, steam, CO_2) have Prandtl numbers of order 1. As a class, they fall right between the limits (a) and (b), which were discussed until now. It is instructive to verify that these two limits, Eqs. (5.72) and (5.78), predict the same heat transfer result ($Nu_x \sim Re_x^{1/2}$) when the Prandtl number is of order 1.

A characteristic of all the boundary layer thicknesses uncovered so far is their monotonic increase as $x^{1/2}$ in the downstream direction. One property of the $x^{1/2}$ function is that it has infinite slope at $x = 0$. This feature is being stressed in Figures 5.5 and 5.7 because it is almost never respected in the illustrations that appear in heat transfer books and journals. Although as noted in Eq. (5.42), the boundary layer approximation does not hold near the tip of the boundary layer region, the shape represented by $x^{1/2}$ has infinite slope at $x = 0$.

The heat transfer results (5.72) and (5.78) have been refined on the basis of more exact methods of solution. The integral method and samples of the Nu_x estimates produced by this method are illustrated in Problem 5.11. The exact solution based on the similarity formulation was carried out by Pohlhausen [9]. Of interest are the "low-Pr" and "high-Pr" limits of this solution [2]:

$$Nu_x \cong 0.564 Pe_x^{1/2} \quad (Pr \lesssim 0.5) \tag{5.79a}$$

$$Nu_x \cong 0.332 Pr^{1/3} Re_x^{1/2} \quad (Pr \gtrsim 0.5) \tag{5.79b}$$

The *total* heat transfer rate between the x-long wall and the adjacent flow per unit length in the direction normal to the plane of Figure 5.7 is

$$\int_0^x q''_{w,x}\, dx = x\, \overline{q}''_{w,x} \tag{5.80}$$

Equation (5.80) is the definition of the x-averaged wall heat flux $\overline{q}''_{w,x}$; this can be calculated by substituting Eqs. (5.79a, 5.79b) into the Nu_x definition (5.71) and using the resulting $q''_{w,x}$ formulas in the integrand on the left side of Eq. (5.80). The average heat flux $\overline{q}''_{w,x}$ obtained in this manner can be summarized in dimensionless form

$$\overline{Nu}_x = \frac{\overline{q}''_{w,x}}{\Delta T}\frac{x}{k} = \frac{\overline{h}_x x}{k} \tag{5.81}$$

where $\overline{Nu}_x$ is the average or overall Nusselt number and $\overline{h}_x$ is the average heat transfer coefficient. The $\overline{Nu}_x$ formulas that correspond to the local Nusselt number asymptotes (5.79a) and (5.79b) are

$$\overline{Nu}_x = 1.128 Pe_x^{1/2} \quad (Pr \lesssim 0.5) \tag{5.82a}$$

$$\overline{Nu}_x = 0.664 Pr^{1/3} Re_x^{1/2} \quad (Pr \gtrsim 0.5) \tag{5.82b}$$

in which, it is worth noting, the numerical coefficients are twice as large as the coefficients of Eqs. (5.79a) and (5.79b). An average Nusselt number expression that covers the entire Prandtl number range was recommended by Churchill and Ozoe [10]:

$$\overline{Nu}_x = \frac{0.677 Pr^{1/3} Re_x^{1/2}}{[1 + (0.0468/Pr)^{2/3}]^{1/4}} \tag{5.83}$$

It is valid when the Péclet number $Pe_x = U_\infty x/\alpha$ is greater than approximately 100.

5.3.3 Nonisothermal Wall

The similarity solutions (5.79a, 5.79b) for heat transfer in laminar boundary layer flow refer to the heat flux from an isothermal wall (T_w) to an isothermal free stream (T_∞). There are many other wall-heating conditions that occur in practical situations, and, for some of these, heat transfer solutions are available. Four results of this kind [2, 11, 12] are shown in Table 5.4. They were derived based on the integral method and applied to fluids with Pr values greater than approximately 0.5.

The first solution in Table 5.4 shows the heat flux through the isothermal section of a wall with unheated starting length. The effect of the unheated length x_0 is to decrease the heat flux to values below those of the fully isothermal wall, Eq. (5.79b). The isothermal wall problem is the special case $x_0 = 0$ of this solution.

Table 5.4   Heat transfer results for laminar boundary layer flows near walls with various heating conditions ($Pr \gtrsim 0.5$).

Unheated starting length:

$$q''_{w,x} = 0.332 \frac{k(T_w - T_\infty)}{x} Pr^{1/3} Re_x^{1/2} \left[1 - \left(\frac{x_0}{x} \right)^{3/4} \right]^{-1/3}$$

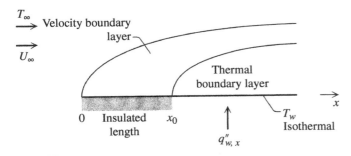

Nonuniform wall temperature:

$$q''_{w,x} = 0.332 \frac{k}{x} Pr^{1/3} Re_x^{1/2} \int_0^x \frac{(dT_w/d\xi)d\xi}{[1 - (\xi/x)^{3/4}]^{1/3}}$$

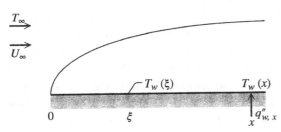

Uniform heat flux:

$$T_w(x) - T_\infty = \frac{q''_w x}{0.453k \, Pr^{1/3} Re_x^{1/2}}$$

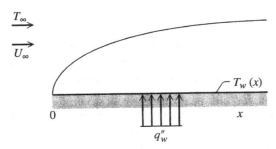

(Continued)

Table 5.4 (Continued)

Nonuniform heat flux:

$$T_w(x) - T_\infty = \frac{0.623}{k\, Pr^{1/3} Re_x^{1/2}} \int_{\xi=0}^{x} \left[1 - \left(\frac{\xi}{x}\right)^{3/4} \right]^{-2/3} q_w''(\xi) d\xi$$

The second solution in Table 5.4 shows the wall heat flux distribution in the general case where the wall temperature is a certain function of longitudinal position, $T_w(\xi)$. The heat flux at the downstream position x is an integral result of the interaction between the wall and the fluid all along the wall section that precedes x, namely, $0 < \xi < x$. The wall with unheated starting length is a simple case of the second solution.

The last two entries in Table 5.4 refer to walls with specified heat flux. The uniform heat flux solution is a special case of the nonuniform wall heating configuration shown at the bottom of the table. In problems with specified wall heat flux, the unknown quantity is the wall–fluid temperature difference, $T_w(x) - T_\infty$, that is, the temperature variation along the wall. For example, the temperature of the wall with constant heat flux increases as $x^{1/2}$ in the downstream direction. Note also that the $T_w(x) - T_\infty$ formula listed for the wall with uniform heat flux agrees within a numerical factor of order 1 with the scale-analysis result (5.78). In other words, the order-of-magnitude conclusions of Section 5.3.2 apply to a laminar boundary layer whose temperature T_w is not necessarily a constant.

5.3.4 Film Temperature

The wall friction and heat transfer results of Section 5.3 are based on the constant property model. In real situations, fluid properties such as k, ν, μ, and α are not constant, as they depend primarily on the local temperature in the flow field. It turns out that the constant property formulas describe sufficiently accurately the actual convective flows encountered in applications, provided the maximum temperature variation experienced by the fluid ($T_w - T_\infty$) over the flow region is small relative to the thermodynamic temperature of the fluid (T_w, or T_∞, expressed in degrees Kelvin). In such cases the properties needed for calculating the various dimensionless groups (Re_x, Pe_x, Pr, $C_{f,x}$, Nu_x) can be evaluated at the average temperature of the fluid in the thermal boundary layer:

$$T = \frac{1}{2}(T_w + T_\infty) \tag{5.84}$$

This average is known as the *film temperature* of the fluid and is generally recommended for use in formulas of the constant property type. Worth keeping in mind is that there are special correlations in which the effect of temperature-dependent properties is taken into account by means of explicit correction factors.

Example 5.1   *The Thermal Laminar Boundary Layer*
A thin plate swept by laminar flow serves as a plate fin attached to a distant wall. The flow is parallel to the wall. The length of the plate in the flow direction is 1 cm. The plate is isothermal, $T_w = 35\,°C$, and the water stream is colder, $T_\infty = 25\,°C$. Calculate the local heat flux at the trailing edge of the plate, the plate-averaged heat flux and heat transfer coefficient, and the total heat transfer rate between the plate and the water stream.

Solution

By evaluating the properties of water at the film temperature,

$$T = \frac{1}{2}(T_w + T_\infty) = \frac{1}{2}(35°C + 25°C) = 30°C$$

we first calculate the Reynolds number, to be sure that the entire boundary layer is laminar (see Eq. (5.85)):

$$Re_x = \frac{U_\infty x}{\nu} = 0.6\,\frac{cm}{s}\,1\,cm\,\frac{s}{0.008\,cm^2} = 75\;(laminar)$$

The local heat flux at $x = 1$ cm can be estimated based on Eq. (5.79b):

$$Nu_x = 0.332 Pr^{1/3} Re_x^{1/2}$$
$$= 0.332(5.49)^{1/3}(75)^{1/2} = 5.07$$

$$Nu_x = \frac{q''_{w,x}}{T_w - T_\infty}\frac{x}{k} = 5.07$$

$$q''_{w,x} = 5.07(T_w - T_\infty)\frac{k}{x}$$
$$= 5.07\,(35 - 25)\,K\,0.61\,\frac{W}{m\cdot K}\,\frac{1}{0.01\,m} = 3094\;W/m^2$$

The value of the heat flux averaged over the entire length x is furnished by Eq. (5.82b):

$$\overline{Nu}_x = 0.664 Pr^{1/3} Re_x^{1/2} = 10.14$$

$$\overline{q}''_{w,x} = 10.14\,(T_w - T_\infty)\frac{k}{x} = 6188\;W/m^2$$

The same values could have been calculated more rapidly by comparing Eqs. (5.82b) and (5.79b) and noting that in the laminar boundary layer the x-averaged heat flux is twice the local heat flux evaluated at x, namely, $\overline{Nu}_x = 2 Nu_x = 10.14$, $\overline{q}''_{w,x} = 2 q''_{w,x} = 6188\;W/m^2$, and

$$\overline{h}_x = \frac{\overline{q}''_{w,x}}{T_w - T_\infty} = \frac{6188\;W/m^2}{10\;K} \cong 620\;W/m^2\cdot K$$

The total heat transfer rate out of the plate fin follows after the observation that both sides of the plate are bathed by the water stream:

$$q' = 2x\,\overline{q}''_{w,x} = 2\times 0.01\,m\,6188\;W/m^2 = 124\;W/m$$

This heat transfer rate is expressed per unit length in the direction perpendicular to the plane of Figure 5.7, that is, away from the wall that supports the fin.

5.4 Turbulent Boundary Layer

5.4.1 Transition from Laminar to Turbulent Flow

The laminar boundary layer flow prevails in the leading section of the flow, that is, at small enough longitudinal distances from the leading edge (x), so that the x-based Reynolds number does not exceed a critical value:

$$Re_x \lesssim 5 \times 10^5 \quad \text{(laminar)}$$
$$Re_x \gtrsim 5 \times 10^5 \quad \text{(turbulent)} \tag{5.85}$$

The critical Reynolds number 5×10^5 is an approximate and much abbreviated way of recording the empirical observations of the laminar – turbulent transition along a plane wall. As illustrated in Figure 5.8, the *transition section* occupies a finite range of x values (or a finite Re_x range), marked by a "beginning" and an "end" of transition. Furthermore, the position of the transition section along the wall depends strongly on the degree of smoothness of the free stream, that is, on the presence of flow disturbances (eddies) in the fluid outside the laminar boundary layer. The critical Reynolds number on the right side of Eq. (5.85) can be as low as 2×10^4 when the free stream is strongly disturbed, and as high as 10^6 and even higher when the outer flow is extremely smooth.

The transition criterion can be expressed in terms of a *local Reynolds number* [2], which is based on the local longitudinal velocity scale (U_∞) and the local thickness of the flow region. Taking the momentum thickness θ of Eq. (5.59) as a measure of the transversal length scale of the velocity boundary layer, the critical local Reynolds number ($U_\infty \theta / \nu$) that corresponds to the observed transition range $2 \times 10^4 < Re_x < 10^6$ is

$$\frac{U_\infty \theta}{\nu} \sim 94 - 660 \tag{5.86}$$

In conclusion, at transition the local Reynolds number has a value of order 10^2. Appendix F shows that the local Reynolds number criterion governs all the laminar – turbulent transitions of the flows treated in the remainder of this book.

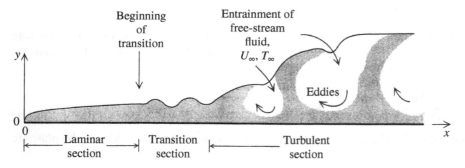

Figure 5.8 Laminar, transition, and turbulent sections in the boundary layer over a plane wall.

Figure 5.9 The laminar section and the beginning of transition in the air boundary layers over a 1.25-m piece of toilet paper falling to the ground: $U_\infty = 2.93$ m/s, ribbon width = 11.4 cm, ribbon mass = 3.21 g.

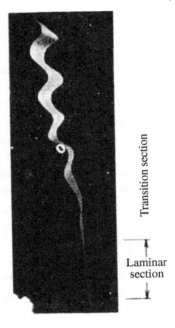

Transition section

Laminar section

Students are often confused that the boundary layer critical Reynolds number 5×10^5 is so much greater than the critical Reynolds number of only about 2000 for duct flow (see Eq. (6.59) in the next chapter). The reason for this apparently huge discrepancy is that the two Reynolds numbers are defined differently. In boundary layer flow the critical *Re* is based on the *longitudinal* distance to the transition region, whereas in duct flow the critical *Re* is based on an appropriate *transversal* length scale of the duct (see "hydraulic diameter," Eq. (6.28)). When the boundary layer Reynolds number is rewritten in terms of the transversal length scale, it becomes the *local* Reynolds number and its critical value agrees much better with the critical value known for duct flow. The order-of-magnitude correlation of seemingly divergent critical numbers for transition in different flow configurations is one of the contributions of the local Reynolds number criterion unveiled in (Appendix F).

One of the simplest and least expensive ways to visualize the transition to turbulence in the laminar boundary layer is by dropping a front-loaded piece of toilet paper through the air. Figure 5.9 shows the instantaneous shape of the paper ribbon [13]. The leading section is straight, as it is lined by laminar boundary layers on both sides. The transition region follows. The tissue paper is flexible enough to mimic the meandering, or buckling, shape of the flow in its first stages of transition. The transition section in Figure 5.9 appears two-dimensional only because of the narrowness of the paper ribbon. On a wide and rigid flat surface, the transition is marked by local, three-dimensional deformations of the laminar flow.

5.4.2 Time-Averaged Equations

In the boundary layer section situated immediately downstream from the transition section, Figure 5.8, the elbows of the sinusoidal deformation become exaggerated, as they are sucked into the surrounding free stream. The deformed boundary layer is then rolled into eddies that tumble as they are swept down the wall. This tumbling flow constitutes the turbulent section of the boundary layer.

One way to visualize the difference between the laminar section and the turbulent section is to recall that the laminar section is ruled by a perfect balance between fluid inertia and by the transversal viscous diffusion of the friction effect of the wall, Eq. (5.35). It can be said that in the laminar section the fluid is fully penetrated in the y direction by the effect of viscous diffusion. This effect requires a certain amount of time, and, as shown in Appendix F, in the transition section the boundary layer has become too thick to allow viscous diffusion to traverse it completely.

In the transition section and, especially, in the subsequent turbulent section, the thickness of the flow regions penetrated by viscous diffusion is of the same order as the thickness of the laminar boundary layer just upstream of the transition section. Consequently, the turbulent section becomes a conglomerate of viscous layers rolled around pockets of inviscid fluid inhaled from the free stream. In Figure 5.8 these inviscid regions are shown as white areas; they travel longitudinally with a velocity of order U_∞; therefore, when they make contact with the solid wall, they generate local (and temporary) laminar layers similar to the leading laminar section of the boundary layer flow.

The classical approach to the analytical study of turbulent flows started from a point of view that differs from Figure 5.8 and Appendix F. It began with a time-averaging process – a smoothing out of the kinks – of the admittedly complicated turbulent flow field. It recognized that a flow variable such as the longitudinal velocity component is a function of spatial position and time, $u(x, y, z, t)$. At a fixed location in the turbulent flow field, the u value fluctuates about a mean value $\bar{u}$ defined by the time-averaging operation

$$\bar{u} = \frac{1}{p} \int_0^p u \, dt \tag{5.87}$$

which is illustrated in Figure 5.10. The value of $\bar{u}$ is independent of time when the period p of the time-averaging operation exceeds the period of the slowest fluctuation exhibited by the actual u. The end result of this time-averaging procedure is the decomposition of the flow variable into a mean value plus a time-dependent correction (fluctuation) labeled u':

$$u(x, y, z, t) = \bar{u}(x, y, z) + u'(x, y, z, t) \tag{5.88}$$

The same decomposition applies to the remaining variables of the flow field:

$$v = \bar{v} + v' \qquad P = \bar{P} + P'$$
$$w = \bar{w} + w' \qquad T = \bar{T} + T' \tag{5.89}$$

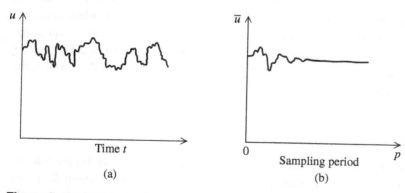

(a) (b)

Figure 5.10 The behavior of an instantaneous quantity (u) in turbulent flow and the calculation of the time-independent mean value $\bar{u}$ by using a long enough sampling period p.

The next step is the time averaging of the mass conservation, momentum, and energy equations assembled in Table 5.1. This consists of substituting Eqs. (5.88)–(5.89) into the governing equations and then applying the time-averaging (5.87) to every term of the resulting equations. This analysis relies on a special set of algebraic rules that follow from Eq. (5.87):

$$\overline{u'} = 0 \qquad \overline{\left(\frac{\partial u}{\partial x}\right)} = \frac{\partial \overline{u}}{\partial x}$$

$$\overline{u + v} = \overline{u} + \overline{v}$$

$$\overline{\overline{u} u'} = 0 \qquad \frac{\partial \overline{u}}{\partial t} = 0$$

$$\overline{uv} = \overline{u}\,\overline{v} + \overline{u'v'} \qquad \overline{\left(\frac{\partial u}{\partial t}\right)} = 0$$

$$\overline{u^2} = (\overline{u})^2 + \overline{(u')^2} \tag{5.90}$$

The governing equations reduce to the following [2]:

$$(m) \quad \frac{\partial \overline{u}}{\partial x} + \frac{\partial \overline{v}}{\partial y} + \frac{\partial \overline{w}}{\partial z} = 0 \tag{5.91}$$

$$(M_x) \quad \overline{u}\frac{\partial \overline{u}}{\partial x} + \overline{v}\frac{\partial \overline{u}}{\partial y} + \overline{w}\frac{\partial \overline{u}}{\partial z} = -\frac{1}{\rho}\frac{\partial \overline{P}}{\partial x} + \nu\,\nabla^2\overline{u}$$

$$- \frac{\partial}{\partial x}\overline{(u'^2)} - \frac{\partial}{\partial y}\overline{(u'v')} - \frac{\partial}{\partial z}\overline{(u'w')} \tag{5.92}$$

$$(M_y) \quad \overline{u}\frac{\partial \overline{v}}{\partial x} + \overline{v}\frac{\partial \overline{v}}{\partial y} + \overline{w}\frac{\partial \overline{v}}{\partial z} = -\frac{1}{\rho}\frac{\partial \overline{P}}{\partial y} + \nu\nabla^2\overline{v}$$

$$- \frac{\partial}{\partial x}\overline{(u'v')} - \frac{\partial}{\partial y}\overline{(v'^2)} - \frac{\partial}{\partial z}\overline{(v'w')} \tag{5.93}$$

$$(M_z) \quad \overline{u}\frac{\partial \overline{w}}{\partial x} + \overline{v}\frac{\partial \overline{w}}{\partial y} + \overline{w}\frac{\partial \overline{w}}{\partial z} = -\frac{1}{\rho}\frac{\partial \overline{P}}{\partial z} + \nu\nabla^2\overline{w}$$

$$- \frac{\partial}{\partial x}\overline{(u'w')} - \frac{\partial}{\partial y}\overline{(\overline{v}w')} - \frac{\partial}{\partial z}\overline{(w'^2)} \tag{5.94}$$

$$(E) \quad \overline{u}\frac{\partial \overline{T}}{\partial x} + \overline{v}\frac{\partial \overline{T}}{\partial y} + \overline{w}\frac{\partial \overline{T}}{\partial z} = \alpha\nabla^2\overline{T}$$

$$- \frac{\partial}{\partial x}\overline{(u'T')} - \frac{\partial}{\partial y}\overline{(v'T')} - \frac{\partial}{\partial z}\overline{(w'T')} \tag{5.95}$$

So far, we treated the turbulent flow field as three-dimensional (u, v, w), because by their very nature turbulent flows are instantaneously three-dimensional, regardless of the simple shapes of the walls that may bound them. In the turbulent section of the boundary layer of Figure 5.8, for example, the eddies can rotate not only in the plane of the figure but also into and out of that plane.

The turbulent boundary layer near a flat wall is two-dimensional only as a time-averaged flow field. This is why beginning with Eqs. (5.91)–(5.95) we set $\overline{w} = 0$ and $\partial()/\partial z = 0$. The first consequence of this decision is that the z-momentum Eq. (5.94) requires that $\partial P/\partial z = 0$, in other words, that $P = P(x, y)$. Boundary layer theory and the y-momentum Eq. (5.93) indicate that $\overline{P}$ is mainly a function of x, which means that

$\overline{P} = \overline{P}(x) = P_\infty$. And, since P_∞ is a constant in the free stream near a flat wall, the term $\partial\overline{P}/\partial x$ of Eq. (5.92) is zero.

The second group of simplifications recommended by boundary layer theory (Section 5.3.1) is the neglect of the longitudinal derivatives $\partial(\ )/\partial x$ and $\partial^2(\ )/\partial x^2$ on the right side of Eqs. (5.92)–(5.95). The final form of the boundary layer-simplified time-averaged equations for the turbulent section of the boundary layer is therefore

$$(m) \quad \frac{\partial \overline{u}}{\partial x} + \frac{\partial \overline{v}}{\partial y} = 0 \tag{5.96}$$

$$(M) \quad \overline{u}\frac{\partial \overline{u}}{\partial x} + \overline{v}\frac{\partial \overline{u}}{\partial y} = \frac{\partial}{\partial y}\left(\nu\frac{\partial \overline{u}}{\partial y} - \overline{u'v'}\right) \tag{5.97}$$

$$(E) \quad \overline{u}\frac{\partial \overline{T}}{\partial x} + \overline{v}\frac{\partial \overline{T}}{\partial y} = \frac{\partial}{\partial y}\left(\alpha\frac{\partial \overline{T}}{\partial y} - \overline{v'T'}\right) \tag{5.98}$$

The momentum and energy, Eqs. (5.97) and (5.98), must be compared with their laminar flow counterparts, Eqs. (5.32) and (5.62), to appreciate the new features introduced by the time averaging of the turbulent flow. The products $\overline{u'v'}$ and $\overline{v'T'}$ survive the time-averaging operation and increase by two the number of unknowns that could be determined by solving the system (5.96)–(5.98). The five unknowns in this three-equation system are $\overline{u}, \overline{v}, \overline{T}, \overline{u'v'}$, and $\overline{v'T'}$. The search for the two additional equations that are required to determine the five unknowns uniquely is activity recognized as turbulence modeling.

5.4.3 Eddy Diffusivities

It is customary to replace the time-averaged products $\overline{u'v'}$, and $\overline{v'T'}$ with

$$-\overline{u'v'} = \varepsilon_M\frac{\partial \overline{u}}{\partial y} \tag{5.99}$$

$$-\overline{v'T'} = \varepsilon_H\frac{\partial \overline{T}}{\partial y} \tag{5.100}$$

where the coefficient ε_M (m²/s) is recognized as the *momentum eddy diffusivity* and ε_H (m²/s) as the *thermal eddy diffusivity*, or the eddy diffusivity for heat. The momentum and energy equations become

$$(M) \quad \overline{u}\frac{\partial \overline{u}}{\partial x} + \overline{v}\frac{\partial \overline{u}}{\partial y} = \frac{\partial}{\partial y}\left[(\nu + \varepsilon_M)\frac{\partial \overline{u}}{\partial y}\right] \tag{5.101}$$

$$(E) \quad \overline{u}\frac{\partial \overline{T}}{\partial x} + \overline{v}\frac{\partial \overline{T}}{\partial y} = \frac{\partial}{\partial y}\left[(\alpha + \varepsilon_H)\frac{\partial \overline{T}}{\partial y}\right] \tag{5.102}$$

The ε_M and ε_H notation is recommended by the view that on the right side of Eqs. (5.97)–(5.98) the time-averaged groups $(-\overline{u'v'})$ and $(-\overline{v'T'})$ are eddy contributions that augment the effects of molecular diffusion, which are represented, respectively, by $\nu(\partial \overline{u}/\partial y)$ and $\alpha(\partial \overline{T}/\partial y)$. Consider first the quantity placed inside the square brackets on the right side of Eq. (5.101):

$$\nu\frac{\partial \overline{u}}{\partial y} + \varepsilon_M\frac{\partial \overline{u}}{\partial y} = \frac{1}{\rho}\left(\underbrace{\mu\frac{\partial \overline{u}}{\partial y}}_{\tau_{mol}} + \underbrace{\rho\varepsilon_M\frac{\partial \overline{u}}{\partial y}}_{\tau_{eddy}}\right) \tag{5.103}$$

where the terms indicated by τ_{mol} and τ_{eddy} represent the usual (molecular) shear stress and the shear stress contribution made by the time-averaged effect of the eddies that act at the point to which Eq. (5.101) applies. The sum of the molecular and eddy shear stresses is the apparent shear stress

$$\tau_{app} = \tau_{mol} + \tau_{eddy} \tag{5.104}$$

Similarly, the right side of the energy (Eq. (5.102)) can be decomposed to see the molecular heat flux $-k(\partial \overline{T}/\partial y)$ and the eddy heat flux $-\rho c_P \varepsilon_H (\partial \overline{T}/\partial y)$:

$$\alpha \frac{\partial \overline{T}}{\partial y} + \varepsilon_H \frac{\partial \overline{T}}{\partial y} = -\frac{1}{\rho c_P} \left[\underbrace{\left(-k \frac{\partial \overline{T}}{\partial y} \right)}_{q''_{mol}} + \underbrace{\left(-\rho c_P \varepsilon_H \frac{\partial \overline{T}}{\partial y} \right)}_{q''_{eddy}} \right] \tag{5.105}$$

Both q''_{mol} and q''_{eddy} are defined as positive when pointing in the positive y direction, that is, away from the wall. Their sum represents the apparent heat flux

$$q''_{app} = q''_{mol} + q''_{eddy} \tag{5.106}$$

In conclusion, the momentum and energy, Eqs. (5.101) and (5.102), can be written in terms of the gradients of apparent shear stress and apparent heat flux:

$$(M) \quad \overline{u} \frac{\partial \overline{u}}{\partial x} + \overline{v} \frac{\partial \overline{u}}{\partial y} = \frac{\partial}{\partial y} \left(\frac{\tau_{app}}{\rho} \right) \tag{5.107}$$

$$(E) \quad \overline{u} \frac{\partial \overline{T}}{\partial x} + \overline{v} \frac{\partial \overline{T}}{\partial y} = \frac{\partial}{\partial y} \left(\frac{q''_{app}}{-\rho c_P} \right) \tag{5.108}$$

The net contribution of the analysis described between Eqs. (5.99) and (5.108) is the replacement of the unknown quantities $\overline{u'v'}$ and $\overline{v'T'}$ with two alternative unknowns, ε_M and ε_H. These require the use of two additional equations (two assumptions), the simplest of which is described next.

To determine ε_M, imagine a packet of fluid that at some point in time is situated at a distance y away from the wall, in the turbulent boundary layer. The average longitudinal velocity at that location is $\overline{u}(x, y)$. Imagine, next, that this fluid packet rides on an eddy and migrates toward the wall to the new location $y - l$, where the mean velocity is $\overline{u}(x, y - l)$. The distance l is called the *mixing length* over which the fluid packet maintains its identity; this length is of the same order as the eddy diameter. The u' fluctuation induced by this migration at the new level $y - l$ is of the same order as

$$|u'| \sim \overline{u}(x, y) - \overline{u}(x, y - l) \sim l \frac{\partial \overline{u}}{\partial y} \tag{5.109}$$

Because of the wheel-like motion of the eddy, it can be argued [2] that the v' fluctuation is of the same order as u', that is, the same as the peripheral speed of the eddy:

$$|v'| \sim l \frac{\partial \overline{u}}{\partial y} \tag{5.110}$$

Substituting these u' and v' estimates into the ε_M definition (5.99), we obtain

$$\varepsilon_M \sim l^2 \left| \frac{\partial \overline{u}}{\partial y} \right| \tag{5.111}$$

Measurements of the $\bar{u} - y$ profile suggest that the mixing length l is proportional to the distance to the wall (like the diameter of the eddy that fits across the distance y):

$$l = \kappa y \tag{5.112}$$

where $\kappa \cong 0.4$ is known as von Karman's constant. In summary, the mixing length model for the momentum eddy diffusivity reads [14]:

$$\varepsilon_M = \kappa^2 y^2 \left| \frac{\partial \bar{u}}{\partial y} \right| \tag{5.113}$$

This is the simplest of many eddy diffusivity models that have been proposed [2]. Regarding the thermal eddy diffusivity ε_H, the simplest model is the assumption that ε_H is approximately the same as ε_M. By analogy with the Prandtl number notation $Pr = \nu/\alpha$, the eddy diffusivity ratio

$$\frac{\varepsilon_M}{\varepsilon_H} = Pr_t \tag{5.114}$$

is called the turbulent Prandtl number. Therefore, the ε_H model consists of writing that Pr_t is a constant approximately equal to 1. Measurements of the temperature distribution in turbulent boundary layers of $Pr \gtrsim 1$ fluids recommend the value $Pr_t \cong 0.9$.

5.4.4 Wall Friction

The writing of the momentum equation as Eq. (5.101) or (5.107) allows us to view the turbulent boundary layer as a sandwich of two distinct regions. In the outer region the inertia of the flow (the left side of Eq. (5.107)) is finite and negative, and the apparent shear stress decreases to zero as y reaches into the free stream. This feature is illustrated in Figure 5.11. The outer layer is also called the wake region of the turbulent boundary layer.

Sufficiently close to the wall, the inertia effect becomes negligible, and both sides of the momentum, Eq. (5.107), approach zero. Integrating in y, we reach the conclusion that in this inner layer τ_{app} is practically independent of y:

$$\tau_{\mathrm{app}} = \mathrm{constant} = \tau_{w,x} \tag{5.115}$$

where the wall shear stress $\tau_{w,x}$ is the value reached by τ_{app} right at the wall. Note that ε_M and τ_{eddy} are zero at $y = 0$. The inner layer is recognized as the constant-τ_{app} region of the boundary layer.

Working now with Eq. (5.115), in which τ_{app} is equal to the quantity contained between brackets on the right side of Eq. (5.103), it is possible to derive an analytical expression for the velocity profile $\bar{u}$ in the constant-τ_{app} region. Figure 5.11 and Eq. (5.113) show that immediately close to the wall, ε_M approaches zero and can be neglected in favor of ν. The region where $\nu \gg \varepsilon_M$ is called the viscous sublayer of the constant-τ_{app} region.

Immediately outside the viscous sublayer (and still inside the constant-τ_{app} region) resides the fully turbulent sublayer, in which ε_M is much greater than ν. In summary, the velocity distribution that is obtained for the constant-τ_{app} region by using Eq. (5.115) and the mixing length model (5.113) is

$$u^+ = \begin{cases} y^+ & \text{(viscous sublayer, } \nu \gg \varepsilon_M) \\ \dfrac{1}{\kappa} \ln y^+ + B & \text{(fully turbulent sublayer, } \varepsilon_M \gg \nu) \end{cases} \tag{5.116}$$

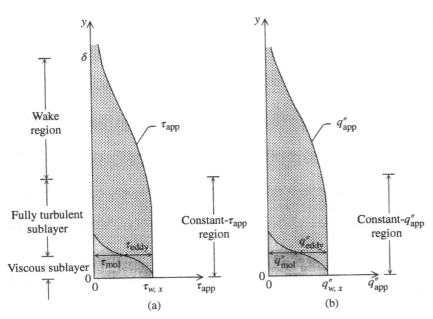

Figure 5.11 The structure of the apparent shear stress τ_{app} and the apparent heat flux q''_{app}.

in which u^+ and y^+ are the dimensionless "wall coordinates" defined by

$$u^+ = \frac{\bar{u}}{(\tau_{w,x}/\rho)^{1/2}} \qquad y^+ = \frac{y}{\nu}\left(\frac{\tau_{w,x}}{\rho}\right)^{1/2} \tag{5.117}$$

Measurements of the $u^+(y^+)$ distribution in the fully turbulent sublayer recommend the value $B \cong 5.5$ for the constant of integration appearing in the second of Eq. (5.116). By equating the u^+ values indicated by the two expressions stacked on the right side of Eq. (5.116), we learn that the interface between the viscous sublayer and the fully turbulent sublayer is located at $y^+ \cong 11.6$. It has been shown [2] that this viscous sublayer thickness (namely, the y^+ thickness of order 10) is a manifestation of the local Reynolds number criterion for transition to turbulence (Eq. (5.86) and Appendix F).

A simpler empirical $u^+(y^+)$ expression that approximates most of the curve represented by Eq. (5.116) is Prandtl's $\frac{1}{7}$th power law [14]:

$$u^+ = 8.7\,(y^+)^{1/7} \tag{5.118}$$

An equally compact formula for the wall shear stress $\tau_{w,x}$ is obtained by stretching the validity of Eq. (5.118) all the way to the edge of the outer (wake) region, where $\bar{u} = U_\infty$ and $y = \delta$:

$$\frac{U_\infty}{(\tau_{w,x}/\rho)^{1/2}} = 8.7\left[\frac{\delta}{\nu}\left(\frac{\tau_{w,x}}{\rho}\right)^{1/2}\right]^{1/7} \tag{5.119}$$

This equation relates $\tau_{w,x}$ to the outer thickness of the turbulent boundary layer, δ. A second relation between $\tau_{w,x}$ and δ is the integral form of the full momentum, Eq. (5.101):

$$\frac{d}{dx}\int_0^\delta \bar{u}(U_\infty - \bar{u})\,dy = \frac{\tau_{w,x}}{\rho} \tag{5.120}$$

Using the $\frac{1}{7}$th power law (5.118) for $\bar{u}$ in the integrand, the system (5.119)–(5.120) can be solved for the wall shear stress and the boundary layer thickness:

$$\frac{\tau_{w,x}}{\rho U_\infty^2} = \frac{1}{2}C_{f,x} = 0.0296\left(\frac{U_\infty x}{\nu}\right)^{-1/5} \tag{5.121}$$

$$\overline{\tau}_{w,L} = 0.037\rho U_\infty^2 Re_L^{-1/5} \tag{5.121'}$$

$$\frac{\delta}{x} = 0.37\left(\frac{U_\infty x}{\nu}\right)^{-1/5} \tag{5.122}$$

The derivation of the *average* shear stress $\overline{\tau}_{w,L}$, which is averaged over a wall of length L, is presented in Eqs. (2) and (3) in Example 5.3 at the end of this section. Multiplied by the total area of the swept wall, the average shear stress permits the calculation of the total tangential drag force exerted by the flow on the plane wall.

The local skin friction coefficient (5.121) may be used at Reynolds numbers as high as $Re_x \cong 10^8$. This formula shows that $\tau_{w,x}$ decreases as $x^{-1/5}$ in the downstream direction, that is, at a much slower rate than in the laminar section (Eq. (5.53)). This trend is illustrated in Figure 5.12.

According to Eq. (5.122), the boundary layer thickness increases as $x^{4/5}$, that is, almost linearly. This increase is considerably steeper than in the leading laminar section, where δ increases as $x^{1/2}$. This trend is presented also in Figure 5.12. Unlike their purely theoretical (predictive) counterparts in the laminar boundary layer, the analytical forms of the turbulent layer, specifically the exponents in $\tau_{w,x} \sim x^{-1/5}$ and $\delta \sim x^{4/5}$, are empirical (descriptive), simply the result of having adopted the empirical curve fit (5.118) at the start of the analysis. In other words, different $u^+(y^+)$ expressions (e.g. Eq. (5.116)) lead to different formulas for $\tau_{w,x}$, $\overline{\tau}_{w,L}$, and δ.

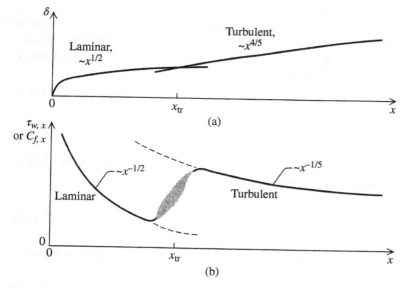

Figure 5.12 The behavior of the boundary layer thickness and wall shear stress in the laminar and turbulent sections of the boundary layer over a plane wall.

5.4.5 Heat Transfer

In $Pr \gtrsim 1$ fluids, the structure of the time-averaged temperature distribution in the turbulent boundary layer mimics that of the longitudinal velocity distribution. Figure 5.11 shows that sufficiently close to the wall the apparent heat flux q''_{app} is independent of the distance to the wall, y. This region is approximately as thick as the constant-τ_{app} region. In it, both sides of the energy, Eqs. (5.102) or (5.108), are zero; therefore, integrating in y we obtain

$$q''_{app} = \text{constant} = q''_{w,x} \tag{5.123}$$

The local wall heat flux $q''_{w,x}$ is the special value of q''_{app} at $y = 0$, because at the wall ε_H and q''_{eddy} are zero (review Eq. (5.106) and the definitions under Eq. (5.105)). The constant-q''_{app} statement can be rewritten as

$$-(k + \rho c_P \varepsilon_H)\frac{\partial \overline{T}}{\partial y} = q''_{w,x} \tag{5.124}$$

It is possible to obtain an expression for the local heat transfer coefficient $h_x = q''_{w,x}/(T_w - T_\infty)$ by combining Eq. (5.124) with the analogous equation recommended by the constant-τ_{app} condition (5.115):

$$(\mu + \rho \varepsilon_M)\frac{\partial \overline{u}}{\partial y} = \tau_{w,x} \tag{5.125}$$

Dividing Eqs. (5.125) and (5.124) side by side,

$$\frac{\rho(\nu + \varepsilon_M)}{\rho c_P(\alpha + \varepsilon_H)}\frac{d\overline{u}}{d\overline{T}} = -\frac{\tau_{w,x}}{q''_{w,x}} \tag{5.126}$$

and assuming that $\nu = \alpha$ and $\varepsilon_M = \varepsilon_H$ (i.e. that $Pr = 1$ and $Pr_t = 1$), we obtain

$$\frac{1}{c_P}\frac{d\overline{u}}{d\overline{T}} = -\frac{\tau_{w,x}}{q''_{w,x}} \qquad (Pr = Pr_t = 1) \tag{5.127}$$

Integrating from the wall ($\overline{u} = 0, \overline{T} = T_w$ at $y = 0$) to a large y where $\overline{u} \cong U_\infty$ and $\overline{T} \cong T_\infty$ yields

$$\frac{U_\infty}{c_P(T_\infty - T_w)} = -\frac{\tau_{w,x}}{q''_{w,x}} \tag{5.128}$$

Rearranged, Eq. (5.128) states that the local heat transfer coefficient $h_x = q''_{w,x}/(T_w - T_\infty)$ is proportional to the local wall shear stress $\tau_{w,x}$. This result can be nondimensionalized by defining the local Stanton number[8]:

$$St_x = \frac{h_x}{\rho c_P U_\infty} = \frac{q''_{w,x}}{\rho c_P U_\infty(T_w - T_\infty)} \tag{5.129}$$

or

$$St_x = \frac{Nu_x}{Pe_x} = \frac{Nu_x}{Re_x Pr} \tag{5.129'}$$

and by recognizing the $\frac{1}{2}C_{f,x}$ definition listed as the first of Eqs. (5.121). The dimensionless form of Eq. (5.128)

$$St_x = \frac{1}{2}C_{f,x} \qquad (Pr = Pr_t = 1) \tag{5.130}$$

---

[8] Thomas Edward Stanton (1865–1931) was a professor of engineering at the Bristol University College. He devoted his research to the relationship between heat transfer and fluid flow, the aerodynamic loads on solid structures, and the air cooling of internal combustion engines.

is known best as the Reynolds analogy between wall friction and heat transfer, in recognition of Osborne Reynolds' essay [15], in which he argued that a relationship between wall shear stress and heat flux must exist.

The preceding analysis shows that the Reynolds analogy holds strictly for $Pr = 1$ and $Pr_t = 1$. For fluids with molecular Prandtl numbers (Pr) different than 1, Colburn [16] proposed an empirical correlation that recommends $Pr^{2/3}$ as a factor:

$$St_x \, Pr^{2/3} = \frac{1}{2} C_{f,x} \qquad (Pr \gtrsim 0.5) \tag{5.131}$$

This can be restated in terms of Nu_x and Re_x, in accord with Eqs. (5.121) and (5.129'):

$$Nu_x = \frac{1}{2} C_{f,x} \, Re_x Pr^{1/3}$$
$$= 0.0296 Re_x^{4/5} Pr^{1/3} \qquad (Pr \gtrsim 0.5) \tag{5.131'}$$

Equation (5.131) is known as the Colburn[9] analogy between wall friction and heat transfer, relative to which the Reynolds analogy (5.130) represents the special case $Pr = 1$.

Although Eq. (5.131') was developed for an isothermal wall, it works satisfactorily when the wall heat flux is uniform. The Nu_x value for a wall with uniform heat flux is only 4% greater than the value furnished by Eq. (5.131'). Note further that when the wall heat flux is uniform, the local Nusselt number is defined by $Nu_x = q_w'' x / k[T_w(x) - T_\infty]$. Regardless of the thermal boundary condition that may exist at the wall, the reference temperature for evaluating the properties in Eqs. (5.131) and (5.131') is the average film temperature $(\overline{T}_w + T_\infty)/2$, where $\overline{T}_w$ is the x-averaged wall temperature.

Colburn's empirical formula (5.131) was predicted theoretically in 1984 (Ref. [2], first edition), based on the view that the time-averaged wall shear and heat flux are dominated by an array of discrete regions of direct contact between the free-stream fluid (U_∞, T_∞) and the wall. According to this theory, the shear flow under each contact spot is *laminar*, with characteristics similar to the leading laminar section of the boundary layer. An additional contribution of the theory is the proof that the Colburn analogy (5.131) applies only to $Pr \gtrsim 0.5$ fluids. For low-Pr fluids the factor $Pr^{2/3}$ is replaced on the left side of Eq. (5.131) by the group $cPr^{1/2}$, where c is a constant of order 10.

In summary, the local heat transfer coefficient h_x (or St_x) may be calculated using Eq. (5.131), for which $\frac{1}{2}C_{f,x}$ is given by Eq. (5.121). The proportionality between St_x and $\frac{1}{2}C_{f,x}$ means that the x dependence of the local heat transfer coefficient is similar to that of the wall shear stress. Figure 5.13 shows that in the turbulent section h_x decreases as $x^{-1/5}$ in the downstream direction. This behavior departs significantly from the $h_x \sim x^{-1/2}$ decrease that prevails in the laminar section of the boundary layer.

Consider finally the *average* heat transfer coefficient $\overline{h}_L$ for the wall of total length L shown in Figure 5.13. This quantity is needed to calculate the total heat transfer rate through the wall, per unit length in the direction normal to the (x, y) plane, namely, $q' = \overline{h}_L(T_w - T_\infty)L$. The value of $\overline{h}_L$ depends on how the transition length

---

9 Allan Philip Colburn (1904–1955) was a professor of chemical engineering at the University of Delaware. His research focused on the condensation of water vapor and the analogy between heat, mass, and momentum transfer.

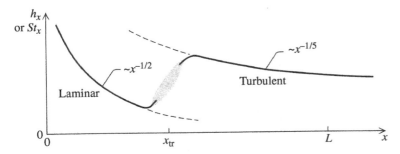

Figure 5.13 The behavior of the local heat transfer coefficient in the laminar and turbulent sections of the boundary layer. (Compare this figure with Figure 5.12.)

x_{tr} compares with the total length L:

$$\bar{h}_L = \frac{1}{L}\left(\int_0^{x_{tr}} \underset{\substack{\text{eq. (5.79b)}}}{h_{x,\text{ laminar}}} \; dx + \int_{x_{tr}}^L \underset{\substack{\text{Eqs. (5.131)}\\ \text{and (5.121)}}}{h_{x,\text{ turbulent}}} \; dx \right) \tag{5.132}$$

Using the h_x formulas indicated under each integrand, we find that the Nusselt number based on $\bar{h}_L$ and L depends on the longitudinal transition Reynolds number ($Re_{x,tr} = U_\infty x_{tr}/\nu$):

$$\overline{Nu}_L = \frac{\bar{h}_L L}{k} = 0.664 Pr^{1/3} Re_{x,\text{tr}}^{1/3} + 0.037 Pr^{1/3}(Re_L^{4/5} - Re_{x,\text{tr}}^{4/5}) \tag{5.133}$$

Taking the right side of Eq. (5.85) as representative of the transition Reynolds number, $Re_{x,\,tr} \cong 5 \times 10^5$, the above $\overline{Nu}_L$ formula becomes

$$\overline{Nu}_L = 0.037 Pr^{1/3}(Re_L^{4/5} - 23\,550) \quad (Pr \gtrsim 0.5) \tag{5.134}$$

In view of the ingredients used in its derivation, this formula is valid for $5 \times 10^5 < Re_L < 10^8$ and $Pr \gtrsim 0.5$. At Reynolds numbers Re_L smaller than 5×10^5, the entire length L is covered by laminar boundary layer flow, and the $\overline{Nu}_L$ formula is given by Eq. (5.82b).

Example 5.2 *Turbulent Boundary Layer: Heat Transfer*

The flat (tabular) iceberg shown in Figure E5.2 drifts over the ocean, as it is driven by the wind that blows over the top. The iceberg may be modeled as a block of frozen fresh water at $0\,°C$. The temperature of the surrounding seawater is $10\,°C$, and the relative velocity between it and the iceberg is $10\,cm/s$. The length of the iceberg in the direction of drift is $L = 100\,m$.

The relative motion between the seawater and the flat bottom of the iceberg produces a boundary layer of length L. The $10\,°C$ temperature difference across this boundary layer drives a heat flux into the bottom surface of the iceberg. This heating effect causes the steady thinning of the flat piece of ice. If $H(t)$ is the instantaneous height of the ice slab, calculate the ice melting rate dH/dt averaged over the swept length of the iceberg.

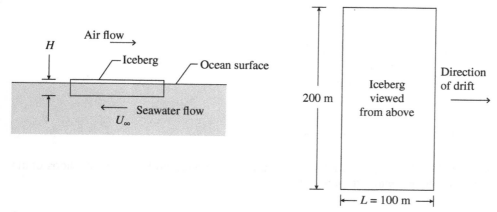

Figure E5.2

Solution

The following calculations are approximate because they are based on the crude model that the bottom surface of the iceberg is plane and that the drift is steady. This is why it is sufficient to approximate the properties of the seawater as those of fresh water at the film temperature of $5\,°C$:

$$v \cong 0.015 \text{ cm}^2/\text{s} \quad k \cong 0.57 \text{W/m·K} \quad Pr \cong 11.13$$

First, we calculate the Reynolds number, to determine the prevailing flow regime:

$$Re_L = \frac{U_\infty L}{v} = 10\frac{\text{cm}}{\text{s}}100 \text{ m}\frac{\text{s}}{0.015 \text{ cm}^2}$$
$$= 6.7 \times 10^6 \quad \text{(turbulent flow)}$$

Next, starting with Eq. (5.134) we calculate, in order, the average Nusselt number, the L-averaged heat transfer coefficient, and the L-averaged heat flux that lands on the bottom of the iceberg:

$$\overline{Nu}_L = 0.037 Pr^{1/3}(Re_L^{4/5} - 23\,550)$$
$$= 0.037(11.13)^{1/3}(2.89 \times 10^5 - 23\,550) = 2.19 \times 10^4$$

$$\overline{h}_L = \frac{k}{L}\overline{Nu}_L = 0.57\frac{\text{W}}{\text{m·K}}\frac{1}{100 \text{ m}}2.19 \times 10^4 =$$
$$\cong 125 \text{ W/m}^2\text{·K}$$

$$\overline{q}''_{w,L} = \overline{h}_L \Delta T = 125\frac{\text{W}}{\text{m}^2\text{·K}}10\,°C$$
$$\cong 1250 \text{ W/m}^2$$

The average heat flux $\overline{q}''_{w,L}$ is absorbed by the melting process (review Section 4.7),

$$\overline{q}''_{w,L} = \rho h_{sf}\frac{dH}{dt}$$

where h_{sf} is the latent heat of melting for ice, $h_{sf} = 333.4 \text{ kJ/kg}$. The average melting rate is therefore

$$\frac{dH}{dt} = \frac{\overline{q}''_{w,L}}{\rho h_{sf}} = 1250\frac{\text{W}}{\text{m}^2}\frac{(0.1 \text{ m})^3}{1 \text{ kg}}\frac{\text{kg}}{333.4 \text{ kJ}}$$
$$= 3.75 \times 10^{-6}\text{m/s} = 1.34 \text{ cm/h}$$

Example 5.3 *Turbulent Boundary Layer: Wall Friction*

Seen from above, the flat iceberg described in the preceding example and Figure E5.2 drifts stably in the direction of its shorter side, $L = 100$ m. The side oriented across the direction of drift is 200 m long. Calculate the time-averaged total drag force exerted by the water on the bottom surface of the iceberg.

Solution

The total drag force is the area integral of all the shear stresses of type $\tau_{w,x}$,

$$F_D = \overline{\tau}_{w,L} A \tag{1}$$

where A is the swept area, $(100\,\text{m})(200\,\text{m}) = 2 \times 10^4\,\text{m}^2$, and $\overline{\tau}_{w,L}$ is the L-averaged shear stress

$$\overline{\tau}_{w,L} = \frac{1}{L} \int_0^L \tau_{w,x}\, dx \tag{2}$$

Assuming that the Reynolds number is high enough so that L is covered almost entirely by turbulent flow, we can use Eq. (5.121) in the above integral; the result is

$$\overline{\tau}_{w,L} = 0.037 \rho U_\infty^2 Re_L^{-1/5} \tag{3}$$

In the present example this formula yields

$$\overline{\tau}_{w,L} = 0.037 \frac{1\,\text{kg}}{(0.1\,\text{m})^3} \left(10 \frac{\text{cm}}{\text{s}}\right)^2 (6.7 \times 10^6)^{-1/5}$$

$$\cong 0.016\,\text{kg/m·s}^2 = 0.016\,\text{N/m}^2$$

Therefore, the total drag force on the bottom surface is

$$F_D = 0.016 \frac{\text{N}}{\text{m}^2} (2 \times 10^4\,\text{m}^2) = 320\,\text{N} \cong 72\,\text{lbf}$$

For steady drift this bottom drag force must be balanced by a driving drag force associated with the air flow over the top of the iceberg.

5.5 Other External Flows

5.5.1 Single Cylinder

In the preceding two sections we learned the most basic aspects of laminar and turbulent convection by focusing on the simplest wall geometry – the flat plate. These aspects are present in more complicated configurations and account for the main body of the external convection literature and handbooks. Here are a few examples:

Consider the heat transfer between a long cylinder oriented across a fluid stream of uniform velocity U_∞ and temperature T_∞, Figure 5.14. The temperature of the cylindrical surface is uniform, T_w. There are many heat transfer correlations for this configuration but, generally speaking, they are not in very good agreement

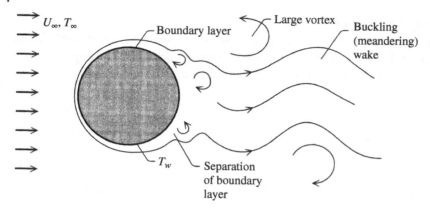

Figure 5.14 Single cylinder (or sphere) in cross-flow and the features of the surrounding flow.

with the experimental data. One correlation that is based on data from many independent sources is [17]

$$\overline{Nu}_D = 0.3 + \frac{0.62 Re_D^{1/2} Pr^{1/3}}{[1 + (0.4/Pr)^{2/3}]^{1/4}} \left[1 + \left(\frac{Re_D}{282\,000} \right)^{5/8} \right]^{4/5} \tag{5.135}$$

where $\overline{Nu}_D = \overline{h}D/k$. This formula holds for all values of Re_D and Pr, provided the Péclet number $Pe_D = Re_D Pr$ is greater than 0.2. In the intermediate Reynolds number range $7 \times 10^4 < Re_D < 4 \times 10^5$, Eq. (5.135) predicts $\overline{Nu}_D$ values that can be 20% smaller than those furnished by direct measurement. The physical properties needed for calculating Nu_D, Pr, and Re_D are evaluated at the film temperature $(T_\infty + T_w/2)$. Equation (5.135) applies also to a cylinder with uniform heat flux, in which case the average heat transfer coefficient $\overline{h}$ is based on the perimeter-averaged temperature difference between the cylindrical surface and the free stream. In flows slow enough so that $Pe_D < 0.2$, more accurate than Eq. (5.135) is the formula [18]

$$\overline{Nu}_D = \frac{1}{0.8237 - 0.5\,\ln(Pe_D)} \tag{5.136}$$

This agrees well with experimental measurements conducted in air; however, it has not been tested for a wide range of the Prandtl numbers.

The single cylinder in cross-flow has been studied extensively from a fluid mechanics standpoint. The first portion of the surrounding flow that becomes turbulent as the Reynolds number $Re_D = U_\infty D/\nu$ increases is the wake. When the Reynolds number is of order 1 or smaller, the flow is nearly symmetric about the transversal diameter of the cylinder.

Eddies of size comparable with the cylinder diameter begin to shed from the cylinder periodically when $Re_D \sim 40$: this is another manifestation of the local Reynolds number criterion of transition to turbulence (Appendix F). The wake becomes increasingly turbulent as Re_D increases; however, the shedding of vortices of diameter comparable with D remains a feature of the wake over most of the $10^2 < Re_D < 10^7$ range. The frequency $f_v(\text{s}^{-1})$ with which the vortices shed is roughly proportional to the flow velocity [19]:

$$f_v \sim 0.28 \frac{U_\infty}{D} \tag{5.137}$$

The vortex-shedding frequency is important in the design of cylinders in cross-flow (e.g. tubes in a heat exchanger) because it may induce and sustain vibrations in the mechanical structure.

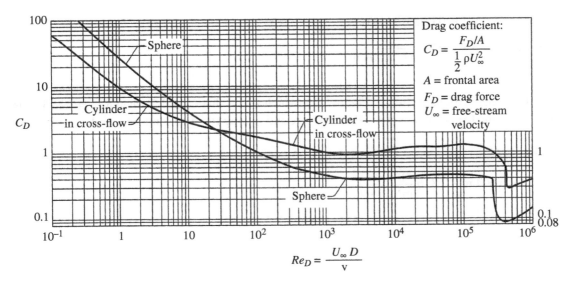

Figure 5.15 Drag coefficients of a smooth sphere and a single smooth cylinder in cross-flow. Source: Bejan [2].

The total, time-averaged drag force F_D exerted by the flow on the cylinder can be calculated with the help of Figure 5.15. Plotted on the ordinate is the *drag coefficient* C_D, which is a dimensionless way of expressing the drag force:

$$C_D = \frac{F_D/A}{\frac{1}{2}\rho U_\infty^2} \tag{5.138}$$

The area A is the frontal area of the cylinder (as seen by the approaching stream), namely, $A = LD$, if L is the length of the cylinder. The drag coefficients of several other body shapes have been compiled by Simiu and Scanlan [20].

Example 5.4 *Cylinder in Cross-Flow*
A pin fin of average temperature 100 °C extends into a 1-m/s uniform air stream of temperature 200 °C. The pin fin diameter is 4 mm. Calculate the average heat transfer coefficient and the drag force between the pin fin and the surrounding air.

Solution

The air properties are evaluated at the film temperature, $(100\,°C + 200\,°C)/2 = 150\,°C$.

$$v = 0.288 \text{ cm}^2/\text{s} \qquad k = 0.036 \text{ W/m·K}$$
$$\rho = 0.846 \text{ kg/m}^3 \qquad Pr = 0.69$$

The average heat transfer coefficient can be calculated by using Eq. (5.135):

$$Re_D = \frac{U_\infty D}{v} = 1\,\frac{\text{m}}{\text{s}}\,0.004 \text{ m}\,\frac{\text{s}}{0.288 \times 10^{-4} \text{ m}^2}$$
$$= 138.9$$

$$\overline{Nu}_D = 0.3 + 5.66\,(1 + 0.0086)^{4/5} \cong 6$$

$$\overline{h} = \overline{Nu}_D \frac{k}{D} = 6 \times 0.036\ \frac{W}{m\cdot K}\,\frac{1}{0.004\ m}$$

$$= 54\ W/m^2{\cdot}K$$

For the drag force calculation, we turn to Figure 5.15, which shows that $C_D \cong 1.7$ when $Re_D = 140$. If L is the length of the fin, the frontal area is $A = DL$, and the resulting drag force per unit fin length is

$$\frac{F_D}{L} = C_D D \frac{1}{2}\rho U_\infty^2$$

$$= 1.7 \times 0.004\ m \frac{1}{2}\,0.846\ \frac{kg}{m^3}\left(1\frac{m}{s}\right)^2$$

$$\cong 0.003\ N/m$$

5.5.2 Sphere

For the average heat transfer coefficient between an isothermal spherical surface (T_w) and an isothermal free stream (U_∞, T_∞), Figure 5.18, Whitaker [21] proposed the correlation

$$\overline{Nu}_D = 2 + (0.4Re_D^{1/2} + 0.06Re_D^{2/3})Pr^{0.4}\left(\frac{\mu_\infty}{\mu_w}\right)^{1/4} \tag{5.139}$$

which has been tested for $0.71 < Pr < 380$, $3.5 < Re_D < 7.6 \times 10^4$, and $1 < (\mu_\infty/\mu_w) < 3.2$. All the physical properties in Eq. (5.139) are evaluated at the free-stream temperature T_∞, except μ_w, which is the viscosity evaluated at the surface temperature, $\mu_w = \mu(T_w)$. It is worth noting that the no-flow limit of this formula, $\overline{Nu}_D = 2$, agrees with the pure conduction estimate for steady radial conduction between the spherical surface and the motionless, infinite conducting medium that surrounds it (Section 2.3). Equation (5.139) also applies to spherical surfaces with uniform heat flux, with the understanding that in such cases $\overline{Nu}_D$ is based on the surface-averaged temperature difference between the sphere and the surrounding stream, $\overline{Nu}_D = \overline{h}D/k = q_w''D/k(\overline{T}_w - T_\infty)$.

Figure 5.15 shows the dimensionless drag coefficient for a sphere suspended in a uniform stream. The total time-averaged drag force F_D experienced by the holder of the sphere can be calculated with Eq. (5.138), in which the frontal area this time is $A = (\pi/4)D^2$. The regimes that are exhibited by the flow around the sphere are similar to those encountered in the case of a cylinder in cross-flow.

5.5.3 Other Body Shapes

The single cylinder and the sphere discussed in Sections 5.5.1 and 5.5.2 are the simplest geometries of bodies immersed in a uniform flow of different temperature. They are the simplest because they have a single length scale, the diameter. The heat transfer literature contains a wealth of results for bodies of other shapes. Some of these formulas have been reviewed critically by Yovanovich [22], who also proposed a universal correlation for spheroids, that is, bodies of nearly spherical shape.

As illustrated in the lower part of Figure 7.11, a spheroid can be obtained by rotating an ellipse about one of the semiaxes. The spheroid geometry is characterized by two dimensions (C, B), where C is the semiaxis aligned with the free stream U_∞. As length scale for the definition of the Reynolds and Nusselt numbers, Yovanovich chose the square root of the spheroid surface A:

$$\mathcal{L} = A^{1/2} \tag{5.140}$$

therefore,

$$Re_{\mathcal{L}} = \frac{U_\infty \mathcal{L}}{\nu} \quad \text{and} \quad \overline{Nu}_{\mathcal{L}} = \frac{\overline{h}\mathcal{L}}{k} \tag{5.141}$$

The correlation developed by Yovanovich [22] is

$$\overline{Nu}_{\mathcal{L}} = \overline{Nu}_{\mathcal{L}}^0 + \left[0.15\left(\frac{p}{\mathcal{L}}\right)^{1/2} Re_{\mathcal{L}}^{1/2} + 0.35 Re_{\mathcal{L}}^{0.566} \right] Pr^{1/3} \tag{5.142}$$

where p is the maximum (equatorial) perimeter of the spheroid, in the plane perpendicular to the flow direction U_∞. The constant $\overline{Nu}_{\mathcal{L}}^0$ is the overall Nusselt number in the no-flow (pure conduction) limit; representative values of this constant are listed in Table 7.1, which is to be read in conjunction with Figure 7.11. Equation (5.142) is recommended for $0 < Re_{\mathcal{L}} < 2 \times 10^5$, $Pr > 0.7$, and $0 < C/B < 5$. The physical properties involved in the definitions of $\overline{Nu}_{\mathcal{L}}$ and $Re_{\mathcal{L}}$ are evaluated at the film temperature.

5.5.4 Arrays of Cylinders

A considerably more complicated geometry is that of a large number of regularly spaced (parallel) cylinders in cross-flow. As indicated in Figure 5.16, this geometry is characterized by the cylinder diameter D, the longitudinal spacing of two consecutive rows (longitudinal pitch, X_l), and the transversal spacing of two consecutive cylinders (transversal pitch, X_t). It is assumed that the array is wide enough, that is, enough cylinders exist in each row, so that the top and bottom boundaries (the shroud) do not affect the overall flow and heat transfer characteristics of the array.

In the field of heat exchanger design, arrays occur when tube banks or tube bundles are placed perpendicularly to a stream of fluid. Assume, for example, that this stream is warm. A second stream of cold fluid flows through the many tubes of the bundle, and, because of the heat transfer interaction between each tube and the external cross-flow, the second stream picks up the enthalpy lost by the warm stream. This two-stream

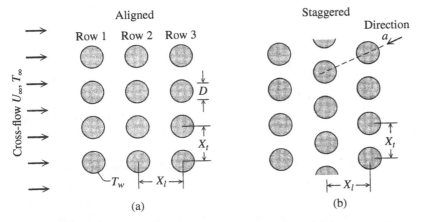

Figure 5.16 Banks of cylinders in cross-flow: aligned (a) versus staggered (b).

arrangement, and the hardware that allows the streams to experience heat transfer without mixing with one another, is an example of a *heat exchanger* (see Chapter 9).

The most common types of arrays are two, where the cylinders are aligned one behind the other in the direction of flow (Figure 5.16a) and where the cylinders are *staggered* (Figure 5.16b). Aligned cylinders form rectangles with the centers of their cross sections, whereas staggered cylinders form isosceles triangles. Significant work has been published on the heat transfer performance of banks of cylinders in cross-flow. The overall Nusselt number formulas presented below are based on Zukauskas' recommendations [23], where

$$\overline{Nu}_D = \frac{\overline{h}D}{k} \tag{5.143}$$

and $\overline{h}$ is the heat transfer coefficient averaged over all the cylindrical surfaces in the array. The total area of these surfaces is $nm\pi DL$, where n is the number of rows, m is the number of cylinders in each row (across the flow direction), and L is the length of the array in the direction perpendicular to the plane of Figure 5.16.

For *aligned* arrays of cylinders, the array-averaged Nusselt number is anticipated within $\pm15\%$ by

$$\overline{Nu}_D = 0.9C_n Re_D^{0.4} Pr^{0.36} \left(\frac{Pr}{Pr_w}\right)^{1/4}, \qquad Re_D = 1 - 10^2$$

$$= 0.52C_n Re_D^{0.5} Pr^{0.36} \left(\frac{Pr}{Pr_w}\right)^{1/4}, \qquad = 10^2 - 10^3$$

$$= 0.27C_n\, Re_D^{0.63} Pr^{0.36} \left(\frac{Pr}{Pr_w}\right)^{1/4}, \qquad = 10^3 - (2 \times 10^5)$$

$$= 0.033C_n Re_D^{0.8} Pr^{0.4} \left(\frac{Pr}{Pr_w}\right)^{1/4}, \qquad = (2 \times 10^5) - (2 \times 10^6) \tag{5.144}$$

where C_n is a function of the total number of rows in the array (Figure 5.17). The Reynolds number Re_D is based on the average velocity through the narrowest cross section formed by the array, that is, the maximum average velocity U_{max}:

$$Re_D = \frac{U_{max} D}{\nu} \tag{5.145}$$

In the case of aligned cylinders, the narrowest flow cross section forms in the plane that contains the centers of all the cylinders of one row. The conservation of mass through such a plane (see Figure 5.16a) requires that

$$U_\infty X_t = U_{max} (X_t - D) \tag{5.146}$$

All the physical properties except Pr_w in Eqs. (5.144) are evaluated at the mean temperature of the fluid that flows through the spaces formed between the cylinders. The mean (or "bulk") temperature is a concept defined in Section 6.2.2, in which the cross-flow of Figure 5.16 can be viewed as a stream that flows through the "duct" constituted by all the spaces between cylinders. The denominator Pr_w is the Prandtl number evaluated at the temperature of the cylindrical surface, $Pr_w = Pr(T_w)$.

Figure 5.17 shows that the number of rows has an effect on the array-averaged Nusselt number only when n is less than approximately 16. In the $n < 16$ range, C_n and $\overline{Nu}_D$ increase as more rows are added to the array. This effect is analogous to the observation that the individual $\overline{h}$ value of a cylinder positioned in the front row is lower than that of the cylinder situated behind it. The front row cylinder is coated by a relatively smooth boundary layer formed by the undisturbed incoming stream U_∞, whereas a downstream cylinder

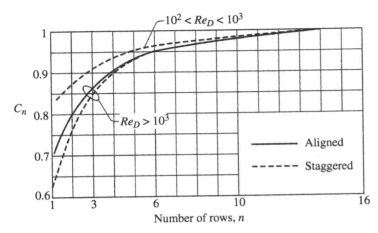

Figure 5.17 The effect of the number of rows on the array-averaged Nusselt number for banks of cylinders in cross-flow. Source: Zukauskas [23].

benefits from the heat transfer "augmentation" effect provided by the eddies of the turbulent wake created by the preceding cylinder.

For staggered arrays, these relations approximate $\overline{Nu}_D$ within ±15%:

$$\overline{Nu}_D = 1.04 C_n Re_D^{0.4} Pr^{0.36} \left(\frac{Pr}{Pr_w}\right)^{1/4}, \qquad Re_D = 1 - 500$$

$$= 0.71 C_n Re_D^{0.5} Pr^{0.36} \left(\frac{Pr}{Pr_w}\right)^{1/4}, \qquad = 500 - 10^3$$

$$= 0.35 C_n Re_D^{0.6} Pr^{0.36} \left(\frac{Pr}{Pr_w}\right)^{1/4} \left(\frac{X_t}{X_l}\right)^{0.2}, \qquad = 10^3 - (2 \times 10^5)$$

$$= 0.031 C_n Re_D^{0.8} Pr^{0.36} \left(\frac{Pr}{Pr_w}\right)^{1/4} \left(\frac{X_t}{X_l}\right)^{0.2}, \qquad = (2 \times 10^5) - (2 \times 10^6) \qquad (5.147)$$

The observations made in connection with Eqs. (5.144) apply here as well. A new effect is the role played by the aspect ratio of the isosceles triangle, X_t/X_l, which is felt at relatively large Reynolds numbers. The Reynolds number continues to be based on U_{max}, Eq. (5.145); however, which flow cross section is the narrowest depends on the slenderness of the isosceles triangle. For example, in the staggered array shown on the right side of Figure 5.16, the narrowest flow area occurs in the vertical plane drawn through the centers of one row of cylinders. In the other extreme, where X_l is considerably smaller than X_t, the strangling of the flow may occur through the area aligned with the sloped direction labeled a.

Means for calculating the drag experienced by a bundle of cylinders in cross-flow are presented in Section 9.5.4. The aggregate effect of the drag forces of all the cylinders is the pressure drop experienced by the stream that bathes the bundle.

5.5.5 Turbulent Jets

There are many flows and convection phenomena in which the solid boundaries are totally absent or irrelevant. Prime examples are the momentum-driven jets discussed below and the buoyancy-driven plumes treated

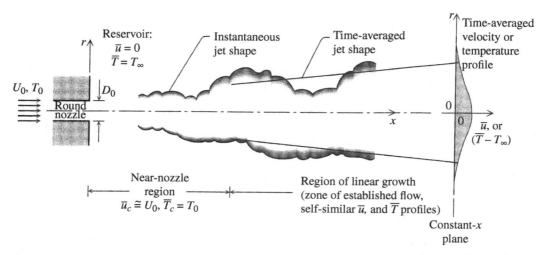

Figure 5.18 Turbulent, nonbuoyant round jet mixing with a stagnant reservoir of different temperature.

in Ref. [2]. These flows are particularly important in environmental applications, where, instead of the usual heat transfer problem (Section 1.2), the chief question is to what extent the stream mixes with and spreads through the surrounding fluid. Since the flow regime is usually turbulent and the solid boundaries are absent, these flows are recognized under the general title of *free turbulent flows*.

In this section we consider the simplest and most basic of the nonbuoyant free turbulent flows, namely, the round turbulent jet (Figure 5.18). A more extensive coverage of this class of convection problems can be found in Chapter 9 of Ref. [2]. The jet of Figure 5.18 issues from a round nozzle of inner diameter D_0 with a cross section-averaged velocity U_0 and a uniform nozzle temperature T_0. The jet flows into a stationary reservoir containing the same type of fluid as the jet itself. The temperature of the reservoir fluid, T_∞, is different than the bulk temperature in the plane of the nozzle, T_0.

Experiments with turbulent jets indicate that the time-averaged velocity field $\bar{u}(r, x)$ has two distinct regions. In the near-nozzle region of axial length of approximately six nozzle diameters, the centerline velocity $\bar{u}_c = \bar{u}(0, x)$ is practically independent of x. In this first region the eddy mixing that operates across the imaginary cylindrical surface between jet and reservoir has not had time to reach the jet centerline. The time-averaged centerline temperature $\bar{T}_c = \bar{T}(0, x)$ is also constant and equal to T_0 in this region.

The jet velocity and temperature distributions described below refer to the second region of the jet, which develops at x values greater than approximately $6D_0$. In this downstream region it is observed that the radial thickness of the $\bar{u}(r, x)$ and $\bar{T}(r, x)$ profiles increases almost linearly with x. This key feature is illustrated in Figure 5.18 along with the characteristic radial variation (bell-shaped profile) of both $\bar{u}$ and $\bar{T}$.

An analytical solution for the time-averaged flow and temperature field can be developed by exploiting the observation that the downstream region of the jet is shaped like a cone. This observation leads, among other things, to the conclusion that the eddy diffusivity ε_M is a constant inside this region of the jet [2]. More practical than the analytical solution is a set of Gaussian expressions that correlate the experimental measurements furnished by many investigators [2]:

$$\bar{u} = \bar{u}_c \exp\left[-\left(\frac{r}{b}\right)^2\right], \qquad b \cong 0.107x \tag{5.148}$$

$$\overline{T} - \overline{T}_\infty = (\overline{T}_c - T_\infty) \exp\left[-\left(\frac{r}{b_T}\right)^2\right], \qquad b_T \cong 0.127x \tag{5.149}$$

The linear growth of the jet mixing region is visible in the empirical expressions listed above for the transversal length scales $b(x)$ and $b_T(x)$. The numerical coefficients 0.107 and 0.127 are accurate within $\pm 3\%$ and are based on experiments conducted mainly in air and water, that is, in $Pr \sim 1$ fluids.

The solution, Eqs. (5.148) and (5.149), is complete only when the x-dependent *centerline* quantities $\overline{u}_c$ and $\overline{T}_c - T_\infty$ are known. These follow from two important theorems, which are proposed as problems at the end of this chapter. The first theorem states that the total longitudinal momentum in any constant-x cut through the jet is conserved (i.e. is independent of x):

$$2\pi \int_0^\infty \rho \overline{u}^2 r \, dr = \text{constant} = \rho U_0^2 \frac{\pi}{4} D_0^2 \tag{5.150}$$

The value of the constant was evaluated from the left-side integral in the plane of the nozzle, where $\overline{u} = U_0$. Combined, Eqs. (5.148) and (5.150) deliver the centerline velocity:

$$\overline{u}_c \cong 6.61 \frac{U_0 D_0}{x} \tag{5.151}$$

The centerline velocity decreases as x^{-1} in the downstream direction, and it is equal to the mean nozzle velocity U_0 at $x \cong 6.61 D_0$ downstream from the nozzle. This location marks the beginning of the linear growth region of the turbulent jet, Figure 5.18.

The second theorem is that the energy flow through a constant-x plane of the jet is conserved:

$$2\pi \int_0^\infty \rho c_p \overline{u} \, (\overline{T} - T_\infty) \, r \, dr = \text{constant} = \rho c_p U_0 \, (T_0 - T_\infty) \frac{\pi}{4} D_0^2 \tag{5.152}$$

The constant listed on the right side is determined by performing the integral in the plane of the nozzle, where $\overline{u} = U_0$ and $\overline{T} = T_0$. Using Eqs. (5.148) and (5.149) in the integrand on the left side of Eq. (5.152), we obtain the expression for the temperature difference between the centerline and the reservoir:

$$\overline{T}_c - T_\infty \cong 5.65 \frac{(T_0 - T_\infty)D_0}{x} \tag{5.153}$$

This result shows that the excess centerline temperature decreases as x^{-1} in the downstream direction. Together, Eqs. (5.149) and (5.153) describe the extent to which the hotness or coldness of the jet has spread into the isothermal reservoir.

When a jet is oriented perpendicular to a solid surface, the spot that is struck by the jet is characterized by a relatively high heat transfer coefficient. Arrays of circular or two-dimensional (curtain) impinging jets can be used to generate an enhanced heat transfer coefficient over a wider area [24].

5.6 Evolutionary Design

5.6.1 Size of Object with Heat Transfer

Traditionally, the discipline of heat transfer is about the relationships between heat currents and temperature differences or gradients. This view is useful as reference. The world, nature, and our own contrivances are things (objects) that in addition to flow and "transfer" have shape, structure, and configurations that evolve

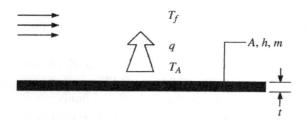

Figure 5.19 Plate surface with specified heat transfer rate.

in a discernible direction in time [25]. Even more basic is the fact that every object has a size, and its size is "characteristic," common, and recognizable. Big heat exchangers belong on big vehicles, small lungs on small mammals. Why?

The place of "size" in thermal science is now firmly established as a new and growing chapter known as constructal design, or evolutionary design [1, 26]. An introductory example is Figure 5.19. The surface of size A and temperature T_A transfers heat to its surroundings of temperature T_f. The heat transfer coefficient is h, therefore

$$q = hA(T_A - T_f) \tag{5.154}$$

This proportionality holds for all modes of heat transfer (convection, conduction, radiation) with more specific definitions of what the "h" factor represents in each case and how it depends on the two temperatures. For now, think of Figure 5.19 as a configuration with convection, forced or natural.

Known from Section 1.6 is that heat transfer is a mechanism for the destruction of useful power. The destroyed power (w_1) is the product $w_1 = T_f s_{gen}$, where the entropy generation rate is (cf., Eq. (1.57))

$$s_{gen} = \frac{q}{T_f} - \frac{q}{T_A} \tag{5.155}$$

Combining this with Eq. (5.154) and assuming that $T_A - T_f \ll (T_A, T_f)$, we obtain

$$s_{gen} = \frac{q^2}{T^2 hA} \tag{5.156}$$

where T [K] stands for either T_A or T_f. The power destroyed by heat transfer is $w_1 = q^2/(ThA)$.

This is not the only way in which power is being destroyed. Power is also destroyed in order to carry on earth the heat transfer surface A on the larger system that uses A. On a vehicle or animal, the power needed to transport an incremental mass (m) is proportional to m, approximately this way [25]: $w_2 = 2mgV$. Here V is the constructal speed of the vehicle, which is proportional to $M^{1/6}$, where M is the total mass of the vehicle. We treat V as a known parameter, which is dictated by the total vehicle mass M, which is fixed. In other words, the mass of the heat transfer device A is negligible relative to the mass of the whole mover.

In the simplest model, the mass m is equal to $\rho_w tA$, where ρ_w and t are the density and thickness of the wall of heat transfer area A. Adding w_1 and w_2 we obtain the total power destroyed because of A:

$$w_1 + w_2 = \frac{q^2}{ThA} + 2gV\rho_w tA \tag{5.157}$$

The area size for minimal power loss is

$$A = \frac{q}{(2gV\rho_w th)^{1/2}} \tag{5.158}$$

This size is proportional to the heat transfer duty q and inversely proportional to other parameters, such as $V^{1/2}$ and $h^{1/2}$. In other words, there is a characteristic size A and heat flux q/A that increase in proportion with $(V\rho_w th)^{1/2}$.

This demonstration of the physics of size was based on the simplest model. More complex heat transfer surfaces, such as the surfaces employed in heat exchangers and power plants of all kinds (stationary, mobile), can be deduced based on the same argument: the trade-off between reduced heat transfer irreversibility and reduced mass.

5.6.2 Evolution of Size

No matter how good, the size of the system with flows and freedom to morph does not remain the same forever. The size changes over time, and the direction of its "movie" tape is discernible and predictable. The big picture of the evolution phenomenon illustrated in Section 5.6.1 is summarized in Figure 5.20 and Ref. [25], on which the following passages are based.

To start, instead of flow object, flow device, or flow component, let us speak of "organ," because all the sections devoted to performance in this book are equally applicable to the organs of animate bodies.

The organ is alive with currents that flow one-way by overcoming resistances, obstacles, and all kinds of "friction." In thermodynamics, this universal phenomenon is called irreversibility, destruction of useful energy, power loss, and entropy generation. The fuel penalty associated with irreversibility is smaller when the organ is larger, because wider ducts and larger heat transfer surfaces pose less resistance to fluid and heat currents. In this limit, larger is better; see the descending curve in Figure 5.20.

Second, the moving body must consume fuel (or food) in order to carry the organ on earth. Likewise, the stationary power plant requires more fuel in order to have a larger component made, transported, installed, and maintained. This fuel penalty increases with the size of the organ and teaches that smaller is better; see the rising line in Figure 5.20.

The second penalty is in conflict with the first, and from this conflict emerges the idea – the prediction, the purely theoretical discovery – that the organ should have a characteristic size that is finite, not too large, not too small, but just right, for that particular vehicle. This was demonstrated in Section 5.6.1. When this trade-off is reached, the total fuel required by the vehicle is proportional to the total weight of its organs.

If the organ is constructed so as to be the size at which the two lines intersect, the sum of the two penalties will be minimal. This trade-off means that large organs (pipes, surfaces, wall material, heat exchangers, etc.)

Figure 5.20 Large organs belong on large vehicles and animals. Every flow component has a characteristic size, which emerges from two conflicting trends. The useful energy dissipated because the imperfection of the component decreases as the component size increases. The useful energy spent by the greater system (vehicle, animal) increases with the component size. The sum of the two penalties is minimal when the component size is finite, at the intersection between the two penalties. In time, the component evolves toward smaller sizes, because it improves and its penalty (the descending curve) slides downward.

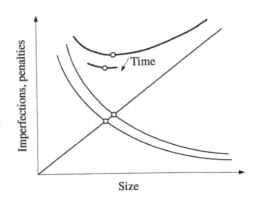

belong on large vehicles and small organs belong on small vehicles. This prediction is in accord with the evolution of all animals and vehicle technologies.

It also means that all organs should be imperfect, because each has a finite size, not an infinite size. The whole (the vehicle) is a construct of organs that are "imperfect" only when examined in isolation. The vehicle design evolves over time and becomes a better construct for moving the vehicle weight on the world map. Here we learn that "better" means moving more weight farther per unit of fuel consumed.

This idea is the principle of evolutionary organization everywhere, the constructal law. It is important to know, useful, and easy to learn. It is also timely, because of the fashionable claim that "biomimetics" helps technology. Biomimetics does not, but the principle does. Biomimicry is the act of seeing something in nature, copying it, and pasting it as an add-on to the evolving design of the human and machine species. This practice is successful only if the observer knows the principle that accounts for the observed natural object in the first place. Had this not been the case, then all the eyed animals and troglodytes would have marched well ahead of our science-based civilization. Those who claim success with biomimetics are unwittingly (intuitively) relying on physics and the constructal law.

In time, the organ evolves toward designs that flow more easily. This means that the descending curve in Figure 5.20 will shift downward over time, as will its intersection with the rising line. The bottom of the bucket-shaped curve follows suit and slides downward and to the left. The discovery is that the future organ must be better in an evolutionary manner, and also *smaller*. This discovery is about the future, and the name for this future is *miniaturization*.

Now we know why miniaturization should happen. It is the natural tendency in each of us to move our body, vehicle, and clan more easily and for longer periods and distances. Miniaturization happens, and it did not start with nanotechnology. Before nanotechnology we had microelectronics, and before microelectronics we had compact heat exchangers, with high density of heat transfer.

5.6.3 Visualization: Heatlines

One cannot photograph heat flow. The ability to "see" an invisible flow is essential to a scientist's ability to learn from experience and in his way to improve his or her technique. In convection problems it is important to visualize the flow of energy. For example, in the two-dimensional Cartesian configuration of Figure 5.2, it is common practice to define a stream function $\psi(x, y)$ as in Eq. (5.45) such that the mass continuity equation for incompressible flow, Eq. (5.6), is satisfied identically. It is easy to verify that the actual flow is locally parallel to the ψ = constant line passing through the point of interest. Therefore, although there are no substitutes for the velocity components u and v as bearers of information regarding the local flow, the family of ψ = constant streamlines provides a bird's-eye view of the entire flow field and its main characteristics.

In convection, the transport of energy through the flow field is a combination of both thermal diffusion and enthalpy flow (cf. Eq. (5.20)). For any such field, Kimura and Bejan [27] and the 1984 edition of Ref. [2] introduced a new function $H(x, y)$ such that the net flow of energy (thermal diffusion and enthalpy flow) is parallel to each H = constant line. The mathematical definition of the *heat function H* follows in the steps of Eqs. (5.45) with the aim of satisfying the energy equation identically. For steady-state two-dimensional convection through a constant property homogeneous fluid, Eq. (5.20) becomes

$$u\frac{\partial T}{\partial x} + v\frac{\partial T}{\partial y} = \alpha \left(\frac{\partial^2 T}{\partial x^2} + \frac{\partial^2 T}{\partial y^2} \right) \tag{5.159}$$

or

$$\frac{\partial}{\partial x}\left(\rho c_{\mathrm{p}}uT - k\frac{\partial T}{\partial x}\right) + \frac{\partial}{\partial y}\left(\rho c_{\mathrm{p}}vT - k\frac{\partial T}{\partial y}\right) = 0 \tag{5.160}$$

The heat function is defined as follows. The net energy flow in the x and y directions are, respectively,

$$\frac{\partial H}{\partial y} = \rho c_{\mathrm{p}}u(T - T_{\mathrm{ref}}) - k\frac{\partial T}{\partial x} \tag{5.161}$$

$$-\frac{\partial H}{\partial x} = \rho c_{\mathrm{p}}v(T - T_{\mathrm{ref}}) - k\frac{\partial T}{\partial y} \tag{5.162}$$

so that the heat function $H(x, y)$ satisfies Eq. (5.159) identically. Note that the definition above also applies to convection through a fluid-saturated porous medium, where Eq. (5.159) accounts for energy conservation [2].

The reference temperature T_{ref} is, in principle, an arbitrary constant that can be selected based on convention. The reference value of steam function is also an arbitrary constant. Patterns of $H = $ constant *heatlines* are instructive when T_{ref} is the lowest temperature that occurs in the heat transfer configuration. For example, if the wall shown in Figure 5.2 is warmer than the free stream, $T_w > T_\infty$, the choice of reference temperature is $T_{\mathrm{ref}} = T_\infty$. For a meaningful comparison of the heatlines of one flow with the heatlines of another flow, I proposed that T_{ref} always be set equal to the lowest temperature of the flow field.

If the fluid flow subsides ($u = v = 0$), the heatlines become identical to the *heat flux lines* employed frequently in the study of conduction phenomena. Therefore, as a heat transfer visualization technique, the use of heatlines is the convection counterpart or generalization of a technique (heat flux lines) used in conduction. The common use of $T = $ constant lines is not a proper way to visualize convective heat transfer; isotherms are a proper heat transfer visualization tool only in the field of conduction (where, in fact, they have been invented) because it is only there that they are locally orthogonal to the true direction of energy flow. The use of $T = $ constant lines to visualize convection heat transfer makes as much sense as using $P = $ constant lines to visualize fluid flow.

The heatline method for the visualization of convective heat transfer was popularized in the first edition (1984) of Ref. [2], along with a first numerical application to natural convection in an enclosure heated from the side [27]. The method has since been adopted and extended in many ways in the post-1984 heat transfer literature [28–63]. Note that there have been attempts to publish the heatlines method as "new" (under new names), as pointed out in Refs. [64–66].

References

1 Bejan, A. (2016). *Advanced Engineering Thermodynamics*, 4e. Hoboken, NJ: Wiley.
2 Bejan, A. (2013). *Convection Heat Transfer*, 4e. Hoboken, NJ: Wiley.
3 Navier, M. (1827). Mémoire sur les lois du movement des fluides. In: *Mémoires de l'Académie des Sciences de l'Institut de France*, vol. 6, 389–440.
4 Fourier, J.B.J. (1833). Mémoire d'analyse sur le mouvement de la chaleur dans les fluides. In: *Mémoires de l'Academie Royale des Sciences de l'Institut de France*, 507–530. Didot, Paris: (presented on September 4, 1820).
5 Fourier, J.B.J. (1890). *Oevres de Fourier*, vol. 2 (ed. G. Darboux), 595–614. Paris: Gauthier-Villars.
6 Poisson, S.D. (1835). *Théorie Mathématique de la Chaleur*, Chapter 4, 86. Paris.

7 Prandtl, L. (1904). Über Flüssigkeitsbewegung bei sehr kleiner Reibung. *Proceedings of the 3rd International Congress of Mathematicians*, Heidelberg; also NACA TM 452, 1928.

8 Blasius, H. (1908). Grenzschichten in Flüssigkeiten mit kleiner Reibung. *Z. Math. Phys.* 56: 1–37; also NACA TM 1256.

9 Pohlhausen, E. (1921). Der Wärmeaustausch zwischen festen Körpern und Flüssigkeiten mit kleiner Reibung und kleiner Wärmeleitung. *Z. Angew. Math. Mech.* 1: 115–121.

10 Churchill, S.W. and Ozoe, H. (1973). Correlations for laminar forced convection in flow over an isothermal flat plate and in developing and fully developed flow in an isothermal tube. *J. Heat Transfer* 95: 416–419.

11 Kays, W.M. and Crawford, M.E. (1980). *Convective Heat and Mass Transfer*, 151. New York: McGraw-Hill.

12 Eckert, E.R.G. (1959). *Introduction to the Transfer of Heat and Mass*. New York: McGraw-Hill.

13 Bejan, A. (1982). The meandering fall of paper ribbons. *Phys. Fluids A* 25: 741–742.

14 Prandtl, L. (1969). *Essentials of Fluid Dynamics*, 117. London: Blackie & Son.

15 Reynolds, O. (1874). On the extent and action of the heating surface for steam boilers. *Proc. Manchester Lit. Philos. Soc.* 14: 7–12.

16 Colburn, A.P. (1933). A method for correlating forced convection heat transfer data and a comparison with fluid friction. *Trans. Am. Inst. Chem. Eng.* 29: 174–210; reprinted in (1964). *Int. J. Heat Mass Transfer* 7: 1359–1384.

17 Churchill, S.W. and Bernstein, M. (1977). A correlating equation for forced convection from gases and liquids to a circular cylinder in crossflow. *J. Heat Transfer* 99: 300–306.

18 Nakai, S. and Okazaki, T. (1975). Heat transfer from horizontal circular wire at small Reynolds and Grashof numbers—I Pure convection. *Int. J. Heat Mass Transfer* 18: 387–396.

19 Bejan, A. (1982). *Entropy Generation through Heat and Fluid Flow*. New York: Wiley.

20 Simiu, E. and Scanlan, R.H. (1986). *Wind Effects on Structures*, 2e, 143–152. New York: Wiley.

21 Whitaker, S. (1972). Forced convection heat transfer correlations for flow in pipes, past flat plates, single cylinders, single spheres, and flow in packed beds and tube bundles. *AIChE J.* 18: 361–371.

22 Yovanovich, M.M. (1988). General expression for forced convection heat and mass transfer from isopotential spheroids. Paper No. AIAA 88-0743, *AIAA 26th Aerospace Sciences Meeting*, Reno, Nevada (11–14 January 1988).

23 Zukauskas, A.A. (1987). Convective heat transfer in cross flow. In: *Handbook of Single-Phase Convective Heat Transfer*, Chapter 6 (eds. S. Kakac, R.K. Shah and W. Aung). New York: Wiley.

24 Martin, H. (1977). Heat and mass transfer between impinging gas jets and solid surfaces. *Adv. Heat Transfer* 13.

25 Bejan, A. (2016). *The Physics of Life: The Evolution of Everything*. New York: St. Martin's Press.

26 Bejan, A. and Lorente, S. (2008). *Design with Constructal Theory*. Hoboken, NJ: Wiley.

27 Kimura, S. and Bejan, A. (1983). The "heatline" visualization of convective heat transfer. *J. Heat Transfer* 105: 916–919.

28 Littlefield, D. and Desai, P. (1986). Buoyant laminar convection in a vertical cylindrical annulus. *J. Heat Transfer* 108: 814–821.

29 Trevisan, O.V. and Bejan, A. (1987). Combined heat and mass transfer by natural convection in a vertical enclosure. *J. Heat Transfer* 109: 104–109.

30 Bello-Ochende, F.L. (1987). Analysis of heat transfer by free convection in tilted rectangular cavities using the energy analogue of the stream function. *Int. J. Mech. Eng. Ed.* 15: 91–98.

31 Bello-Ochende, F.L. (1988). A heat function formulation for thermal convection in a square cavity. *Int. Commun. Heat Mass Transfer* 15: 193–202.

32 Morega, A.M. (1988). The heat function approach to the thermo-magnetic convection of electroconductive melts. *Rev. Roum. Sci. Tech. Ser. Electrotech. Energ.* 33: 33–39.

33 Aggarwal, S.K. and Manhapra, A. (1989). Use of heatlines for unsteady buoyancy-driven flow in a cylindrical enclosure. *J. Heat Transfer* 111: 576–578.

34 Aggarwal, S.K. and Manhapra, A. (1989). Transient natural convection in a cylindrical enclosure nonuniformly heated at the top wall. *Numer. Heat Transfer, Part A* 15: 341–356.

35 Ho, C.J., Lin, Y.H., and Chen, T.C. (1989). A numerical study of natural convection in concentric and eccentric horizontal cylindrical annuli with mixed boundary conditions. *Int. J. Heat Fluid Flow* 10: 40–47.

36 Ho, C.J. and Lin, Y.H. (1989). Thermal convection heat transfer of air/water layers enclosed in horizontal annuli with mixed boundary conditions. *Wärme und Stoffübertrag* 24: 211–224.

37 Ho, C.J. and Lin, Y.H. (1990). Natural convection of cold water in a vertical annulus with constant heat flux on the inner wall. *J. Heat Transfer* 112: 117–123.

38 Morega, A.M. and Bejan, A. (1993). Heatline visualization of forced convection boundary layers. *Int. J. Heat Mass Transfer* 36: 3957–3966.

39 Morega, A.M. and Bejan, A. (1994). Heatline visualization of forced convection in porous media. *Int. J. Heat Fluid Flow* 15: 42–47.

40 Costa, V.A.F. (1997). Double diffusive natural convection in a square enclosure with heat and mass diffusive walls. *Int. J. Heat Mass Transfer* 40: 4061–4071.

41 Costa, V.A.F. (1998). Double diffusive natural convection in enclosures with heat and mass diffusive walls. In: *Proceedings of the International Symposium on Advances in Computational Heat Transfer (CHT'97)* (eds. G. De Vahl Davis and E. Leonardi), 338–344. New York: Begell House.

42 Wang, H.Y., Penot, F., and Saulnier, J.B. (1997). Numerical study of a buoyancy-induced flow along a vertical plate with discretely heated integrated circuit packages. *Int. J. Heat Mass Transfer* 40: 1509–1520.

43 Costa, V.A.F. (1999). Unification of the streamline, heatline and massline methods for the visualization of two-dimensional transport phenomena. *Int. J. Heat Mass Transfer* 42: 27–33.

44 Kim, S.J. and Jang, S.P. (2001). Experimental and numerical analysis of heat transfer phenomena in a sensor tube of a mass flow controller. *Int. J. Heat Mass Transfer* 44: 1711–1724.

45 Deng, Q.-H. and Tang, G.-F. (2002). Numerical visualization of mass and heat transport for conjugate natural convection/heat conduction by streamline and heatline. *Int. J. Heat Mass Transfer* 45: 2375–2385.

46 Deng, Q.-H. and Tang, G.-F. (2002). Numerical visualization of mass and heat transport for mixed convective heat transfer by streamline and heatline. *Int. J. Heat Mass Transfer* 45: 2387–2396.

47 Mukhopadhyay, A., Qin, X., Aggarwal, S.K., and Puri, I.K. (2002). On extension of "heatline" and "massline" concepts to reacting flows through the use of conserved scalars. *J. Heat Transfer* 124: 791–799.

48 Costa, V.A.F. (2003). Comment on the paper by Qi-Hong Deng and Guang-Fa Tang, Numerical visualization of mass and heat transport for conjugate natural convection/heat conduction by streamline and heatline. *Int. J. Heat Mass Transfer* 46: 185–187.

49 Costa, A.F. (2003). Unified streamline, heatline and massline methods for the visualization of two-dimensional heat and mass transfer in anisotropic media. *Int. J. Heat Mass Transfer* 46: 1309–1320.

50 Mukhopadhyay, A., Qin, X., Puri, I.K., and Aggarwal, S.K. (2003). Visualization of scalar transport in nonreacting and reacting jets through a unified "heatline" and "massline" formulation. *Numer. Heat Transfer, Part A* 44: 683–704.

51 Dalal, A. and Das, M.K. (2008). Heatline method for the visualization of natural convection in a complicated cavity. *Int. J. Heat Mass Transfer* 51: 263–272.

52 Zhao, F.-Y., Liu, D., and Tang, G.-F. (2007). Application issues of the streamline, heatline and massline for conjugate heat and mass transfer. *Int. J. Heat Mass Transfer* 50: 320–334.

53 Basak, T. and Roy, S. (2008). Role of "Bejan's heatlines" in heat flow visualization and optimal thermal mixing for differentially heated square enclosures. *Int. J. Heat Mass Transfer* 51: 3486–3503.

54 Hakyemez, E., Mobedi, M., and Oztop, H.F. (2008). Effects of wall-located heat barrier on conjugate conduction/natural-convection heat transfer and fluid flow in enclosures. *Numer. Heat Transfer, Part A* 54: 197–220.

55 Mahmud, S. and Fraser, R.A. (2007). Visualizing energy flows through energy streamlines and pathlines. *Int. J. Heat Mass Transfer* 50: 3990–4002.

56 Basak, T., Roy, S., and Pop, I. (2009). Heat flow analysis for natural convection within trapezoidal enclosures based on heatline concept. *Int. J. Heat Mass Transfer* 52: 2471–2483.

57 Basak, T., Aravind, G., and Roy, S. (2009). Visualization of heat flow due to natural convection within triangular cavities using Bejan's heatline concept. *Int. J. Heat Mass Transfer* 52: 2824–2833.

58 Kaluri, R.S., Basak, T., and Roy, S. (2009). Bejan's heatlines and numerical visualization of heat flow and thermal mixing in various differentially heated porous square cavities. *Numer. Heat Transfer, Part A* 55: 487–516.

59 Waheed, M.A. (2009). Heatfunction formulation of thermal convection in rectangular enclosures filled with porous media. *Numer. Heat Transfer, Part A* 55: 185–204.

60 Varol, Y., Oztop, H.F., Mobedi, M., and Pop, I. (2008). Visualization of natural convection heat transport using heatline method in porous non-isothermally heated triangular cavity. *Int. J. Heat Mass Transfer* 51: 5040–5051.

61 Basak, T., Roy, S., and Aravind, G. (2009). Analysis of heat recovery and thermal transport within entrapped fluid based on heatline approach. *Chem. Eng. Sci.* 64: 1673–1686.

62 Kaluri, R.S., Basak, T., and Roy, S. (2010). Heatline approach for visualization of heat flow and efficient thermal mixing with discrete heat sources. *Int. J. Heat Mass Transfer* 53: 3241–3261.

63 Lima, T.P. and Ganzarolli, M.M. (2016). A heatline approach on the analysis of the heat transfer enhancement in a square enclosure with an internal conducting solid body. *Int. J. Therm. Sci.* 105: 45–56.

64 Bejan, A. (2015). Heatlines (1983) versus synergy (1998). *Int. J. Heat Mass Transfer* 81: 654–658.

65 Lukose, L. and Basak, T. (2019). Heatlines vs energy flux vectors: tools for heat flow visualization. *Int. Commun. Heat Mass Transfer* 108: 104265.

66 Bejan, A. (2002). *Freedom and Evolution: Hierarchy in Nature, Society and Science*. Switzerland AG: Springer Nature.

67 Bejan, A. (2020). Boundary layers from constructal law. *Int. Commun. Heat Mass Transfer* 116: 104672.

Problems

Basic Principles of Convection

5.1 Retrace the steps between Eqs. (5.8) and (5.11). Show that the derivation of Eq. (5.11) requires the use of the constant-μ and nearly constant-ρ models.

5.2 Perform the two-step analysis outlined in connection with Eqs. (5.16) and (5.17), and show that the left side of the energy equation, Eq. (5.14), can be expressed in terms of temperature and pressure gradients, as in Eq. (5.19).

Laminar Boundary Layers

5.3 The free-stream water temperature in a laminar boundary layer is $T_\infty = 25\,°C$. The temperature of the wall is $T_w = 24\,°C$. The free-stream velocity is 0.6 cm/s. Use the properties of 25 °C water, and calculate the local wall heat flux at $x = 20$ cm and the heat flux averaged over the wall section of length x.

5.4 Assume that the entire swept length of a thin plate is $x = 10$ cm. The water free-stream temperature is 30 °C, and the uniform temperature of the plate is 20 °C. The free-stream velocity is 0.6 cm/s. By using the properties of distilled water at the appropriate film temperature, calculate the following:
a) The average heat transfer coefficient, the average heat flux, and the total heat transfer rate received by the plate.
b) The total drag force experienced by the plate.

5.5 Assume that a thin plate is an electrical resistance that is being cooled on both sides by the laminar boundary layer flow of water. The width of the resistance (the length swept by the flow) is $x = 10$ cm. The free-stream velocity is 0.6 cm/s. Assume further that the water is pure and that its temperature is $T_\infty = 20\,°C$. The heat flux is distributed uniformly over the swept length, $q''_w = 1000\,W/m^2$. Calculate the following:
a) The plate temperature at its trailing edge (x).
b) The x-averaged temperature of the plate.

5.6 The wind blows at 0.5 m/s parallel to the short side of a flat roof with the rectangular area 10 m × 20 m. The roof temperature is 40 °C, and the temperature of the air free stream is 20 °C. Calculate the total force experienced by the roof. Estimate also the total heat transfer rate by laminar forced convection, from the roof to the atmosphere.

5.7 Make a qualitative sketch of how the local heat flux varies along an isothermal wall bathed by a laminar boundary layer of total length L. Use $q''_{w,x}$ on the ordinate and x on the abscissa. On the same sketch, draw a horizontal line at the level that would correspond to the heat flux averaged over the entire length of the plate, $\overline{q}''_{w,L}$. Determine analytically the following:
a) The position x where the local heat flux matches the value of the L-averaged heat flux.
b) The relationship between the midpoint local flux and the L-averaged value, that is, the ratio $q''_{w,L/2}/\overline{q}''_{w,L}$.

5.8 Consider the sharp-edged entrance to a round duct of diameter D (Figure P5.8). The laminar boundary layer that forms over the duct length L is much thinner than the duct diameter. The temperature difference between the duct wall (isothermal) and the inflowing stream is ΔT. The longitudinal inlet velocity of the stream is U_∞. Derive expressions for the total force F experienced by the duct section of length L and the total heat transfer rate from the duct wall to the stream, q. In the end, show that q and F are proportional:

$$\frac{q}{F} = Pr^{-2/3}\frac{c_P\Delta T}{U_\infty} \qquad (Pr \gtrsim 0.5)$$

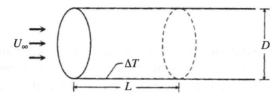

Figure P5.8

5.9 A stream of 20 °C water enters a duct (Figure P5.9). The wall temperature is uniform and equal to 50 °C. The water inlet velocity is 5 cm/s. The duct cross section is a 20 cm × 20 cm. Assume that the thickness of the boundary layer that lines the inner surface of the duct is much smaller than 20 cm, and calculate the following:

a) The local heat transfer coefficient at $x = 1$ m downstream from the mouth.

b) The total heat transfer rate between the duct section of length $x = 1$ m and the water stream.

c) The velocity boundary layer thickness (δ_{99}) at $x = 1$ m. Verify in this way the validity of the assumption that δ_{99} is much smaller than the duct width.

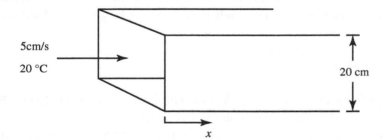

5cm/s
20 °C
20 cm
x

Figure P5.9

5.10 Perform the integral analysis of the laminar velocity boundary layer illustrated in Figure 5.5 and follow these steps:

a) Show that the momentum equation, Eq. (5.32), is equivalent to

$$\frac{\partial(u^2)}{\partial x} + \frac{\partial(uv)}{\partial y} = v\frac{\partial^2 u}{\partial y^2}$$

b) Integrate this momentum equation across the boundary **layer** and find

$$\frac{d}{dx}\int_0^\delta u(U_\infty - u)\,dy = v\left(\frac{\partial u}{\partial y}\right)_{y=0}$$

where $\delta(x)$ is the unknown thickness of the boundary layer.

c) Assume a reasonable shape for the $u - y$ profile,

$$\frac{u}{U_\infty} = m(n) \text{ where } n = \frac{y}{\delta(x)}$$

and $m(n)$ is one of the functions listed in the left column of the table below. Show that the integral analysis produces the following $\delta(x)$ and $C_{f,x}$ estimates [2]:

Assumed profile	$\dfrac{\delta}{x}Re_x^{1/2}$	$C_{f,x}Re_x^{1/2}$
Linear: $m = n$	3.46	0.577
Parabolic: $m = \dfrac{n}{2}(3 - n^2)$	4.64	0.646
Sinus arc: $m = \sin\left(\dfrac{\pi n}{2}\right)$	4.8	0.654

5.11 Carry out the integral analysis of the thermal boundary layer near the isothermal wall of Figure 5.9 (bottom), that is, when the fluid has a large Prandtl number. Execute the following steps:

a) Show that the boundary layer energy equation, Eq. (5.62), is equivalent to

$$\frac{\partial(uT)}{\partial x} + \frac{\partial(vT)}{\partial y} = \alpha\frac{\partial^2 T}{\partial y^2}$$

b) Integrate the energy equation across the thermal boundary layer, and arrive at the integral **energy** equation

$$\frac{d}{dx}\int_0^{\delta_T} u(T_\infty - T)dy = \alpha\left(\frac{\partial T}{\partial y}\right)_{y=0}$$

c) Assume the parabolic temperature profile

$$\frac{T_w - T}{T_w - T_\infty} = \frac{p}{2}(3 - p^2) \quad \text{where } p = \frac{y}{\delta_T}$$

and note that the corresponding velocity boundary layer thickness $\delta(x)$ is listed in the middle column of the table in the preceding problem statement. Note that since $Pr \gg 1$, you can use the approximation $\delta \gg \delta_T$. Show that the local Nusselt number is $Nu_x = 0.331 Pr^{1/3}Re_x^{1/2}$.

5.12 Consider the boundary layers δ_T and δ near an isothermal wall, Figure 5.9 (top). In the low-Pr limit δ is negligible relative to δ_T, therefore the flow in the δ_T-thick region is uniform and parallel:

$$u = U_\infty \quad \text{and} \quad v = 0$$

a) Show that in this limit the energy equation for the thermal boundary layer has the same form as Eq. (4.22) of Section 4.3.1. Recognize further the analogy between the boundary conditions in your problem and those of the unsteady conduction problem of Section 4.3.1.

b) Note the different notations employed in the two problems, and deduce from Eq. (4.42) the closed-form expression for the temperature distribution $T(x, y)$ in the low-Pr thermal boundary layer.

c) Determine analytically the local heat flux and the local Nusselt number; in other words, prove the validity of Eq. (5.79a).

5.13 The average Nusselt number $\overline{Nu_x}$ is not the result of averaging the local Nusselt number Nu_x over the downstream length of the boundary layer,

$$\overline{Nu_x} \neq \frac{1}{x}\int_0^x Nu_{x'}dx'$$

The proper definition of $\overline{Nu}_x$ is Eq. (5.81). It is the average wall heat flux $\overline{q}''_{w,x}$ (or the average heat transfer coefficient $\overline{h}_x$), which is the result of averaging the local quantity $q''_{w,x}$ (or h_x) over the distance x (review Eq. (5.80)). Keep this observation in mind as you derive the $\overline{Nu}_x$ expressions (5.82a, 5.82b), by starting with the local Nusselt number formulas listed as Eqs. (5.79a, 5.79b).

5.14 The plane wall shown in the Figure P5.14 is swept by the laminar boundary layer flow of an isothermal fluid (T_∞, U_∞) with Prandtl number greater than 0.5. Deposited on the surface of this wall is a narrow strip of metallic film that runs parallel to the leading edge of the wall, that is, in the direction normal to the plane of the figure. As part of an electrical circuit, the strip generates Joule heating at the rate q'_w (W/m), or as the heat flux $q''_w = q'_w/\Delta x$. The width of the strip is much smaller than the distance to the leading edge, $\Delta x \ll x_1$.

Consult Table 5.4 and determine analytically the wall temperature distribution over the unheated downstream portion $x > x_1$. In other words, determine the "thermal wake" effect of the strip conductor. Assume that the entire Joule heating effect q'_w can escape through the fluid side of the metallic strip. In other words, assume that the wall side of the strip is adiabatic.

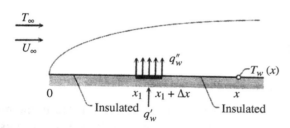

Figure P5.14

5.15 Use the formula for the nonuniform heat flux case (the fourth in Table 5.4), set $q''_w(\xi) = $ constant, and derive the temperature distribution on a wall with uniform heat flux (the third in Table 5.4).

5.16 A plane wall of length L is cooled by the laminar boundary layer flow of a fluid with $Pr \gtrsim 0.5$. The wall is heated electrically so that it releases the uniform heat flux q'' over the front half of its swept length, $0 < x < L/2$. The trailing half $L/2 < x < L$ is without heat transfer. Consult Table 5.4 and determine analytically the wall–fluid temperature difference at the trailing edge, $T_w(L) - T_\infty$.

A simpler (approximate) approach would be to assume that the total heat transfer rate described above ($q'' L/2$) is distributed uniformly over the entire length L. Determine the trailing edge temperature difference, compare it with the previous estimate, and comment on the accuracy of this approximate approach.

5.17 An isothermal flat strip is swept by a parallel stream of water with a temperature of 20 °C and free-stream velocity of 0.5 m/s. The width of the strip, $L = 1$ cm, is parallel to the flow. The temperature difference between the strip and the free stream is $\Delta T = 1$ °C. Calculate the L-averaged shear stress $\overline{\tau}_{w,L}$ and the L-averaged heat flux $\overline{q}''_{w,L}$ between the strip and the water flow.

5.18 An unspecified fluid with $Pr = 17.8$ flows parallel to a flat isothermal wall and develops a laminar boundary layer along the wall. At the location x along the wall, the local skin friction coefficient is equal to 0.008. Calculate the value of the local Nusselt number at the same location.

5.19 It is proposed to estimate the uniform velocity U_∞ of a stream of air of temperature 20 °C, by measuring the temperature of a thin metallic blade that is heated and inserted parallel to U_∞ in the air stream (Figure P5.19). The width of the blade (i.e. the dimension aligned with U_∞) is $L = 2$ cm. The blade is considerably longer in the direction normal to the attached figure; therefore, the boundary layer flow that develops is two-dimensional. The blade is heated volumetrically by an electric current, so that 0.03-W electrical power is dissipated in each square centimeter of metallic blade. It is assumed that the blade is so thin that the effect of heat conduction through the blade (in the x direction) is negligible. A temperature sensor mounted on the trailing edge of the blade reads $T_w = 30$ °C. Calculate the free-stream velocity U_∞ that corresponds to this reading.

Figure P5.19

Turbulent Boundary Layers

5.20 Demonstrate that the time-averaged version of the mass conservation equation, Eq. (m) of Table 5.1, is Eq. (5.91). In the derivation of Eq. (5.91), use the velocity decomposition relationships (5.88) and (5.89) and the appropriate algebraic rules listed as Eqs. (5.90).

5.21 Derive the time-averaged form of the energy equation in Cartesian coordinates, Eq. (5.95), by starting with Eq. (E) of Table 5.1, in which $\dot{q} = 0$ and $\mu\Phi = 0$. In this analysis, use the decomposition formulas (5.88) and (5.89) and the time-averaging rules (5.90).

5.22 Derive the two-curve expression for the longitudinal velocity profile $u^+(y^+)$ in turbulent boundary layer flow near a flat wall, Eq. (5.116). Begin with the constant-τ_{app} condition, Eq. (5.115), and in the fully turbulent sublayer ($\varepsilon_M \gg v$) use the mixing-length model (5.113).

5.23 Derive the formulas for skin friction coefficient and boundary layer thickness for turbulent flow over a flat plate, Eqs. (5.121) and (5.122). Base this derivation on the $\frac{1}{7}$th power law profile (5.118). As starting condition for evaluating the left-side integral of Eq. (5.120), use $\delta = 0$ at $x = 0$.

5.24 a) Show that these relationships exist between the Stanton, Nusselt, Reynolds, Péclet, and Prandtl numbers in boundary layer flow:

$$St_x = \frac{Nu_x}{Re_x Pr} = \frac{Nu_x}{Pe_x}$$

b) Show that the Colburn analogy (5.131) applies also to the *laminar* section of the boundary layer near an isothermal wall if the fluid has a Prandtl number in the range $Pr \gtrsim 0.5$.

5.25 Evaluate the average heat transfer coefficient for a flat wall of length L, with laminar and turbulent flow on it, Eq. (5.132). Show that in $Pr \gtrsim 0.5$ fluids the average Nusselt number $\overline{Nu_L}$ is given by Eq. (5.133).

5.26 The laminar section of the paper ribbon falling in Figure 5.9 has the approximate length $x \cong 0.3$ m. The air temperature is 20 °C.
  a) Calculate the Re_x value for the end of the laminar section, and compare it with the transition criterion recommended in the text.
  b) Imagine that the ribbon is straight (rigid) and covered with laminar boundary layers over its entire length. Assume also that the terminal velocity of free fall has been reached. Calculate this velocity, and compare it with the measured value listed in the caption of Figure 5.9. Why is the calculated value larger?

5.27 Examine again the information listed in the caption of Figure 5.9, and imagine that the paper ribbon remains straight and vertical as it falls to the ground. The air temperature is 20 °C.
  a) Calculate the terminal free-fall velocity of the ribbon, $U_{\infty,\,turb}$, by assuming that the boundary layers are turbulent over the entire length L.
  b) Had we assumed instead that the flow is laminar over L, the answer to part (a) would have been $U_{\infty,\,lam} = 11.7$ m/s. Compare your $U_{\infty,\,turb}$ estimate with the measured value (Figure 5.9 caption) and $U_{\infty,\,lam}$, and comment on which method of calculation is more accurate and why.

5.28 Water flows with the velocity $U_\infty = 0.2$ m/s parallel to a plane wall. The following calculations refer to the position $x = 6$ m measured downstream from the leading edge. The water properties can be evaluated at 20 °C.
  a) A probe is to be inserted in the viscous sublayer to the position represented by $y^+ = 2.7$. Calculate the actual spacing y (mm) between the probe and the wall.
  b) Calculate the boundary layer thickness δ, and compare this value with the estimate based on the assumption that the length x is covered by turbulent boundary layer flow.
  c) Calculate the heat transfer coefficient averaged over the length x.

5.29 The flat iceberg described in Examples 5.2 and 5.3 in the text is driven by the air drag felt by its upper surface. The iceberg drifts steadily when the air drag is balanced by the bottom-surface seawater drag calculated in Example 5.3. The relative speed between seawater and iceberg is 10 cm/s. Calculate the corresponding wind velocity when the atmospheric air temperature is 40 °C.

5.30 Air flows with velocity 3.24 m/s over the top surface of the flat iceberg discussed in Example 5.2. The air temperature outside the boundary layer is 40 °C, while the ice surface temperature is 0 °C. The length of the iceberg in the direction of air flow is $L = 100$ m. The ice latent heat of melting is $h_{sf} = 333.4$ kJ/kg. Calculate the L-averaged heat flux deposited by the air flow into the upper surface of the iceberg (model this surface as flat). Calculate, in millimeters per hour, the rate of melting caused by this heat flux, that is, the erosion (thinning) of the ice slab.

5.31 The large faces of the tall building shown in Figure P5.31 are swept by a parallel wind with $U_\infty = 15\,\text{m/s}$ and $T_\infty = 0\,°\text{C}$. Calculate the heat transfer coefficient for the outer surface of a window situated on a large face, at $x = 60\,\text{m}$ downstream from the leading surface of the building. Evaluate all the air properties at $0\,°\text{C}$. Do you expect the calculated heat transfer coefficient to be smaller or larger than the true value?

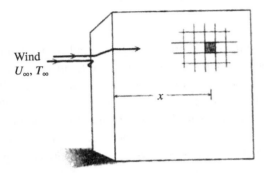

Figure P5.31

Other External Flows

5.32 The cold wind forces the air to flow with a velocity of 4 cm/s through the hair strands in the fur of a bear. The hair strand is approximately cylindrical, with a diameter of $21 \times 10^{-6}\,\text{m}$. The mean temperature of the air that sweeps the hair strands is $10\,°\text{C}$. Calculate the average coefficient for heat transfer between the individual hair strand and the surrounding air.

5.33 Imagine that you are fishing while wading in a 75-cm-deep river that flows at 1 m/s. The immersed portion of each bare leg can be approximated as a vertical cylinder with a diameter of 15 cm. The river velocity is approximately uniform, and the water temperature is $10\,°\text{C}$.
a) Calculate the horizontal drag force experienced by one leg.
b) Compare the drag force with the weight (in air) of the immersed portion of one leg.
c) Determine the average heat transfer coefficient between the wetted skin and the river water.
d) After a long enough wait, the temperature of the wetted skin drops to $11\,°\text{C}$. Calculate the instantaneous heat transfer rate that escapes into the river through one leg.

5.34 The baseballs used in the major leagues have an average diameter of 7.4 cm and an average weight of 145 g. The distance between the pitcher's mound and home plate is 18.5 m. The pitcher throws an 80 miles/h fastball. The rotation of the ball and its motion in the vertical direction are negligible.
a) Calculate the drag force experienced by the fastball.
b) Show that the calculated drag force is comparable with the weight of the ball.
c) Estimate the horizontal velocity of the ball as it reaches the catcher's glove.

5.35 An electrical light bulb for the outdoors is approximated well by a sphere with a diameter of 6 cm. It is being swept by a 2 m/s wind, while its surface temperature is $60\,°\text{C}$. The air temperature is $10\,°\text{C}$.

Calculate the rate of forced convection heat transfer from the outer surface of the light bulb to the atmosphere.

5.36 Rely on the results of Section 2.3 to prove that in the pure conduction limit the overall Nusselt number for a sphere immersed in a motionless conducting medium is $\overline{Nu}_D = 2$. In other words, demonstrate that the $Re_D = 0$ limit of Eq. (5.139) is correct. Show also that in the same limit $\overline{Nu}_{\mathcal{L}} = 3.545$, for which the $\mathcal{L}$ scale is defined in Eq. (5.140).

5.37 During the cooling and hardening phase of the manufacturing process, a glass bead with a diameter of 0.5 mm is dropped from a height of 10 m. The bead falls through still air of temperature 20 °C. The properties of the bead material are the same as those listed for window glass in Appendix B.
   a) Calculate the terminal velocity of the free-falling bead, that is, the velocity when its weight is balanced by the air drag force. Calculate the approximate time that is needed by the bead to achieve this velocity, and show that the bead travels most of the 10-m height at terminal velocity.
   b) Calculate the average heat transfer coefficient between the bead surface and the surrounding air, when the bead travels at terminal velocity and when its surface temperature is 500 °C. Treat the bead as a lumped capacitance, and estimate its temperature at the end of the 10-m fall. Assume that its initial temperature was 500 °C.

5.38 Combine the empirical turbulent jet velocity profile (5.148) with the momentum conservation theorem (5.150) to determine the centerline jet velocity (5.151). In this way, prove that the centerline velocity decreases as x^{-1}, from a distance $x \cong 6.61 D_0$ downstream from the nozzle, where $\overline{u}_c = U_0$.

5.39 Demonstrate that the time-averaged excess temperature on the centerline of a turbulent jet decreases as x^{-1}, in accordance with Eq. (5.153). In this derivation, rely on the Gaussian profiles (5.148) and (5.149), the centerline velocity (5.151), and the energy conservation theorem (5.152).

5.40 Guided by Table 5.2, it would not be difficult to show that the boundary layer-simplified mass and momentum equations for a time-averaged turbulent jet (Figure 5.28) are

$$(m) \quad \frac{\partial \overline{u}}{\partial x} + \frac{1}{r}\frac{\partial}{\partial r}(r\overline{v}) = 0$$

$$(M) \quad \overline{u}\frac{\partial \overline{u}}{\partial x} + \overline{v}\frac{\partial \overline{u}}{\partial r} = \frac{1}{r}\frac{\partial}{\partial r}\left[r(v + \varepsilon_M)\frac{\partial \overline{u}}{\partial r}\right]$$

Rely on these equations to prove the correctness of Eq. (5.150), which states that the cross-section integrated longitudinal momentum of the jet is conserved (independent of x). Use part (a) of Problem 5.10 as a hint. Does the longitudinal momentum theorem (5.150) apply also to a laminar jet?

5.41 The boundary layer-simplified energy equation for a time-averaged turbulent round jet (Figure 5.18) is

$$(E) \quad \overline{u}\frac{\partial \overline{T}}{\partial x} + \overline{v}\frac{\partial \overline{T}}{\partial r} = \frac{1}{r}\frac{\partial}{\partial r}\left[r(\alpha + \varepsilon_H)\frac{\partial \overline{T}}{\partial r}\right]$$

Rely on this equation and the mass conservation Eq. (m) listed in the preceding problem to prove the validity of the theorem (5.152). Step (a) of Problem 5.12 can be used as a guide. The theorem (5.152) states that the flow of energy through any constant-x cross section is conserved. Does this theorem apply also to a laminar round jet?

Evolutionary Design

5.42 Consider the laminar boundary layer formed by the flow of 10°C water over a 10°C flat wall of length L. Show that the total shear force experienced by the wall and the mechanical power P spent on dragging the wall through the fluid is proportional to $v^{1/2}$.

a) The dissipated drag power described above refers to the case in which the wall is as cold as the free-stream water. Show that if the wall is heated isothermally so that its temperature rises to 90°C, the dissipated power decreases by 35%. In other words, show that $P_h/P_c = 0.65$, where h and c refer to the hot-wall and cold-wall conditions.

b) Compare the power savings due to heating the wall ($P_c - P_h$) with the electrical power needed to heat the wall to 90°C. How fast must the water flow be that the savings in fluid-friction power dissipation become greater than the electrical power invested in heating the wall? How short must the swept length L be so that the boundary layer remains laminar while the power savings $P_c - P_h$ exceed the heat input to the wall?

5.43 It is proposed here to reduce the drag experienced by the hull of a ship by heating the hull to a high enough temperature so that the viscosity of the water in the boundary layer decreases. Evaluate the merit of this proposal in the following steps:

a) Model the hull as a flat wall of length L that is swept by turbulent boundary layer flow. Show that the power spent on dragging the wall through the water is proportional to $v^{1/5}$.

b) Assume that the hull temperature is raised to 90°C, while the water free-stream temperature is 10°C. Calculate the relative decrease in the drag power, a decrease that is caused by the heating of the wall.

c) Compare the decrease (savings) in drag power with the electrical power that would be needed to maintain a wall temperature of 90°C. Show that when the ship speed is of order 10 m/s, the savings in drag power are much smaller than the power used for heating the wall.

5.44 By holding and rubbing the ball in his hand, the pitcher warms the leather cover of the baseball to 30°C. The outside air temperature is 20°C and the ball diameter is 7 cm. The pitcher throws the ball at 50 miles/h to the catcher, who is stationed 18.5 m away.

a) Assume that the ball surface temperature remains constant, and calculate the heat transferred from the ball to the surrounding air during the throw.

b) Calculate the temperature drop experienced by the leather cover to account for the heat transfer calculated in part (a). Assume that the thickness of the layer of leather that experiences the air cooling effect is comparable with the conduction penetration depth $\delta \sim (\alpha t)^{1/2}$ where α is the thermal diffusivity of leather. Validate in this way the correctness of the constant surface temperature assumption made in part (a).

5.45 Hot air with an average velocity $U_\infty = 2$ m/s flows across a bank of 4 cm-diameter cylinders in an array with $X_l = X_t = 7$ cm. Assume that the air bulk temperature is 300 °C and that the cylinder wall temperature is 30 °C. The array has 20 rows in the direction of flow. Calculate the average heat transfer coefficient when the cylinders are as follows:

(a) Aligned

(b) Staggered

Comment on the relative heat transfer augmentation effect associated with staggering the cylinders.

5.46 An object falls vertically at its terminal velocity (V = constant) through still air. The object is roundish (a clump) that can be modeled as having a single length scale (D), which may be regarded as an approximate diameter. The local Reynolds number is high enough ($VD/\nu > 10^2$) so that the wake behind the object meanders as a sequence of shed vortices of diameter D and peripheral velocity V. The vertical distance between the centers of successive vortices is half the buckling wavelength λ, which is of order $\lambda \sim 2D$, cf., Appendix F. The weight of the object is Mg, where M is its mass.

a) Determine the terminal velocity V by invoking the first law of thermodynamics for the process executed by the whole system (falling object, plus surrounding air) as the object falls to a vertical distance of order D. During the same process, the air immediately behind the object is set in motion as a ball of diameter D and peripheral velocity V. The energy of the whole system is conserved; the potential energy lost by the falling object is equal to the gain in kinetic energy experienced by the vortex created behind the object.

b) Determine the drag coefficient C_D that corresponds to the terminal velocity obtained above. Recall that during free fall at terminal velocity the drag force experienced by the object is balanced by its weight. Compare your C_D estimate with C_D measurements found in Figure 5.15. Note that the success of your theoretical prediction of C_D is due to the correctness of two principles, the first law of thermodynamics and the constructal law (in this case, the predicted flow configuration in the wake of the object when the local Reynolds number is of order 10^2 or greater).

5.47 A small rigid plate falls steadily through still air. Its mass and weight are M and Mg. The plate is almost horizontal; its plane makes a small angle (α) with the horizontal plane, and, consequently it slides parallel to itself as it is being pulled by the component of Mg aligned with its plane. Model the plate as a rectangle of length L aligned with its sliding motion, and width W. Label the weight of the plate per unit of surface area as m''. Both sides of the plate are lined by laminar boundary layers. Which configuration will the plate favor as it falls? Will it become perfectly horizontal or vertical? Determine the future configuration in two ways:

a) Determine the order of magnitude of the velocity with which the plate slides in its plane, slightly downward on the incline of angle α. If this velocity is to increase toward greater access (in accord with the constructal law), will the future configuration be more inclined or less?

b) Determine the terminal speed during vertical fall in two extreme configurations: horizontal plate and vertical plate. For the horizontal plate the order of magnitude of C_D is indicate by the measurements assembled in Figure 5.15. For the vertical plate the drag force is the total skin friction force experienced by the plate along its swept surfaces of length L. The boundary layers are laminar. Show that the ratio of terminal speeds $V_{\text{vertical}}/V_{\text{horizontal}}$ has the same order of magnitude as the product $Re^{1/3} C_D^{2/3}$, where Re is the local Reynolds number based on $V_{\text{horizontal}}$, namely $Re = LV_{\text{horizontal}}/\nu$. Finally, assume that Re is of order 10^2 or greater (so that a buckling turbulent wake forms behind the horizontal plate), and conclude which configuration (horizontal or vertical) offers greater access to the falling plate.

5.48 The natural phenomenon of economies of scale can be predicted by analyzing the force and power needed to drag an object through a fluid reservoir. Would the power needed to drag two identical river barges be smaller or greater than the power needed to drag one large barge? The size of large barge equals the sum of the two smaller barges.

   The simple model that helps you answer this question consists of two balls of diameter D_1 that are dragged with the speed U through a stagnant fluid. The drag force on one ball is F_1. The competing design is a single ball of diameter D_2 and drag force F_2, which has the same volume as the two smaller balls combined. Calculate the ratio $2F_1/F_2$, and discover the physics: economies of scale are a manifestation of the constructal law [66]. For simplicity, assume that in both designs the Reynolds number is greater than 10^2 so that the drag coefficient C_D is the same constant of order 1.

5.49 Freight of total size M (kg) is being transported on water on two ships linked together, one with motor (the tug boat) and the other without motor (the barge). The travel speed of the ensemble is V. The freight M is divided and loaded as M_1 on the tug boat and M_2 on the barge. Assume that the size of each empty vessel ($M_{e,1}$, $M_{e,2}$) is proportional to its load (M_1, M_2). In addition, the tug boat carries a motor of size M_m. The body of water displaced by each vessel is proportional to its size, namely, ($M_{e,1} + M_1 + M_m$) versus ($M_{e,2} + M_2$).

   Determine the relation between the power spent by the tug boat, as a function of the weight distribution ratio M_1/M_2. How should M be divided into M_1 and M_2 such that the power required by the ensemble is minimal?

   Hint: the length scale of the body of displaced water is proportional to the 1/3 power of the volume of that body; the frontal cross-sectional area of that body (for drag force calculations) is proportional to the square of the length scale of the body.

5.50 Figure P5.50 shows one of the most common observations of fluid flow behavior. When the water jet from the faucet impinges on the sink bottom, it first spreads smoothly as a wall jet, radially. Suddenly, the wall jet "jumps" and forms a thicker layer that continues to flow radially, with rolls and bulges that indicate a three-dimensional flow, not a purely radial flow. This phenomenon is commonly known as the "hydraulic jump," based on the popular view that at the transition from radial flow to three-dimensional flow the water jumps against gravity and acquires an amount of potential energy that the original radial jet did not have.

   In this problem, you are invited to question [66] the established view by modeling and analyzing the radial jet flow while neglecting gravity. Assume that the mass flowrate and velocity of the faucet jet ($\dot{m}$, V) are fixed. Construct the analysis in the following steps, and employ scale analysis all the way:

   First, consider the initial (smooth, radial) wall jet, and determine the scales of its thickness (δ) and radial velocity (u). Model the flow as inertial, i.e. frictionless.

   Next, model the radial wall jet as viscid, with viscous diffusion having penetrated the δ thickness (as in laminar boundary layer flow), and determine the corresponding δ and u scales of this flow regime.

Intersect the two solutions determined above, and deduce the radial distance where the inertial jet ceases to be inviscid. As a numerical example, calculate the radial distance (r) for the case where faucet jet consists of water at room temperature, with $\dot{m} \sim 0.01 \ kg/s$ and $V \sim 1$ m/s.

Finally, estimate the transition distance (r) by invoking the local Reynolds number criterion dictated by the constructal law (Appendix F), $u\delta/v \sim 10^2$, for which the scales u and δ are from the analysis of the viscid radial jet. Show that the estimated transition distance (r) has the same scale as in the preceding numerical example and in observations (Figure P5.50). Comment on the role of gravity in your model and analysis.

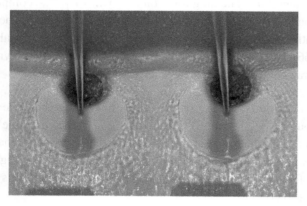

Figure P5.50

5.51 Motionless fluid is bounded by a very long plane wall. At the time $t = 0$ the wall starts moving along itself at constant speed. The thickness of the entrained layer increases in time. Because the moving wall is very long and the entrained fluid layer is very thin, the fluid flows in the same direction (x) as the wall, and its local speed is $u(y, t)$, where y is the distance in the direction perpendicular to the wall. Let δ be the order of magnitude of the distance from the wall where the speed decreased from u_0 at the wall to essentially zero far from the wall. Determine $\delta(t)$ by scale analysis; start with Eq. (5.11), keep only the terms that represent the $u(y, t)$ flow field, recognize the boundary and initial conditions that accompany the simplified Eq. (5.11), and solve the scale-analysis version of that equation. How fast is the thickness δ increasing as t increases?

5.52 Consider a two-dimensional volume of motionless fluid of length L and thickness H. The fluid is set in motion by a very thin solid blade that enters (at $t = 0$ and speed V) along the bottom side of the $H \times L$ frame (Figure P5.52). The flow regime is laminar. The fluid volume is fixed ($HL = A$, constant), but its shape H/L is free to vary. The time required by the entire fluid volume to start moving depends on both L and H. Invoke the constructal law, and determine the shape H/L such that the required time is minimum. Compare this shape with the shape of the laminar boundary layer (δ/x) determined in Eq. (5.36). What is the fundamental significance of this coincidence? To learn more, read Ref. [67].

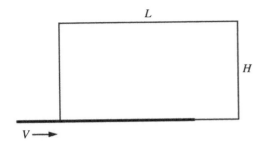

Figure P5.52

5.53 Historians of the game of soccer are discussing rankings of players who are famous for having kicked the ball the fastest during competition. The kicks are shown on videos. In one such report, the 10th fastest was one of the best players in the world, with a ball velocity of over 100 km/h. The fastest of the ten players in the video was credited with a ball velocity well over 200 km/h.

I became curious about this huge difference in performance because I had never heard of the player with the 200 km/h shot. My suspicion was that the no. 10 player kicked the ball from much farther than the no. 1 player.

Analyze this proposed explanation by determining the relation between ball velocity and distance traveled. Obviously, because of air drag the ball velocity decreases after the kick. Consult Figure 5.15, and assume that the drag coefficient is constant, independent of instantaneous velocity. Keep in mind that the ball velocity reported in soccer commentaries is the average velocity, which is calculated as the traveled distance divided by the time of travel. By what factor does the average velocity decrease if the distance to the goal is doubled?

5.54 A rogue rocket from a rogue nation falls at terminal speed through the atmosphere and impacts the ground. The terminal speed is reached when the weight of the falling object is matched by the air drag force that opposes the fall. Assume that the damage caused on the ground is measurable in terms of the kinetic energy of the body that hits the ground.

Can the damage be decreased by fragmenting the body before impact? Assume that the rocket has the mass M_1 and the length scale D_1. In other words, model the falling object as a lump of length scale D_1. What is the relation between its terminal kinetic energy (KE_1) and D_1?

Next, M_1 is fragmented into two equal pieces of size M_2 and length scale D_2. Compare with KE_1 the total kinetic energy of the two pieces ($2KE_2$). By how much has the impact been reduced?

The falling rocket threatens a country where the life of the individual is valued. Buildings are designed to withstand a total impact that does not exceed $\frac{1}{2}KE_1$. How many stages of fragmentation must be designed into the falling rocket ($M_1 \rightarrow 2M_2 \rightarrow 4M_3 \rightarrow \ldots$) such that the total KE of the fragments falls below $\frac{1}{2}KE_1$?

6

Internal Forced Convection

6.1 Laminar Flow Through a Duct

6.1.1 Entrance Region

In this chapter we turn our attention to forced convection configurations in which the flow is "internal," that is, surrounded by the wall of a duct at a different temperature. Consider the laminar flow through the round tube of diameter D shown in Figure 6.1. The mouth is located in the plane $x = 0$, and x is measured in the downstream direction.

Throughout this chapter we shall write U for the *mean velocity*, that is, the longitudinal velocity averaged over the duct cross section A:

$$U = \frac{1}{A} \int_A u \, dA \tag{6.1}$$

In Figure 6.1, A is equal to πr_o^2, where r_o is the outer radius of the flow region (the inner radius of the tube wall). Equation (6.1) is a general definition of U. Note that regardless of the shape of the duct cross section, U is also the uniform (sluglike) distribution of longitudinal velocity through the entrance to the duct.

Assume that the mean velocity U is constant or nearly constant along the duct. According to the principle of mass conservation, this assumption is valid when the fluid density ρ is constant or nearly constant along the duct. It is particularly good in the case of liquids (e.g. water and oils) and fairly good for gases that experience only moderate temperature changes while flowing through the duct. "Moderate" means that the change in temperature from inlet to outlet is at least ten times smaller than the absolute (Kelvin) temperature of the gas. On the other hand, if the density variation is significant, the gas stream accelerates or decelerates along the duct, depending on whether it is heated or cooled. The complications that arise when ρ and U are not sufficiently constant along the duct are treated in Section 9.5.3.

The uniform velocity assumed across the $x = 0$ plane in Figure 6.1 is also an approximation. For example, if the flow originates from a large reservoir positioned to the left of the $x = 0$ plane, the entering fluid flows longitudinally only if it is close to the tube centerline. Closer to the rim, the velocity of the entering fluid has a negative radial component. These and other entrance effects form the subject of Section 9.5.2.

Immediately downstream from the mouth, the flow develops a boundary layer that coats the inner surface of the wall. The thickness of this velocity boundary layer, δ, initially grows as $x^{1/2}$. For this reason we must identify the longitudinal position X where the velocity boundary layer thickness δ has grown to the same size as the tube radius r_o. In other words, X is the longitudinal position where the inflowing stream has had enough time to feel the effect of wall friction (transversal viscous diffusion) all the way to its centerline.

Heat Transfer: Evolution, Design and Performance, First Edition. Adrian Bejan.
© 2022 John Wiley & Sons, Inc. Published 2022 by John Wiley & Sons, Inc.
Companion website: www.wiley.com/go/bejan/heattransfer

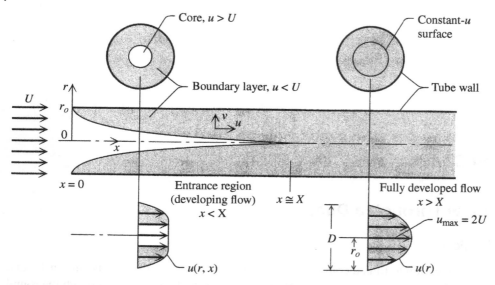

Figure 6.1 Entrance region and fully developed flow region of laminar flow in a tube.

The length X divides the flow into two distinct regions: the flow entrance region situated between the mouth and $x \sim X$, and the fully developed flow region located downstream from $x \sim X$. The flow entrance region and the fully developed flow region are known also as the *hydrodynamic* entrance region and the *hydrodynamically* fully developed region. The size of X can be estimated using the results of Section 5.3.1. As a measure of the thickness of the boundary layer present in the entrance region, we can use an approximate version of Eq. (5.57):

$$\delta(x) \cong 5x\left(\frac{Ux}{\nu}\right)^{-1/2} \tag{6.2}$$

The order-of-magnitude definition of the entrance-length X is

$$\delta \cong r_0 \quad \text{when} \quad x \cong X \tag{6.3}$$

Combining this with Eq. (6.2), we conclude that the geometric aspect ratio of the flow entrance region, X/D, is proportional to the Reynolds number based on tube diameter ($Re_D = UD/\nu$):

$$\frac{X}{D} \cong 10^{-2} Re_D \tag{6.4}$$

The more exact and more frequently used estimate is

$$\frac{X}{D} \cong 0.05 Re_D \tag{6.4'}$$

In the case of a duct whose cross section is not circular, in Eq. (6.4'), the diameter D is replaced by the hydraulic diameter D_h, which will be defined later in Eq. (6.28).

The flow that develops in the entrance region has two subregions of its own – the annular boundary layer and the core that surrounds the centerline. The fluid that entered the tube with the velocity U is slowed down by the wall in the boundary layer ($u < U$). The conservation of the flowrate $\rho \pi r_o^2 U$ in each cross section requires longitudinal velocities larger than U in the core region. In Section 6.1.2, we will learn that the centerline fluid accelerates from $u = U$ at the mouth to $u = 2U$ on the centerline of the fully developed flow region.

6.1.2 Fully Developed Flow Region

The flow in the fully developed region can be determined based on a relatively simple analysis. We start with the equations that govern the conservation of mass and momentum in the cylindrical system of coordinates (r, θ, x) attached to the tube centerline. Assuming a flow with nearly constant density (ρ) and viscosity (μ) and replacing z with x, v_z with u, and v_r with v in the equations of Table 5.2, we find the three equations that govern the flow in the steady state:

$$\frac{1}{r}\frac{\partial}{\partial r}(rv) + \frac{\partial u}{\partial x} = 0 \tag{6.5}$$

$$\rho\left(v\frac{\partial v}{\partial r} + u\frac{\partial v}{\partial x}\right) = -\frac{\partial P}{\partial r} + \mu\left(\frac{\partial^2 v}{\partial r^2} + \frac{1}{r}\frac{\partial v}{\partial r} - \frac{v}{r^2} + \frac{\partial^2 v}{\partial x^2}\right) \tag{6.6}$$

$$\rho\left(v\frac{\partial u}{\partial r} + u\frac{\partial u}{\partial x}\right) = -\frac{\partial P}{\partial x} + \mu\left(\frac{\partial^2 u}{\partial r^2} + \frac{1}{r}\frac{\partial u}{\partial r} + \frac{\partial^2 u}{\partial x^2}\right) \tag{6.7}$$

In writing these equations, we left out the derivatives with respect to θ, because the tube wall is such that the flow field does not change as θ changes. It is said that the flow exhibits "θ symmetry." Equations (6.5)–(6.7) assume a dramatically simpler form if we recognize that in the flow region of length x and thickness D, the mass conservation Eq. (6.5) requires

$$\frac{v}{D} \sim \frac{U}{x} \tag{6.8}$$

The transversal velocity scale v is therefore

$$v \sim \frac{UD}{x} \tag{6.9}$$

This scale shows that v becomes negligible relative to U as x increases:

$$v = 0 \quad \text{(fully developed flow)} \tag{6.10}$$

From Eq. (6.10) follow two additional conclusions. First, from Eq. (6.5) we learn that in the fully developed region, $\partial u/\partial x$ must also be zero, which means that

$$u = u(r) \tag{6.11}$$

Second, by substituting $v = 0$ into Eq. (6.6), we find that $\partial P/\partial r = 0$; therefore,

$$P = P(x) \tag{6.12}$$

In view of the last three conclusions, Eqs. (6.10)–(6.12), the longitudinal momentum Eq. (6.7) assumes the much simpler form (note the disappearance of the partial derivative signs)

$$\frac{dP}{dx} = \mu\left(\frac{d^2 u}{dr^2} + \frac{1}{r}\frac{du}{dr}\right) \tag{6.13}$$

where the left side is, at best, a function of x and the right side is a function of r. In conclusion, both sides of Eq. (6.13) must be equal to the same constant:

$$\frac{dP}{dx} = \text{constant} = \mu\left(\frac{d^2 u}{dr^2} + \frac{1}{r}\frac{du}{dr}\right) \tag{6.14}$$

Two equations (two equal signs) appear in Eq. (6.14). The first indicates that the pressure varies linearly along the fully developed section of the tube flow. If the entrance-length X is much smaller than the overall length of the tube, L, and if ΔP is the pressure difference (pressure "drop") maintained between the mouth and the downstream end of the tube, then the constant alluded to in Eq. (6.14) is

$$\frac{dP}{dx} = -\frac{\Delta P}{L} \tag{6.15}$$

The equation associated with the second equal sign in Eq. (6.14) can be solved for the fully developed velocity distribution $u(r)$. The two boundary conditions are

$$u = 0 \quad \text{at} \quad r = r_o \quad \text{(no slip)} \tag{6.16a}$$

$$\frac{du}{dr} = 0 \quad \text{at} \quad r = 0 \quad \text{(symmetry)} \tag{6.16b}$$

The solution can be derived as an exercise (Problem 6.1):

$$u = \frac{r_o^2}{4\mu}\left(-\frac{dP}{dx}\right)\left[1 - \left(\frac{r}{r_o}\right)^2\right] \tag{6.17}$$

Substituted into the U definition (6.1), this solution yields

$$U = \frac{r_o^2}{8\mu}\left(-\frac{dP}{dx}\right) \tag{6.18}$$

which shows that the average velocity (or the flowrate $\rho\pi r_o^2 U$) is proportional to the longitudinal pressure gradient. The $u(r)$ solution (6.17) can then be rewritten as

$$u = 2U\left[1 - \left(\frac{r}{r_o}\right)^2\right] \tag{6.19}$$

to show that the velocity along the centerline of the tube is exactly twice as large as the tube-averaged velocity U. The fully developed flow distribution (6.19) is known as Hagen–Poiseuille flow, after the first two investigators who reported it [1, 2].

Similarly, the laminar flow into the space of thickness D formed between two parallel walls (Figure 6.2) can be divided into an entrance region followed by a fully developed region. With reference to the (x, y) system of coordinates and boundary conditions defined in Figure 6.2, the governing equations for the fully

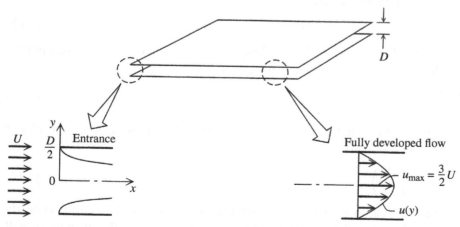

Figure 6.2 Laminar flow through a parallel-plate channel.

developed region reduce to

$$\frac{dP}{dx} = \text{constant} = \mu \frac{d^2u}{dy^2} \qquad (6.20)$$

The resulting velocity profile $u(y)$ is parabolic (Problem 6.2):

$$u = \frac{3}{2}U \left[1 - \left(\frac{y}{D/2} \right)^2 \right] \qquad (6.21)$$

$$U = \frac{D^2}{12\mu} \left(-\frac{dP}{dx} \right) \qquad (6.22)$$

and the maximum (midplane) velocity is only 50% larger than the cross section-averaged value U. The flowrate is proportional to the longitudinal pressure gradient $-dP/dx$, which drives the flow.

6.1.3 Friction Factor and Pressure Drop

The fully developed flow solutions exhibited in Eqs. (6.19) and (6.21) show that the longitudinal velocity u is not a function of the longitudinal position x. For this reason, the wall shear effect τ_ω is also independent of x. For Hagen–Poiseuille flow through a tube, Figure 6.1 (right side), the τ_ω constant is evaluated from Eq. (6.19):

$$\tau_\omega = \mu \left(-\frac{du}{dr} \right)_{r=r_o} = 4\mu \frac{U}{r_o} \qquad (6.23)$$

It is customary to express τ_ω in dimensionless form as the Fanning friction factor:

$$f = \frac{\tau_\omega}{\frac{1}{2}\rho U^2} \qquad (6.24)$$

The Fanning friction factor f should not be confused with the Darcy–Weisbach [4] friction factor f_D, which is $f_D = 4f$. Note the similarities between f and the local skin friction coefficient defined in Eq. (5.39). This time, the role of velocity unit is played by the tube-averaged velocity U, and, unlike in Eq. (5.39), the resulting dimensionless group (f) is x-independent. Combining Eqs. (6.23) and (6.24), we find that the friction factor for fully developed tube flow is inversely proportional to the Reynolds number based on mean velocity and tube diameter ($Re_D = UD/v$);

$$f = \frac{16}{Re_D} \quad \text{(fully developed tube flow)} \qquad (6.25)$$

Either in f or τ_ω form, the wall friction information can be used to deduce the overall pressure difference ΔP that must be maintained across the duct of length L to drive the flow of average velocity U. Consider the flow through the straight duct shown in the upper-left corner of Figure 6.5. The unspecified geometry of the duct cross section is described by the flow cross-sectional area A and by the wetted perimeter of the cross section, p.

When the length L is much larger than the entrance length estimated in Section 6.1.1, the inner surface of the duct is covered by a wall shear stress τ_ω that does not vary with the longitudinal position. In the round-tube and parallel-plate examples, τ_ω is also uniformly distributed around the wetted perimeter of the cross section. In a duct with a less regularly shaped cross section (e.g. triangular), τ_ω varies around the cross section periphery in such a way that its smallest values occur in the corners. For this reason, in the force balance suggested by the first drawing of Figure 6.3, the term τ_ω represents the wall shear stress *averaged* around the periphery of length p. The product $\tau_\omega pL$ is, therefore, the total friction force experienced by the tube wall.

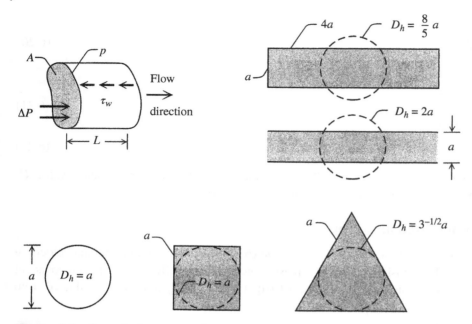

Figure 6.3 Force balance over the flow control volume (the upper-left corner) and five duct cross-sectional shapes and their hydraulic diameters. Source: After Ref. [3].

The force balance on the control volume AL requires

$$\Delta P \, A = \tau_\omega p L \tag{6.26}$$

In terms of the friction factor f, the pressure drop given by Eq. (6.26) reads

$$\Delta P = f \frac{L}{A/p} \frac{1}{2} \rho U^2 \tag{6.27}$$

Equations (6.26) and (6.27) hold for either laminar or turbulent flow, provided the length L contains fully developed flow. Note further that the denominator A/p has units of length; for example, the A/p value of a circular cross section of diameter D is equal to $D/4$. It makes sense, then, to define $4A/p$ as the hydraulic diameter D_h of a cross section that is not circular:

$$D_h = \frac{4A}{p} \tag{6.28}$$

and to regard D_h as a characteristic transversal length scale of the flow. The pressure drop (6.27) becomes

$$\Delta P = f \frac{4L}{D_h} \frac{1}{2} \rho U^2 \tag{6.27'}$$

Figure 6.3 shows several of the more common duct cross sections and their respective hydraulic diameters. These cross sections have been drawn to scale in such a way that they all have the same hydraulic diameter [3]. For example, in the case of a round cross section of diameter D, the hydraulic diameter is equal to the actual diameter, $D_h = 4(\pi D^2/4)(\pi D) = D$. In a parallel-plate channel of spacing S and width W (i.e. flow cross section $S \times W$), the hydraulic diameter is two times greater than the spacing, $D_h = 4(SW)/(2W) = 2S$. Inside regular polygons, the hydraulic diameter is equal to the diameter of the inscribed circle (e.g., square and equilateral triangle in Figure 6.3).

Table 6.1 Friction factors (f), cross section shape numbers (B), and Nusselt numbers (Nu_D) for hydrodynamically and thermally fully developed duct flows [3].

Cross section shape	$f\,Re_{D_h}$	B	$Nu_{D_h} = hD_h/k$ Uniform q_w''	Uniform T_w
60° triangle	13.3	0.605	3	2.35
square	14.2	0.785	3.63	2.89
circle	16	1	4.364	3.66
$a \times 4a$ rectangle	18.3	1.26	5.35	4.65
thin rectangle	24	1.57	8.235	7.54
thin rectangle, One side insulated	24	1.57	5.385	4.86

The first column of Table 6.1 shows the friction factor formulas for fully developed laminar flow through the cross sections of Figure 6.3. The general form of these formulas is

$$f = \frac{C}{Re_{D_h}} \tag{6.29}$$

where Re_{D_h} is the Reynolds number based on hydraulic diameter, UD_h/ν. The constant C appearing in the numerator depends on the shape of the duct cross section. It was pointed out in Ref. [3] that C increases monotonically with a new dimensionless parameter, the cross section shape number:

$$B = \frac{\pi D_h^2/4}{A} \tag{6.30}$$

whose purpose is to measure the deviation of the actual shape from the circular shape. Note that $B = 1$ corresponds to a round tube. The C values of Table 6.1 are approximated within 4% by the expression

$$C \cong 16\ \exp(0.294B^2 + 0.068B - 0.318) \tag{6.31}$$

Example 6.1 *Laminar Flow: Round Tube*
Consider the laminar flow of 20 °C water through a tube with the diameter $D = 2.7$ cm. The mean velocity is $U = 6$ cm/s. Calculate the pressure drop per unit length $\Delta P/L$ in the fully developed region. Determine the flow entrance length.

Solution

We begin with the calculation of the Reynolds number to verify that the flow is indeed laminar:

$$Re_D = \frac{UD}{\nu}$$

$$= 6\frac{\text{cm}}{\text{s}} 2.7\ \text{cm}\ \frac{\text{s}}{0.01\ \text{cm}^2} = 1620 \quad \text{(laminar)} \tag{6.25}$$

$$f = \frac{16}{Re_D} = \frac{16}{1620} = 0.0099$$

$$\frac{\Delta P}{L} = f\frac{4}{D}\frac{1}{2}\rho U^2$$

$$= 0.0099\frac{4}{2.7\ \text{cm}}\frac{1}{2}1\frac{\text{g}}{\text{cm}^3}\left(6\ \frac{\text{cm}}{\text{s}}\right)^2$$

$$= 0.264\frac{\text{g}}{\text{cm}^2\text{s}^2} = 2.64\frac{\text{N/m}^2}{\text{m}} \tag{6.27'}$$

$$X \cong 0.05DRe_D$$

$$= (0.05 \times 2.7\ \text{cm})\,(1620) = 2.2\ \text{m}$$

6.2 Heat Transfer in Laminar Flow

6.2.1 Thermal Entrance Region

The fluid mechanics covered in the preceding section is a prerequisite for the analysis of the heat transfer between the duct wall and the fluid stream. We continue to assume that the duct is long enough so that the flow is fully developed over most of the length L. This is why in Figure 6.4 the flow entrance region is not shown and the fully developed velocity profile u is drawn to emphasize that it does not change along the duct.

The entering stream has a uniform initial temperature T_0 at $x = 0$. Heat transfer occurs between the duct wall and the fluid, and this effect is represented by the local wall heat flux q''_w and the local wall temperature T_w. In this opening subsection, we say nothing more about q''_w and T_w, except that, in general, they can both be functions of longitudinal position. Two special types of wall heating conditions will be analyzed soon, namely, uniform heat flux and uniform temperature.

The heating effect penetrates the flow gradually. It is felt first in a thin thermal boundary layer that coats the wall. The thermal boundary layer grows in the downstream direction and eventually becomes as thick as the duct itself. The approximate longitudinal position $x \sim X_T$ where the thermal boundary layer loses its identity divides the duct length into two distinct regions: (i) a thermal entrance region ($0 < x < X_T$) in which the shape of the transversal temperature profile "develops" (i.e. where the T profile changes from one x to the next) and (ii) a thermally fully developed region ($x > X_T$) where the shape of the temperature profile is preserved. In tube flow, for example, the unchanging shape of the fully developed temperature profile is similar to that of the meniscus at the end of a liquid column in a capillary. This shape is preserved (i.e. it remains a "meniscus") even though its amplitude may vary in the longitudinal direction. In Figure 6.4, the amplitude is the difference between the temperature at the wall and the minimum or maximum temperature on the duct centerline. The figure shows that in the thermally fully developed region ($x > X_T$), the amplitude changes while the shape – the meniscus – is preserved.

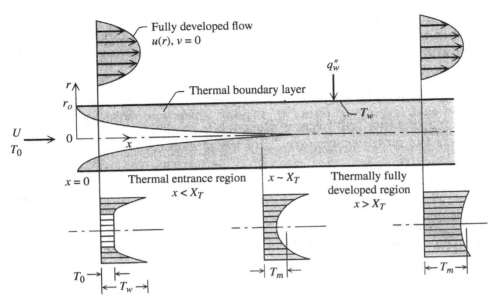

Figure 6.4 Thermal entrance region and the thermally fully developed flow in a tube.

A scale analysis similar to the one used in the derivation of the flow entrance-length X, Eq. (6.4), can be used to estimate the thermal entrance-length X_T [3]:

$$\frac{X_T}{D_h} \cong 0.05 Re_{D_h} Pr \tag{6.32}$$

Reference [3] showed that this estimate holds for both high-Pr and low-Pr fluids and is insensitive to whether the flow entrance-length X is longer or shorter than the thermal entrance-length X_T. In fluids with Prandtl numbers that do not differ greatly from 1 (e.g. air), the lengths X and X_T are of the same order of magnitude: if X is negligible relative to L, then X_T is negligible.

6.2.2 Thermally Fully Developed Region

For reasons that will become clear while reading Eq. (6.38), it is convenient to define the mean temperature T_m of the stream as

$$T_m = \frac{1}{UA} \int_A uT \, dA \tag{6.33}$$

This is also known as the bulk temperature of the stream, and it is an integral feature of the bulk flow model that is used routinely in the thermodynamic analysis of open (flow) systems [5]. The temperature T_m is a cross-sectional *weighted* average of the local fluid temperature T: The role of weighting factor is played by the longitudinal velocity u, which is zero at the wall and large in the central portion of the duct cross section.

The mean temperature difference between the wall and the stream is

$$\Delta T = T_w - T_m \tag{6.34}$$

in which both T_w and T_m can vary in the longitudinal direction. In problems of internal convection (duct flow), the mean temperature difference is used as temperature difference scale in the definition of the wall–fluid heat transfer coefficient h, which generally can be a function of x:

$$h = \frac{q''_w}{T_w - T_m} = \frac{q''_w}{\Delta T} \tag{6.35}$$

Consider now the problem of determining the heat transfer coefficient for a hydrodynamically and thermally fully developed flow through a straight duct of unspecified cross-sectional shape (A, p). Hydrodynamically developed means that the u profile is x-independent and that the order of magnitude of u is the mean value U. The thermal development of the temperature field means that the transversal length scale of the temperature profile is the same as the duct size D_h. For example, in a tube of radius r_o full thermal development means that

$$\left(\frac{\partial T}{\partial r}\right)_{r=r_o} \sim \frac{\Delta T}{r_o} \tag{6.36}$$

The relationship between the local wall heat flux q''_w and the local change in mean fluid temperature, dT_m/dx, follows from the first law of thermodynamics, as applied to the steady flow through the duct control volume of length dx and cross-sectional area A:

$$\int_A \rho u \left(i_{x+dx} - i_x\right) dA = q''_w p \, dx \tag{6.37}$$

The integral expresses the difference between the enthalpy outflow (at $x + dx$) and the enthalpy inflow (at x). The right side of Eq. (6.37) represents the rate of heat transfer into the control volume, through the lateral area $p \, dx$. In accordance with the assumption of negligible local pressure changes, which was invoked in the derivation of the energy equation (review Eq. (5.17)), we can replace the specific enthalpy change, $i_{x+dx} - i_x$, with $c_p dT$:

$$\rho c_P \, d\left(\int_A u T \, dA\right) = q''_w p \, dx \tag{6.38}$$

On the left side of Eq. (6.38), we recognize the integral used in the definition of the mean temperature; in this way, we arrive at the conclusion that the local mean temperature gradient is proportional to the local wall heat flux:

$$\frac{dT_m}{dx} = \frac{p}{A} \frac{q''_w}{\rho c_P U} \tag{6.39}$$

In the special case of a round tube, this conclusion reads

$$\frac{dT_m}{dx} = \frac{2}{r_o} \frac{q''_w}{\rho c_P U} \qquad \text{(round tube)} \tag{6.39'}$$

It is instructive to show that the first-law statement (6.39) and the condition of full thermal development, Eq. (6.36), are sufficient for estimating the order of magnitude of the heat transfer coefficient h. According to Table 5.2 and the cylindrical coordinates and velocity components defined in Figure 6.4, the steady-state energy equation for fully developed flow and temperature in a round tube is

$$\rho c_P u \frac{\partial T}{\partial x} = k \frac{1}{r} \frac{\partial}{\partial r}\left(r \frac{\partial T}{\partial r}\right) \tag{6.40}$$

This equation expresses the balance between the heat conducted from the wall radially into the fluid (the right side) and the enthalpy carried away longitudinally by the stream (the left side). The orders of magnitude of the two sides of this balance are evaluated using Eqs. (6.39′) and (6.36):

$$\rho c_P U \frac{1}{r_o} \frac{q_w''}{\rho c_P U} \sim k \frac{\Delta T}{r_o^2} \tag{6.41}$$

From this we conclude that the heat transfer coefficient, $h = q_w''/\Delta T$, must be x-independent:

$$h \sim \frac{k}{r_o} \quad \text{(constant)} \tag{6.42}$$

The internal flow heat transfer coefficient is commonly nondimensionalized as a Nusselt number based on hydraulic diameter:

$$Nu_{D_h} = \frac{h D_h}{k} \tag{6.43}$$

In a round tube, D_h is the same as the tube diameter D; therefore, the Nu_D estimate that corresponds to Eq. (6.42) is

$$Nu_D \sim 1 \quad \text{(constant)} \tag{6.44}$$

The Nusselt number for hydrodynamically and thermally fully developed flow is a constant whose order of magnitude is 1. This conclusion means also that the local heat transfer coefficient is independent of longitudinal position, even though q_w'' and ΔT may vary with x. It was shown in Ref. [3] that the constant-h conclusion (6.42) means that the fluid temperature $T(r, x)$ obeys a function of the type

$$T(r,x) = T_w(x) - [T_w(x) - T_m(x)]\phi\left(\frac{r}{r_o}\right) \tag{6.45}$$

where ϕ is a function of r only. Equation (6.45) is usually postulated as the definition of what is meant by thermally fully developed flow, that is, the condition that must be met by the temperature profile to be "fully" developed. In this subsection we started from physics, Eq. (6.36), and predicted not only Eq. (6.45) but also the Nusselt number (6.44).

6.2.3 Uniform Wall Heat Flux

The Nusselt number for fully developed flow through a tube with uniform wall heat flux can be determined analytically by solving the energy Eq. (6.40). This equation acquires a simpler form based on the observation that when q_w'' is constant, Eqs. (6.35) and (6.42) require

$$\frac{dT_w}{dx} = \frac{dT_m}{dx} \tag{6.46}$$

Differentiating with respect to x the fully developed temperature profile (6.45), and using Eq. (6.46), we learn further that

$$\frac{\partial T}{\partial x} = \frac{dT_w}{dx} = \frac{dT_m}{dx} \tag{6.47}$$

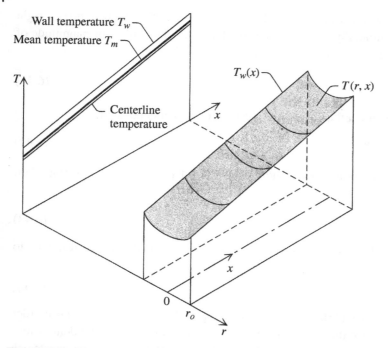

Figure 6.5 Fully developed temperature distribution in a tube with uniform heat flux.

In conclusion, instead of $\partial T/\partial x$ on the left side of the energy Eq. (6.40), we use dT_w/dx, which according to Eq. (6.39′) is a constant. This conclusion is illustrated graphically by the three-dimensional surface $T(r, x)$ in Figure 6.5. Finally, on the right side of the energy Eq. (6.40), we replace T with the expression listed in Eq. (6.45), so that in the end the energy equation reads

$$2\frac{u}{U}\frac{q_w''}{kr_o\Delta T} = -\frac{1}{r}\frac{d}{dr}\left(r\frac{d\phi}{dr}\right) \tag{6.48}$$

In the denominator on the left side, ΔT is the temperature difference $T_w(x) - T_m(x)$.

The velocity profile u/U is given by Eq. (6.19). Finding $T(r, x)$ reduces to solving Eq. (6.48) for $\phi(r)$, subject to the boundary conditions $\phi(r_o) = 0$ and $\phi'(0) = 0$. The result is

$$\phi(r) = \frac{q_w''r_o}{k\Delta T}\left[\frac{3}{4} - \left(\frac{r}{r_o}\right)^2 + \frac{1}{4}\left(\frac{r}{r_o}\right)^4\right] \tag{6.49}$$

Combined with Eq. (6.45), this $\phi(r)$ solution represents the temperature distribution $T(r, x)$ subject to the still undetermined factor $q_w''r_o/k\Delta T$ that appears in Eq. (6.49). This factor is the unknown of the problem, as it is equal to $Nu_D/2$, where $Nu_D = q_w''D/k\Delta T$. We determine this factor from the condition that $T(r, x)$ must obey the T_m definition (6.33). In terms of ϕ, the mean temperature definition (6.33) reads

$$\frac{1}{UA}\int_A u\phi \, dA = 1 \tag{6.50}$$

where $A = \pi r_o^2$. The substitution of Eqs. (6.19) and (6.49) into this integral condition leads to the conclusion that the Nusselt number is indeed a constant of order 1 (cf. Eq. (6.44)):

$$Nu_D = \frac{48}{11} = 4.364 \quad \text{(constant } q_w'') \tag{6.51}$$

Table 6.2 Friction factors and Nusselt numbers for heat transfer to laminar flow through ducts with regular polygonal cross sections.

		Nu_{D_h}			
		Uniform heat flux		Isothermal wall	
Cross section	$f\,Re_{D_h}$ Fully developed flow	Fully developed flow	Slug flow	Fully developed flow	Slug flow
Square	14.167	3.614	7.083	2.980	4.926
Hexagon	15.065	4.021	7.533	3.353	5.380
Octagon	15.381	4.207	7.690	3.467	5.526
Circle	16	4.364	7.962	3.66	5.769

Source: The data are from Asako et al. [6].

The third column of Table 6.1 lists the corresponding Nu values for five additional duct cross-sectional shapes. Except for the bottom entry (parallel-plate channel with one side insulated), the Nusselt numbers increase monotonically as the shape number B increases [3]. In the triangular, square, and rectangular cases, the wall temperature varies along the perimeter of the cross section when the heat flux is uniform. In such cases the heat transfer coefficient is defined by $h = q_w'' / [\overline{T}_w(x) - T_m(x)]$, where $\overline{T}_w(x)$ is the wall temperature averaged over the entire perimeter. Additional examples of Nu values for fully developed flow through ducts with uniform heat flux can be found in Ref. [4] and, later here, in Table 6.2.

Figure 6.6 shows the complete behavior of the Nusselt number along a tube with uniform heat flux. The figure was drawn based on data reported in Refs. [4, 7], The Nu_D constant determined in Eq. (6.51) is reached when the flow is in the thermally fully developed region. (Note that x is measured starting from the mouth of the tube.) In the thermal entrance region, the local Nusselt number based on tube diameter ($Nu_D = q_w'' D / k\Delta T$) is $q_w'' D / (k\Delta T)$ a function of x, indicating that, unlike in the fully developed region, the temperature difference $\Delta T = T_w - T_m$ is not constant. It can be verified that toward the extreme left of Figure 6.6, the Nu_D (or ΔT) behavior is consistent with the behavior seen in a thermal boundary layer near a plane wall with uniform heat flux, Table 5.4.

A closed-form expression that covers both the entrance and fully developed regions of Figure 6.6 was developed by Churchill and Ozoe [8]:

$$\frac{Nu_D}{4.364\,[1 + (Gz/29.6)^2]^{1/6}}$$
$$= \left[1 + \left(\frac{Gz/19.04}{[1 + (Pr/0.0207)^{2/3}]^{1/2}[1 + (Gz/29.6)^2]^{1/3}} \right)^{3/2} \right]^{1/3} \tag{6.52}$$

The new dimensionless group Gz is the Graetz number[1]:

$$Gz = \frac{\pi D^2 U}{4\alpha x} = \frac{\pi}{4} \left(\frac{x/D}{Re_D Pr} \right)^{-1} \tag{6.53}$$

---

1 Leo Graetz (1856–1941) was a professor of physics at the University of Munich. His writings covered a very wide domain, from heat transfer (conduction, convection, and radiation) to mechanics (elasticity, friction) and electromagnetism.

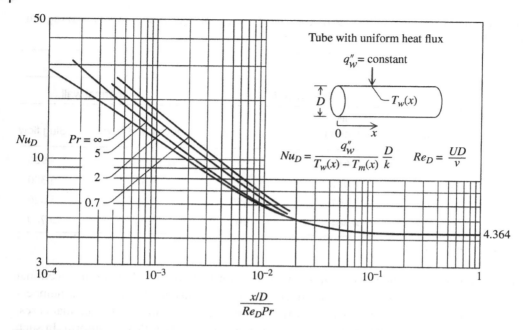

Figure 6.6 The Nusselt number for laminar flow through a tube with uniform wall heat flux.

which is inversely proportional to the abscissa parameter used in Figure 6.6. The formula (6.52) agrees within 5% with experimental and numerical data for $Pr = 0.7$ and 10 and has the correct asymptotic behavior for large and small Gz and Pr.

6.2.4 Isothermal Wall

When the flow is hydrodynamically and thermally fully developed and when the tube wall is isothermal, the mean temperature of the stream varies exponentially toward the plateau value T_w as x increases. See the projection on the vertical plane to the left in Figure 6.7. Such a relationship is required by Eq. (6.39′) because, after writing $q_w'' = h[T_w - T_m(x)]$, that equation reads

$$\frac{dT_m}{dx} = \frac{2h}{r_o \rho c_p U}[T_w - T_m(x)] \tag{6.54}$$

This equation can be integrated from a reference position $x = x_1$ where $T_m = T_{m,1}$, by noting that h must be a constant (cf. Eq. (6.42)):

$$\frac{T_w - T_m(x)}{T_w - T_{m,1}} = \exp\left[-\frac{2h(x - x_1)}{r_o \rho c_p U}\right] \tag{6.55}$$

The heat transfer coefficient h, or the Nusselt number $Nu_D = hD/k$, can be determined by following the analytical steps outlined in the preceding subsection. This time, however, the resulting equation for $\phi(r)$ is more complicated and has to be solved iteratively or numerically [3]. The result is

$$Nu_D = 3.66 \quad (\text{constant } T_w) \tag{6.56}$$

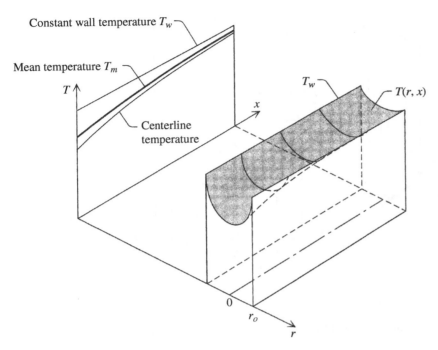

Figure 6.7 Fully developed temperature distribution in a tube with constant wall temperature.

The constancy of Nu_D implies that when the wall is isothermal, both ΔT and q''_w vary exponentially in the downstream direction:

$$q''_w(x) = 3.66 \frac{k}{D}(T_w - T_{m,1}) \exp\left[-3.66 \frac{\alpha(x - x_1)}{r_o^2 U}\right] \tag{6.57}$$

Table 6.1 lists the Nu_D constants for other cross-sectional shapes of ducts with isothermal wall when the laminar flow is both hydrodynamically and thermally fully developed. The table shows again that duct cross sections with large shape numbers B have correspondingly large fully developed flow Nusselt numbers.

Figure 6.8 shows the Nusselt number in the thermal entrance and fully developed regions of a tube with isothermal wall. The curves were drawn based on data furnished by Refs. [4, 7]. The D-based Nusselt number approaches the 3.66 value as the stream proceeds into the thermally fully developed region. In the thermal entrance region, Nu_D varies as $x^{-1/2}$, indicating that the thermal resistance between wall and stream is associated with a thermal boundary layer of the type described by Eq. (5.79). Keep in mind the difference between the D-based Nusselt number, Eq. (6.43), and the Nu_x definition (5.71).

Table 6.2 shows the fully developed values of the friction factor and the Nusselt number in a duct with regular polygonal cross section. Some of the cases in this table are also included in Table 6.1. The less than 1% discrepancies that exist between the data covered by both tables (e.g. the square cross section) represent the discrepancies found between the results reported by different investigators in the literature. Note again that in the noncircular cases with uniform heat flux the wall temperature changes along the

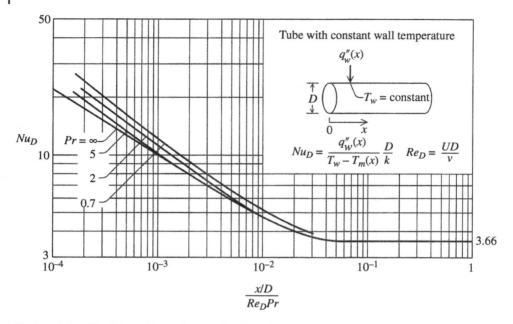

Figure 6.8 The Nusselt number for laminar flow through a tube with constant wall temperature.

perimeter, and the heat transfer coefficient is based on the perimeter-averaged temperature $\overline{T}_w(x)$, namely, $h = q''_w/[\overline{T}_w(x) - T_m(x)]$.

Table 6.2 also shows that the thermally developed Nu_D value is considerably smaller in fully developed flow than in "slug flow." The latter refers to the flow of a fluid with extremely small Prandtl number ($Pr \to 0$), in which the viscosity is so much smaller than the thermal diffusivity that the longitudinal velocity profile remains uniform over the cross section, $u = U$, like the velocity distribution inside a solid slug. The persistence of slug flow in the $Pr \to 0$ limit, even at a large enough x where the temperature profile is fully developed, is expected from the relationship between the flow and the thermal developing lengths, cf. Eqs. (6.4′) and (6.32):

$$\frac{X_T}{X} \cong Pr \tag{6.58}$$

When Pr is very small, the flow entrance distance X is much longer than the thermal entrance-length X_T, and when $X_T < x < X$, the flow distribution in the thermally fully developed region is approximately sluglike.

Example 6.2 *Heat Transfer, Laminar Flow, and Round Tube*

Water at room temperature is being heated as it flows through a pipe with uniform wall heat flux, $q''_w = 0.1 \text{ W/cm}^2$. The flow is hydrodynamically and thermally fully developed. The mass flowrate of the stream is $\dot{m} = 10$ g/s and the pipe radius is $r_o = 1$ cm. The properties of room-temperature water are approximately $\mu = 0.01$ g/cm · s and $k = 0.006$ W/cm · K. Calculate (a) the Reynolds number based on diameter ($D = 2r_o$) and mean velocity (U), (b) the heat transfer coefficient (h), and (c) the difference between the local wall temperature and the local mean (bulk) temperature.

Solution

(a) The D-based Reynolds number is

$$Re_D = \frac{\rho U D}{\mu}$$

for which the product ρU can be determined from the given mass flowrate:

$$\rho U = \frac{\dot{m}}{\pi D^2 / 4} = \frac{10 \text{ g}}{\text{s}} \frac{4}{\pi} \frac{1}{(2 \text{ cm})^2}$$

$$= 3.18 \text{ g/cm}^2 \cdot \text{s}$$

The Reynolds number is therefore

$$Re_D = 3.18 \frac{\text{g}}{\text{cm}^2 \cdot \text{s}} 2 \text{ cm} \frac{\text{cm} \cdot \text{s}}{0.01 \text{ g}}$$

$$= 637 \qquad \text{(laminar regime)}$$

(b) The heat transfer coefficient follows from the definition of the Nusselt number for tube flow:

$$Nu_D = \frac{hD}{k} = 4.364$$

Therefore,

$$h = Nu_D \frac{k}{D} = 4.364 \times 0.006 \frac{\text{W}}{\text{cm} \cdot \text{K}} \frac{1}{2 \text{ cm}} = 0.0131 \text{W/cm}^2 \cdot \text{K}$$

(c) From the definition of h in internal flow, $q''_w = h(T_w - T_m)$, we find that

$$T_w - T_m = \frac{q''_w}{h} = \frac{0.1 \text{ W}}{\text{cm}^2} \frac{\text{cm}^2 \cdot \text{K}}{0.0131 \text{ W}}$$

$$= 7.64 \text{ K} = 7.64 \,^\circ\text{C}$$

6.3 Turbulent Flow

6.3.1 Transition, Entrance Region, and Fully Developed Flow

Consider first the turbulent flow through a straight pipe of inner diameter D. Experimental observations (Appendix F) indicate that the fully developed laminar flow, Eq. (6.19), ceases to exist if the D-based Reynolds number exceeds approximately

$$Re_D \cong 2000 \qquad (6.59)$$

The turbulent regime takes over when Re_D exceeds approximately 2300. It must be noted, however, that the laminar regime can survive at considerably higher Reynolds numbers (for example, in the range $Re_D \sim 10^4 - 10^5$) if the pipe wall is exceptionally smooth and the supply of fluid to the pipe is steady and free of disturbances. As soon as a disturbance occurs, however, the laminar flow disappears and is replaced by the meandering procession of eddies of several sizes known as turbulent duct flow. The usual condition of the internal surface of ducts is sufficiently rough and their operation is sufficiently "noisy" so that the transition is summarized adequately by the criterion (6.59). The same criterion applies to ducts with other cross-sectional shapes, provided Re_D is replaced by the Reynolds number based on hydraulic diameter, Re_{D_h}.

The onset of turbulence in a pipe has been visualized by several methods, such as the dye injection method in ducts and in open channels. Most of the fluid meanders at high speed through the core of the channel while bumping into and rebounding from the walls. These collisions lead to the formation of slower eddy-filled layers of fluid along both walls, which serve as lubricant for the relative motion between the fast core fluid and the stationary walls.

In terms of the time-averaged flow variables discussed in Section 5.4.2, the turbulent pipe flow is described by the mean longitudinal velocity component $\bar{u}$, the radial component $\bar{v}$, the pressure $\bar{P}$, and the temperature $\bar{T}$ (Figure 6.9). The time averaging of the governing equations for a flow in cylindrical coordinates with θ symmetry yields the following system [3]:

$$\frac{\partial \bar{u}}{\partial x} + \frac{1}{r}\frac{\partial}{\partial r}(r\bar{v}) = 0 \tag{6.60}$$

$$\bar{u}\frac{\partial \bar{u}}{\partial x} + \bar{v}\frac{\partial \bar{u}}{\partial r} = -\frac{1}{\rho}\frac{d\bar{P}}{dx} + \frac{1}{r}\frac{\partial}{\partial r}\left[(\nu + \varepsilon_M)r\frac{\partial \bar{u}}{\partial r}\right] \tag{6.61}$$

$$\bar{u}\frac{\partial \bar{T}}{\partial x} + \bar{v}\frac{\partial \bar{T}}{\partial r} = +\frac{1}{r}\frac{\partial}{\partial r}\left[(\alpha + \varepsilon_H)r\frac{\partial \bar{T}}{\partial r}\right] \tag{6.62}$$

The eddy diffusivities ε_M and ε_H emerge after the same argument as in Section 5.4.3. They contribute to model the apparent shear stress and the apparent radial heat flux, as sums of molecular and eddy contributions:

$$\tau_{\text{app}} = -\mu\frac{\partial \bar{u}}{\partial r} - \rho\varepsilon_M\frac{\partial \bar{u}}{\partial r} \tag{6.63}$$

$$q''_{\text{app}} = \underbrace{k\frac{\partial \bar{T}}{\partial r}}_{\text{Molecular}} + \underbrace{\rho c_P \varepsilon_H \frac{\partial \bar{T}}{\partial r}}_{\text{Eddy}} \tag{6.64}$$

The radial heat flux is positive when oriented toward the centerline.

It is observed that the turbulent flow becomes hydrodynamically and thermally fully developed after a relatively short distance from the entrance to the tube:

$$\frac{X}{D} \cong 10 \cong \frac{X_T}{D} \tag{6.65}$$

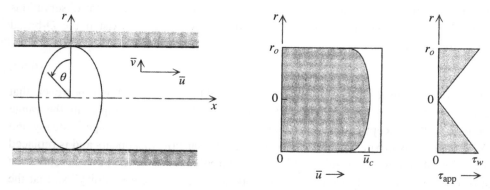

Figure 6.9 Longitudinal velocity profile and apparent shear stress distribution in fully developed turbulent flow through a pipe.

This full-development criterion is particularly applicable to fluids with Prandtl numbers of order 1. It is easy to verify that the turbulent entrance length (6.65) is much shorter than the laminar estimate (6.4′) when $Re_D > 2000$. According to Section 6.1.2, in fully developed flow, $\bar{v} = 0$. Therefore, $\bar{u} = \bar{u}(r)$, $\overline{P} = \overline{P}(r)$, and the momentum and energy Eqs. (6.61) and (6.62) reduce to

$$0 = -\frac{1}{\rho}\frac{d\overline{P}}{dx} - \frac{1}{\rho r}\frac{d}{dr}(r\tau_{\mathrm{app}}) \tag{6.66}$$

$$\bar{u}\frac{\partial \overline{T}}{\partial x} = \frac{1}{\rho c_P r}\frac{\partial}{\partial r}(rq''_{\mathrm{app}}) \tag{6.67}$$

In the remainder of this subsection, we study the implications of the momentum Eq. (6.66). The longitudinal pressure gradient is a constant proportional to the wall shear stress τ_w:

$$-\frac{d\overline{P}}{dx} = 2\frac{\tau_w}{r_o} \tag{6.68}$$

This is a rewriting of the force balance (6.26), in which $\Delta P/L = -d\overline{P}/dx$, $A = \pi r_o^2$, and $p = 2\pi r_o$. Eliminating the pressure gradient between Eqs. (6.66) and (6.68) and integrating the resulting equation away from the centerline (where symmetry requires $\tau_{\mathrm{app}} = 0$), we obtain

$$\frac{\tau_{\mathrm{app}}}{\tau_w} = \frac{r}{r_o} \tag{6.69}$$

In conclusion, the apparent shear stress increases linearly toward the wall value τ_w, and, in a sufficiently thin flow region close to the wall, we may write

$$\tau_{\mathrm{app}} \cong \tau_w \tag{6.70}$$

In this way we rediscover the constant-τ_{app} wall region that was discussed in some detail in Section 5.4.4. All the features of the constant-τ_{app} region (Figure 5.11) apply here as well. The velocity distribution obeys the universal law $u^+ = u^+(y^+)$ presented in Eq. (5.116), where

$$u^+ = \frac{\bar{u}}{(\tau_w/\rho)^{1/2}} \qquad y^+ = \frac{y}{v}\left(\frac{\tau_w}{\rho}\right)^{1/2} \tag{6.71}$$

and $y = r_o - r$ is the short distance measured away from the wall. The velocity profile is remarkably flat over most of the central portion of the cross section: this feature was intentionally exaggerated in the $\bar{u}(r)$ diagram of Figure 6.9. Of use in the friction factor derivation presented as follows is the 1/7th power law (5.118), which approximates well the $u^+(y^+)$ function:

$$u^+ \cong 8.7(y^+)^{1/7} \tag{6.72}$$

6.3.2 Friction Factor and Pressure Drop

In problems of internal convection, we use the friction factor f defined by Eq. (6.24). The cross section-averaged velocity U is (cf. Eq. (6.1))

$$U = \frac{1}{\pi r_o^2}\int_0^{2\pi} d\theta \int_0^{r_o} \bar{u}r\, dr \tag{6.73}$$

A compact friction factor formula can be derived by applying Eq. (6.72) at the pipe centerline ($r = 0$), where $\bar{u}$ reaches its centerline value $\bar{u}_c$:

$$\frac{\bar{u}_c}{(\tau_w/\rho)^{1/2}} \cong 8.7 \left[\frac{r_o}{\nu} \left(\frac{\tau_w}{\rho} \right)^{1/2} \right]^{1/7} \tag{6.74}$$

The f definition (6.24) provides an additional relationship between f and τ_w/ρ:

$$\left(\frac{\tau_w}{\rho} \right)^{1/2} = U \left(\frac{f}{2} \right)^{1/2} \tag{6.75}$$

The relation between the cross section-averaged velocity U and the centerline velocity $\bar{u}_c$ follows from the definition (6.73), for which $\bar{u}$ is provided by Eq. (6.72). The result of combining Eqs. (6.72) through (6.75) is the friction factor formula

$$f \cong 0.079 Re_D^{-1/4} \qquad 2 \times 10^3 < Re_D < 2 \times 10^4 \tag{6.76}$$

Compared with the curve drawn for smooth pipes in Figure 6.10, the friction factor (6.76) is fairly accurate in the specified range, $2 \times 10^3 < Re_D < 2 \times 10^4$, where $Re_D = UD/\nu$. An empirical relation that holds at higher Reynolds numbers is

$$f \cong 0.046 Re_D^{-1/5} \qquad 2 \times 10^4 < Re_D < 10^6 \tag{6.77}$$

Figure 6.10 also shows the laminar flow friction factor formula $f = 16/Re_D$, which was derived in Eq. (6.25). The use of $4f$ is worth noting on the ordinate of this figure, which is known as the Moody chart [9] even though it was originally reported in 1933 by Nikuradse [10, 11] based on his own measurements of f_D (or $4f$) in rough pipes. See Figure 3.46 in Prandtl's textbook [11].

Across the $Re_D \cong 2000$ transition, the friction factor executes a jump of a factor of order 10 from the laminar level to the corresponding turbulent level. A similar observation can be made regarding the behavior of the local skin friction coefficient $C_{f,x}$ along a plane wall with boundary layer flow, Figure 5.12. For fully developed turbulent flow through ducts with cross sections other than round, Eqs. (6.76) and (6.77) and Figure 6.10 continue to apply, provided Re_D is replaced by the Reynolds number based on hydraulic diameter, Re_{D_h}. In a duct with noncircular cross section, the time-averaged wall shear stress τ_w does not have a constant value around the periphery of the cross section. Consequently, in the friction factor definition (6.24), τ_w represents the wall shear stress averaged over the perimeter of the duct cross section.

Ducts with rough internal surfaces have higher friction factors and pressure drops than their counterparts with perfectly polished walls. The effect of wall roughness is presented in Figure 6.10 by means of the dimensionless ratio k_s/D, where the length scale k_s (mm) is Nikuradse's "sand roughness" size [10, 11]. Experimentally, a k_s value can be attached to a given type of surface, as illustrated in the table in Figure 6.10. In general, the friction factor for fully developed turbulent flow through a duct with rough internal surface approaches a constant in the "fully rough" limit; that is, as Re_D becomes sufficiently large,

$$f = \left[1.74 \ln \left(\frac{D}{k_s} \right) + 2.28 \right]^{-2} \tag{6.78}$$

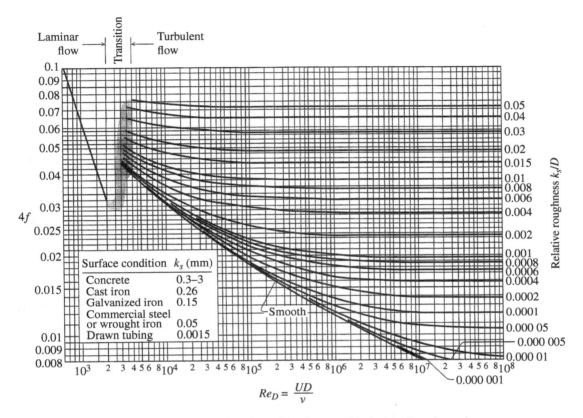

Figure 6.10 Friction factor for fully developed laminar and turbulent flow in a pipe.

In this subsection we focused on the calculation of the friction factor f in the case of fully developed flow in a straight pipe, or a duct with different cross-sectional shape. The overall pressure drop registered between the ends of a duct of length L can then be calculated by applying Eqs. (6.27) and (6.27′).

6.3.3 Heat Transfer Coefficient

In fully developed turbulent flow through a straight duct, the heat transfer coefficient can be calculated using a Stanton number formula analogous to the Colburn relationship (5.131). This analogy exists because sufficiently close to the duct wall the apparent heat flux is nearly constant, that is, similar to the distribution plotted in Figure 5.11 for a turbulent boundary layer.

The existence of the constant-q''_{app} layer near the wall of the duct can be demonstrated by using the pipe flow energy Eq. (6.62). Noting that in fully developed flow $\bar{v} = 0$, and using Eq. (6.64), the energy equation becomes

$$\rho c_p \bar{u} \frac{\partial \bar{T}}{\partial x} = \frac{1}{r}\frac{\partial}{\partial r}(r q''_{\text{app}}) \tag{6.79}$$

This can be integrated over a central cross-sectional disc of radius r:

$$2\pi \int_0^r \rho c_p \bar{u}\frac{-\partial \bar{T}}{\partial x} r\, dr = 2\pi r q''_{app} \tag{6.80}$$

where q''_{app} is the apparent heat flux at the radial distance r. Note that in accordance with Eq. (6.64), the heat flux is positive when pointing toward the centerline (i.e. away from the pipe wall). The special form of Eq. (6.81) at the wall ($r = r_o$) is

$$\int_0^{r_o} \rho c_p \bar{u}\frac{-\partial \bar{T}}{\partial x}\, r\, dr = r_o q''_w \tag{6.81}$$

Dividing Eqs. (6.80) and (6.81) side by side, we obtain a relation that resembles Eq. (6.69):

$$\frac{q''_{app}}{q''_w} = M\frac{r}{r_o} \tag{6.82}$$

except for the factor M, which is shorthand for

$$M = \frac{\dfrac{1}{r^2}\displaystyle\int_0^r \bar{u}\frac{\partial \bar{T}}{\partial x} r\, dr}{\dfrac{1}{r_o^2}\displaystyle\int_0^{r_o} \bar{u}\frac{-\partial \bar{T}}{\partial x} r\, dr} \tag{6.83}$$

In Section 6.2.3 we learned that when the wall heat flux q''_w is independent of x, the longitudinal temperature gradient ($\partial \bar{T}/\partial x$ in this case) is independent of r; therefore, M assumes the simpler form

$$M = \frac{\dfrac{1}{r^2}\displaystyle\int_0^r \bar{u}\, r\, dr}{\dfrac{1}{r_o^2}\displaystyle\int_0^{r_o} \bar{u}\, r\, dr} \tag{6.84}$$

Furthermore, since the turbulent flow $\bar{u}$ profile is relatively flat (Figure 6.13), the effect of $\bar{u}(r)$ is weak on the right side of Eq. (6.84). In conclusion, the M factor is a weak function of radius and does not deviate much from the value 1. This means that the apparent heat flux increases almost linearly in the radial direction (cf. Eq. (6.82)):

$$\frac{q''_{app}}{q''_w} \cong \frac{r}{r_o} \tag{6.85}$$

This distribution can be visualized by replacing τ_{app} with q''_{app} and τ_w with q''_w on the extreme-right drawing shown in Figure 6.9. Equation (6.85) shows that sufficiently close to the wall ($r \lesssim r_o$), the apparent heat flux is nearly constant:

$$q''_{app} \cong q''_w \tag{6.86}$$

This conclusion is the starting point – the basis – of analyses [3] that lead to Stanton number formulas similar to Eq. (5.131), except that in the case of duct flow, the longitudinal velocity scale is the cross section-averaged velocity U, not U_∞:

$$St = \frac{h}{\rho c_p U} \tag{6.87}$$

Another distinction to be made between Eqs. (6.87) and (5.129) is that in fully developed flow both h and St are independent of x. Furthermore, in duct flow we speak of friction factors f, Eq. (6.24), instead of local skin friction coefficients; therefore, the equivalent Colburn formula for the Stanton number is [12]

$$St = \frac{\frac{1}{2}f}{Pr^{2/3}} \tag{6.88}$$

This formula holds for $Pr \gtrsim 0.5$ and is to be used in conjunction with the Moody chart (Figure 6.10), which supplies the value of the friction factor. It applies to ducts of various cross-sectional shapes, with wall surfaces having various degrees of roughness. For example, in the special case of a pipe with smooth internal surface, we can combine Eq. (6.88) with Eq. (6.77) to derive the Nusselt number formula

$$Nu_D = \frac{hD}{k} = 0.023 Re_D^{4/5} Pr^{1/3} \tag{6.89}$$

which, in accordance with Eq. (6.77), holds in the range $2 \times 10^4 < Re_D < 10^6$.

There are many formulas that, in one way or another, improve on the accuracy with which Colburn's Eq. (6.88) predicts actual measurements [13]. Popular is the formula correlation due to Dittus and Boelter [14]:

$$Nu_D = 0.023 Re_D^{4/5} Pr^n \tag{6.90}$$

which was developed for $0.7 \leq Pr \leq 120$, $2500 \leq Re_D \leq 1.24 \times 10^5$, and $L/D > 60$. The Prandtl number exponent is $n = 0.4$ when the fluid is being heated ($T_w > T_m$), and $n = 0.3$ when the fluid is being cooled ($T_w < T_m$). All the physical properties needed for the calculation of Nu_D, Re_D, and Pr are to be evaluated at the bulk temperature T_m. The maximum deviation between experimental data and values predicted using Eq. (6.90) is of the order of 40%.

For applications in which the temperature influence on properties is significant, Sieder and Tate's [15] modification of Eq. (6.89) is recommended:

$$Nu_D = 0.027 Re_D^{4/5} Pr^{1/3} \left(\frac{\mu}{\mu_w} \right)^{0.14} \tag{6.91}$$

The correlation is valid for $0.7 < Pr < 16\,700$ and $Re_D > 10^4$. The effect of temperature-dependent properties is taken into account by evaluating all the properties (except μ_w) at the mean temperature of the stream, T_m. The viscosity μ_w is evaluated at the wall temperature, $\mu_w = \mu(T_w)$.

The most accurate of the correlations of type (6.89) through (6.91) is [16]

$$Nu_D = \frac{(f/2)(Re_D - 10^3)Pr}{1 + 12.7(f/2)^{1/2}(Pr^{2/3} - 1)} \tag{6.92}$$

in which the friction factor is supplied by Figure 6.10. It is accurate within $\pm 10\%$ in the range $0.5 < Pr < 10^6$ and $2300 < Re_D < 5 \times 10^6$. Like Eqs. (6.90)–(6.92), the correlation (6.92) can be used both in constant-q_w'' and constant-T_w applications. Two simpler alternatives to Eq. (6.92) are [16]

$$Nu_D = 0.0214 \left(Re_D^{0.8} - 100 \right) Pr^{0.4} \quad 0.5 \leq Pr \leq 1.5, \ 10^4 \leq Re_D \leq 5 \times 10^6 \tag{6.93}$$

$$Nu_D = 0.012 \left(Re_D^{0.87} - 280 \right) Pr^{0.4} \quad 1.5 \leq Pr \leq 500, \ 3 \times 10^3 \leq Re_D \leq 10^6 \tag{6.93'}$$

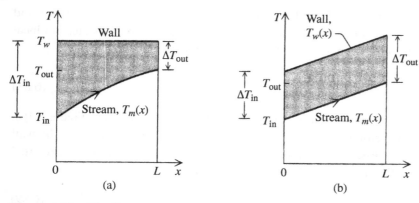

Figure 6.11 Distribution of temperature along a duct: (a) isothermal wall and (b) wall with uniform heat flux.

The preceding results refer to gases and liquids, that is, to the range $Pr \gtrsim 0.5$. For liquid metals, the most accurate formulas are those of Notter and Sleicher [17]:

$$Nu_D = 6.3 + 0.0167 Re_D^{0.85} Pr^{0.93} \quad (q_w'' = \text{constant}) \tag{6.94}$$

$$Nu_D = 4.8 + 0.0156\, Re_D^{0.85} Pr^{0.93} \quad (T_w = \text{constant}) \tag{6.95}$$

These are valid for $0.004 < Pr < 0.1$ and $10^4 < Re_D < 10^6$, and are based on both computational and experimental data. All the properties used in Eqs. (6.94) and (6.95) are evaluated at the mean temperature T_m.

One peculiarity of the mean temperature of the stream is that it varies with the position along the duct, $T_m(x)$. This variation is linear in the case of constant-q_w'' and exponential when the duct wall is isothermal (review Figures 6.6 and 6.8). To simplify the recommended evaluation of the physical properties at the T_m temperature, it is convenient to choose as representative mean temperature the average value:

$$T_m = \frac{1}{2}(T_{in} + T_{out}) \tag{6.96}$$

In this definition, T_{in} and T_{out} are the bulk temperatures of the stream at the duct inlet and outlet, respectively (Figure 6.11).

Example 6.3 *Pressure Drop, Turbulent Flow, and Annular Space*

The inner pipe of a downhole coaxial heat exchanger used for geothermal energy extraction has a diameter of 16 cm. The pipe material is commercial steel. At a certain location along this pipe, the mean temperature of the water stream that flows through it is 80 °C. The water flowrate is 100 tons/h. Calculate the frictional pressure drop per unit pipe length ($\Delta P/L$) experienced by the water stream at that location.

Solution

The frictional pressure drop per unit length can be calculated with Eq. (6.27), which can be rewritten as

$$\frac{\Delta P}{L} = \frac{f}{D_h} 2\rho U^2 \tag{6.27'}$$

The right side of this expression asks us to evaluate, in order,

$$U = \frac{\dot{m}}{\rho A} = 100 \frac{10^3\,\text{kg}}{3600\,\text{s}} \frac{\text{cm}^3}{0.9718\,\text{g}} \frac{1}{(\pi/4)(16)^2\,\text{cm}^2}$$

$$= 1.42\,\text{m/s}$$

$$Re = \frac{UD}{\nu} = 142\frac{\text{cm}}{\text{s}}16\text{ cm}\frac{\text{s}}{0.00\,366\text{ cm}^2}$$
$$= 6.2 \times 10^5 \quad \text{(turbulent flow)}$$

The properties ρ and ν have been evaluated at the mean temperature of $80\,°C$.

Next, Figure 6.10 delivers the value of the friction factor, provided we also know the roughness parameter k_s/D. In the case of a commercial steel surface, this is

$$\frac{k_s}{D} = \frac{0.05\text{ mm}}{160\text{ mm}} = 3.1 \times 10^{-4}$$

Therefore, Figure 6.10 recommends $4f \cong 0.016$, or $f \cong 0.004$. The right side of Eq. (6.27′) can now be evaluated numerically:

$$\frac{\Delta P}{L} = \frac{0.004}{16\text{ cm}}2 \times 0.9718\frac{\text{g}}{\text{cm}^3}(142)^2\,\frac{\text{cm}^2}{\text{s}^2}$$
$$= 9.8\frac{\text{g}}{\text{cm}^2 \cdot \text{s}^2} = 98\frac{\text{N/m}^2}{\text{m}} = 9.7 \times 10^{-4}\text{atm/m}$$

This $\Delta P/L$ value may not look impressive; however, for a pipe that reaches to a depth of 1 km into the Earth's crust, it indicates a frictional pressure drop of approximately 1 atm.

6.4 Total Heat Transfer Rate

The primary objective of Sections 6.2 and 6.3 was the evaluation of the heat transfer coefficient between the duct wall and the stream. We learned that in fully developed laminar and turbulent duct flows, h is independent of longitudinal position. The heat transfer coefficient is essential in the calculation of the *total* heat transfer rate $q(W)$ that is received by the stream as it travels the entire length L of the duct. The heat transfer rate q can be expected to be proportional to the h constant, to the total surface swept by the stream ($A_w = pL$), and to an "effective" temperature difference labeled for the time being ΔT_{lm}:

$$q = hA_w\Delta T_{lm} \tag{6.97}$$

6.4.1 Isothermal Wall

The magnitude of ΔT_{lm} depends on how the wall-stream temperature difference varies along the duct, $T_w(x) - T_m(x)$. Consider first the case where the wall temperature T_w is constant, as shown earlier in Figure 6.6 and now in Figure 6.11a. In fully developed laminar or turbulent flow, the temperature difference

$$\Delta T = T_w - T_m \tag{6.98}$$

decreases exponentially in the downstream direction, between a value at the tube inlet and a smaller value at the tube outlet:

$$\Delta T_{in} = T_w - T_{in}, \qquad \Delta T_{out} = T_w - T_{out} \tag{6.99}$$

The effective temperature difference ΔT_{lm} falls somewhere between the extremes ΔT_{in} and ΔT_{out}. Its precise value can be determined by deriving the formula (6.97) in a rigorous manner, based on thermodynamic analysis. From the point of view of the stream as an elongated control volume, note that the total heat transfer rate through the duct wall is

$$q = \dot{m}c_P(T_{out} - T_{in}) \tag{6.100}$$

Figure 6.11a shows that the bulk temperature excursion $T_{out} - T_{in}$ is the same as the difference $\Delta T_{in} - \Delta T_{out}$; therefore, an alternative to Eq. (6.100) is

$$q = \dot{m}c_P(\Delta T_{in} - \Delta T_{out}) \tag{6.101}$$

It remains to determine the relationship between the heat capacity flowrate $\dot{m}c_P$ and the group hA_w that appears on the right side of Eq. (6.97). For this we use Eq. (6.39), in which $q''_w = h(T_w - T_m)$; therefore,

$$\frac{dT_m}{T_w - T_m} = \frac{hp}{A\rho c_P U}dx \tag{6.102}$$

The special form of this equation for a duct of circular cross section is Eq. (6.54). Assuming constant A, p, and c_P, we integrate Eq. (6.102) from the inlet ($T_m = T_{in}$ at $x = 0$) to the outlet ($T_m = T_{out}$ at $x = L$) and obtain

$$\ln \frac{T_m - T_{in}}{T_w - T_{out}} = \frac{hpL}{\rho AU c_P} \tag{6.103}$$

In this equation we recognize the inlet and outlet temperature differences (6.99), the mass flowrate $\rho AU = \dot{m}$, and the total duct area $pL = A_w$. Therefore, an alternative form of Eq. (6.103) is

$$\ln \frac{\Delta T_{in}}{\Delta T_{out}} = \frac{hA_w}{\dot{m}c_P} \tag{6.103'}$$

The proper definition of the ΔT_{lm} factor adopted in Eq. (6.97) becomes clear as we eliminate $\dot{m}c_P$ between Eqs. (6.103') and (6.101):

$$q = hA_w \frac{\Delta T_{in} - \Delta T_{out}}{\ln\left(\dfrac{\Delta T_{in}}{\Delta T_{out}}\right)} \tag{6.104}$$

in other words,

$$\Delta T_{lm} = \frac{\Delta T_{in} - \Delta T_{out}}{\ln\left(\dfrac{\Delta T_{in}}{\Delta T_{out}}\right)} \tag{6.105}$$

Because of the ΔT-averaging operation prescribed by Eq. (6.105), the ΔT_{lm} factor is recognized as the *log-mean temperature difference* between the wall and the stream. When the wall and inlet temperatures are specified, Eq. (6.104) expresses the relationship between the total heat transfer rate q, the total duct surface conductance hA_w, and the outlet temperature of the stream. Alternatively, Eqs. (6.103')–(6.105) can be combined to express the total heat transfer rate in terms of the inlet temperatures, mass flowrate, and duct surface thermal conductance hA_w:

$$q = \dot{m}c_P\Delta T_{in}\left[1 - \exp\left(-\frac{hA_w}{\dot{m}c_P}\right)\right] \tag{6.105'}$$

Note that it was not necessary to assume that h is independent of x while integrating Eq. (6.102). In cases where the heat transfer coefficient varies longitudinally, $h(x)$, the h factor on the right side of Eqs. (6.103) and (6.103′) represents the L-averaged heat transfer coefficient, namely, $\int_L h(x)dx/L$.

6.4.2 Uniform Wall Heating

It will be demonstrated in Section 9.3 that the applicability of Eq. (6.104) is considerably more general than what is suggested by Figure 6.11a. For example, when the heat transfer rate q is distributed uniformly along the duct, the temperature difference ΔT does not vary with the longitudinal position (review Section 6.2.3). This case is illustrated in Figure 6.11b, where it was again assumed that A, p, h, and c_P are independent of x. It is evident that the value labeled ΔT_{lm} is the same as the constant ΔT along the duct:

$$\Delta T_{lm} = \Delta T_{in} = \Delta T_{out} \tag{6.106}$$

Equation (6.106) is a special case of Eq. (6.105), namely, the limit $\Delta T_{in}/\Delta T_{out} \rightarrow 1$.

6.5 Evolutionary Design

6.5.1 Size of Duct with Fluid Flow

The size of a system subjected to observation and analysis (and even to design) is usually not questioned. It is assumed, not discovered, predicted, or recommended. This is a failing of scientific method, particularly in view of the obvious fact that in nature every object has size. The same holds for every object drawn in art, design, and scientific publications. In Section 5.6.1 we made a first corrective step by predicting the size of an object exposed to external heat transfer. In the present section, we continue this line by questioning the size of a duct with internal fluid flow. The presentation follows the analysis put forth in Ref. [5].

Consider a round tube of diameter D and mass flowrate $\dot{m}$, which serves as flow component on a larger system (heat exchanger, or lung) that is itself an organ of a moving body (vehicle, or animal). What should be the size D?

Figure 5.20 showed that, in general, a flow component destroys useful power in two ways. Internally, it destroys power (w_1) because of the irreversibility of the flow of fluid (friction, heat transfer). Externally, it destroys power (w_2) as it rides on the vehicle to which it belongs. In the present analysis, the internal irreversibility is due entirely to fluid friction. If the flow regime is laminar and fully developed, the pressure loss per unit of duct length is

$$\frac{\Delta P}{L} = \frac{32\mu}{D^2}U \tag{6.107}$$

where U is the mean fluid velocity, $\dot{m}/(\rho\pi D^2/4)$. The pumping power required to force $\dot{m}$ to flow through the tube is (Figure 6.12)

$$\frac{w_1}{L} = \frac{\dot{m}\Delta P}{\eta_p \rho L} \tag{6.108}$$

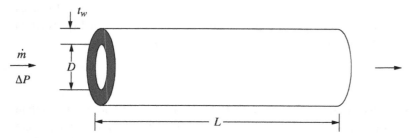

Figure 6.12 Round tube with specified mass flowrate.

where η_p is the pump isentropic efficiency and ρ is the density of the single-phase fluid. The power used to transport the duct is expressed per unit of duct length:

$$\frac{w_2}{L} \cong 2\frac{m}{L}gV \tag{6.109}$$

where m/L is the duct mass per unit length. Assume that the competing designs have various D sizes and cross sections that are geometrically similar; that is, the thickness of the duct wall (t_w) is a fixed fraction of the duct diameter, $t_w = \varepsilon D$ where $\varepsilon \ll 1$, constant. In this case the duct mass per unit length is $m/L = \rho_w\pi\varepsilon D^2$. The total destruction of power is

$$\frac{w_1 + w_2}{L} \cong \frac{c_1}{D^4} + c_2D^2 \tag{6.110}$$

where $c_1 = 128\ \mu\dot{m}^2/\pi\eta_p\rho^2$ and $c_2 = 2\pi\varepsilon\rho_wgV$. We see the competition between the two destruction mechanisms: A large D is attractive from the point of view of avoiding large flow friction irreversibility, but it is detrimental because of the large duct mass. The tube size for which the total power loss is minimal is

$$D = \left(2\frac{c_1}{c_2}\right)^{1/6} = \left(\frac{128\ \mu\dot{m}^2}{\pi^2\eta_p\varepsilon\rho_w^2gV}\right)^{1/6} \tag{6.111}$$

This diameter increases with the flowrate as $\dot{m}^{1/3}$ and decreases weakly as the vehicle or animal speed increases.

This analysis is just the beginning of what can be done in the direction of discovering the size of every flow component in a larger installation. For example, the size of the round tube can be determined similarly for duct flow in the turbulent regime and for geometries where the tube wall thickness is not necessarily a fraction of the tube diameter. Fittings, junctions, bends, and all the other geometric features that impede flow and add mass to vehicles can have their sizes discovered in the same manner. The flow passages of heat exchangers are next in line, although in their case the destruction of power is due to three mechanisms: heat transfer, flow friction, and the transportation of the heat exchanger mass (cf. Section 9.6.3).

6.5.2 Tree-Shaped Ducts

In Sections 3.4.2 and 4.8.2, we learned that when heat flows between one point and an infinity of points (area or volume), the flow architecture that offers greater access is shaped as a tree. The shape is also known as dendritic or arborescent. In the present section we consider the fluid flow counterpart of the conduction tree discovery. When the fluid flows from the point (the base of the trunk) to the volume (the tree canopy), the tree architecture "distributes" the flow. When the flow is oriented the other way, from canopy to trunk,

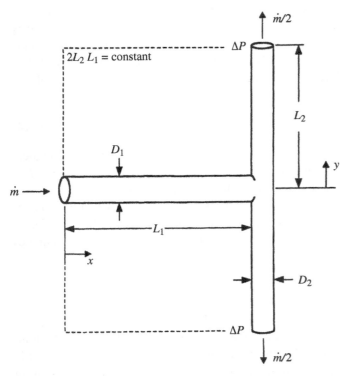

Figure 6.13 T-shaped construct of round tubes.

the tree "collects" the flow. As we saw earlier, the direction of the actual flow is not the issue. The physics phenomenon is the evolution of the design – the emergence of the tree flow architecture.

Consider the simplest tree, which is the T-shaped construct of round tubes shown in Figure 6.13. The flow connects one point (source or sink) with two points. There are two finite quantities:

1. Total duct volume ($D_1^2 L_1 + 2D_2^2 L_2 = $ constant)
2. Total space allocated to the construct ($2L_2 L_1 = $ constant)

If the flow is laminar and fully developed, the minimization of the flow resistance subject to constraint 1 yields the ratio of tube diameters [3, 5]

$$\frac{D_1}{D_2} = 2^{1/3} \tag{6.112}$$

This ratio is a result known as the Hess–Murray law [18, 19]. This result is remarkable for its robustness: The ratio D_1/D_2 is independent of the assumed tube lengths and the angles between the three tubes. It is independent of layout. The result that corresponds to Eq. (6.112) for channels that are formed between parallel plates of spacings D_1 and D_2 is $D_1/D_2 = 2^{1/2}$.

New is the second degree of freedom, which consists of selecting the ratio L_1/L_2 subject to the space constraint 2. The result

$$\frac{L_1}{L_2} = 2^{1/3} \tag{6.113}$$

shows that at the junction the tube lengths must change in the same proportion as the tube diameters. Equations (6.112) and (6.113) are a condensed summary of the geometric proportions found more laboriously in the evolution of three-dimensional flow constructs [20], where the tube lengths increase by factors in the cyclical sequence 2, 1, 1, 2, 1, 1, and so on. The average of this factor for one step of construction is $2^{1/3}$.

If the flow in the T-shaped construct is fully developed and turbulent, the preceding results are replaced by [21]

$$\frac{D_1}{D_2} = 2^{3/7}, \qquad \frac{L_1}{L_2} = 2^{1/7} \tag{6.114}$$

Unlike in the laminar case, where the ratio D/L was preserved in going from each tube to its stem or branch, in turbulent flow the geometric ratio that is preserved is D/L^3. Note that Eq. (6.114) yield $D_1/L_1^3 = D_2/L_2^3$.

The assembly of three tubes can be improved further by giving the morphing drawing more degrees of freedom. One way is to allow the angle of confluence to vary so that the T-shaped construct acquires a better Y shape. Even better can be the Y construct in which the two branches are not identical (Problem 6.34).

More rewards come from more freedom. The research literature on tree-shaped flow architectures continues to flourish [22–28].

6.5.3 Spacings

The size of a duct (diameter, or plate-to-plate spacing) also depends on the volume in which many ducts of the same size are packed [29]. In forced convection, the whole architecture also depends on how the fluid is forced to flow through the coarse porous structure. How to determine the coarseness (the duct size) is challenging, which is why the solution was not available until recently [30]. Here we illustrate the solution with reference to Figure 6.14 and Ref. [3].

Of principal interest is the maximum heat transfer rate, that is, the maximum density of heat-generating electronics that can be fitted in a package of specified volume. In Figure 6.14 the coolant inlet temperature T_∞ and the pressure difference established by the compressor (fan or pump) ΔP are fixed. The solution method was first proposed for natural convection spacings in the 1984 edition of my convection book [3], as shown here in Section 7.5.1. The analysis was based on the *intersection-of-asymptotes method*. The flow is assumed laminar, and the board temperature is assumed uniform at the safe level T_w. Each board has a thickness t that is sufficiently smaller than D. To determine the board-to-board spacing, D is the same as determining the number of boards ($n \gg 1$) that fill a volume of thickness H, namely, $n \simeq H/D$:

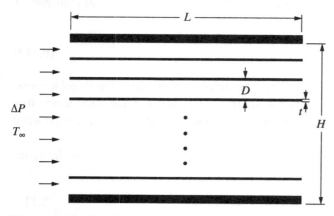

Figure 6.14 Stack of parallel heat-generating plates cooled by forced convection.

(a) Consider the limit $D \to 0$ when each channel is slender enough so that the flow is fully developed all along L, and the fluid outlet temperature approaches the board temperature T_w. The average fluid velocity in each channel is

$$U = \frac{D^2}{12\mu}\frac{\Delta P}{L} \tag{6.115}$$

and the total mass flowrate $\dot{m}'$ through the stack of thickness H is

$$\dot{m}' = \rho U H = \rho H \frac{D^2}{12\mu}\frac{\Delta P}{L} \tag{6.116}$$

The mass flowrate $\dot{m}'$ is expressed per unit length in the direction perpendicular to Figure 6.14. The total heat transfer rate removed from the stack by the $\dot{m}'$ stream is

$$q'_a \simeq \dot{m}' c_P (T_w - T_\infty) = \rho H \frac{D^2}{12\mu}\frac{\Delta P}{L} c_P (T_w - T_\infty) \tag{6.117}$$

In conclusion, in the limit $D \to 0$, the total cooling rate (or total rate of allowable Joule heating in the package) decreases as D^2. This trend is illustrated qualitatively as curve (a) in Figure 6.15. The direction toward small D values is not good for design performance.

(b) In the limit $D \to \infty$, the boundary layer that lines one surface is distinct, thin. The overall pressure drop is fixed at ΔP; therefore, what is the free-stream velocity U_∞ that sweeps these boundary layers? The overall force balance on the control volume $H \times L$ requires that

$$\Delta P H = n \cdot 2\bar{\tau}_w L \tag{6.118}$$

in which $\bar{\tau}_w$ is the L-averaged wall shear stress

$$\bar{\tau}_w = 1.328 Re_L^{-1/2}\frac{1}{2}\rho U_\infty^2 \tag{6.119}$$

Combined, Eqs. (6.118) and (6.119) yield

$$U_\infty = \left(\frac{1}{1.328}\frac{\Delta P}{nL^{1/2}}\frac{H}{\rho\nu^{1/2}} \right)^{2/3} \tag{6.120}$$

The total heat transfer rate from one of the L-long surfaces (q'_1) is calculated from the overall Nusselt number for $Pr > 0.5$:

$$\frac{\bar{h}L}{k} = \frac{\bar{q}''}{T_w - T_\infty}\frac{L}{k} = 0.664 Pr^{1/3}\left(\frac{U_\infty L}{\nu} \right)^{1/2} \tag{6.121}$$

Figure 6.15 Intersection-of-asymptotes method: the recommended spacing as the intersection of the small-D asymptote with the large-D asymptote.

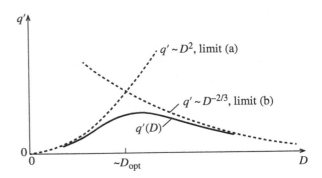

which leads to

$$q_1' = \overline{q}''L = k(T_w - T_\infty)0.664Pr^{1/3}\left(\frac{U_\infty L}{\nu}\right)^{1/2} \tag{6.122}$$

The total heat transfer rate released by the stack is $2n$ times larger than q_1'. In view of the n and U_∞ expressions presented earlier, the total heat transfer rate becomes

$$q_b' = 1.208k(T_w - T_\infty)H\frac{Pr^{1/3}L^{1/3}\Delta P^{1/3}}{\rho^{1/3}\nu^{2/3}D^{2/3}} \tag{6.123}$$

In the large-D limit, the total heat transfer rate decreases as $D^{-2/3}$. This trend has been added as curve (b) in Figure 6.15, to suggest that the peak of the actual (unknown) curve $q'(D)$ occurs at a spacing D that is of the same order as the D value obtained by intersecting the asymptotes (6.117) and (6.123). The order-of-magnitude statement $q_a' \sim q_b'$ yields the spacing

$$D \simeq 2.73L\,Be_L^{-1/4} \tag{6.124}$$

where

$$Be_L = \frac{\Delta PL^2}{\mu\alpha} \tag{6.125}$$

is the pressure drop number that in 1988 Bhattacharjee and Grosshandler [31] termed the *Bejan number*. The role of this dimensionless group in forced convection is discussed by Petrescu [32] and Zimparov et al. [33]. Equation (6.124) agrees very well with the more exact result obtained by locating the maximum of the actual $q'(D)$ curve.

The order of magnitude of the maximum package heat transfer rate that corresponds to the spacing discovered in Eq. (6.124) is

$$q' \lesssim 0.62\left(\frac{\rho\,\Delta P}{Pr}\right)^{1/2}Hc_P(T_w - T_\infty) \tag{6.126}$$

The inequality sign is a reminder that the actual q' maximum is lower than the q' value obtained by intersecting asymptotes (a) and (b) in Figure 6.15.

6.5.4 Packaging for Maximum Heat Transfer Density

The fundamentals identified in this chapter are driving new currents in heat transfer science and technology. One example is the spacing D discovered in Eq. (6.124), which is the same scale as the thickness of the laminar boundary layer at the end of the duct length L. The discovery [34, 35] is that the package of channels that has maximum heat transfer rate per unit volume must be such that each channel must have a flow length that matches the entrance length of its developing flow. The package then is a volume filled by entrance-length channels. This concept is now explored in the evolutionary design of new heat exchanges (Section 9.6.1).

The heat transfer density objective also governs the evolutionary design perpendicular to the duct flow length. How should the ducts be stacked in the bundle that fills the available volume? In other words, what should be the arrangement of the array of channels when seen in cross section? The question is particularly interesting in view of the fact that the cross section is also occupied by the solid walls between the channels. The solid walls are necessary for mechanical strength and heat transfer between channels. This concept is outlined in Refs. [29, 36].

References

1 Hagen, G. (1839). Über die Bewegung des Wassers in engen zylindrischen Röhren. *Pogg. Ann.* 46: 423–442.

2 Poiseuille, J. (1840). Récherches expérimentales sur le mouvement des liquides dans les tubes de très petits diamètres. *C. R. Acad. Sci.* 11: 961–967, 1041–1048; Vol. 12, 1841, 112–115.

3 Bejan, A. (2013). *Convection Heat Transfer*, 4e. Hoboken, NJ: Wiley.

4 Shah, R.K. and London, A.L. (1978). *Laminar Flow Forced Convection in Ducts*, 40. New York: Academic Press.

5 Bejan, A. (2016). *Advanced Engineering Thermodynamics*, 4e. Hoboken, NJ: Wiley.

6 Asako, Y., Nakamura, H., and Faghri, M. (1988). Developing laminar flow and heat transfer in the entrance region of regular polygonal ducts. *Int. J. Heat Mass Transfer* 31: 2590–2593.

7 Hornbeck, R.W. (1965). An all-numerical method for heat transfer in the inlet of a tube. *American Society of Mechanical* Engineers, Paper No. 65-WA/HT-36.

8 Churchill, S.W. and Ozoe, H. (1973). Correlations for forced convection with uniform heating in flow over a plate and in developing and fully developed flow in a tube. *J. Heat Transfer* 95: 78–84.

9 Moody, L.F. (1944). Friction factors for pipe flows. *Trans. ASME* 66: 671–684.

10 Nikuradse, J. (1933). Strömungsgesetze in rauhen Rohren. *VDI-Forschungsheft* 361: 1–22.

11 Prandtl, L. (1952). *Essentials of Fluid Dynamics*, 3e. London: Blackie & Son.

12 Colburn, A.P. (1933). A method for correlating forced convection heat transfer data and a comparison with fluid friction. *Trans. Am. Inst. Chem. Eng.* 29: 174–210; reprinted in (1964). *Int. J. Heat Mass Transfer* 7: 1359–1384.

13 Bejan, A. and Kraus, A.D. (2003). *Heat Transfer Handbook*. Hoboken, NJ: Wiley.

14 Dittus, P.W. and Boelter, L.M.K. (1985). Heat transfer in automobile radiators of the tubular type, *Univ. California Pub. Eng.*, Vol. 2, No. 13, pp. 443–461, Oct. 17, 1930; reprinted in. *Int. Commun. Heat Mass Transfer* 12: 3–22.

15 Sieder, E.N. and Tate, G.E. (1936). Heat transfer and pressure drop of liquids in tubes. *Ind. Eng. Chem.* 28: 1429–1436.

16 Gnielinski, V. (1976). New equations for heat and mass transfer in turbulent pipe and channel flow. *Int. Chem. Eng.* 16: 359–368.

17 Notter, R.H. and Sleicher, C.A. (1972). A solution to the turbulent Graetz problem III. Fully developed and entry region heat transfer rates. *Chem. Eng. Sci.* 27: 2073–2093.

18 Hess, W.R. (1914). Das Prinzip des kleinsten Kraftverbrauches im Dienste hämodynamischer Forschung. *Arch. Anat. Physiol.*: 1–62.

19 Murray, C.D. (1926). The physiological principle of minimal work, in the vascular system, and the cost of blood volume. *Proc. Natl. Acad. Sci. U.S.A.* 12: 207–214.

20 Bejan, A. (1997). Constructal tree network for fluid flow between a finite-size volume and one source or sink. *Rev. Gén. Therm.* 36: 592–604.

21 Bejan, A., Rocha, L.A.O., and Lorente, S. (2000). Thermodynamic optimization of geometry: T- and Y-shaped constructs of fluid streams. *Int. J. Therm. Sci.* 39: 949–960.

22 Miguel, A.F. (2018). Constructal branching design for fluid flow and heat transfer. *Int. J. Heat Mass Transfer* 122: 204–211.

23 Clemente, M.R. and Panão, M.R.O. (2018). Introducing flow architecture in the design and optimization of mold inserts cooling systems. *Int. J. Therm. Sci.* 127: 288–293.

24 Kamps, T., Biedermann, M., Seidel, C., and Reinhart, G. (2018). Design approach for additive manufacturing employing Constructal Theory for point-to-circle flows. *Addit. Manuf.* 20: 111–118.

25 Rubio-Jimenez, C.A., Hernandez-Guerrero, A., Cervantes, J.G. et al. (2016). CFD study of constructal microchannel networks for liquid-cooling of electronic devices. *Appl. Therm. Eng.* 95: 374–381.

26 Silva, C. and Reis, A.H. (2015). Scaling relations of branching pulsatile flows. *Int. J. Therm. Sci.* 88: 77–83.

27 Chen, Y. and Deng, Z. (2015). Gas flow in micro tree-shaped hierarchical network. *Int. J. Heat Mass Transfer* 80: 163–169.

28 Rastogi, P. and Mahulikar, S.P. (2018). Optimization of micro-heat sink based on theory of entropy generation in laminar forced convection. *Int. J. Therm. Sci.* 126: 96–104.

29 Bejan, A., Almerbati, A., Lorente, S. et al. (2016). Arrays of flow channels with heat transfer embedded in conducting walls. *Int. J. Heat Mass Transfer* 99: 504–511.

30 Bejan, A. and Sciubba, E. (1992). The optimal spacing of parallel plates cooled by forced convection. *Int. J. Heat Mass Transfer* 35: 3259–3264.

31 Bhattacharjee, S. and Grosshandler, W.L. (1988). The formation of a wall jet near a high temperature wall under microgravity environment. *ASME HTD* 96: 711–716.

32 Petrescu, S. (1994). Comments on the optimal spacing of parallel plates cooled by forced convection. *Int. J. Heat Mass Transfer* 37: 1283.

33 Zimparov, V.D., Angelov, M.S., and Hristov, J.Y. (2020). New insight into the definitions of the Bejan number. *Int. Commuun. Heat Mass Transfer* 116: 104637.

34 Bejan, A. (2000). *Shape and Structure, from Engineering to Nature*. Cambridge: Cambridge University Press.

35 Bejan, A. and Lorente, S. (2008). *Design with Constructal Theory*. Hoboken, NJ: Wiley.

36 Bejan, A., Alalaimi, M., Lorente, S. et al. (2016). Counterflow heat exchanger with core and plenums at both ends. *Int. J. Heat Mass Transfer* 99: 622–629.

37 Bejan, A. (2020). *Freedom and Evolution: Hierarchy in Nature, Society and Science*. Switzerland AG: Springer Nature.

Problems

Laminar Flow

6.1 Derive the parabolic velocity distribution (6.17)–(6.18) for fully developed laminar flow through a tube of radius r_o. In other words, solve Eq. (6.14) subject to the boundary conditions (6.16a and 6.16b).

6.2 Consider the fully developed laminar flow through the parallel-plate channel shown in Figure 6.2. Determine the velocity profile (6.21)–(6.22) by solving Eq. (6.20) subject to the boundary conditions identified on the figure.

6.3 The friction factor for fully developed laminar flow through a straight duct with regular hexagonal cross section (Table 6.2) is $f = 15.065/Re_{D_h}$. Compare the constant 15.065 with the approximate value that can be anticipated using Eq. (6.31).

6.4 Consider the laminar flow of 5 °C water through a parallel-plate channel. The plate-to-plate spacing is $D = 2$ cm and the mean velocity is $U = 3.2$ cm/s. Calculate the pressure drop per unit length $\Delta P/L$ in the fully developed region. Estimate also the flow entrance-length X.

6.5 Water flows with the mean velocity 6 cm/s through a 2.7-cm-diameter pipe. The pipe wall is isothermal, $T_w = 10\,°C$, and the water inlet temperature is 30 °C. The total length of the pipe is $L = 10$ m. Calculate the outlet mean temperature of the stream, by assuming that the flow is fully developed and that the water properties can be evaluated at 20 °C. Verify that the flow is in the laminar regime.

6.6 Water is heated as it flows through a stack of parallel metallic blades. The blade-to-blade spacing is $D = 1$ cm and the mean velocity through each channel is $U = 3.2$ cm/s. Each blade is heated electrically so that both sides of the blade release together 1600 W/m^2 into the water. Assuming that the water properties can be evaluated at 50 °C and that the flow is thermally fully developed, do the following:
a) Verify that the flow is laminar.
b) Calculate the mean temperature difference between the blade and the water stream.
c) Calculate the rate of temperature increase along the channel.
d) Develop a feeling for how long the channel must be so that the assumption that the flow is thermally fully developed is valid.

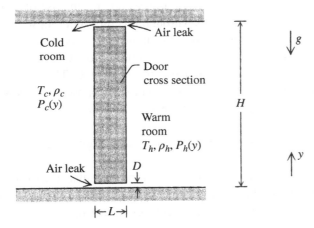

Figure P6.7

6.7 The air flow through the gaps formed at the top and bottom of a closed door is driven by the local air pressure difference between the two sides of the door (Figure P6.7). The door separates two isothermal rooms at different temperatures, T_c and T_h. In each room the pressure distribution is purely hydrostatic, $P_c(y)$ and $P_h(y)$, and the height-averaged pressure is the same on both sides of the door.
a) Assume that the air flow through each gap is laminar and fully developed. In terms of the geometric parameters indicated in the figure, show that the air flowrate through one gap is $\dot{m} = (\rho_c - \rho_h)gD^3WH/(24\nu L)$, where W is the door width in the direction perpendicular to the plane of the figure. Show further that the net convection heat transfer rate from the warm room to the cold room, through the two gaps, is $q = \dot{m}c_P(T_h - T_c)$.
b) Given are $T_c = 10\,°C$, $T_h = 30\,°C$, $D = 0.5$ mm, $L = 5$ cm, $H = 2.2$ m, and $W = 1.5$ m. Calculate $\dot{m}$ and q, and comment on how these quantities react to an increase in the gap thickness D.

6.8 A stream of water is heated with uniform heat flux in a pipe with a diameter of 2 cm and length of 2.86 m. The stream enters the pipe with the uniform velocity 10 cm/s and temperature $T_{in} = 10\,°C$. The mean temperature at the pipe outlet is $T_{out} = 20\,°C$. The water properties can be estimated at the average bulk temperature $(T_{in} + T_{out})/2$.

a) Calculate the thermal entrance length and compare it with the length of the pipe. Is the stream thermally fully developed at the pipe outlet?

b) Estimate the local Nusselt number Nu_D at $x = 2.86$ m in four ways: (i) graphically, based on Figure 6.6, (ii) by assuming that the temperature profile is fully developed, (iii) by assuming that the thermal boundary layer extends over the entire length of the pipe wall, and (iv) based on the correlation (6.52)–(6.53). Comment on the relative merit (accuracy) of the four methods.

c) Calculate the wall heat flux and the local wall temperature at the pipe outlet. Estimate the wall temperature averaged over the entire length of the pipe.

6.9 Determine the Nusselt number for fully developed laminar flow through a tube with uniform heat flux. Begin with Eq. (6.48), integrate that equation twice, and invoke finally the mean temperature definition (6.50).

6.10 The design of the cooling jacket for the apparatus shown in the figure calls for the use of a copper tube of inner diameter $D = 0.5$ cm and total length $L = 4$ m. Through this tube, cooling water flows with the mean velocity $U = 10$ cm/s. The tube is soldered to the side of the apparatus, the temperature of which is $30\,°C$. Neglecting the thermal resistance due to pure conduction across the copper wall of the tube, we can model the tube wall as isothermal with the temperature $T_w = 30\,°C$. The water stream enters the tube with the mean temperature $T_1 = 20\,°C$. In the calculations proposed as follows, the physical properties of water can be evaluated at $25\,°C$. Assume that the tube is sufficiently straight, that is, neglect the effect of bends.

a) Verify that the flow is hydrodynamically fully developed over most of the tube length. Calculate the pressure drop across the tube.

b) Verify that the flow is thermally fully developed along most of the tube. Calculate the mean outlet temperature of the water stream and the total heat transfer rate extracted by the stream from the tube wall (i.e. from the apparatus to which the tube is attached) (Figure P6.10).

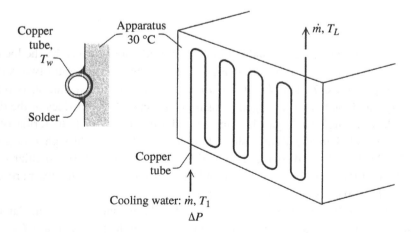

Figure P6.10

6.11 A highly viscous fluid is forced to flow through a straight pipe of inner radius r_o. The effect of friction (viscous shearing) tends to warm up the fluid as it advances through the pipe. This effect is offset by the cooling provided all along the pipe wall, which is isothermal (T_w = constant). The flow is hydrodynamically and thermally fully developed. The energy equation for a fluid with constant properties reduces in this case to

$$0 = k \frac{1}{r} \frac{d}{dr} \left(r \frac{dT}{dr} \right) + \mu \Phi$$

where the viscous dissipation function is $\Phi = (du/dr)^2$, and $u(r)$ is the Hagen–Poiseuille velocity profile.

a) Determine the temperature distribution through the fluid, $T(r)$.

b) If q is the total heat transfer rate (i.e. the cooling effect) through the wall of a pipe of length L, show that $q = \dot{m}\Delta P/\rho$, where ΔP is the pressure difference that drives the mass flowrate $\dot{m}$ through the pipe.

Turbulent Flow

6.12 The criterion for transition to turbulence in duct flow, Eq. (6.59), is essentially the same as the criterion for transition in boundary layer flow, Eq. (5.85). Demonstrate this equivalence by noting that when the flow is at the laminar/turbulent threshold in the fully developed region, it is also undergoing transition at the end of the entrance region (i.e. at $x \sim X$: see Figure P6.12). Rely on this observation as you start with Eq. (6.59) and derive a critical Re_x criterion that reproduces approximately Eq. (5.85). The equivalence of these and other transition criteria forms the subject of Appendix F.

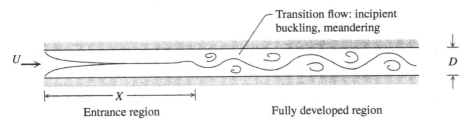

Figure P6.12

6.13 Consider the flow through a tube of fixed diameter D and length L. The mass flowrate $\dot{m}$ is fixed. The only change that may occur is the switch from laminar to turbulent flow, because the Reynolds number Re_D happens to be in the vicinity of 2000. In either regime, the flow is fully developed. Calculate the change in the required pumping power as the laminar flow is replaced by turbulent flow.

6.14 A stream of air ($Pr = 0.72$) is heated in fully developed flow through a pipe of diameter D (fixed) and uniform heat flux q''_w (fixed). Since the Reynolds number Re_D is 2500, there is some uncertainty with regard to the flow regime that prevails in the air stream. Calculate the change experienced by the local temperature difference $T_w - T_m$ as the flow regime switches from laminar to turbulent.

6.15 Water flows at the rate of 0.5 kg/s through a 10-m-long pipe with an inside diameter of 2 cm. It is being heated with uniform wall heat flux at the rate of 5×10^4 W/m^2. Evaluate the water properties at 20 °C, assume that the flow and temperature fields are fully developed, and calculate the following:

a) The pressure drop over the entire pipe length

b) The heat transfer coefficient based on the Colburn analogy (6.89)

c) The heat transfer coefficient based on the Dittus–Boelter correlation (6.91)

d) The difference between the wall temperature and the local mean water temperature

e) The temperature increase experienced by the mean water temperature in the longitudinal direction, from the inlet to the outlet

6.16 Atmospheric pressure steam condenses on the outside of a metallic tube and maintains the tube wall temperature uniform at 100 °C. The interior of the tube is cooled by a stream of 1 atm air with a mean velocity of 5 m/s and an inlet temperature of 30 °C. The tube inside diameter is 4 cm. Assume that the flow and temperature distributions across the tube are fully developed, and calculate the following:

a) The heat transfer coefficient

b) The length of the tube, if the outlet mean temperature of the air stream is 90 °C

c) The flow and thermal entrance lengths. Is the assumption of fully developed flow and heat transfer justified?

6.17 Water is being heated in a straight pipe with an inside diameter of 2.5 cm. The heat flux is uniform, $q''_w = 10^4 \, \text{W/m}^2$, and the flow and temperature fields are fully developed. The local difference between the wall temperature and the mean temperature of the stream is 4 °C. Calculate the mass flowrate of the water stream, and verify that the flow is turbulent. Evaluate the properties of water at 20 °C.

6.18 Derive Eq. (6.69), which shows that the apparent shear stress in fully developed turbulent pipe flow increases linearly in the radial direction.

6.19 Tap water of temperature 20 °C flows through a straight pipe with a 1 cm diameter, and a mean velocity of 1 m/s. Verify that the flow regime is turbulent, and calculate the friction factor for fully developed flow. If the dimensionless thickness of the viscous sublayer of the constant-τ_{app} region of the flow is equal to $y^+ \cong 11.6$, what is the actual thickness y (mm) of the viscous sublayer?

6.20 Show that in fully developed turbulent flow through a pipe, in which the velocity distribution is given by Eq. (6.72), the relationship between the centerline velocity $\bar{u}_c$ and the mean velocity U is $\bar{u}_c = 1.224\,U$. Rely on this result and Eqs. (6.74)–(6.75) to derive your version of the friction factor (6.76).

6.21 Derive Colburn's (6.89) for fully developed flow in a straight tube by combining Eqs. (6.88) and (6.77). This derivation is based on the assumption that Eq. (6.89) is accurate above $Re_D \cong 2 \times 10^4$. Derive an alternative Nu_D formula that is more accurate at lower Reynolds numbers, in the range $2 \times 10^3 < Re_D < 2 \times 10^4$.

Total Heat Transfer Rate

6.22 A water chiller circulates a water stream of 0.1 kg/s through a pipe immersed in a bath of crushed ice and water. For this reason, the pipe wall temperature may be modeled as constant, $T_w = 0\,°C$. The inlet temperature of the water stream is $T_1 = 40\,°C$. The pipe is long enough so that the flow is hydrodynamically and thermally fully developed over most of the pipe length L. The mean temperature at the pipe outlet is $T_L = 6\,°\text{C}$.

a) Calculate the total heat transfer rate between the stream and the $0\,°\text{C}$ bath.

b) Let D be the inner diameter of the pipe discussed until now. Consider the option of using a different pipe with the diameter $D_1 = D/2$. The mass flowrate, and the inlet and outlet temperatures of the water stream are fixed by design. What is the required length of the new pipe (L_1)?

6.23 Consider the definition of the log-mean temperature difference, Eq. (6.105), in an application in which the end differences ΔT_{in} and ΔT_{out}, are equal (e.g. Figure 6.15b). Show that in the limit $\Delta T_{\text{in}}/\Delta T_{\text{out}} \to 1$, Eq. (6.105) approaches the ΔT_{lm} values listed in Eq. (6.106).

6.24 One method of extracting the energy contained in a geothermal reservoir consists of using the "downhole coaxial heat exchanger" shown in Figure P6.24. The underground temperature increases almost linearly with depth. The stream $\dot{m} = 100$ tons/h of cold water is pumped downward through the annular space of outer diameter $D_o = 22$ cm and inner diameter $D_i = 16$ cm. In this portion of its circuit, the stream is heated by contact with the increasingly warmer rock material, across the wall of diameter D_o.

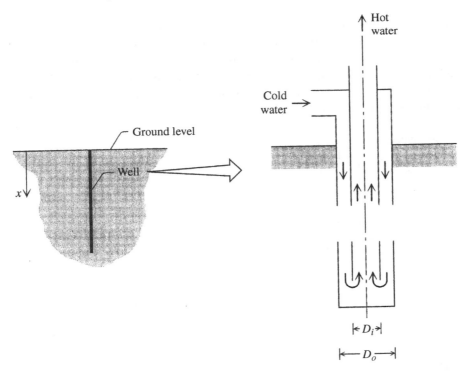

Figure P6.24

After reaching the lower extremity of the well, the heated stream returns to the surface by flowing through the inner pipe. A very effective layer of insulation is built into the wall of diameter D_i, which separates the downflowing cold stream from the upflowing hot stream.

a) Consider only the downflow through the annular space, and assume that the depth (x) to which your calculations apply is such that the mean temperature of the stream is $80\,°\text{C}$. The wetted surfaces of the annular space are made of commercial steel. Calculate the frictional pressure drop per unit length experienced by the stream at that depth.

b) Calculate the temperature difference ΔT between the outer wall of the annulus and the mean temperature of the stream. Again, the depth x is such that the mean temperature of the stream is $80\,°C$. Known is the mean temperature gradient $dT_m/dx = 200\,°C/km$, that is, the rate of temperature increase with depth.

6.25 Review the method of extracting energy from hot dry rock, which is illustrated in Problem 4.24. The fracture can be modeled as a parallel-plate channel of thickness $D = 1$ cm, length parallel to the flow $L = 100$ m, and width $W = 100$ m. The wetted rock surfaces are isothermal at $T_w = 200\,°C$. The water inlet temperature is $T_{in} = 30\,°C$, and the flowrate is $\dot{m} = 10$ kg/s. The water properties are approximately the same as those of saturated liquid at $200\,°C$. Calculate the following:

a) The flow and thermal entrance lengths
b) The pressure drop across the length L
c) The total heat transfer rate between the rock surfaces and the water stream, that is, the rate of energy extraction from the hot rock

6.26 Air at $300\,°C$ and 2 m/s approaches a bundle of 4 cm-diameter tubes arranged in a staggered array ($X_t = X_l = 7$ cm), with 21 rows and 6 or 5 tubes per row (i.e. across the flow). Each tube is 3 m long, and its wall temperature is maintained at $30\,°C$ by water flowing in the tube (Figure P6.26). The bundle-averaged heat transfer coefficient between tubes and air stream is 62 W/m² · K. The parallel side walls of the air duct are insulated. Calculate the outlet temperature of the air stream, and the total heat transfer rate absorbed by the tube bundle.

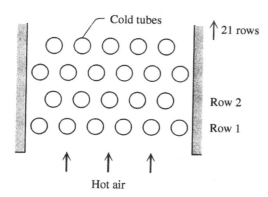

Figure P6.26

Evolutionary Design

6.27 The metallic blade shown in Figure P6.27 is an electric conductor that must be cooled by forced convection in a channel with insulated walls, with spacing D and length L. The blade and the channel are sufficiently long in the direction perpendicular to the figure. The pressure difference across the arrangement is fixed, ΔP, and the flow on either side of the blade is laminar and fully developed. The inlet temperature of the coolant is T_0.

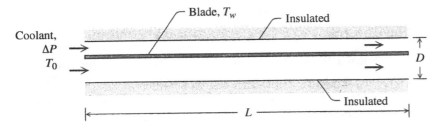

Figure P6.27

The objective is to lower the blade temperature as much as possible. The total heat transfer rate from the blade to the fluid, through both sides of the blade, is fixed by electrical design. What is the best position that the blade should occupy in the channel – right in the middle or closer to one of the side walls?

For simplicity, assume that the blade is isothermal (T_w). Assume that on either side of the blade, the group $hA_w/\dot{m}c_P$ is sufficiently greater than 1 so that the outlet temperature of the stream is approximately equal to T_w. The blade thickness is negligible relative to D.

6.28 Figure P6.28 shows a model of an electronic circuit board cooled by a laminar fully developed flow in a parallel-wall channel of fixed length L. The walls of the channel are insulated. The board substrate has a sufficiently high thermal conductivity so that the board temperature T_w may be assumed uniform in the longitudinal direction. The pressure difference that drives the flow is fixed, ΔP, and the fluid inlet temperature is T_0.

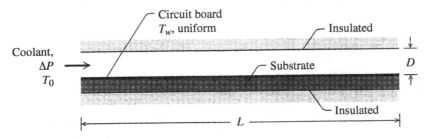

Figure P6.28

The channel spacing D must be selected in such a way that the thermal conductance $q/(T_m - T_0)$ is maximum. In this ratio, q is the total heat transfer rate removed by the stream from the board. Show that the optimal spacing is given by

$$\frac{D_{opt}}{L} = 2.7\left(\frac{\mu\alpha}{\Delta P \cdot L^2}\right)^{1/4}$$

and that the corresponding maximum thermal conductance or average heat transfer coefficient is

$$\left(\frac{\overline{q''}}{T_w - T_0}\right)_{max} \frac{L}{k} = 0.7\left(\frac{\Delta P \cdot L^2}{\mu\alpha}\right)^{1/4}$$

The dimensionless group that emerged on the right side is the pressure drop number $Be = \Delta P \cdot L^2/\mu\alpha$, cf. Eq. (6.125).

6.29 The electronic circuit board shown in Figure P6.29 is thin and long enough to be modeled as a surface with uniform heat flux q''. The heat generated by the circuitry is removed by the fully developed laminar flow channeled by the board and a parallel wall above it. That wall and the underside of the board are insulated. The length L is specified, and the inlet temperature of the coolant is T_0.

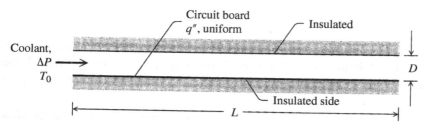

Figure P6.29

The board reaches its highest temperature (T_h) at the trailing edge, that is, in the plane of the outlet. That temperature ceiling is fixed by electrical design, otherwise the performance of the electronic components incorporated in the board will deteriorate. The designer would like to build as much circuitry and as many components into the board as possible. This objective is equivalent to seeking a board and channel design that ensures the removal of the largest rate of heat generated by the board ($q''L$). The lone degree of freedom is the selection of the spacing D.

a) Maximize the heat transfer rate removed by the stream, and show that the design is characterized by

$$\frac{D_{opt}}{L} = 3.14\left(\frac{\mu\alpha}{\Delta P \cdot L^2}\right)^{1/4}$$

$$\left(\frac{q''}{T_h - T_0}\right)_{max}\frac{L}{k} = 0.644\left(\frac{\Delta P \cdot L^2}{\mu\alpha}\right)^{1/4}$$

b) Compare this design with the results of the preceding problem (with $T_w = T_h$), in which the board was made isothermal by bonding it to a high conductivity substrate. Why is the maximum heat transfer rate higher when the board is isothermal? Is the increase in heat transfer significant enough to justify the use of a high-conductivity substrate?

6.30 Flow strangulation is not good for performance. Consider a long duct with Poiseuille flow (Figure P6.30). The duct has two sections, a narrow one of length L_1 and cross-sectional area A_1, followed by a wider one of length L_2 and cross-sectional area A_2. The total length L, the mass flowrate, and the total duct volume are fixed. Show that the duct with minimal global flow resistance is the one with uniform cross-section ($A_1 = A_2$). To demonstrate this analytically, write that the pressure drop along each duct section is proportional to the flow length and inversely proportional to the cross-sectional area squared.

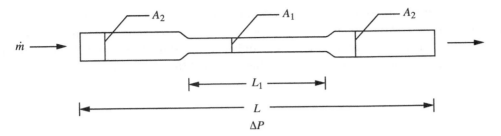

Figure P6.30

6.31 The Y-shaped construct of three round tubes shown in Figure P6.31 stretches on a square of side L. The trunk touches the midpoint of the left side, and the branches touch the opposite two corners. Variable is the location of the branching point or the trunk length L_1. The flow in every tube is in the Poiseuille regime, and the pressure drops are due to fully developed laminar flow distributed along the tubes. The total volume (V) occupied by the three tubes is fixed. The ratio D_1/D_2 has already been selected in accord with the Hess–Murray rule for laminar flow, Eq. (6.112). Minimize the overall flow resistance $\Delta P/\dot m_1$ by selecting the position of the branching point (L_1/L) and show that in the resulting design the angle between the two branches is close to $75°$.

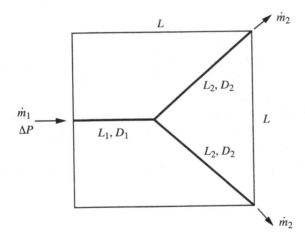

Figure P6.31

6.32 Water is pumped in at one location along the beach, and it is delivered to users at the two ends of the beach by two pipes of lengths L_1 and L_2, and diameters D_1 and D_2 (Figure P6.32). The water flowrates through each pipe ($\dot m_1$, $\dot m_2$) depend on the pipe dimensions, which are free to vary. The objective is to determine the complete configuration in which the pumping power requirement is minimal: the location of the water inlet, and the relative dimensions of the two pipes. The total volume occupied by the pipes is fixed, and so is the total length of the beach, $L = L_1 + L_2$. For analytical ease, introduce the dimensionless fractions $x = L_1/L$ and $y = \dot m_1/\dot m$, and determine analytically the desired configuration

$(x, y, D_1/D_2)$. For simplicity, start with the assumptions that $y = 1$ and the flow is laminar and fully developed in both pipes. After obtaining the solution for x and D_1/D_2, comment on how you would start the analysis when the flow regime is turbulent, fully developed and fully rough. Finally, comment on how you would approach the problem when y has a value different than 1.

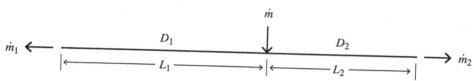

Figure P6.32

6.33 The natural phenomenon of economies of scale can be predicted by analyzing the flow of water through two identical pipes, each of length L, mass flowrate $\dot{m}_1$, diameter D_1, and end to end pressure drop ΔP_1. The pipes are long enough so that the flow is fully developed, and the effect of entrance and other local pressure losses can be neglected. What happens to the pressure drop if the two $\dot{m}_1$ streams and pipe volumes are replaced by one stream and volume, in a pipe of diameter D_2, length L, mass flowrate $\dot{m}_2$, and pressure drop ΔP_2? Will ΔP_2 be smaller or greater than ΔP_1? Calculate the ratio $\Delta P_2/\Delta P_1$ in two extreme regimes: (a) laminar fully developed flow and (b) laminar fully developed and fully rough regime.

6.34 A round tube of diameter D_1 and length L_1 splits into two tubes of diameters D_2 and D_3 and lengths L_2 and L_3. The tube length L_2 is not equal to the tube length L_3. The three tubes form a Y-shaped construct: The layout of this Y on a plane or in the three-dimensional space is not an issue in this problem. The tube lengths are given. The tube diameters may vary, but the total volume occupied by the tubes is fixed. The flow is in the Poiseuille regime in every duct. Each tube is *svelte* enough so that the pressure drop across the entire Y construct, from the free end of L_1 to the free ends of L_2 and L_3, is dominated by fluid friction along the straight portions of the ducts. In other words, the pressure loss at the Y function is negligible. Minimize the overall flow resistance and show that the optimal distribution of tube diameters is

$$\frac{D_1}{D_2} = \left[1 + \left(\frac{L_3}{L_2} \right)^3 \right]^{1/3}, \qquad \frac{D_2}{D_3} = \frac{L_2}{L_3}$$

Show that when $L_2 = L_3$, these results confirm the Hess–Murray rule for D_1/D_2. In other words, the results obtained in this problem represent a generalization of the Hess–Murray ratio for the entire range where $L_2 \neq L_3$.

6.35 Rivers do not flow straight and smooth like in the textbook sketch and the laboratory water channel. They meander, and they fall. The falls are known as cataracts, cascades, and water falls. They do this repeatedly with a particular periodicity. It is as if the natural tendency of the flow is to make its course more difficult, with obstacles not shortcuts. In this problem you are encouraged to question the obvious [37].

To predict the necessary occurrence of cataracts and their periodicity, compare the two ways to flow: straight down an incline and stepwise (Figure P6.35). The straight channel is a river bed with rough bottom and a cross section with width (W) proportional to depth (D), cf. page 657 in Ref. [5]. The flow is turbulent in the fully rough regime, with constant C_f. The alternative is the stepped path consisting of pools (L) interrupted by water in free fall (z). The mass flowrate is the same along both paths.

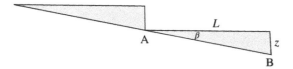

Figure P6.35

Calculate the time of water travel from A to B along both paths, invoke the constructal law of flow configuration evolution toward greater access [37], and predict which path should occur naturally. You will discover that the grade (the angle β) is key. Show that when the grade is small, the density of steps should be greater, and when the river is big (large W), the water falls should be sparse.

7

Natural Convection

7.1   What Drives Natural Convection?

In natural convection, or free convection, the fluid flows "naturally" (by itself), as it is driven by a heat engine, as shown in the 1984 edition of Ref [1]. This effect is distributed throughout the fluid and is associated with the general tendency of fluids to expand (or, in special cases,[1] to contract), when heated at constant pressure. The layer that feels the warm vertical wall shown in Figure 7.1 becomes lighter than the rest of the fluid. Its lightness forces it to flow upward, sweeping the wall in a manner that reminds us of the boundary layer shown in Figure 5.5. This time, however, the flow is a vertical jet parallel to the wall, whereas the fluid situated far from the wall is stagnant.

Remembering the impossibility of a perpetual motion machine of the first kind, it is instructive to show what thermodynamic principle is responsible for the fluid motion in natural convection. Consider the evolution of a small fluid packet Δm (a closed thermodynamic system) as it proceeds clockwise along the circuit drawn in Figure 7.1. While it is in the close vicinity of the wall, the fluid packet is *heated* by thermal diffusion from the wall. At the same time, Δm *expands* because it rises to altitudes where the hydrostatic pressure due to the distant fluid reservoir is lower.

Mass conservation requires that the upflow of the wall jet be complemented by a downflow of the cold reservoir fluid. The fluid packet Δm returns eventually to the heated wall, by flowing downward (and very slowly) through the reservoir. In the reservoir, the fluid packet is *cooled* and *compressed* while it descends to higher pressure.

In summary, the Δm system executes a cycle, which processes are arranged in the following sequence: heating – expansion – cooling – compression. This is the classical sequence of a work-producing cycle. The finite work delivered by Δm is invested in accelerating and increasing the kinetic energy inventory of Δm to the point where any additional work is dissipated fully by the effect of friction between Δm and the other fluid packets with which Δm comes in contact.

Looking at the entire flow field as an ensemble of fluid packets, we see the "wheel" of a heat engine driven by the temperature difference $T_w - T_\infty$. The Carnot work potential of this heat engine is dissipated entirely in the "fluid brake" that is distributed throughout the fluid, that is, in the interfaces between all the fluid packets that experience relative movement. The existence of a work-producing potential in natural convection can be demonstrated by inserting a propeller across the path followed by Δm, especially in the wall jet section where Δm travels the fastest.

---

1 One example of this kind is water at nearly atmospheric pressures, in the 0–4 °C temperature range.

Heat Transfer: Evolution, Design and Performance, First Edition. Adrian Bejan.
© 2022 John Wiley & Sons, Inc. Published 2022 by John Wiley & Sons, Inc.
Companion website: www.wiley.com/go/bejan/heattransfer

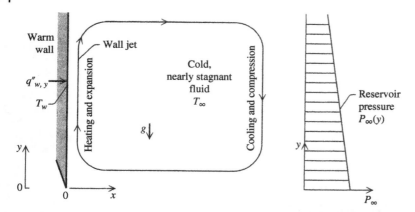

Figure 7.1 Wall jet driven by buoyancy along a heated wall, and pressure distribution in the reservoir of nearly stagnant fluid.

7.2 Boundary Layer Flow on Vertical Wall

7.2.1 Boundary Layer Equations

The challenge is again to determine the local heat transfer coefficient:

$$h_y = \frac{q''_{w,y}}{T_w - T_\infty} = \frac{-k(\partial T/\partial x)_{x=0}}{T_w - T_\infty} \tag{7.1}$$

The two-dimensional frame x–y is oriented in the usual way that Cartesian systems are drawn, with y pointing upward and x pointing horizontally. The gravitational acceleration g points in the negative y direction.

The numerator on the right side of Eq. (7.1) invites us to determine the temperature distribution in the fluid that makes contact with the wall. The four equations that govern the two-dimensional flow and temperature field can be found in Table 5.1, in which the body force components are $X = 0$ and $Y = -\rho g$:

$$(m) \quad \frac{\partial u}{\partial x} + \frac{\partial v}{\partial y} = 0 \tag{7.2}$$

$$(M_x) \quad \rho\left(u\frac{\partial u}{\partial x} + v\frac{\partial u}{\partial y}\right) = -\frac{\partial P}{\partial x} + \mu\left(u\frac{\partial^2 u}{\partial x^2} + \frac{\partial^2 u}{\partial y^2}\right) \tag{7.3}$$

$$(M_y) \quad \rho\left(u\frac{\partial v}{\partial x} + v\frac{\partial v}{\partial y}\right) = -\frac{\partial P}{\partial y} + \mu\left(\frac{\partial^2 v}{\partial x^2} + \frac{\partial^2 v}{\partial y^2}\right) - \rho g \tag{7.4}$$

$$(E) \quad u\frac{\partial T}{\partial x} + v\frac{\partial T}{\partial y} = \alpha\left(\frac{\partial^2 T}{\partial x^2} + \frac{\partial^2 T}{\partial y^2}\right) \tag{7.5}$$

Starting with Figure 7.2, we consider the flow region immediately adjacent to the vertical wall. The only assumption we make in connection with the wall temperature T_w is that it is *different* than the temperature of the fluid reservoir, T_∞. More specialized models of the wall heating mechanism will be adopted in Sections 7.2.3 (isothermal wall) and 7.2.5 (uniform heat flux).

The thickness of the near-wall region, δ_T, is defined as the transversal (horizontal) distance over which the temperature changes from its wall value, T_w, to a value comparable to that of the fluid reservoir, T_∞. The

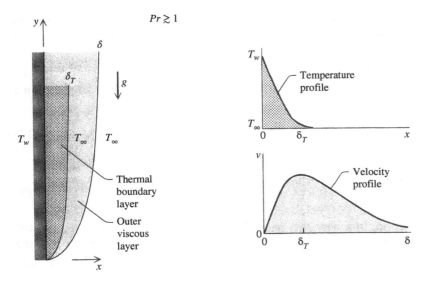

Figure 7.2   High Prandtl number fluids: thermal and velocity boundary layers on a vertical wall.

order-of-magnitude definition of δ_T is

$$\left(\frac{-\partial T}{\partial x}\right)_{x=0} \sim \frac{\Delta T}{\delta_T} \tag{7.6}$$

where $\Delta T = T_w - T_\infty$. Next, we assume that this wall layer of thickness δ_T and height y is slender, that is, a thermal boundary layer in the sense of Section 5.3.2:

$$\delta_T \ll y \tag{7.7}$$

The boundary layer feature (7.7) has the power to simplify the governing equations in two ways. First, all the longitudinal curvature terms $\partial^2/\partial y^2$ can be neglected in favor of the transversal curvature terms of type $\partial^2/\partial x^2$. This simplification occurs on the right side of Eqs. (7.3)–(7.5), that is, in a total of three places.

Second, the transversal momentum Eq. (7.3) implies that the pressure P does not vary appreciably across the δ_T-thin region; in other words,

$$P(x, y) \cong P(y) = P_\infty(y) \tag{7.8}$$

According to the right side of Figure 7.1, the distribution of pressure in the reservoir is hydrostatic:

$$\frac{dp_\infty}{dy} = -\rho_\infty g \tag{7.9}$$

In view of Eqs. (7.8) and (7.9), the longitudinal pressure gradient $\partial P/\partial y$ in Eq. (7.4) can be replaced by $-\rho_\infty g$, in which ρ_∞ is the density of reservoir fluid.

To summarize the two simplifications made possible by the slenderness of the thermal boundary layer, we record the boundary layer simplified forms of the momentum and energy equations:

$$(M)\quad \rho\left(u\frac{\partial v}{\partial x} + v\frac{\partial v}{\partial y}\right) = \mu\frac{\partial^2 v}{\partial x^2} + (\rho_\infty - \rho)g \tag{7.10}$$

$$(E)\quad u\frac{\partial T}{\partial x} + v\frac{\partial T}{\partial y} = \alpha\frac{\partial^2 T}{\partial x^2} \tag{7.11}$$

There is only one momentum equation now – Eq. (7.10) – because the other one, Eq. (7.3), was invoked to write Eq. (7.8), that is, to eliminate P as one of the unknowns of the problem. Note that in the original system of four equations, Eqs. (7.2)–(7.5), the four unknowns were u, v, P, and T.

The temperature T appears explicitly in the momentum equation if we make use of the thermodynamic equation of state $\rho = \rho(T, P)$. Expanding the density function as a Taylor series for small departures from a reference density,

$$\rho_\infty = \rho(T_\infty, P_0) \tag{7.12}$$

we obtain

$$\rho \cong \rho_\infty + \left(\frac{\partial \rho}{\partial T}\right)_P (T - T_\infty) + \left(\frac{\partial \rho}{\partial P}\right)_T (P - P_0) + \cdots \tag{7.13}$$

where P_0 is a reference pressure level (e.g. the pressure at the bottom of the reservoir, $P_{\infty,0}$). In most of the natural convection configurations that are encountered at ground level, the pressure correction term on the right side of Eq. (7.13) is negligible relative to the temperature correction term; therefore, Eq. (7.13) reduces to [1]

$$\rho \cong \rho_\infty - \rho_\infty \beta(T - T_\infty) + \cdots$$
$$\cong \rho_\infty[1 - \beta(T - T_\infty) + \cdots] \tag{7.14}$$

where β is the coefficient of volumetric thermal expansion defined in Eq. (5.18). In an ideal gas, for example, β is equal to $1/T$, where T is the *absolute* temperature expressed in degrees Kelvin or Rankine. Also worth noting is the fact that Eq. (7.14) is an adequate linear approximation of the $\rho = \rho(T, P)$ equation of state only when the departures from the reference density $\rho_\infty(T_\infty, P_0)$ are sufficiently small, that is, when the temperature correction is small enough so that

$$\beta(T - T_\infty) \ll 1 \tag{7.15}$$

The writing of v as shorthand for μ/ρ_∞ does not mean that in an actual application the fluid properties are to be evaluated at the reservoir temperature T_∞. When the fluid properties vary with the temperature across the boundary layer region, reasonable agreement between predictions based on this theory and actual measurements is achieved when the fluid properties are evaluated at the film temperature $(T_w + T_\infty)/2$.

Substituting the ρ approximation (7.14) on both sides of the momentum Eq. (7.10), we obtain

$$\rho_\infty[1 - \beta(T - T_\infty)]\left(u\frac{\partial v}{\partial x} + v\frac{\partial v}{\partial y}\right) = \mu\frac{\partial^2 v}{\partial x^2} + \rho_\infty \beta(T - T_\infty)g \tag{7.16}$$

In the square brackets on the left side, the $\beta(T - T_\infty)$ term can be neglected (left out) in favor of 1, in accordance with the observation (7.15). Dividing both sides of the equation by ρ_∞, and writing $v = \mu/\rho_\infty$ for the kinematic viscosity of the fluid, we arrive at a momentum equation with T in the buoyancy term:

$$(M) \quad u\frac{\partial v}{\partial x} + v\frac{\partial v}{\partial y} = v\frac{\partial^2 v}{\partial x^2} + g\beta(T - T_\infty)$$

$$\text{Inertia} \qquad \text{Friction} \;\; \text{Buoyancy} \tag{7.17}$$

The linearization of the $\rho = \rho(T, P)$ equation of state, Eq. (7.14), or the rewriting of the momentum Eq. (7.10) as Eq. (7.17), is recognized as the *Oberbeck–Boussinesq approximation* [2, 3] or, more succinctly, as the Boussinesq[2] approximation. The presence of the temperature in the buoyancy term of the momentum Eq. (7.17) "couples" the flow to the temperature field, and vice versa. In summary, there are three boundary layer-simplified equations: the mass conservation Eq. (7.2), which remained unchanged, the momentum Eq. (7.17), and the energy Eq. (7.11).

7.2.2 Scale Analysis of the Laminar Regime

The simplest and most direct approach to calculate the local heat transfer coefficient h_y is the order-of-magnitude analysis. According to the second h_y expression in Eq. (7.1), and in view of Eq. (7.6), the chief unknown is

$$h_y \sim \frac{k}{\delta_T} \tag{7.18}$$

The problem reduces to figuring out the size of the thermal boundary layer thickness δ_T. Let u and v represent the orders of magnitude (the scales) of the horizontal and vertical velocity components *inside* the flow region of thickness δ_T and height y. The mass, momentum, and energy Eqs. (7.2), (7.17), and (7.11) require, in order,

$$(m) \quad \frac{u}{\delta_T} \sim \frac{v}{y} \tag{7.19}$$

$$(M) \quad u\frac{v}{\delta_T}, \quad v\frac{v}{y} \sim v\frac{v}{\delta_T^2}, g\beta\Delta T \tag{7.20}$$

$$(E) \quad u\frac{\Delta T}{\delta_T}, \quad v\frac{\Delta T}{y} \sim \alpha\frac{\Delta T}{\delta_T^2} \tag{7.21}$$

The mass conservation proportionality (7.19) shows that the first two scales (the convection scales) on the left side of the energy Eq. (7.21) are of the same order of magnitude, $v\Delta T/y$. Therefore, the energy Eq. (7.21) reduces to

$$(E) \quad \underset{\text{Convection}}{v\frac{\Delta T}{y}} \sim \underset{\substack{\text{Transversal}\\\text{conduction}}}{\alpha\frac{\Delta T}{\delta_T^2}} \tag{7.22}$$

that is, a balance between the heat conducted horizontally, from the wall to the thermal boundary layer, and the enthalpy carried upward by the vertical stream.

Similarly, the mass conservation proportionality $u/\delta_T \sim v/y$ shows that the first two terms (the inertia scales) on the left side of the momentum Eq. (7.20) are of the same order of magnitude, v^2/y. The momentum equation is the competition between the three scales

$$(M) \quad \underset{\text{Inertia}}{\frac{v^2}{y}} \quad \underset{\text{Friction}}{v\frac{v}{\delta_T^2}} \quad \underset{\text{Buoyancy}}{g\beta\Delta T} \tag{7.23}$$

among which the driving force (the buoyancy scale) is never negligible (without it, there would be no flow). The remaining effects – inertia and friction – oppose the driving effect of buoyancy. Which flow-restraining effect plays the dominant role, inertia or friction? To answer this question, consider the following two limits:

---

2 (Valentin-)Joseph Boussinesq (1842–1929) was a professor of physics and experimental mechanics at the Faculty of Sciences in Paris. He made many contributions to the theory of elasticity and magnetism, and was the first to attack the subject of turbulent flow in a fundamental theoretical manner. In 1877 he pioneered the use of the time-averaged Navier–Stokes equations in the analytical study of turbulent flows.

(a) **Buoyancy Balanced by Friction.** In this limit, the momentum Eq. (7.23) reduces to

$$(M) \quad v\frac{v}{\delta_T^2} \sim g\beta\Delta T \tag{7.24}$$

Equations (7.24), (7.22), and (7.19) form a system of approximate algebraic equations containing three unknowns, u, v, and δ_T. The solution is

$$u \sim \frac{\alpha}{y}Ra_y^{1/4} \tag{7.25a}$$

$$v \sim \frac{\alpha}{y}Ra_y^{1/2} \tag{7.25b}$$

$$\delta_T \sim yRa_y^{-1/4} \tag{7.25c}$$

in which the dimensionless group labeled Ra_y is the Rayleigh number[3]:

$$Ra_y = \frac{g\beta(T_w - T_\infty)y^3}{\alpha v} \tag{7.26}$$

Equation (7.25c) shows that Ra_y is a number of the same order of magnitude as the slenderness ratio y/δ_T raised to the fourth power. This is why Ra_y has characteristically large numerical values in calculations. Equation (7.25c) also delivers the answer to the question that motivated this analysis, namely, Eq. (7.18):

$$h_y \sim \frac{k}{y}Ra_y^{1/4} \tag{7.27}$$

This conclusion is nondimensionalized by defining the local Nusselt number:

$$Nu_y = \frac{h_y y}{k} \tag{7.28}$$

so that Eq. (7.27) is the same as

$$Nu_y \sim Ra_y^{1/4} \tag{7.29}$$

When are the results (7.25) and (7.29) valid? They are valid when the assumed buoyancy–friction balance holds, namely, whenever the inertia scale v^2/y is negligible relative to either buoyancy or friction, for example,

$$\frac{v^2}{y} < v\frac{v}{\delta_T^2} \tag{7.30}$$

After using the v and δ_T scales shown in Eqs. (7.25b) and (7.25c), this inequality becomes $\alpha < v$, or $1 < Pr$. In conclusion, the results (7.25) and (7.29) are valid in fluids with Prandtl numbers of order 1 or greater. That is, the thermal boundary layer of a $Pr \gtrsim 1$ fluid is ruled by the momentum balance between the effects of buoyancy and friction.

When the kinematic viscosity is greater than the thermal diffusivity, the thermal boundary layer flow of thickness δ_T and vertical velocity v entrains (drags along) an additional layer that contains nearly isothermal fluid. It can be shown using the method discussed in Section 5.3.1 wherein the thickness of this outer layer is of order [1]

$$\delta \sim yRa_y^{-1/4}Pr^{1/2} \tag{7.31}$$

---

3 Lord Rayleigh (John William Strutt, 1842–1919) was a professor of natural philosophy in the Royal Institution of Great Britain. Next to his seminal work in sound theory, optics, electrodynamics, and elasticity, he is known for his work on hydrodynamic stability and cellular convection in a fluid layer heated from below (Bénard convection, Section 7.4.3).

The outer layer is thicker than the thermal boundary layer because, according to Eqs. (7.25c) and (7.31),

$$\frac{\delta}{\delta_T} \sim Pr^{1/2} > 1 \tag{7.32}$$

The structure and relative position of the thermal and vertical velocity profiles in $Pr \geq 1$ fluids are illustrated in Figure 7.2. Note that the velocity profile has two length scales, the distance from the wall to the velocity peak (δ_T) and the distance to the nearly stagnant reservoir (δ). Combining Eqs. (7.25c) and (7.31) with slenderness criteria of type (7.7), we learn that the δ_T and δ layers are indeed "boundary layers" if, respectively,

$$Ra_y^{1/4} > 1 \qquad Ra_y^{1/4} Pr^{-1/2} > 1 \tag{7.33}$$

In $Pr \gtrsim 1$ fluids, the second of these criteria is more stringent than the first, as $Ra_y^{1/4}$ must exceed the order of magnitude of $Pr^{1/2}$. This is why natural convection boundary layer flows are often referred to as "high Rayleigh number" flows.

(b) **Buoyancy Balanced by Inertia**. When the contribution made by the friction term is negligible in the momentum balance (7.23), what is left reads

$$(M) \quad \frac{v^2}{y} \sim g\beta\Delta T \tag{7.34}$$

This equation and Eqs. (7.19) and (7.22) are sufficient for determining the unknown scales of the δ_T-thin region:

$$u \sim \frac{\alpha}{y}(Ra_y Pr)^{1/4} \tag{7.35a}$$

$$v \sim \frac{\alpha}{y}(Ra_y Pr)^{1/2} \tag{7.35b}$$

$$\delta_T \sim y(Ra_y Pr)^{-1/4} \tag{7.35c}$$

These results bring to light the new dimensionless group $(Ra_y\ Pr)$, which does not contain the kinematic viscosity v in the denominator:

$$Ra_y Pr = \frac{g\beta(T_w - T_\infty)y^3}{\alpha^2} \tag{7.36}$$

This group is recognized as the Boussinesq number, Bo_y. In the present treatment, the $Ra_y\ Pr$ label will be used to stress that the shift from the results (7.25a–c) to the new set of results (7.35a–c) occurs when the Prandtl number changes.

The local Nusselt number Nu_y follows from Eqs. (7.18), (7.28), and (7.35c):

$$Nu_y \sim (Ra_y Pr)^{1/4} \tag{7.37}$$

The results (7.35a–c) and (7.37) apply when the friction effect is negligible in the momentum balance, that is, when the opposite of the inequality (7.30) prevails. Substituting the v and δ_T scales (7.35b) and (7.35c) into the inequality leads to $\alpha > v$ or to the conclusion that the present results apply to $Pr \lesssim 1$ fluids.

The structure of the temperature and vertical velocity profiles across the boundary layer flow of a low-Pr fluid is illustrated in Figure 7.3. The wall jet is as thick as the temperature profile (δ_T). Immediately adjacent

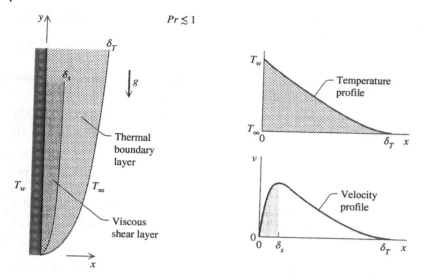

Figure 7.3 Low Prandtl number fluids: thermal and velocity boundary layers on a vertical wall.

to the wall, there is a thin shear layer in which the effect of friction is important. The thickness of this shear layer (i.e. the distance to the velocity peak) is of the order [1]

$$\delta_s \sim y\left(\frac{Ra_y}{Pr}\right)^{-1/4}$$
(7.38)

which shows that δ_s is smaller than δ_T:

$$\frac{\delta_s}{\delta_T} \sim Pr^{1/2} < 1$$
(7.39)

The dimensionless group (Ra_y/Pr) is known as the Grashof number[4]:

$$Gr_y = \frac{Ra_y}{Pr} = \frac{g\beta(T_w - T_\infty)y^3}{\nu^2}$$
(7.40)

Contrary to the impression left by the presence of Pr when the Grashof number is written as Ra_y/Pr, the Grashof number – the right side of Eq. (7.40) – does not contain the thermal diffusivity α in its denominator.

The results developed in this limit of low Prandtl numbers are valid provided that the δ_s and δ_T layers are slender. These requirements lead, respectively, to the following inequalities:

$$\left(\frac{Ra_y}{Pr}\right)^{1/4} > 1 \qquad (Ra_y Pr)^{1/4} > 1$$
(7.41)

The second of these inequalities is more restrictive because Pr is a number of order 1 or smaller. The slenderness of the boundary layer structure is assured if $Ra_y > Pr^{-1}$, that is, when the Rayleigh number is sufficiently large.

The limits (**a**) and (**b**) analyzed in this section correspond to the Prandtl number limits $Pr \gtrsim 1$ and $Pr \lesssim 1$. At the intersection of these two limiting sets of results lies the class of fluids with Prandtl numbers of order

---

4 Franz Grashof (1826–1893) was a professor of mechanical engineering at the University of Karlsruhe.

1 (e.g. air). Note that the two limiting Nu_y formulas, Eqs. (7.29) and (7.37), give the same result when Pr is of order 1 and fixed.

7.2.3 Isothermal Wall

More exact methods of solution have the power to predict the dimensionless coefficients of order 1 that are now missing from the right side of the Nu_y formulas (7.29) and (7.37). For the flow near an isothermal wall, the similarity solution can be determined by first defining the similarity variable [1]

$$\eta = \frac{x}{y Ra_y^{-1/4}} \tag{7.42}$$

in which the reference transversal length scale is the thermal boundary layer thickness of a high-Pr fluid, Eq. (7.25c). The similarity vertical velocity profile recommended by Eq. (7.25b) is

$$G(\eta, Pr) = \frac{v}{(a/y)Ra_y^{1/2}} \tag{7.43}$$

Introducing the streamfunction $\psi(x, y)$ defined by

$$u = \frac{\partial \psi}{\partial y} \quad \text{and} \quad v = -\frac{\partial \psi}{\partial x} \tag{7.44}$$

it can be shown that the similarity velocity profile (7.43) requires a similarity streamfunction of the form

$$F(\eta, Pr) = \frac{\psi}{\alpha Ra_y^{1/4}} \tag{7.45}$$

where $G = -dF/d\eta$. Finally, the similarity temperature profile is the function

$$\theta(\eta, Pr) = \frac{T - T_\infty}{T_w - T_\infty} \tag{7.46}$$

in which the denominator $T_w - T_\infty = \Delta T$ is a constant.

The similarity form of the energy and momentum Eqs. (7.11) and (7.17) is obtained in two steps: first, by eliminating u and v in favor of the respective ψ derivatives (7.44) and, second, by eliminating x, y, ψ, and T in favor of η, F, and θ using Eqs. (7.42), (7.45), and (7.46). The two equations that emerge from this analysis are [1]

$$(E) \quad \frac{3}{4}F\theta' = \theta'' \tag{7.47}$$

$$(M) \quad \frac{1}{Pr}\left(\frac{1}{2}F'^2 - \frac{3}{4}FF''\right) = -F''' + \theta \tag{7.48}$$

The boundary conditions that must be satisfied by the functions F and θ are

$$F = 0 \text{ at } \eta = 0 \quad \text{(impermeable wall, } u = 0) \tag{7.49a}$$

$$F' = 0 \text{ at } \eta = 0 \quad \text{(no slip, } v = 0) \tag{7.49b}$$

$$\theta = 1 \text{ at } \eta \to \infty \quad \text{(isothermal wall, } T = T_w) \tag{7.49c}$$

$$F' \to 0 \text{ as } \eta \to \infty \quad (v = 0 \text{ in the fluid reservior}) \tag{7.49d}$$

$$\theta \to 0 \text{ as } \eta \to \infty \quad (T = T_\infty \text{ in the fluid reservior}) \tag{7.49e}$$

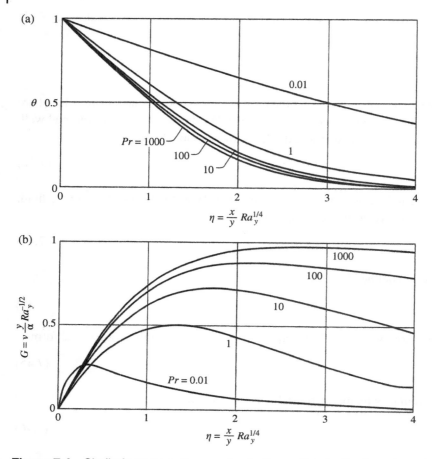

Figure 7.4 Similarity temperature and velocity profiles for laminar natural convection on a vertical isothermal wall. Source: Bejan [1].

Figure 7.4 shows the resulting similarity profiles for temperature and vertical velocity near the isothermal wall [1]. The upper graph shows that the temperature profiles fall on the same curve when the Prandtl number is large. The local Nusselt number revealed by the similarity solution of Figure 7.4 is [cf. Eqs. (7.28) and (7.1)]

$$Nu_y = \frac{h_y y}{k} = \frac{y}{k \Delta T} \left(-k \frac{\partial T}{\partial x} \right)_{x=0}$$
$$= \left(-\frac{d\theta}{d\eta} \right)_{\eta=0} Ra_y^{1/4} \tag{7.50}$$

The proportionality between Nu_y and $Ra_y^{1/4}$ indicates that when the vertical wall is isothermal, the local heat flux $q_{w,y}''$ varies as $y^{-1/4}$. The similarity temperature gradient at the wall, $(d\theta/d\eta)_{\eta=0}$, is a function of Prandtl number because Pr appears as a parameter in the momentum Eq. (7.48). That function is approximated with 0.5% by [4]

$$Nu_y = 0.503 \left(\frac{Pr}{Pr + 0.986 Pr^{1/2} + 0.492} \right)^{1/4} Ra_y^{1/4} \tag{7.51}$$

This expression covers the entire range of Prandtl numbers. Its two asymptotes

$$Nu_y = 0.503\, Ra_y^{1/4} \quad (Pr \gg 1) \tag{7.52a}$$

$$Nu_y = 0.6(Ra_y Pr)^{1/4} \quad (Pr \ll 1) \tag{7.52b}$$

confirm the validity of the order-of-magnitude conclusions (7.29) and (7.37), respectively.

The *average* or *overall* Nusselt number $\overline{Nu}_y$ for a wall of height y is defined as

$$\overline{Nu}_y = \frac{\overline{h}_y y}{k} = \frac{\overline{q}''_{w,y}}{T_w - T_\infty} \frac{y}{k} \tag{7.53}$$

In this expression, $\overline{q}''_{w,y}$ is the wall heat flux averaged from $y = 0$ to y, and

$$q'_{w,y} = \overline{q}''_{w,y}\, y \tag{7.54}$$

is the total heat transfer rate through the wall, per unit length in the direction perpendicular to the plane of Figure 7.2. Or, if W is the width of the wall in the direction perpendicular to Figure 7.2, the total heat transfer rate through the wall $q_{w,y}$ (watts) is $q_{w,y} = q'_{w,y} W = q''_{w,y} Wy$. The wall-averaged Nusselt number $\overline{Nu}_y$ is a dimensionless version of the total heat transfer rate $q'_{w,y}$, namely, $\overline{Nu}_y = q'_{w,y}/k\Delta T$, and the formula that corresponds to Eq. (7.51) is

$$\overline{Nu}_y = 0.671 \left(\frac{Pr}{Pr + 0.986\, Pr^{1/2} + 0.492} \right)^{1/4} Ra_y^{1/4} \tag{7.55}$$

For air or any other gas with $Pr = 0.72$, Eq. (7.55) yields $\overline{Nu}_y = 0.517\, Ra_y^{1/4}$. An alternative correlation for the entire Pr range in the laminar regime will be presented in Eq. (7.62), that is, after the discussion of the transition from laminar flow to turbulent flow, Eq. (7.56).

Looking back at the material covered in Sections 7.2.1–7.2.3, we focused in detail on natural convection along a vertical isothermal wall because it is the most basic (and oldest) of all the natural convection boundary layer configurations. The history of the study of this flow was recounted by Martin [5]. The first similarity formulation was published by Schmidt and Beckmann [6]. The scales and double-layer structures presented in Section 7.2.2 were first reported by Kuiken [7]. The scale analysis of the same problem was developed independently in the first (1984) edition of Ref. [1].

Example 7.1 *Laminar Natural Convection Boundary Layer*
The door of a kitchen oven is a vertical rectangular area 0.5 m tall and 0.65 m wide. The external surface of the oven door is at 40 °C, while the room air is at 20 °C. Calculate the natural convection heat transfer rate from the door to the ambient air.

Solution

The film temperature of the air in the boundary layer is $(T_w + T_\infty)/2 = 30\,°C$, and the air properties that will be needed are

$$Pr = 0.72 \quad k = 0.026\, \frac{W}{m \cdot k} \quad \frac{g\beta}{\alpha\nu} = \frac{90.7}{cm^3 \cdot k}$$

The calculation begins with the Rayleigh number based on the door height $H = 50$ cm:

$$Ra_H = \frac{g\beta}{\alpha\nu} H^3 (T_w - T_\infty)$$

$$= \frac{90.7}{cm^3 \cdot K}(50\ cm)^3(40-20)\ K = 2.27 \times 10^8 \quad \text{(laminar)}$$

$$\overline{Nu}_H = 0.517 Ra_H^{1/4}$$

$$= 0.517(2.27 \times 10^8)^{1/4} = 63.44$$

$$\overline{Nu}_H = \frac{\overline{q''}_w H}{\Delta T k}$$

$$\overline{q''}_w = (63.44 \times 20\ K)\left(0.026\frac{W}{m \cdot K}\right)\frac{1}{0.5\ m} = 66\ W/m^2$$

The door area is $A = 0.5\ m \times 0.65\ m = 0.325\ m^2$, and this means that the natural convection heat transfer rate to the room air is

$$q = \overline{q''}_w A$$

$$= 66\frac{W}{m^2}(0.325\ m^2) = 21.4\ W$$

7.2.4 Transition and the Effect of Turbulence

The boundary layer flow discussed until now remains laminar if y is small enough so that the Rayleigh number Ra_y does not exceed a certain critical value (Figure 7.5). The transition to turbulent flow was thought to occur at the y position where $Ra_y \sim 10^9$, regardless of the value of the Prandtl number. Bejan and Lage [8] showed that it is the Grashof number of order 10^9 that marks the transition in all fluids:

$$Gr_y \sim 10^9 \quad (10^{-3} \le Pr \le 10^3) \tag{7.56}$$

This universal transition criterion can be expressed in terms of the Rayleigh number, by recalling that $Ra_y = Gr_y Pr$:

$$Ra_y \sim 10^9 Pr \quad (10^{-3} \le Pr \le 10^3) \tag{7.57}$$

It is supported very well by numerous experimental observations reviewed in Ref. [8]. The $Gr_y \sim 10^9$ transition criterion coincides with the traditional criterion $Ra_y \sim 10^9$ only in the case of fluids with Prandtl numbers of order 1 (e.g. air). In the liquid metal range of Prandtl numbers ($Pr \sim 10^{-3}$ to 10^{-2}), for example, the Grashof number criterion (7.56) means that the actual transition Rayleigh number is of order 10^6 to 10^7, which is well below the often mentioned threshold of 10^9. When the Rayleigh number exceeds the order of magnitude of $10^9\ Pr$, the Nusselt number can be calculated with Eq. (7.61), which is discussed later in this section.

The transition criterion (7.56) is a manifestation of the constructal law through local Reynolds number criterion for transition to turbulence, which is universally applicable (Appendix F). For example, if the present flow is a Prandtl number of order 1, its local Reynolds number has the δ_T of Eq. (7.25c) as transversal length scale and the v of Eq. (7.25b) as longitudinal velocity scale:

$$Re \sim \frac{\delta_T v}{\nu} \sim \frac{y Ra_y^{-1/4}}{\nu}\frac{\alpha}{y}Ra_y^{1/2}$$

$$\sim Ra_y^{1/4}/Pr \tag{7.58}$$

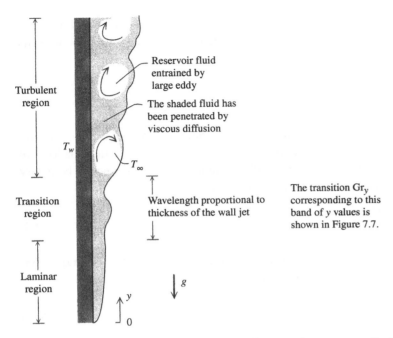

Figure 7.5 Laminar, transition, and turbulent sections on a wall with natural convection.

In this last expression, Ra_y can be replaced by Gr_y because $Pr \sim 1$:

$$Re \sim Gr_y^{1/4} \quad (Pr \sim 1) \tag{7.59}$$

From Eq. (7.56) we learn that at transition the local Reynolds number is of order 10^2:

$$Re \sim (10^9)^{1/4} = 178 \ (Pr \sim 1) \tag{7.60}$$

The heat transfer rate from a vertical wall in the presence of turbulence in the boundary layer has been measured experimentally and correlated as a function $\overline{Nu}_y(Ra_y, Pr)$. It was found that in the turbulent regime $\overline{Nu}_y$ is approximately proportional to $Ra_y^{1/3}$, which differs from the $\overline{Nu}_y \sim Ra_y^{1/4}$ proportionality for laminar flow in high-Pr fluids. An empirical correlation that reports the wall-averaged Nusselt number $\overline{Nu}_y$ for the entire Rayleigh number range (laminar, transition, turbulent) is [9]

$$\overline{Nu}_y = \left\{ 0.825 + \frac{0.387 Ra_y^{1/6}}{[1 + (0.492/Pr)^{9/16}]^{8/27}} \right\}^2 \tag{7.61}$$

This correlation holds true for $10^{-1} < Ra_y < 10^{12}$ and for all Prandtl numbers. The physical properties used in the definition of $\overline{Nu}_y$, Ra_y, and Pr are evaluated at the film temperature $(T_w + T_\infty)/2$. In the case of air $(Pr = 0.72)$, Eq. (7.61) reduces to $\overline{Nu}_y = (0.825 + 0.325 \, Ra_y^{1/6})^2$. In the laminar range, $Gr_y < 10^9$, a correlation that represents the experimental data more accurately than Eq. (7.61) is [9]

$$\overline{Nu}_y = 0.68 + \frac{0.67 Ra_y^{1/4}}{[1 + (0.492/Pr)^{9/16}]^{4/9}} \tag{7.62}$$

which for air ($Pr = 0.72$) yields $\overline{Nu}_y = 0.68 + 0.515\, Ra_y^{1/4}$. This correlation is an alternative to Eq. (7.55) especially in the low Rayleigh number limit where the boundary layer (slender flow) approximation loses its appeal.

7.2.5 Uniform Heat Flux

When the vertical wall is heated uniformly, $q''_w = $ constant, the wall temperature T_m increases monotonically in the y direction. In the laminar regime the temperature difference $T_w - T_\infty$ increases as $y^{1/5}$; this relationship follows from Eqs. (7.29) and (7.37), which have general applicability in laminar natural convection boundary layer flow. In the case of a high-Pr fluid, for example, Eq. (7.29) states that

$$\frac{q''_w}{[T_w(y) - T_\infty]}\frac{y}{k} \sim \left[\frac{g\beta(T_w - T_\infty)y^3}{\alpha\nu}\right]^{1/4} \tag{7.63}$$

This result can be rearranged to show that $T_w - T_\infty$ is indeed proportional to $y^{1/5}$. Alternatively, Eq. (7.63) can be rewritten so that the right side no longer contains the temperature difference $T_w - T_\infty$:

$$\frac{q''}{[T_w(y) - T_\infty]}\frac{y}{k} \sim \left(\frac{g\beta q''_w y^4}{\alpha\nu k}\right)^{1/5} \tag{7.64}$$

The group formed inside the brackets on the right side is the Rayleigh number based on constant heat flux,

$$Ra_y^* = \frac{g\beta q''_w y^4}{\alpha\nu k} \tag{7.65}$$

while the left side of Eq. (7.64) continues to show the local Nusselt number defined in Eq. (7.28). In conclusion, in the laminar high-Pr flow, Nu_y is of the same order as $Ra_y^{*1/5}$. This is confirmed by the similarity solution for the uniform heat flux case [10], which is fitted well by the expression

$$Nu_y \cong 0.616\left(\frac{Pr}{Pr + 0.8}\right)^{1/5} Ra_y^{*1/5} \tag{7.66}$$

In fluids of the air–water Prandtl number range, the transition to turbulence occurs in the vicinity of $Ra_y^* \sim 10^{13}$. For calculating the local and wall-averaged Nusselt numbers, recommended are the following formulas [11]:

$$\left.\begin{array}{l} Nu_y = 0.6 Ra_y^{*1/5} \\ \overline{Nu}_y = 0.75 Ra_y^{*1/5} \end{array}\right\} \quad \text{laminar,} \quad 10^5 < Ra_y^* < 10^{13} \tag{7.67}$$

$$\left.\begin{array}{l} Nu_y = 0.568 Ra_y^{*0.22} \\ \overline{Nu}_y = 0.645 Ra_y^{*0.22} \end{array}\right\} \quad \text{turbulent,} \quad 10^{13} < Ra_y^* < 10^{16} \tag{7.68}$$

The average Nusselt number $\overline{Nu}_y$ is based on the wall-averaged temperature difference $T_w - T_\infty$. In particular, for calculations of heat transfer to *air*, recommended are [12]

$$Nu_y = 0.55 Ra_y^{*1/5} \quad \text{(laminar)} \tag{7.69a}$$

$$Nu_y = 0.17 Ra_y^{*1/4} \quad \text{(turbulent)} \tag{7.69b}$$

A correlation valid for all Rayleigh and Prandtl numbers is [9]

$$\overline{Nu}_y = \left\{ 0.825 + \frac{0.387 Ra_y^{1/6}}{[1 + (0.437/Pr)^{9/16}]^{8/27}} \right\}^2 \tag{7.70}$$

In this expression, Ra_y is based on the y-averaged temperature difference, $\overline{T}_w - T_\infty$. This correlation is almost identical to the one recommended for isothermal walls, Eq. (7.61). For air at room conditions, the Eq. (7.70) reduces to

$$\overline{Nu}_y = (0.825 + 0.328 Ra_y^{1/6})^2 \quad (Pr = 0.72) \tag{7.70'}$$

The high Rayleigh number asymptote of this last formula is $\overline{Nu}_y \cong 0.107\, Ra_y^{1/3}$, for $Pr = 0.72$ and $Ra_y > 10^{10}$. Equation (7.70) can be restated in terms of the flux Rayleigh number Ra_y^* by noting the substitution $Ra_y = Ra_y^*/\overline{Nu}_y$. For example, the high Rayleigh number asymptote for air becomes $\overline{Nu}_y \cong 0.187 Ra_y^{*1/4}$ $(Pr = 0.72$ and $Ra_y^* > 10^{12})$.

7.3 Other External Flows

7.3.1 Thermally Stratified Reservoir

There are many instances in which the fluid reservoir that bathes the heated or cooled object is not isothermal, as was assumed throughout Section 7.2. Quiescent fluid reservoirs often exhibit a temperature variation – a stratification – in the vertical direction. The core fluid in an enclosure heated from the side (Section 7.4.2) and the bulk of the air in a sealed room are just two examples. The inset in the upper-right corner of Figure 7.6 shows the assumed linear distribution of temperature in a fluid reservoir adjacent to an *isothermal* vertical wall of height H. The dimensionless stratification parameter

$$b = \frac{\Delta T_{max} - \Delta T_{min}}{\Delta T_{max}} \tag{7.71}$$

varies between the isothermal reservoir limit ($b = 0$) and the limit of maximum stratification ($b = 1$), where the uppermost layer of the reservoir is as warm as the heated wall. The temperature differences ΔT_{min} and ΔT_{max} are the wall–reservoir differences measured at $y = H$ and $y = 0$, respectively.

Figure 7.6 shows the average Nusselt number in the laminar regime. Both $\overline{Nu}_H$ and Ra_H are based on the height and the *maximum* temperature difference:

$$\overline{Nu}_H = \frac{\overline{q}_{w,H}''}{\Delta T_{max}} \frac{H}{k} \tag{7.72a}$$

$$Ra_H = \frac{g\beta \Delta T_{max} H^3}{\alpha \nu} \tag{7.72b}$$

In the laminar range, the Nusselt number obeys the relationship

$$\overline{Nu}_H = f(b, Pr)\, Ra_H^{1/4} \tag{7.73}$$

for which the factor $f(b, Pr)$ is displayed in Figure 7.6. The factor f decreases significantly as the stratification parameter b increases from 0 to 1. The Prandtl number has a less pronounced effect especially when it is considerably greater than 1. The curves drawn for $Pr = 0.7$ and $Pr = 6$ are based on results [13], obtained using the local nonsimilarity method [14]. The $Pr \to \infty$ curve was developed based on the integral method in the 1984 edition of Ref. [1].

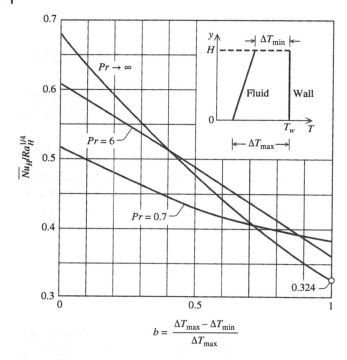

Figure 7.6 Average Nusselt number for laminar flow along an isothermal wall in contact with a linearly stratified fluid reservoir.

7.3.2 Inclined Walls

Figure 7.7 shows a plane wall inclined relative to the vertical direction. The angle between the plane and the vertical direction, ϕ, is restricted to the range $-60° < \phi < 60°$ (horizontal walls are discussed in subsection 7.3.3). In cases (*a*) and (*d*) – heated wall tilted upward, and cooled wall tilted downward – the effect of the angle ϕ thickens the tail end of the boundary layer and gives the wall jet a tendency to separate from the wall. The opposite effect is illustrated in cases (*b*) and (*c*), where the wall jet is squeezed against the wall until it spills over the trailing edge.

In the boundary layer analysis of the flows of Figure 7.7, it is found that the momentum equation is analogous to Eq. (7.17), except that $g \cos \phi$ replaces g in the buoyancy term. The group $g \cos \phi$ is the gravitational acceleration component oriented parallel to the wall. For this reason, the heat transfer rate in the *laminar* regime along an isothermal wall can be calculated with Eq. (7.62), provided the Rayleigh number Ra_y is based on g cos ϕ:

$$Ra_y = \frac{(g \cos \phi)\beta(T_w - T_\infty)y^3}{\alpha \nu} \tag{7.74}$$

Similarly, for laminar flow over a plate with uniform heat flux, the Nusselt number can be calculated with Eqs. (7.67) and (7.69a), in which

$$Ra_y^* = \frac{(g \cos \phi)\beta q_w'' y^4}{\alpha \nu k} \tag{7.75}$$

For the *turbulent* regime, it was found that the heat transfer measurements are correlated better using g instead of $g \cos \phi$ in the Rayleigh number group [15]. Therefore, Eq. (7.61) is recommended for isothermal

Figure 7.7 Plane walls inclined relative to the vertical direction.

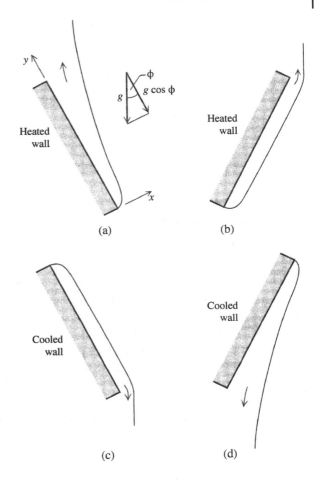

(a) (b)

(c) (d)

plates, in which Ra_y follows the usual definition, Eq. (7.26). For inclined walls with uniform heat flux, the Nusselt number is given by Eqs. (7.68) and (7.69b), and the flux Rayleigh number Ra_y^* is defined by Eq. (7.65).

The tilt angle ϕ has a noticeable effect on the location of the laminar–turbulent transition when the uniform flux wall is oriented as in cases (a) and (d) [15]. The flux Rayleigh numbers tabulated below mark the beginning and the end of the transition region in water experiments ($Pr \cong 6.5$):

φ	Ra_y^*
0°	$5 \times 10^{12} - 10^{14}$
30°	$3 \times 10^{10} - 10^{12}$
60°	$6 \times 10^7 - 6 \times 10^9$

The Ra_y^* spread between the beginning and end of transition covers almost two orders of magnitude. This observation reinforces the approximate character of the threshold value $Ra_y^* \sim 10^{13}$ used in Eqs. (7.67) and (7.68). The effect of wall inclination on transition along an isothermal wall was measured for water ($Pr \sim 6$) [16]. The transition Rayleigh number $Ra_y = g\beta y^3 \Delta T / \alpha v$ decreases as the tilt angle ϕ increases:

φ	Ra_y
0°	8.7×10^8
20°	2.5×10^8
45°	1.7×10^7
60°	7.7×10^5

7.3.3 Horizontal Walls

The flow changes its character as the tilt angle ϕ increases beyond the moderate values considered in Figure 7.7. Two flow types are encountered in the extreme where the plane wall becomes horizontal (Figure 7.8). When the wall is heated and faces upward, or when it is cooled and faces downward, the flow leaves the boundary layer as a vertical plume rooted in the central region of the wall. When the temperature difference is sufficiently large, the heated fluid rises from all over the surface: the flow is intermittent and consists of balls of heated fluid (called *thermals*) that rise and meander through the colder fluid.

In cases where the surface is hot and faces downward, or when it is cold and faces upward, the boundary layer covers the entire surface and the flow spills over the edges, Figure 7.8. A two-sided plate, hot or cold, will have flow of one type on the top side and flow of the other type on the bottom side.

Average Nusselt number measurements for several horizontal plate configurations have been correlated by defining the characteristic length of the plane surface [17]:

$$L = \frac{A}{p} \tag{7.76}$$

where A is the area of the plane surface and p is the perimeter of A. For example, for a disc of diameter D, $L = D/4$. The $\overline{Nu}_L$ formulas listed below are valid for Prandtl numbers greater than 0.5. The average Nusselt number $\overline{Nu}_L$ and the Rayleigh number Ra_L are both based on L. For hot surfaces facing upward, or cold

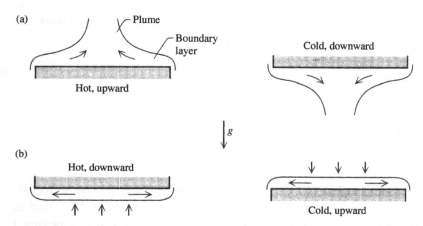

Figure 7.8 Horizontal surfaces with central plume (a) and without plume flow (b).

surfaces facing downward (Figure 7.8a), the Nusselt number is [18]

$$\overline{Nu_L} = 0.54 Ra_L^{1/4} \quad (10^4 < Ra_L < 10^7) \tag{7.77a}$$

$$\overline{Nu_L} = 0.15 Ra_L^{1/3} \quad (10^7 < Ra_L < 10^9) \tag{7.77b}$$

The corresponding correlation for hot surfaces facing downward or cold surfaces facing upward (Figure 7.8b) is (cf. Ref. [19], p. 548)

$$\overline{Nu_L} = 0.27 Ra_L^{1/4} \quad (10^5 < Ra_L < 10^{10}) \tag{7.78}$$

Equations (7.77)–(7.78) are for isothermal surfaces. The same correlations can be used for uniform flux surfaces, noting that in those cases $\overline{Nu_L}$ and Ra_L would be based on the *L-averaged* temperature difference between the surface and the surrounding fluid. The flux Rayleigh number Ra_L^* can be made to appear on the right side of these correlations by noting one more time the substitution $Ra_L = Ra_L^*/\overline{Nu_L}$.

Example 7.2 *Hot Plate Facing Upward in a Colder Fluid*
The temperature of a horizontal plate in a water pool is $T_w = 43.1\,°C$, while the temperature of the water pool is $T_\infty = 23.6\,°C$. The plate is a square with side $a = 8.9\,cm$. Calculate the rate at which the plate must be heated electrically to maintain its temperature.

Solution

The film temperature is $(43.1\,°C + 23.6\,°C)/2 = 33.4\,°C$ and, after interpolating linearly in Appendix C, we find that the relevant water properties are

$$k = 0.62 \frac{W}{m \cdot K} \qquad \frac{g\beta}{\alpha v} = \frac{28\,982}{cm^3 \cdot K}$$

The characteristic length of the upward facing square of side a is

$$L = \frac{A}{p} = \frac{a^2}{4a} = \frac{8.9\,cm}{4} = 2.23\,cm$$

with the corresponding Rayleigh number

$$Ra_L = \frac{g\beta}{\alpha v} L^3 (T_w - T_\infty)$$
$$= \frac{28\,982}{cm^3 \cdot K} (2.23\,cm)^3 (43.1 - 23.6)K = 6.23 \times 10^6$$

The appropriate heat transfer correlation is Eq. (7.77a):

$$\overline{Nu_L} = 0.54 Ra_L^{1/4}$$
$$= 0.54 (6.23 \times 10^6)^{1/4} = 26.97$$

$$\overline{h} = \overline{Nu_L} \frac{k}{L}$$
$$= 26.97 \times 0.62 \frac{W}{m \cdot K} \frac{1}{0.0223\,m} = 750\,W/m^2 \cdot K$$

$$q = \overline{h} a^2 (T_w - T_\infty)$$
$$= 750 \frac{W}{m^2 \cdot K} (0.089\,m)^2 (43.1 - 23.6)\,K = 116\,W$$

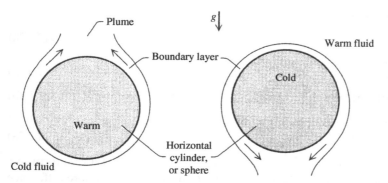

Figure 7.9 Horizontal cylinder or sphere immersed in a fluid at a different temperature.

7.3.4 Horizontal Cylinder

The natural flow around an isothermal cylinder positioned horizontally in a fluid reservoir, Figure 7.9, is similar to the flow along a vertical surface. The only difference is the fact that the wall is curved, and instead of the wall height y (or H), the vertical dimension is the cylinder diameter D. These similarities explain why the heat transfer correlation for horizontal cylinders [9],

$$\overline{Nu}_D = \left\{ 0.6 + \frac{0.387\, Ra_D^{1/6}}{[1 + (0.559/Pr)^{9/16}]^{8/27}} \right\}^2 \tag{7.79}$$

has the same form as the vertical wall correlation (7.61). Equation (7.79) is valid for $10^{-5} < Ra_D < 10^{12}$ and the entire Prandtl number range. The average Nusselt number and the Rayleigh number are both based on diameter:

$$\overline{Nu}_D = \frac{\overline{q}''_{w,D}}{\Delta T} \frac{D}{k} \qquad Ra_D = \frac{g\beta\Delta T D^3}{\alpha\nu} \tag{7.80}$$

7.3.5 Sphere

The flow around a sphere suspended in a pool at a different temperature has the general features outlined in Figure 7.9. The vertical dimension of the spherical body is its diameter D, on which both $\overline{Nu}_D$ and Ra_D are based (cf. Eqs. (7.80)). Heat transfer measurements are correlated for $Pr \gtrsim 0.7$ and $Ra_D < 10^{11}$ by the formula [20]

$$\overline{Nu}_D = 2 + \frac{0.589 Ra_D^{1/4}}{[1 + (0.469/Pr)^{9/16}]^{4/9}} \tag{7.81}$$

7.3.6 Vertical Cylinder

There are three surfaces for heat transfer in the vertical cylinder geometry, the top and bottom discs, which can be handled according to Section 7.3.3, and the lateral surface. The boundary layer flow that develops over the lateral surface is illustrated in Figure 7.10. When the boundary layer thickness δ_T is much smaller than the cylinder diameter D, the curvature of the lateral surface does not play a role, and the Nusselt number

Figure 7.10 Vertical cylinders with boundary layer flow on the lateral surface.

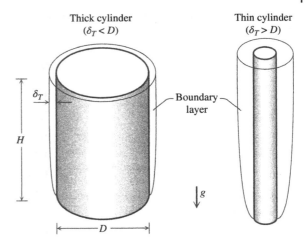

can be calculated with the vertical wall formulas (7.61)–(7.62). Note that if H is the height of the cylinder and if $Pr \gtrsim 1$, the $\delta_T < D$ criterion requires

$$\frac{D}{H} > Ra_H^{-1/4} \tag{7.82}$$

This inequality and the simplified heat transfer calculation recommended by it may be regarded as the "thick cylinder" limit. The opposite limit is illustrated on the right side of Figure 7.10. An integral solution that accounts for the effect of wall curvature in the laminar regime is [21]

$$\overline{Nu}_H = \frac{4}{3}\left[\frac{7Ra_H Pr}{5(20 + 21Pr)}\right]^{1/4} + \frac{4(272 + 315\,Pr)H}{35(64 + 63\,Pr)D} \tag{7.83}$$

where $\overline{Nu}_H = \bar{h}H/k, Ra_H = g\beta\Delta T H^3/\alpha\nu, \bar{h}$ is the wall-averaged heat transfer coefficient, and ΔT is the temperature difference between the surface and the fluid reservoir.

7.3.7 Other Immersed Bodies

In Section 7.3, we have reviewed several of the simplest and most common body shapes that are encountered in calculations of external natural convection. The heat transfer from bodies with less regular shapes follows $\overline{Nu}$ that are similar to what we have seen thus far. The simplest generalization was proposed by Lienhard [22]:

$$\overline{Nu}_l \cong 0.52 Ra_l^{1/4} \tag{7.84}$$

where $\overline{Nu}_l = \bar{h}l/k, Ra_l = g\beta\Delta T l^3/\alpha\nu$, and $\bar{h}$ is the heat transfer coefficient averaged over the entire surface of the body. The length l, on which both Nu_l and Ra_l are based, is the distance traveled by the boundary layer fluid while in contact with the body. In the case of a horizontal cylinder, for example, $l = \pi D/2$, Eq. (7.84) should be accurate within 10%, provided that $Pr \gtrsim 0.7$ and the Rayleigh number is sufficiently large so that the boundary layer is thin. Equation (7.84) was tested in an experiment in which a vertical cylinder of equal height and diameter ($H = D$, hence $l = 2D$) was suspended in air [23].

Yovanovich [24] developed a correlation that covers the entire laminar range, $0 < Ra_L < 10^8$, that is, including the limit of pure conduction $Ra_L \to 0$. As length scale, he used the square root of the entire surface of the

Table 7.1 Constants in Eq. (7.86) for laminar natural convection on immersed bodies.

Body shape	$\overline{Nu}_{\mathcal{L}}^{0}$	$G_{\mathcal{L}}$
Sphere	3.545	1.023
Bisphere	3.475	0.928
Cube 1	3.388	0.951
Cube 2	3.388	0.990
Cube 3	3.388	1.014
Vertical cylinder[a]	3.444	0.967
Horizontal cylinder[a]	3.444	1.019
Cylinder[a] at 45°	3.444	1.004
Prolate spheroid ($C/B = 1.93$)	3.566	1.012
Oblate spheroid ($C/B = 0.5$)	3.529	0.973
Oblate spheroid ($C/B = 0.1$)	3.342	0.768

a) Short cylinder, $H = D$.

immersed body

$$\mathcal{L} = A^{1/2} \tag{7.85}$$

and defined $\overline{Nu}_{\mathcal{L}} = \overline{h}\mathcal{L}/k$ and $Ra_{\mathcal{L}} = g\beta\Delta T\mathcal{L}^3/\alpha v$. The correlation has two constants,

$$\overline{Nu}_{\mathcal{L}} = \overline{Nu}_{\mathcal{L}}^{0} + \frac{0.67 G_{\mathcal{L}} Ra_{\mathcal{L}}^{1/4}}{[1 + (0.492/Pr)^{9/16}]^{4/9}} \tag{7.86}$$

namely, the conduction limit Nusselt number $\overline{Nu}_{\mathcal{L}}^{0}$ and the geometric parameter $G_{\mathcal{L}}$. The latter is a weak function of body shape, aspect ratio, and orientation in the gravitational field. Table 7.1 lists the two constants for the bodies and orientations shown in Figure 7.11. These values do not vary appreciably; therefore, a general expression based on the average values of $\overline{Nu}_{\mathcal{L}}^{0}$ and $G_{\mathcal{L}}$, and valid for $Pr \gtrsim 0.7$, is

$$\overline{Nu}_{\mathcal{L}} \cong 3.47 + 0.51 Ra_{\mathcal{L}}^{1/4} \tag{7.87}$$

A more extensive correlation that covers the conduction, laminar, and turbulent regimes was developed by Hassani and Hollands [25]. This correlation employs two length scales, one of which is $\mathcal{L}$, Eq. (7.85). It covers the Rayleigh number range 0–10^{14}. Experimental and numerical results for heat transfer from horizontal discs and rings were published by Sahraoui et al. [26].

Example 7.3 *Horizontal Cylinder: Laminar Natural Convection*

A horizontal cylinder has the diameter $D = 6$ cm and length $L = 60$ cm. Its surface is isothermal at $T_w = 34\,°C$, while the surrounding air has the temperature $T_\infty = 25\,°C$. Verify numerically that the boundary layer is laminar. Calculate the heat transfer rate released by the cylindrical surface.

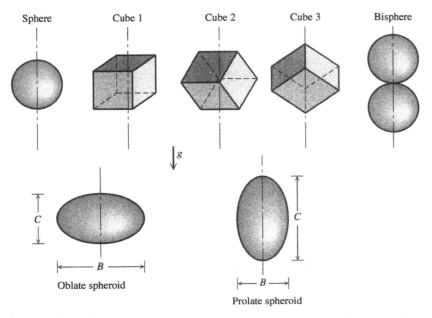

Sphere   Cube 1   Cube 2   Cube 3   Bisphere

g

C B
Oblate spheroid

C B
Prolate spheroid

Figure 7.11   Shapes and orientations of isothermal bodies immersed in a fluid at a different temperature (Table 7.1).

Solution

The film temperature of the air boundary layer is $(34\,°C + 25\,°C)/2 \cong 30\,°C$; therefore, the physical properties that will be needed are

$$Pr = 0.72 \quad k = 0.026\frac{W}{m \cdot K} \qquad \frac{g\beta}{\alpha v} = \frac{90.7}{cm^3 \cdot K}$$

The Rayleigh and Grashof numbers based on diameter

$$Ra_D = \frac{g\beta}{\alpha v}D^3(T_w - T_\infty)$$
$$= \frac{90.7}{cm^3 \times K}(6\,cm)^3(34 - 25)K = 1.76 \times 10^5$$
$$Gr_D = \frac{Ra_D}{Pr} = \frac{1.76 \times 10^5}{0.72} = 2.45 \times 10^5 < 10^9$$

confirm that the boundary layer is laminar. The thermal boundary layer thickness is approximately

$$\delta_T \sim DRa_D^{-1/4} = 6\,cm(1.76 \times 10^5)^{-1/4}$$
$$= 0.3\,cm$$

which shows that the slenderness ratio δ_T/D is of order 1/20. To calculate the total heat transfer rate released by the cylinder, we turn our attention to Eq. (7.79), in which we substitute $Pr = 0.72$:

$$\overline{Nu}_D = \left(0.6 + 0.322Ra_D^{1/6}\right)^2 = 9.06$$
$$\overline{q}''_{w,D} = \overline{Nu}_D\frac{\Delta T}{D}k$$

$$= 9.06 \frac{9\text{K}}{0.06 \text{ m}} 0.026 \frac{\text{W}}{\text{m} \cdot \text{K}} = 35.3 \frac{\text{W}}{\text{m}^2}$$

$$A = \pi DL$$

$$= \pi(0.06 \text{ m})(0.6 \text{ m}) = 0.113 \text{ m}^2$$

$$q = \overline{q''}_{w,D} A$$

$$= 35.3 \frac{\text{W}}{\text{m}^2}(0.113 \text{ m}^2) = 4 \text{ W}$$

7.4 Internal Flows

7.4.1 Vertical Channels

Consider situations where the walls surround the buoyant fluid. One example is the vertical gap formed between two heated plates, Figure 7.12, which is a model for the air space between two vertically mounted circuit boards in an electronic package or for the space between two vertical plate fins. The plates are modeled as isothermal (T_w), and so is the fluid situated outside the channel (T_∞). It is assumed that the plates are heated ($T_w > T_\infty$) and that the fluid rises through the chimney formed between them.

The channel has two length scales: the height H and the plate-to-plate spacing L. When the thermal boundary layers that coat each plate (δ_T) are considerably thinner than the plate-to-plate spacing, the heat transfer rate from the plates to the channel fluid can be calculated with the single-wall formulas (7.61)–(7.62) or, if they are modeled as uniform flux, with Eqs. (7.67)–(7.70). In $Pr \gtrsim 1$ fluids, the "wide channel" limit represented by $\delta_T < L$ is represented by

$$\frac{L}{H} > Ra_H^{-1/4} \quad \text{or} \quad \frac{L}{H} > Ra_L^{-1} \tag{7.88}$$

Of special interest is the "narrow channel" limit in which the above inequalities break down. Figure 7.12 shows that when the channel is narrow enough, the velocity profiles merge into a single profile that resembles that of Hagen–Poiseuille flow (Figure 6.3, right side). Even though the temperature of the fluid that enters through the bottom opening is equal to T_∞, when the channel is narrow and tall enough, the fluid temperature $T(x, y)$ is nearly the same as the plate temperature T_w. We record this observation as an inequality between temperature differences:

$$T_w - T(x, y) < T_w - T_\infty \tag{7.89}$$

This inequality means that the duct flow that proceeds vertically through the channel is thermally fully developed over most of the channel height H. Assuming further that the Prandtl number is of order 1 (e.g. air) or greater, the inequality (7.89) means also that the channel flow is hydrodynamically fully developed. Therefore, with reference to the momentum Eq. (7.17), which applies at every point in the channel, the hydrodynamic development of the flow means the flow is vertical ($u = 0$) and, consequently, $\partial v/\partial y = 0$. [Review Eqs. (6.10)–(6.14), and keep in mind that in the present problem the longitudinal direction is y.] In conclusion, Eq. (7.17) reduces to

$$0 = \nu \frac{d^2 u}{dx^2} + g\beta(T - T_\infty) \tag{7.90}$$

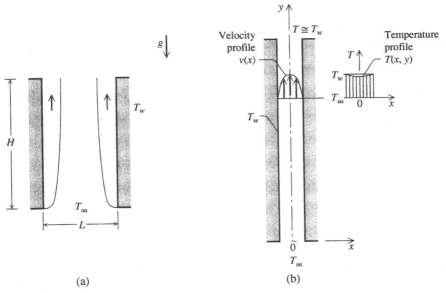

Figure 7.12 Vertical channel with isothermal walls; the top and bottom ends communicate with a reservoir of isothermal fluid. (a) Wide channel and (b) narrow channel.

and, in view of Eq. (7.89),

$$\frac{d^2v}{dx^2} \cong -\frac{g\beta}{\nu}(T_w - T_\infty) = \text{constant} \tag{7.91}$$

Note that Eq. (7.90) follows from Eq. (7.17), which was developed for an external flow with $dP_\infty/dy = -\rho_\infty g$, Eq. (7.9). In the present geometry, $\partial P/\partial y$ is a constant equal to dP/dy because the flow is assumed to be fully developed. [Review Eqs. (6.12) and (6.14).] It is because both ends of the channel are open that the dP/dy constant inside the channel is equal to (i.e. controlled by) the hydrostatic pressure gradient of the surroundings:

$$dP/dy = dP_\infty/dy = -\rho_\infty g.$$

Equation (7.91) is equivalent to the momentum Eq. (6.20) of forced convection through a parallel-plate channel. The role of the forced convection constant $(1/\mu)\, dP/dx$ is now played by the natural convection constant $-g\beta\Delta T/\nu$, where $\Delta T = T_w - T_\infty$. Solving Eq. (7.91) subject to the no-slip conditions that are evident in Figure 7.12, we obtain the parabolic velocity distribution:

$$v = \frac{g\beta\Delta T L^2}{8\nu}\left[1 - \left(\frac{x}{L/2}\right)^2\right] \tag{7.92}$$

and the vertical mass flowrate per unit length in the direction normal to Figure 7.12:

$$\dot{m}' = \int_{-L/2}^{L/2} \rho v\, dx = \frac{\rho g\beta\Delta T L^3}{12\nu} \tag{7.93}$$

An important observation is that in fully developed laminar flow, the vertical velocity and mass flowrate are independent of the channel height H. Although this may seem strange, it is a consequence of the fact that the net longitudinal (i.e. vertical) pressure gradient that drives the chimney flow is independent of H. By reason of the analogy between Eqs. (7.91) and (6.20), that net pressure gradient is $-\rho g \beta (T_w - T_\infty)$.

The total heat transfer rate extracted by the $\dot{m}'$ stream from the two vertical surfaces that touch it is (again, per unit length normal to the plane of Figure 7.17)

$$q' = \dot{m}' c_P (T_w - T_\infty) = \frac{\rho g \beta c_P (\Delta T)^2 L^3}{12 \nu} \tag{7.94}$$

This result can be nondimensionalized by defining the channel-averaged heat flux (note that the channel has two sides):

$$\overline{q''} = \frac{q'}{2H} \tag{7.95}$$

and the average Nusselt number:

$$\overline{Nu_H} = \frac{\overline{q''}}{\Delta T} \frac{H}{k} = \frac{1}{24} Ra_L \tag{7.96}$$

The right side of Eq. (7.96) follows from using Eq. (7.94). This $\overline{Nu_H}$ result is valid when the inequality (7.89) prevails over most of the channel: the two sides of that inequality are, respectively, represented by the scales

$$\frac{\overline{q''} L}{k} < \Delta T \tag{7.97}$$

Dividing by ΔT and using the last of Eq. (7.96), we arrive at

$$Ra_L < \frac{H}{L} \tag{7.98}$$

which is the opposite of the wide-channel criterion (7.88). In conclusion, the $\overline{Nu_H}$ result (7.96) holds in the narrow-channel limit represented by the inequality (7.98). Results similar to Eq. (7.96) have been obtained in the narrow-channel limit of vertical ducts with other cross-sectional shapes (Table 7.2) [1]. In each case the ratio $\overline{Nu_H}/Ra_{D_h}$ is a constant, D_h being the hydraulic diameter of the particular cross section, and $\overline{Nu_H}$ is the channel-averaged Nusselt number defined by the first of Eq. (7.96). The height of the channel in all cases is H.

Table 7.2 Average Nusselt numbers for chimney flow in the narrow-channel limit.

Cross section shape	$\overline{Nu_H}/Ra_{D_h}$
Parallel plates	$\dfrac{1}{192}$
Circular	$\dfrac{1}{128}$
Square	$\dfrac{1}{113.6}$
Equilateral triangle	$\dfrac{1}{106.4}$

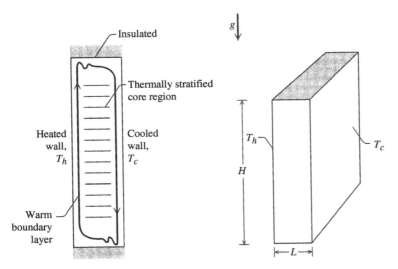

Figure 7.13 Enclosure filled with fluid and heated and cooled along the two lateral walls.

7.4.2 Enclosures Heated from the Side

Another class of internal natural convection configurations are the flows induced in enclosed spaces that are subjected to temperature differences in the horizontal direction. An example is the circulation of air in the slot of a double-pane window. Figure 7.13 shows that the circulation consists of a vertical jet that rises along the heated wall and of a cold jet that descends along the cooled wall. The circulation is completed by two horizontal streams that proceed in counterflow along the top and bottom walls. It is assumed that the flow is practically two-dimensional, in other words, that the enclosure is sufficiently wide in the direction normal to the plane of Figure 7.13.

A significant volume of research has been published on the heat transfer rate between the differentially heated walls of the enclosure (T_h, T_c). The general trends revealed by this research have been sorted out in Ref. [1]. Begin with Figure 7.14, which shows that in $Pr \gtrsim 1$ fluids the flow pattern depends on the Rayleigh number $Ra_H = g\beta(T_h - T_c)H^3/\alpha\nu$ and the geometric aspect ratio H/L.

The cavity is "wide" when the vertical thermal boundary layer thickness δ_T is smaller than the horizontal dimension L. The condition $\delta_T < L$ leads to the inequality (7.88), which is represented by the rising diagonal in Figure 7.14. To the right of this diagonal, the vertical boundary layers are distinct and, to a large degree, the overall heat transfer rate can be predicted with the formulas of Section 7.2.4. (Note that in the present problem the temperature difference between one wall and the "fluid reservoir" would be $\Delta T/2$, where $\Delta T = T_h - T_c$.) Consequently, in the laminar regime, we expect a proportionality between the average Nusselt number and $Ra_H^{1/4}$, but there is no effect due to L. The Berkovsky–Polevikov correlations[5] recommended by Catton [27] deviate only slightly from this rule:

$$\overline{Nu}_H = 0.22\left(\frac{Pr}{0.2 + Pr}Ra_H\right)^{0.28}\left(\frac{L}{H}\right)^{0.09}$$

$$2 < \frac{H}{L} < 10, \qquad Pr < 10^5, \ Ra_H < 10^{13} \tag{7.99}$$

---

5 In Ref. [27], the Berkovsky–Polevikov correlations show a deceptively strong L/H effect, because the Nusselt and Rayleigh numbers in that paper were based on L as length scale, namely, $\overline{Nu}_L$ and Ra_L.

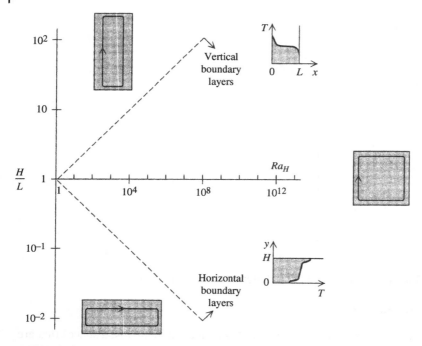

Figure 7.14 Flow regimes for natural convection in enclosures heated from the side $Pr \gtrsim 1$ fluid. Source: Bejan [1].

$$\overline{Nu}_H = 0.18\left(\frac{Pr}{0.2 + Pr}Ra_H\right)^{0.29}\left(\frac{L}{H}\right)^{-0.13}$$

$$1 < \frac{H}{L} < 2, \quad 10^{-3} < Pr < 10^5, \quad 10^3 < \frac{Pr}{0.2 + Pr}Ra_H\left(\frac{L}{H}\right)^3 \tag{7.100}$$

where $\overline{Nu}_H = q''H/k\,\Delta T$. These correlations are valid in the indicated (H/L, Pr, Ra_H) domain and in the "wide" cavity limit, Eq. (7.88).

In the opposite extreme, $L/H < Ra_H^{-1/4}$, the cavity is too narrow to allow distinct vertical thermal boundary layers. Instead, the boundary layers are squeezed together, giving birth to a linear temperature variation in the horizontal direction. This feature occurs at large values of H/L, that is, to the left of the rising diagonal of Figure 7.14. The linear temperature profile means that in this limit $\overline{Nu}_H$ approaches the constant H/L:

$$\overline{Nu}_H = \overline{q}''\frac{H}{k\Delta T} \;\;\rightarrow\;\; \left(k\frac{\Delta T}{L}\right)\frac{H}{k\Delta T} = \frac{H}{L} \tag{7.101}$$

The preceding discussion referred to square or tall enclosures, where $H/L \geq 1$. The flows in shallow enclosures (Figure 7.14, $H/L < 1$) can be divided into flows with distinct horizontal jets and flows without distinct jets along the top and bottom walls. The overall Nusselt number is reported in Figure 7.15, which is based on an integral analysis of the natural circulation in the entire cavity [28]. At sufficiently low Rayleigh numbers Ra_H, the Nusselt number $\overline{Nu}_H$ approaches again the pure conduction constant H/L. This limit corresponds to the wedge-shaped domain situated to the left of the descending diagonal in Figure 7.14.

A theoretical solution for the flow and temperature field is possible [29] when the side walls are heated and, respectively, cooled with uniform heat flux q''. In the boundary layer regime, the temperature increases

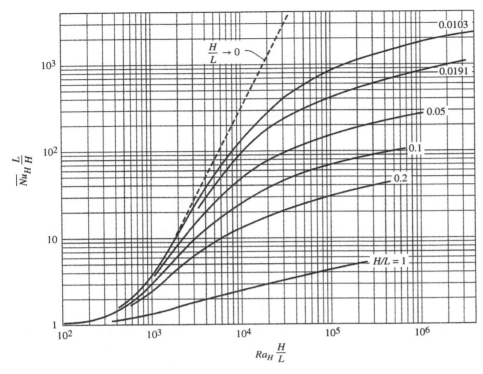

Figure 7.15 Shallow enclosures heated from the side: average Nusselt number for $Pr \gtrsim 1$ fluids.

linearly in the vertical direction along the heated wall, the cooled wall, and in the core region:

$$\frac{\partial T}{\partial y} = 0.0425 \frac{\alpha \nu}{g \beta H^4} \left(\frac{H}{L}\right)^{4/9} Ra_H^{*8/9} \quad \text{(constant)} \tag{7.102}$$

Since the temperature increases at the same rate in the vertical direction along both walls, the wall-to-wall temperature difference is a constant at every level, $T_h(y) - T_c(y) = \Delta T$. The theoretical solution for the average Nusselt number $\overline{Nu}_H = q'' H/(\Delta T k)$ in the boundary layer regime in $Pr \gtrsim 1$ fluids is [29]

$$\overline{Nu}_H = 0.34 \, Ra_H^{*2/9} \left(\frac{H}{L}\right)^{1/9} \tag{7.103a}$$

where $Ra_H^* = g \beta H^4 q''/(\alpha \nu k)$. If, instead of the flux Rayleigh number Ra_H^*, we use the ΔT-based Rayleigh number $Ra_H = Ra_H^*/\overline{Nu}_H = g \beta \Delta T H^3/(\alpha \nu)$, then Eq. (7.103a) becomes

$$\overline{Nu}_H = 0.25 \, Ra_H^{2/7} \left(\frac{H}{L}\right)^{1/7} \tag{7.103b}$$

Because $Ra_H^{2/7} = Ra_H^{0.286}$, this theoretical alternative reproduces almost all the features of the empirical correlations (7.99)–(7.100) recommended for enclosures with isothermal side walls. This reinforces the observation that the heat transfer correlations developed for a system with isothermal walls apply reasonably well to the uniform flux configuration, provided the Ra_H number is based then on the wall-averaged temperature difference, $Ra_H = Ra_H^*/Nu_H$.

There are many applications in which the enclosed fluid departs from the single-enclosure model used throughout this subsection 7.4.2 (e.g. rooms in buildings, solar collectors, lakes). More advanced models for

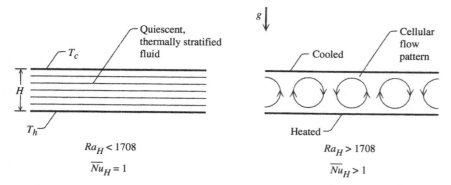

Figure 7.16 Horizontal fluid layer between parallel walls and heated from below.

the description of such systems are partially divided enclosure and two enclosures communicating laterally through a doorway, window, or corridor, or over an incomplete dividing wall [1].

7.4.3 Enclosures Heated from Below

The fundamental difference between enclosures heated from the side (Figure 7.13) and enclosures heated from below (Figure 7.16) is that in the former, a buoyancy-driven flow is present as soon as a very small temperature difference $(T_h - T_c)$ is imposed between the two side walls, while in the latter, the imposed temperature difference must exceed a finite critical value before the first signs of fluid motion and convective heat transfer are detected.

When the enclosure is sufficiently long and wide in the horizontal direction, the condition for the onset of convection is expressed by the critical Rayleigh number [30]:

$$Ra_H \gtrsim 1708 \tag{7.104}$$

where $Ra_H = g\beta(T_h - T_c)H^3/(\alpha\nu)$. As suggested in Figure 7.16, immediately above $Ra_H \cong 1708$, the flow consists of counterrotating two-dimensional rolls, which cross sections are almost square. This flow pattern is commonly recognized as Bénard cells, or Bénard convection, in honor of H. Bénard who reported the first investigation of this phenomenon in 1900. The cellular flow becomes considerably more complicated as Ra_H exceeds by one or more orders of magnitude, which is a critical (convection onset) value. The two-dimensional rolls break up into three-dimensional cells, which appear hexagonal in shape when viewed from above (Figure 7.17). At even higher Rayleigh numbers, the cells multiply (become narrower) and, eventually, the flow becomes oscillatory and turbulent [1].

The heat transfer effect of the cellular flow is to augment the net heat transfer rate in the vertical direction, that is, to increase it above the pure conduction rate that would prevail in the absence of fluid motion. The dimensionless number that measures this augmentation effect is the average Nusselt number based on the vertical dimension H, namely, $\overline{Nu}_H = \overline{q''}H/(k\Delta T)$. Experimental heat transfer measurements in the range $3 \times 10^5 < Ra_H < 7 \times 10^9$ support the correlation [31]

$$\overline{Nu}_H = 0.069\, Ra_H^{1/3} Pr^{0.074} \tag{7.105}$$

The physical properties needed for calculating $\overline{Nu}_H$, Ra_H, and Pr are evaluated at the average fluid temperature $(T_h + T_c)/2$. Equation (7.105) holds when the horizontal layer is sufficiently wide so that the effect of the short vertical sides is minimal.

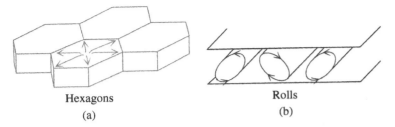

Hexagons
(a)

Rolls
(b)

Figure 7.17 Two-dimensional rolls and three-dimensional hexagonal cells in a fluid layer heated from below. (a) Free surface and (b) rigid surface.

The experimental data correlated by Eq. (7.105) line up in such a way that the exponent of Ra_H in that correlation increases slightly as Ra_H increases. The exponent is closer to 0.29 at the low Ra_H end of the correlation (roughly, when $Ra_H \lesssim 10^8$). Its value becomes practically equal to 1/3 at higher Rayleigh numbers; in other words $\overline{Nu_H} \sim Ra_H^{1/3}$ when Ra_H exceeds approximately 10^8. The proportionality $\overline{Nu_H} \sim Ra_H^{1/3}$ persists as Ra_H increases above the range covered by Eq. (7.105).

The $\overline{Nu_H} \sim Ra_H^{1/3}$ proportionality expressed by Eq. (7.105) means that the actual heat transfer rate $(\overline{q''})$ from T_h to T_c is independent of the layer thickness H. The $\overline{Nu_H} \sim Ra_H^{1/3}$ proportionality was predicted theoretically [1] by performing the scale analysis of a single roll (cell) of the Bénard convection pattern. An even simpler alternative derivation is presented in Example 7.5.

At subcritical Rayleigh numbers, $Ra_H \lesssim 1708$, the fluid is quiescent and the temperature decreases linearly from T_h to T_c. The heat transfer rate across the fluid layer is by pure conduction; therefore, $\overline{Nu_H} = 1$. The convection onset criterion (7.104) – the value 1708 on the right side – refers strictly to an infinite and mathematically horizontal layer with rigid (no-slip) and isothermal top and bottom boundaries. Similar criteria, with critical Rayleigh numbers of the order of 10^3, hold for other horizontal layer configurations (i.e. combinations of top and bottom boundary conditions). It is important to note that as soon as the orientation deviates from the horizontal, natural convection sweeps the long walls even in the limit of an infinitesimally small temperature difference across the small dimension of the enclosure.

Example 7.4 *The Origin of Thermals Rising from a Heated Surface*

We are in a position to estimate the time interval between the formation of two successive balls of heated fluid (thermals) that rise from the same spot on the hot surface, Figure E7.4. As in Example 7.2, assume that the bottom surface and water pool temperatures are 43.1 and 23.6 °C, respectively. The following order-of-magnitude calculation is based on the observation that the surface heats by conduction a thin layer of water. The thickness of this layer increases in time until its thickness-based Rayleigh number reaches the critical level for the onset of convection. Thermals are the aftermath of the onset of convection in the conduction layer that grew over the surface.

Solution

The water properties that will be used are evaluated at the film temperature of 33.4 °C:

$$\alpha = 0.00148 \frac{cm^2}{s} \quad \frac{g\beta}{\alpha \nu} = \frac{28{,}982}{cm^3 \cdot K}$$

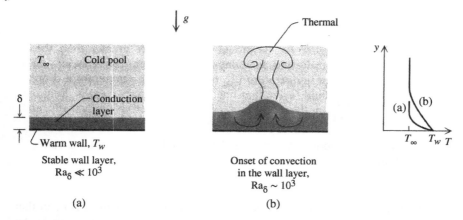

Figure E7.4

Assume that in the beginning the water is motionless. The water that comes in contact with the hot surface develops a conduction layer, which thickness increases as (see Chapter 4)

$$\delta \sim (\alpha t)^{1/2} \tag{1}$$

This δ-tall layer is heated from below, and, in accordance with a criterion similar to Eq. (7.104), it becomes unstable when its Rayleigh number based on height exceeds the order of magnitude 10^3:

$$Ra_\delta \sim 10^3 \tag{2}$$

$$\frac{g\beta}{\alpha\nu}\delta^3(T_w - T_\infty) \sim 10^3 \tag{3}$$

$$\delta^3 \sim 10^3 \frac{\text{cm}^3 \cdot \text{K}}{28\ 982} \frac{1}{(43.1 - 23.6)\text{K}} = 0.0018\,\text{cm}^3 \tag{4}$$

or $\delta \sim 1$ mm. In view of Eq. (1), this result means that

$$\alpha t \sim 0.0146\,\text{cm}^2 \tag{5}$$

$$t \sim \frac{0.0146\,\text{cm}^2}{0.00148\,\text{cm}^2/\text{s}} \cong 10\,\text{s} \tag{6}$$

In conclusion, thermals will rise from the same spot on the heated surface at time intervals of the order of 10 seconds (Figure E7.4).

Example 7.5 *Turbulent Bénard Convection: Why $\overline{Nu}_H$ Increases as $Ra_H^{1/3}$ and Why the Heat Transfer Rate Is Independent of H*

When the Rayleigh number Ra_H is orders of magnitude greater than the critical value, convection in the bottom-heated fluid layer is turbulent. The core of the fluid layer is practically at the average temperature $(T_h + T_c)/2$, while temperature drops of size $(T_h - T_c)/2$ occur across thin fluid layers that line the two horizontal walls (Figure E7.5). Rely on the scales of the thermals analyzed in the preceding example to predict the heat transfer correlation (7.105).

Solution

In the high Rayleigh number regime, the turbulence over most of the horizontal layer of thickness H is caused by the buckling thermals that rise from the heated bottom and those that fall from the cooled top. In this way, the turbulent core of the layer is sandwiched between two δ-thin conduction layers of the type described in the preceding example. There, we learned that the δ thickness is such that, in an order of magnitude sense (Figure E7.5),

$$Ra_\delta \sim 10^3 \tag{1}$$

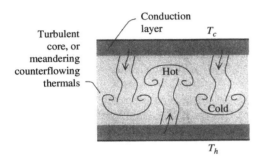
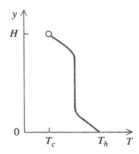

Figure E7.5

This is equivalent to writing

$$Ra_H \sim 10^3 \left(\frac{H}{\delta}\right)^3 \tag{2}$$

where Ra_H is the usual Rayleigh number, $Ra_H = g\beta\Delta T H^3/(\alpha\nu)$, and $\Delta T = T_h - T_c$. The heat transfer from T_h to T_c is impeded by layers of thickness δ (not H); therefore,

$$\overline{q''} \sim k\frac{\Delta T}{\delta} \tag{3}$$

$$\overline{Nu}_H = \frac{\overline{q''}}{k\Delta T/H} \sim \frac{H}{\delta} \tag{4}$$

Equation (3) shows that the heat transfer rate does not depend on H. By eliminating H/δ between Eqs. (2) and (4), we predict that

$$\overline{Nu}_H \sim 10^{-1} Ra_H^{1/3} \tag{5}$$

It is indeed rewarding to see that this simple theoretical result reproduces almost all the features of the empirical correlation recommended for turbulent (high Ra_H) Bénard convection calculations, Eq. (7.105).

7.4.4 Inclined Enclosures

Consider now the tilted enclosure configuration of Figure 7.18, in which L is the distance measured in the direction of the imposed temperature difference $T_h - T_c$. The angle made with the horizontal direction, τ, is defined such that in the range $0° < \tau < 90°$ the heated surface is positioned below the cooled surface. The enclosure heated from the side, Figure 7.13, is the $\tau = 90°$ case of the inclined enclosure of Figure 7.18.

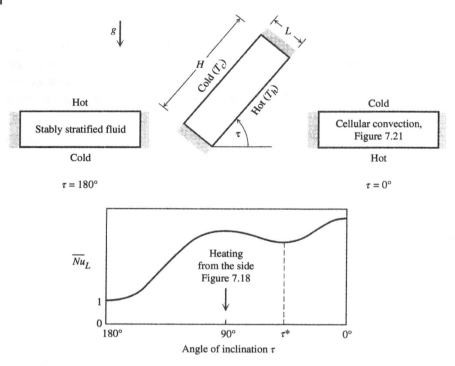

Figure 7.18 The effect of tilt angle on the heat transfer rate and flow pattern in an enclosure.

The $\tau = 0°$ case represents the rectangular enclosure heated from below (Figure 7.16), except that now the distance between the differentially heated walls is labeled L.

The inclination angle τ has a dramatic effect on the flow and heat transfer in the enclosure. As τ decreases from 180° to 0°, the heat transfer mechanism switches from pure conduction at $\tau = 180°$ to single-cell convection at $\tau = 90°$ and, finally, to Bénard convection at $\tau = 0°$. The conduction-referenced Nusselt number $\overline{Nu}_L = q''/(k\Delta T)$ rises from the pure conduction level $\overline{Nu}_L(180°) = 1$ to a maximum $\overline{Nu}_L$ value near $\tau = 90°$. As τ decreases below 90°, the Nusselt number decreases and passes through a local minimum at a special tilt angle $\tau = \tau*$, which is a function of the geometric aspect ratio of the enclosure [32]:

H/L	1	3	6	12	>12
$\tau *$	25°	53°	60°	67°	70°

As τ continues to decrease below $\tau = \tau*$, the heat transfer rate rises toward another maximum associated with the Bénard convection regime, $\overline{Nu}_L(0°)$. The $\overline{Nu}_L(\tau)$ curve is illustrated qualitatively in the lower part of Figure 7.18. The curve is represented well by the formulas [27]

$$\overline{Nu}_L(\tau) = 1 + [\overline{Nu}_L(90°) - 1]\sin\tau \quad 180° > \tau > 90° \tag{7.106a}$$

$$\overline{Nu}_L(\tau) = \overline{Nu}_L(90°)(\sin\tau)^{1/4} \quad 90° > \tau > \tau* \tag{7.106b}$$

$$\overline{Nu}_L(\tau) = \left[\frac{\overline{Nu}_L(90°)}{\overline{Nu}_L(0°)}(\sin\tau*)^{1/4}\right]^{\tau/\tau*} \quad \tau* > \tau > 0° \text{ and } H/L < 10 \tag{7.106c}$$

Figure 7.19 Natural convection in the annular space between horizontal concentric cylinders.

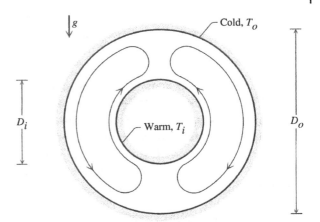

$$\overline{Nu_L}(\tau) = 1 + 1.44\left(1 - \frac{1708}{Ra_L \cos \tau}\right)^*\left[1 - \frac{(\sin 1.8\tau)^{1.6} \times 1708}{Ra_L \cos \tau}\right] +$$

$$\left[\left(\frac{Ra_L \cos \tau}{5830}\right)^{1/3} - 1\right]^* \qquad \tau^* > \tau > 0° \text{ and } H/L > 10 \qquad (7.106d)$$

The quantities contained between the parentheses with asterisk, $()^*$, must be set equal to zero if they become negative. The Rayleigh number is based on the distance between the differentially heated walls, $Ra_L = g\beta(T_h - T_c)L^3/(\alpha v)$.

7.4.5 Annular Space Between Horizontal Cylinders

The flow generated between concentric horizontal cylinders at different temperatures (T_i, T_0) has features similar to the circulation in an enclosure heated from the side. As illustrated in Figure 7.19, in the laminar regime, two counterrotating kidney-shaped cells are positioned symmetrically about the vertical plane drawn through the cylinder centerline. Because the two cells are being heated and cooled from the side, the overall heat transfer correlation should have the features of Eqs. (7.99) and (7.100). The role of vertical dimension in this case can be played by either of the two diameters D_i and D_o: when D_i is sensibly smaller than D_o, the overall heat transfer rate is impeded by the smallness of D_i rather than D_o.

These expectations are confirmed by a theoretical interpretation [1][6] of a correlation developed by Raithby and Hollands [33]:

$$q' \cong \frac{2.425k(T_i - T_o)}{[1 + (D_i/D_o)^{3/5}]^{5/4}}\left(\frac{PrRa_{D_i}}{0.861 + Pr}\right)^{1/4} \qquad (7.107)$$

This q' (W/m) expression refers to the total heat transfer rate between the two cylinders, per unit length in the direction normal to the plane of Figure 7.19. The Rayleigh number is based on the cylinder-to-cylinder temperature difference and on the inner diameter, $Ra_{D_i} = g\beta(T_i - T_o)D_i^3/(\alpha v)$. Equation (7.107) agrees within $\pm 10\%$ with experimental data at moderate (laminar regime) Rayleigh numbers as high as 10^7. The low

---

6 The writing and use of the original correlation [33] are a bit more complicated, because they were patterned after the pure conduction formula (2.32), and its Rayleigh number was based on the thickness of the annular space $(D_0 - D_i)/2$.

Rayleigh number limit of the validity of Eq. (7.107) occurs where the thickest free convection boundary layer is larger than the transversal dimension of the annular cavity:

$$D_o \text{Ra}_{D_o}^{-1/4} > D_o - D_i \tag{7.108}$$

The left side of this inequality shows the thermal boundary layer thickness along (i.e. inside) the outer cylinder, which[7] is greater than the corresponding thickness associated with the inner cylinder, $D_i Ra_{D_i}^{-1/4}$. When the Rayleigh number is small enough so that the inequality (7.108) is satisfied, the heat transfer mechanism approaches the pure conduction regime, for which the recommended formula is Eq. (2.32). Another way of making certain that the Rayleigh number is large enough for Eq. (7.107) to apply is to calculate q' twice, using Eqs. (7.107) and (2.32), and to retain the larger of the two q' values. This alternative is based on the constructal idea that the true heat transfer rate q' cannot be smaller than the pure conduction estimate based on Eq. (2.32).

Equation (7.107) has been tested extensively in the range $Pr > 0.7$. It should also hold in the low Prandtl number range, because the function $[Pr/(0.861 + Pr)]^{1/4}$ is known to account properly for the Pr effect at both ends of the Pr spectrum. Compare, for example, Eq. (7.107) with Eq. (7.55). All the physical properties that appear on the right side of Eq. (7.107) are evaluated at the average temperature $(T_i + T_o)/2$. A lengthier correlation that covers a wider Rayleigh number range, including the turbulent regime, is available in Ref. [34].

7.4.6 Annular Space Between Concentric Spheres

The natural circulation in the annulus between two concentric spheres has the approximate shape of a doughnut. In fact, at small enough Rayleigh numbers in the laminar regime, a vertical cut through the center of the two spheres reveals two kidney-shaped flow patterns similar to what is shown in Figure 7.19. A relatively compact expression for the total heat transfer rate $q(W)$ between the two spherical surfaces can be derived from a correlation developed in Ref. [33]:

$$q \cong \frac{2.325 k D_i (T_i - T_o)}{[1 + (D_i/D_o)^{7/5}]^{5/4}} \left(\frac{Pr Ra_{D_i}}{0.861 + Pr} \right)^{1/4} \tag{7.109}$$

where, again, $Ra_{D_i} = g\beta(T_i - T_o)D_i^3/(\alpha\nu)$. The argument for employing the Rayleigh number based on D_i, is the same as in the opening paragraph to the preceding Section 7.4.6. Equation (7.109) holds over the entire Prandtl number range, and the physical properties are evaluated at the average temperature $(T_i + T_0)/2$. The similarities between Eq. (7.109) and Eq. (7.107) are worth contemplating.

Equation (7.109) is accurate within $\pm 10\%$ in the laminar range, at Rayleigh numbers as high as 10^7. When the Rayleigh number is so low that the inequality (7.108) applies, Eq. (7.109) is no longer valid: the formula to use in this limit is the pure conduction expression (2.39). As a safety check, it is a good idea to calculate q based on both formulas, Eqs. (7.109) and (2.39), and to retain the larger of the two values. Note that the ratio $q_{7.109}/q_{2.39}$ indicates the degree of heat transfer enhancement (augmentation) that is due to the presence of convection inside the cavity. In this sense, the q ratio is similar to the conduction-referenced Nusselt number $\overline{Nu}_H$ encountered in fluid layers heated from below, Eq. (7.105).

---

7 Recall that in the laminar boundary layer natural convection along (or around) a surface (flat or curved) of a certain height, the boundary layer thickness increases as the height raised to the power $\frac{1}{4}$. The boundary layer thickness scale is equal to the height times the Rayleigh number (based on height) raised to the power $-\frac{1}{4}$. This observation stood behind the laminar flow correlations assembled in Section 7.3.

All the topics and configurations assembled in this chapter refer exclusively to pure natural convection, that is, to flows driven solely by the effect of buoyancy. Even more numerous in nature are the flows that result from a combination of natural and forced convection. Think, for example, of the room air flow near a warm window at the time when the air-conditioning system blows a charge of new (cleaner and cooler) air through the room. These more complicated flows are examples of *mixed convection*: their fundamentals are covered in the more advanced courses on convection [1].

7.5 Evolutionary Design

7.5.1 Spacings

Packages of electronics, electric windings, arrays of fins, and many other flow systems (including compact organs from animal design) are bathed by fluids that permeate through a heat-generating structure. For such flows to happen, the solid structure must have spacings (coarse porosity), but not just any spacings. The dimension of the spacing is tied intimately to the performance of the heat-generating package. The architectures of cooled packages evolve toward greater densities of heat-generating entities present in the volume.

This section outlines the fundamental problem and its central role in evolutionary design. The formulation is the simplest, and it comes from the homework problem that I wrote for the first edition of my convection book [1] in 1984, when I discovered not only the spacing for natural convection cooling but also how to determine it most quickly: the most direct method is the intersection of asymptotes. To the history and implications of this theoretical step, I return at the end of this section.

With reference to Figure 7.20, we seek the spacing D so that the heat transfer from the vertical stack to the ambient (T_∞) is most intense. This search is fundamental in the cooling of electronic packages, where the objective is to increase the density of heat-generating electronics that can be fitted in a package of specified volume. In Figure 7.20, the package volume is LHW, where W is the width perpendicular to the plane of Figure 7.20.

The simplest way to solve this problem is by recognizing the limiting regimes ($D \to 0$ versus $D \to \infty$) of the stack cooling phenomenon. Assume that the surface of each board is smooth and isothermal at T_0 and that the board thickness t is negligible with respect to D. The number of boards is $n \sim L/D$, and the parameters H, W, T_0, T_∞, and L are assumed given:

(a) Consider first the limit of vanishingly small plate-to-plate spacing, $D \to 0$. In this limit we can use with confidence Eq. (7.94) for the overall heat transfer rate extracted from the two surfaces of a single channel:

$$q_1 \cong \frac{\rho g \beta c_P (\Delta T)^2 D^3}{12 \nu} W \tag{7.110}$$

where $\Delta T = T_0 - T_\infty$. The total number of channels is L/D; therefore, the total heat transfer rate from the assembly is

$$q_a = q_1 \frac{L}{D} = \frac{\rho g \beta c_P (\Delta T)^2 D^3}{12 \nu} W \frac{L}{D} \tag{7.111}$$

This result and Figure 7.21 show that in the $D \to 0$ limit, the total heat transfer rate decreases as D^2.

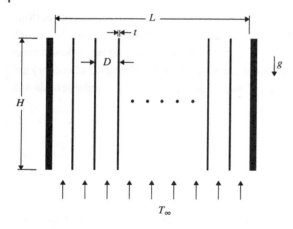

Figure 7.20 Vertical stack of heat-generating plates cooled by natural convection.

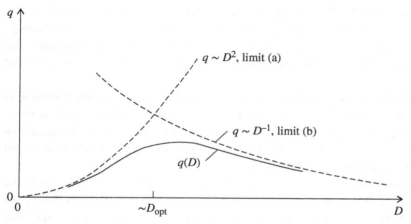

Figure 7.21 Plate-to-plate spacing at the intersection of the small-*D* and the large-*D* asymptotes.

(b) Consider next the limit in which the spacing D is sufficiently large, that is, larger than the thickness of the air boundary layer formed on one of the vertical surfaces, $D > H\,Ra_H^{-1/4}$, where $Ra_H = (g\beta H^3 \Delta T)/\alpha\nu$ and $\Delta T = T_0 - T_\infty$. The boundary layers are distinct (i.e. thin compared with D), and the center of the plate-to-plate spacing is occupied by T_∞ – air. The number of air boundary layers is $2(L/D)$ because there are two boundary layers for each D spacing. The corresponding formula for the total heat transfer rate is

$$q_b = 2\frac{L}{D}q_{\mathrm{BL}} \tag{7.112}$$

where q_{BL} is the heat transfer through one boundary layer (one surface $W \times H$):

$$q_{\mathrm{BL}} = \overline{h}_H WH\,\Delta T \tag{7.113}$$

If we set $Pr = 0.72$ for air, the average heat transfer coefficient is [1]

$$\overline{h}_H = \frac{k}{H}0.516Ra_H^{1/4} \tag{7.114}$$

In conclusion, in the large-*D* limit, the total heat transfer rate is inversely proportional to D:

$$q_b = 2\frac{L}{D}\,WH\,\Delta T\frac{k}{H}0.516Ra_H^{1/4} \tag{7.115}$$

We have determined the two asymptotes of the unknown curve of q versus D. Figure 7.21 shows that the asymptotes intersect above what would be the q maximum of the actual curve. The recommended spacing can be estimated (approximately) as the D value where the two asymptotes intersect:

$$q_a \cong q_b \tag{7.116}$$

which yields

$$Ra_D \sim 12.4 Ra_H^{1/4} \quad \text{and} \quad D \sim 2.3 H Ra_H^{-1/4} \tag{7.117}$$

The peak heat transfer rate that corresponds to this spacing is

$$q \lesssim 2 \frac{L}{D} 0.45 k \Delta T \frac{LW}{H} Ra_H^{1/2} \tag{7.118}$$

The inequality sign is a reminder that the peak of the actual curve q versus D is located under the intersection of the two asymptotes, as shown in Figure 7.21. Despite this inequality, the right side of Eq. (7.118) represents the correct order of magnitude of q. The peak heat generation (or electronics) density q_{max}/HLW is proportional to $H^{-1/2} \Delta T^{3/2}$ because Ra_H is proportional to $H^3 \Delta T$.

I first proposed this problem and solved it intuitively by intersecting the asymptotes in the 1984 edition of Ref. [1], specifically, as Problem 11 on p. 157. The same topic was analyzed by a much lengthier method by Bar-Cohen and Rohsenow [35]. In fact, both versions, my 1984 book and Ref. [35], appeared in print at exactly the same time, in August 1984, displayed side by side in the book exhibit at the 1984 ASME National Heat Transfer Conference in Niagara Falls, New York. I remember my first reaction to this amazing coincidence. Warren Rohsenow had been my heat transfer professor at MIT. I thought that if a full-length article in a top journal was needed to determine what my students could derive on the back of an envelope by solving a homework problem, then my method of intersecting the asymptotes is novel and extremely powerful. History has proven this to be true. The identification of spacings for the internal structures of volumes cooled by natural, forced and mixed convection, laminar and turbulent, has become a very active and distinct research direction in heat transfer and constructal design.

7.5.2 Miniaturization

The chief consequence of identifying the correct spacing for vertical natural convection cooling is the maximum packing of heat-generating electronics, Eq. (7.118). The packing per unit volume is represented by the ratio $q/(HLW)$, which is proportional to $Ra_H^{1/2}/H^2$, which is proportional to $H^{-1/2}$ because $Ra_H^{1/2}$ is proportional to $H^{3/2}$.

In conclusion, the packing density increases as the size of the package in the flow direction (H in Figure 7.20, and L in the lower-right detail of Figure 7.22) decreases. Smaller electronics make better and more efficient vehicles for heat transfer, and this way we arrive again at the discovery that the march toward miniaturization is predictable from the constructal law, and it visualizes the time arrow of evolutionary designs everywhere.

For packages of heat-generating components (electronics, computers, windings), the evolution toward greater density of functionality is illustrated in Figure 7.22. This is a review [36] of four decades of designs for the cooling of electronics. The length scale (L) of the device filled with electronics can vary. Think of the evolution of electronics from phone booths to today's servers and laptops and handheld devices, and how that length scale has been shrinking.

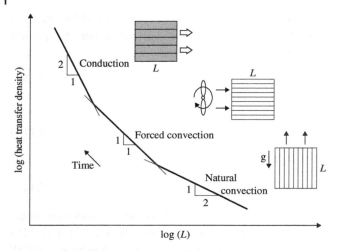

Figure 7.22 The evolution of heat transfer density toward higher values, showing two phenomena: evolution toward smaller sizes (miniaturization) and stepwise (dramatic) changes in cooling technology.

Three cooling technologies are natural convection (NC, flow driven by buoyancy, because warm air is lighter and rises through the body to be cooled), forced convection (FC, flow driven by a fan or pump), and solid-body conduction (C, heat flows from hot to cold, through the casing of the device). The chronological sequence in which these technologies emerged and took over the world is NC → FC → C, not the reverse. The following is why it had to happen this way.

A cooling technology offers the greatest packing density of functionality (heat transfer density) when the spacing between the plates populated by electronics has a certain value. The oldest design of this kind was for NC cooling, and its packing density of heat-generating electronics varies as $L^{-1/2}$, as shown in Section 7.5.1. We see that greater densities of functionality are possible if this old package can be made smaller. The time arrow of technology evolution of NC cooling points to the left, toward smaller sizes.

The second oldest cooling technology is based on forced convection (FC) [1, 37]. In the designs with the highest density of electronics, the spacing between plates is such that the packing density varies as L^{-1}, which is also shown in the middle segment of Figure 7.22. This means that greater cooling densities by forced convection are possible in the direction of smaller elements L, toward miniaturization. This is in accord with the trend exhibited by NC cooling technology. This is a new prediction (correct, in hindsight) that in the pursuit of greater densities in smaller elements, there must be a stepwise *transition* in technology, from natural to forced convection cooling, not from forced to natural convection.

The evolution of volumetric cooling does not end with convection through properly sized spacings, parallel plates, or other packed elements (cylinders, spheres, staggered or aligned, pin fin arrays, etc.) [38, 39]. It is possible to cool the L-scale body by pure conduction, as shown in the upper detail of Figure 7.22. The body generates heat volumetrically at a uniform rate. To facilitate the flow of heat from the volume to be cooled to one or more points on the side, solid inserts (blades, pins, trees) of a material with much higher conductivity are placed inside the original material.

In sum, conduction cooling is facilitated by designing the body as a composite material with two solid organs (high and low conductivity) and with *design,* the organization of the high-conductivity paths on the low-conductivity background. Review Section 3.4.2. The composition is described by the two materials, while the design evolves so that the heat generated by the volume flows more and more easily to the heat sink on the boundary. All such designs tend toward greater densities of heat generation (denser electronics) that increase as L^{-2} when L decreases.

Because the path toward greater density is more direct than in designs with forced convection cooling, and certainly greater than in natural convection cooling, we discover that there must be a transition from forced convection to solid-body conduction throughout the volume. This step change technology evolution occurs in only one direction, from forced convection to conduction, and not the other way around.

This story of evolution indicated that cooling technology must evolve in two ways: toward smaller scales (miniaturization) and, though dramatic, stepwise changes (transitions) in heat flow mechanisms. Changes are ongoing and occur in the same time direction as the morphing toward miniaturization.

References

1 Bejan, A. (2013). *Convection Heat Transfer*, 4e, Hoboken: Wiley 2013 (first edition, 1984).
2 Oberbeck, A. (1879). Über die Wärmeleitung der Flüssigkeiten bei Berücksichtigung der Strömungen infolge von Temperaturdifferenzen. *Ann. Phys. Chem.* 7: 271–292.
3 Boussinesq, J. (1903). *Theorie Analytique de la Chaleur*, vol. 2. Paris: Gauthier-Villars.
4 LeFevre, E.J. (1956). Laminar free convection from a vertical plane surface. *Proceedings of the 9th International Congress for Applied Mechanics*, Brussels, Volume 4, pp. 168–174.
5 Martin, B.W. (1984). An appreciation of advances in natural convection along an isothermal vertical surface. *Int. J. Heat Mass Transfer* 27: 1583–1586.
6 Schmidt, E. and Beckmann, W. (1930). Das Temperatur- und Geschwindigkeitsfeld vor einer Wärme abgebenden senkrechten Platte bei natürlicher Konvektion. *Tech. Mech. Thermo-Dynam.* 1: 391–406.
7 Kuiken, H.K. (1968). An asymptotic solution for large Prandtl number free convection. *J. Eng. Math.* 2: 355–371.
8 Bejan, A. and Lage, J.L. (1990). The Prandtl number effect on the transition in natural convection along a vertical surface. *J. Heat Transfer* 112: 787–790.
9 Churchill, S.W. and Chu, H.H.S. (1975). Correlating equations for laminar and turbulent free convection from a vertical plate. *Int. J. Heat Mass Transfer* 18: 1323–1329.
10 Sparrow, E.M. and Gregg, J.L. (1956). Laminar free convection from a vertical plate with uniform surface heat flux. *Trans. ASME* 78: 435–440.
11 Vliet, G.C. and Liu, C.K. (1969). An experimental study of turbulent natural convection boundary layers. *J. Heat Transfer* 91: 517–531.
12 Vliet, G.C. and Ross, D.C. (1975). Turbulent natural convection on upward and downward facing inclined heat flux surfaces. *J. Heat Transfer* 97: 549–555.
13 Chen, C.C. and Eichhorn, R. (1976). Natural convection from a vertical surface to a thermally stratified medium. *J. Heat Transfer* 98: 446–451.
14 Minkowycz, W.J. and Sparrow, E.M. (1974). Local nonsimilar solutions for natural convection on a vertical cylinder. *J. Heat Transfer* 96: 178–183.
15 Vliet, G.C. (1969). Natural convection local heat transfer on constant-heat-flux inclined surfaces. *J. Heat Transfer* 91: 511–516.
16 Lloyd, J.R. and Sparrow, E.M. (1970). On the instability of natural convection flow on inclined plates. *J. Fluid Mech.* 42: 465–470.

17 Goldstein, R.J., Sparrow, E.M., and Jones, D.C. (1973). Natural convection mass transfer adjacent to horizontal plates. *Int. J. Heat Mass Transfer* 16: 1025–1035.

18 Lloyd, J.R. and Moran, W.R. (1974). Natural Convection Adjacent to Horizontal Surfaces of Various Planforms. *ASME Paper 74-WA/HT-66.*

19 Incropera, F.P. and DeWitt, D.P. (1990). *Fundamentals of Heat and Mass Transfer*, 3e, 539–540. New York: Wiley.

20 Churchill, S.W. (1983). Free convection around immersed bodies. In: *Heat Exchanger Design Handbook*, Section 2.5.7 (ed. E.U. Schlünder). New York: Hemisphere.

21 LeFevre, E.J. and Ede, A.J. (1956). Laminar free convection from the outer surface of a vertical circular cylinder. *Proceedings of the 9th International Congress for Applied Mechanics*, Brussels, Volume 4, pp. 175–183.

22 Lienhard, J.H. (1973). On the commonality of equations for natural convection from immersed bodies. *Int. J. Heat Mass Transfer* 16: 2121–2123.

23 Sparrow, E.M. and Ansari, M.A. (1983). A refutation of King's rule for multi-dimensional external natural convection. *Int. J. Heat Mass Transfer* 26: 1357–1364.

24 Yovanovich, M.M. (1987). On the Effect of Shape, Aspect Ratio and Orientation Upon Natural Convection from Isothermal Bodies of Complex Shape. *ASME HTD-Vol. 82*, pp. 121–129.

25 Hassani, A.V. and Hollands, K.G.T. (1989). On natural convection heat transfer from three-dimensional bodies of arbitrary shape. *J. Heat Transfer* 111: 363–371.

26 Sahraoui, M., Kaviany, M., and Marshall, H. (1990). Natural convection from horizontal disks and rings. *J. Heat Transfer* 112: 110–116.

27 Catton, I. (1979). Natural convection in enclosures. *6th International Heat Transfer Conference*, Toronto, 1978, Volume 6, pp. 13–43.

28 Bejan, A. and Tien, C.L. (1978). Laminar natural convection heat transfer in a horizontal cavity with different end temperatures. *J. Heat Transfer* 100: 641–647.

29 Kimura, S. and Bejan, A. (1984). The boundary layer natural convection regime in a rectangular cavity with uniform heat flux from the side. *J. Heat Transfer* 106: 98–103.

30 Pellew, A. and Southwell, R.V. (1940). On maintained convection motion in a fluid heated from below. *Proc. R. Soc.* A176: 312–343.

31 Globe, S. and Dropkin, D. (1959). Natural convection heat transfer in liquids confined by two horizontal plates and heated from below. *J. Heat Transfer* 81: 24–28.

32 Arnold, N., Catton, I., and Edwards, D.K. (1976). Experimental investigation of natural convection in inclined rectangular regions of differing aspect ratios. *J. Heat Transfer* 98: 67–71.

33 Raithby, G.D. and Hollands, K.G.T. (1975). A general method of obtaining approximate solutions to laminar and turbulent free convection problems. *Adv. Heat Transfer* 11: 265–315.

34 Kuehn, T.H. and Goldstein, R.J. (1976). Correlating equations for natural convection heat transfer between horizontal circular cylinders. *Int. J. Heat Mass Transfer* 19: 1127–1134.

35 Bar-Cohen, A. and Rohsenow, W.M. (1984). Thermally optimum spacing of vertical, natural convection cooled, parallel plates. *J. Heat Transfer* 106: 116–123.

36 Bejan, A. (2013). Technology evolution, from the constructal law. *Adv. Heat Transf.* 45: 183–207.

37 Bejan, A. and Sciubba, E. (1992). The optimal spacing of parallel plates cooled by forced convection. *Int. J. Heat Mass Transfer* 35: 3259–3264.

38 Bejan, A. and Lorente, S. (2008). *Design with Constructal Theory*. Hoboken, NJ: Wiley.

39 Stafford, J. and Egan, V. (2014). Configurations for single-scale cylinder pairs in natural convection. *Int. J. Therm. Sci.* 84: 62–74.

40 Bejan, A. (2016). *The Physics of Life: The Evolution of Everything*, New York: St. Martin's Press.

41 Bejan, A. (2021). Watching Physics at the Olympics. *Academia Letters*, article 2577.

Problems

Laminar Vertical Boundary Layer Flow

7.1 a) Show that in the case of a fluid that can be modeled as an ideal gas, the linearized equation of state (7.13) becomes

$$\frac{\rho}{\rho_\infty} \cong 1 - \frac{T - T_\infty}{T_\infty} + \frac{P - P_0}{P_0} + \cdots$$

  b) Consider an application in which the fluid is air ($T_\infty = 20\,°C$, $P_0 = 1$ atm), the vertical heated wall is 4 m tall, and the wall temperature is approximately $40\,°C$. The pressure excursion $P-P_0$ is of the same order as the hydrostatic pressure difference between floor and ceiling (Figure 7.1, right side). Numerically, show that the term $(P-P_0)/P_0$ is relatively negligible in the ρ/ρ_∞ formula derived in part (a).

7.2 Let $\delta(y)$ be the outer thickness of the vertical velocity profile for laminar natural convection along a vertical wall. Show that the integral form of the boundary layer-simplified momentum Eq. (7.17) is

$$\frac{d}{dy} \int_0^\delta v^2 dx = -v\left(\frac{\partial u}{\partial x}\right)_{x=0} + \int_0^\delta g\beta(T - T_\infty)dx$$

7.3 Let δ_T be the thermal boundary layer thickness of the buoyant jet along a vertical wall. Show that the integral form of the boundary layer-simplified energy Eq. (7.11) is

$$\frac{d}{dy} \int_0^{\delta_T} v(T - T_\infty)\ dx = -\alpha\left(\frac{\partial T}{\partial x}\right)_{x=0}$$

In this derivation, keep in mind that the vertical velocity v at the outer edge of the thermal boundary layer ($x = \delta_T$) is not necessarily zero.

7.4 Consider the boundary layer flow near a vertical isothermal wall (T_w) surrounded by an isothermal fluid reservoir (T_∞). Assume the following temperature and vertical velocity profiles:

$$T - T_\infty = \Delta T\left(1 - \frac{x}{\delta}\right)^2, \quad v = V\frac{x}{\delta}\left(1 - \frac{x}{\delta}\right)^2$$

where $\Delta T = T_w - T_\infty$. Determine the functions $\delta(y)$ and $V(y)$ by solving the integral momentum and energy equations listed in the preceding two problem statements. Show that the local Nusselt number that corresponds to this solution is

$$Nu_y = \frac{q''_{w,y}}{\Delta T}\frac{y}{k} = 0.508\left(1 + \frac{20}{21Pr}\right)^{-1/4} Ra_y^{1/4}$$

7.5 The door of a household refrigerator is 1 m tall and 0.6 m wide. The temperature of its external surface is $20\,°C$, while the room air temperature is $30\,°C$. Verify that the natural convection air boundary layer that descends along the door is laminar, and calculate the heat transfer rate absorbed by the door.

7.6 An immersion heater for water consists of a thin vertical plate shaped as a rectangle with height of 8 cm and length of 15 cm. The plate is heated electrically and maintained at $55\,°C$, while the average temperature in the surrounding water tank is 15 °C. The plate is bathed by water on both sides. Show that the natural convection boundary layer that rises along the plate is laminar. Next, calculate the total heat transfer rate released by the heater into the water pool.

7.7 For the buoyant jet along a vertical wall with uniform heat flux q_w'', assume the following temperature and velocity profiles:

$$T - T_\infty = (T_w - T_\infty)\left(1 - \frac{x}{\delta}\right)^2$$

$$v = V\frac{x}{\delta}\left(1 - \frac{x}{\delta}\right)^2$$

Solve the integral boundary layer-simplified equations for momentum and energy listed in Problems 7.2 and 7.3, and show that the local Nusselt number is

$$Nu_y = \frac{q_w''}{T_w(y) - T_\infty}\frac{y}{k} = 0.616\left(\frac{0.8}{Pr} + 1\right)^{-1/5}Ra_y^{*1/5}$$

7.8 One way to visualize the $y^{1/4}$-dependence of the thickness of the laminar natural convection boundary layer is to perform the experiment shown in Figure P7.8. The vertical isothermal wall, $T_w = 20$ °C, is placed in contact with an isothermal pool of paraffin, $T_\infty = 35$ °C. Since the solidification point of this paraffin is $T_m = 27.5$ °C, the wall becomes covered with a thin layer of solidified paraffin.

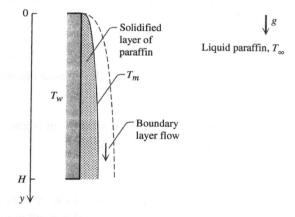

Figure P7.8

Show that under steady-state conditions, the thickness of the solidified layer, L, is proportional to the laminar boundary layer thickness, that is, it increases in the downward direction as $y^{1/4}$. Calculate L numerically and plot to scale the $L(y)$ shape of the solidified layer. The properties of liquid paraffin are $k_f = 0.15$ W/m · K, $\beta = 8.5 \times 10^{-4}$ K^{-1}, $\alpha = 9 \times 10^{-4}$ cm^2/s, and $Pr = 55.9$. The thermal conductivity of solid paraffin is $k_s = 0.36$ W/m · K. The overall height of the isothermal wall is $H = 10$ cm (Figure P7.8).

Turbulent Vertical Boundary Layer Flow

7.9 One 0.5 m × 0.5 m vertical wall of a parallelepipedic water container is heated uniformly by an array of electrical strip heaters mounted on its back. The total heat transfer rate furnished by these heaters is 1000 W. The water temperature is 25 °C. Calculate the height-averaged temperature of the heated surface. Then, verify that most of the boundary layer that rises along this surface is turbulent.

7.10 The size of the small box filled with water must be selected for the purpose of simulating in the laboratory the natural convection of air in a 3 m tall room. The average temperature of the room air is 25 °C, while the temperature of the cold vertical wall is 15 °C. The designer must select a certain vertical dimension (height H) for the water enclosure so that the Rayleigh number $g\beta\Delta T H^3/(\alpha v)$ of the water flow matches the Rayleigh number of the air flow in the actual room. In the water experiment, the average water temperature and the cold-wall temperature are 25 and 15 °C, respectively. What is the required height H of the water apparatus?

7.11 a) Consider the natural convection boundary layer flow of water along a vertical wall with uniform heat flux q''_w. The Rayleigh number based on heat flux, Ra_y^*, is known. Show that $Ra_y = Ra_y^*/\overline{Nu}_y$, where $\overline{Nu}_y$ is the average Nusselt number (based on average temperature difference, $\overline{T}_w - T_\infty$) and Ra_y is the Rayleigh number based on $\overline{T}_w - T_\infty$.
b) It is observed that the transition to turbulence takes place in the vicinity of $Ra_y^* \sim 10^{13}$. Calculate the corresponding value of the Rayleigh number based on temperature difference, Ra_y, and compare the Ra_y value with the one anticipated based on Eq. (7.57).

Other External Flow Configurations

7.12 The downward trend exhibited by the $\overline{Nu}_H/Ra_H$ curves of Figure 7.6 can be anticipated by means of a very simple analysis. Let ΔT_{avg} be the average temperature difference between the isothermal wall and the stratified reservoir. Assume a large Prandtl number ($Pr \to \infty$), and estimate the Nusselt number by using Eq. (7.55), in which both $\overline{Nu}_y$ and Ra_y are based on ΔT_{avg} (note also that $y = H$). Multiply and divide this relationship by ΔT_{max} to reshape it in terms of the $\overline{Nu}_H$ and Ra_H groups defined in Eq. (7.72a, b). Show that this relationship becomes

$$\overline{Nu}_H = 0.671\left(1 - \frac{b}{2}\right)^{5/4} Ra_H^{1/4}$$

and that it approximates within 13% the values suggested by the $Pr \to \infty$ curve of Figure 7.6.

7.13 The air in a room is thermally stratified so that its temperature increases by 5 °C for each 1-m rise in altitude. Facing the room air is a 10 °C vertical window, which is 1 m tall and 0.8 m wide. The room air temperature at the same level as the window midheight is 20 °C. Assume that the natural convection boundary layer that falls along the window is laminar, and calculate the heat transfer rate through the window.

7.14 A horizontal disc with a diameter of 8.7 cm and temperature of 43.1 °C is immersed facing upward in a pool of 23.6 °C water. Calculate the total heat transfer rate released by the surface into the water.

7.15 A long, rectangular metallic blade has width $H = 4$ cm and temperature $T_w = 40$ °C. It is surrounded on both sides by atmospheric air at $T_\infty = 20$ °C. The long side of the blade is always horizontal. Calculate the total heat transfer rate per unit of blade length, when the short side of its rectangular shape (H) is as follows:
a) Vertical
b) Inclined at 45° relative to the vertical
c) Horizontal
Comment on the effect that blade orientation has on the total heat transfer rate.

7.16 The external surface of a spherical container with the diameter $D = 3$ m has a temperature of 10 °C. The container is surrounded by 30 °C air, which is motionless except in the immediate vicinity of the container.
a) Calculate the total heat transfer rate absorbed by the spherical container.
b) It is proposed to replace the spherical container with one shaped as a horizontal cylinder with the diameter $d = 1.5$ m. This new container would have the same volume as the old one. Calculate the total heat transfer rate absorbed by the cylindrical container.
c) Which container design would you choose if your objective is to prevent the warming of the liquid stored inside the container?

7.17 A cubic block of metal is immersed in a pool of 20 °C water and is oriented as "Cube 1" in Figure 7.11. The cube has a 2 cm side and an instantaneous temperature of 80 °C. Calculate the average heat transfer coefficient between the cube and the water, by using Eqs. (7.84), (7.86), and (7.87). Comment on the agreement between these three estimates.

7.18 A block of ice has the shape of a parallelepiped with the dimensions indicated in Figure P7.18. The block is oriented in such a way that two of its long surfaces are horizontal. It is surrounded by 20 °C air from all sides. Calculate the total heat transfer rate between the ambient air and the ice block, q (W), by estimating the heat transfer through each face. Calculate the ice melting rate that corresponds to the total heat transfer rate. The ice latent heat of melting is $h_{sf} = 333.4$ kJ/kg.

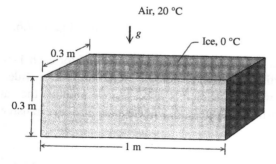

Figure P7.18

7.19 Consider again the natural convection heat transfer from air to the ice block described in Figure P7.18. Obtain a quick estimate of the total heat transfer rate $q(W)$ by using Eq. (7.84). The length l on which $\overline{Nu}_l$ and Ra_l are based can be approximated as the half-perimeter of the smaller ($0.3\,\text{m} \times 0.3\,\text{m}$) cross section of the ice block.

7.20 An electrical wire of diameter $D = 1$ mm is suspended horizontally in air of temperature 20 °C. The Joule heating of the wire is responsible for the heat generation rate $q' = 0.01$ W/cm, per unit length in the axial direction. The wire can be modeled as a cylinder with isothermal surface. Sufficiently far from the wire, the ambient air is motionless. Calculate the temperature difference that is established between the wire and the ambient air. Note: This calculation requires a trial-and-error procedure; expect a relatively small Rayleigh number.

7.21 The large-diameter cylindrical reservoir shown in Figure P7.21 is perfectly insulated and filled with water. Two horizontal tubes with an outer diameter of 4 cm are positioned at the same level in the vicinity of the reservoir centerline. The temperatures of the tube walls are maintained at $T_1 = 30\,°\text{C}$ and $T_2 = 20\,°\text{C}$ by internal water streams of appropriate (controlled) temperature.

Calculate the heat transfer rate from the hot tube to the cold tube via the water reservoir, by assuming that the tube-to-tube spacing is wide enough so that the boundary layers that coat the tubes do not touch (this case is illustrated in Figure P7.21). Assume that the film-temperature properties of the two boundary layers are equal to the properties evaluated at the average temperature of the water reservoir, T_∞. Begin with the calculation of T_∞, and recognize the centrosymmetry of the flow pattern, that is, the symmetry about the centerline of the large reservoir (Figure P7.21).

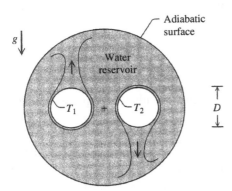

Figure P7.21

7.22 A hot dog is immersed horizontally in a stagnant pool of 95 °C water. The temperature of the hot dog is uniform at 20 °C. The shape of the hot dog is approximated by a cylinder with a diameter of 1.7 cm and length of 10 cm. Assume that the natural convection boundary layer flow has had time to develop around the hot dog surface.

a) Calculate the average heat transfer coefficient and the rate of heat transfer absorbed by the hot dog.

b) As the time increases and the hot dog warms up, does the hot dog temperature remain uniform over the hot dog cross section? In other words, would it be a good idea to model the hot dog as a lumped capacitance?

7.23 Cake batter is poured to fill completely a horizontal shallow pan that is shaped as a square. The side of the square is 46 cm (measured internally, i.e. touched by the batter). The room air is motionless and at 28 °C, while the initial temperature of the batter is 22 °C. Calculate the rate of heat transfer through the top surface, into the batter. Assuming that the pan is not put in the oven too soon, will the calculated heat transfer rate increase or decrease as time passes?

Internal Flow Configurations

7.24 Consider the narrow-channel limit of a vertical duct with isothermal wall T_w, height H, cross-sectional area A, wetted perimeter p, and hydraulic diameter $D_h = 4A/p$. The top and bottom ends of the duct communicate with an isothermal fluid reservoir (T_∞), Figure 7.12. By assuming fully developed laminar flow and reviewing Sections 6.1.2 and 6.1.3, show that the average vertical velocity through the duct is

$$V = \frac{g\beta(T_w - T_\infty)D_h^2}{2\nu f \, Re_{D_h}}$$

where $Re_{D_h} = D_h V / \nu$. Show further that the average Nusselt number $\overline{Nu}_H = \overline{q''}H/(k\Delta T)$ is proportional to the Rayleigh number based on hydraulic diameter:

$$\overline{Nu}_H = \frac{1}{8(f Re_{D_h})} Ra_{D_h}$$

where the group $f \, Re_{D_h}$ is a constant listed in Table 6.1. Use this formula for the purpose of verifying the validity of the numerical values assembled in Table 7.2.

7.25 The single-pane window problem consists of estimating the heat transfer rate through the vertical glass layer shown in Figure P7.25. The window separates two air reservoirs of temperatures T_h and T_c. Assuming constant properties, laminar boundary layers on both sides of the glass, and a uniform glass temperature T_w, show that the average heat flux $\overline{q''}$ from T_h to T_c obeys the relationship

$$\frac{\overline{q''}}{T_h - T_c} \frac{H}{k} = 0.217 \left[\frac{g\beta(T_h - T_c)H^3}{\alpha\nu} \right]^{1/4}$$

Use Eq. (7.55) as starting point in this analysis, and neglect the thermal resistance due to pure conduction across the glass layer itself (Figure P7.25).

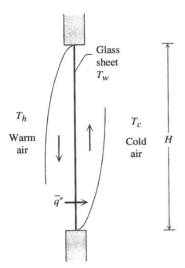

Figure P7.25

7.26 Consider again the single-pane window described in Figure P7.25, and model the heat flux through the glass layer as uniform, q''. Starting with the height-averaged version of Eq. (7.66) for air, show that the relationship between q'' and the overall temperature difference $T_h - T_c$ is

$$\frac{q''}{T_h - T_c}\frac{H}{k} = 0.252\left[\frac{g\beta(T_h - T_c)H^3}{\alpha v}\right]^{1/4}$$

Compare this result with the formula recommended by the uniform-T_w model in Problem 7.25, and you will get a feel for the "certainty" with which you can calculate the total heat transfer rate through the window.

7.27 The double-pane window problem consists of determining the H-averaged heat flux through the system, when the overall temperature difference $T_h - T_c$ is specified (see Figure P7.27). If the glass-to-glass spacing is wide enough to house distinct laminar boundary layers, it is possible to approximate the double-pane system as a sandwich of two of the single-pane windows treated in the preceding problem (Figure P7.27). Use the formula listed above for the single-pane window with uniform heat flux, and show that the average heat flux through the double-pane window system is given approximately by

$$\frac{q''}{T_h - T_c}\frac{H}{k} \cong 0.106\left[\frac{g\beta(T_h - T_c)H^3}{\alpha v}\right]^{1/4}$$

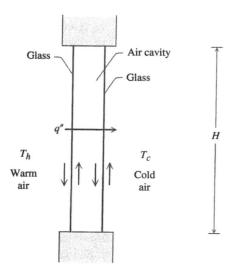

Figure P7.27

7.28 In Figure P4.3, a small bottle containing $60\,°C$ temperature milk is immersed in a pool of $10\,°C$ tap water. The height of the bottle wall wetted on the inside by milk and on the outside by tap water is $H = 6\,cm$. Calculate the H-averaged overall heat transfer coefficient between the warm milk and the cold tap water. In this calculation recognize that the two sides of the vertical wall are lined by natural convection boundary layers and that the overall heat transfer coefficient is related to the individual heat transfer coefficients calculated for the two sides of the vertical surface.

Evolutionary Design

7.29 An interesting application of the air flow in enclosures heated from the side is shown in Figure P7.29. The space between the vertical surfaces of the double wall is divided into many cells by means of inclined partitions. In the house heating arrangement illustrated here, the function of this wall is to allow the transfer of heat from left to right (when the outer surface is heated by the sun) and to prevent the transfer of heat in the opposite direction (during night time).

In the "on" mode, the heated portion of each cell is positioned below the cooled portion. The enclosed air rises at one end and sinks at the other and completes one roll. In the "off" mode, the heated air remains trapped above the cold air, and the circulation is inhibited. Estimate the average heat flux through the double wall in the two modes. Model the flow during the "on" mode as the single-roll circulation inside a square cavity heated from the side, $H = L = 10\,cm$. Assume that the heat transfer during the "off" mode is by pure conduction. In both cases the temperature difference between the vertical surfaces of the double wall is $20\,°C$. Evaluate all the air properties at $20\,°C$ (Figure P7.29).

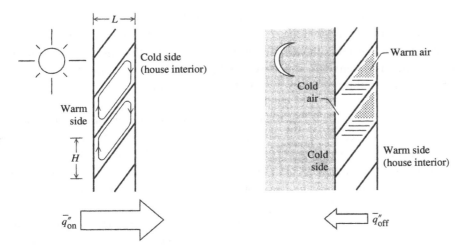

Figure P7.29

7.30 You have a can of cold drink (beer, coke) and you would like it to remain cold as you sip it for 10 minutes or longer. The surrounding air is warm. You also have an empty glass, the diameter of which is a bit larger than the diameter of the can. How would you configure the arrangement (can, glass) and the way you drink the contents over time?

There is plenty of freedom in how to change the design. For example, pour a small amount of liquid in the glass, and drink it. Or you may pour a larger amount in the glass, on the theory that more cold liquid in the glass has a greater thermal inertia and will stay cold longer. Or perhaps there is an optimal amount of liquid to reside in the glass as you sip it and replenish it intermittently. Think, write down your proposed design, and explain why it promises to keep your drink cold the longest. My solution is in the solutions manual, and it will surprise you.

7.31 Here is how to discover that humid air is lighter than dry air. You just washed your cylindrical coffee cup, and you want it to become dry while sitting in still air. The last to become dry are the inner walls of the cup. How fast the inner walls become dry is the performance of the configuration that you choose for the drying process.

There is plenty of freedom in how you configure the drying process, and it all has to do with how you orient the cup. If its mouth faces upward, then the humid air that forms inside the cup rises and leaves the cup, but a layer of liquid accumulates at the bottom. If the cup is rotated 180°, with the mouth flush on a horizontal surface, the humid air is trapped inside the cup. If the cup is rotated 90°, with its side on the table, then the humid air escapes but a string of liquid accumulates at the bottom, along its lowest generator. Think and propose a configuration that offers the shortest drying time. My solution is in the solutions manual, and it will surprise you.

7.32 How much warmer than steam must dry air be so that steam does not rise by buoyancy through dry air? Assume that initially the dry air and the steam are at 100 °C and atmospheric pressure. Consult Appendix D, and you will learn that the density of dry air is greater than the density of steam. Let T be the dry air temperature where the air density matches the density of steam at 100 °C. Calculate T. Will the 100 °C steam rise through dry air when the air temperature exceeds T?

7.33 The density of atmospheric air (ρ) is lower than the density of cold air, cf. Eq. (7.14). The density difference is felt in the drag force (F) that is overcome by the runner in the 100 m dash: the force is proportional to both ρ and U^2, where $U \cong 10$ m/s is the current speed record.

(ΔT) Determine the relative reduction in drag force ($\Delta F/F$) associated with the reduction in air density ($\Delta \rho$) when the air temperature changes from 10 to 40 °C. For the average air density in this comparison, use the density at 300 K.

(ΔU) Recognized in track and field is the advantage due to tail wind. The maximum tail wind speed allowed for certifying records in the 100 m dash is 2 m/s. Determine the relative reduction in drag force ($\Delta F/F$) due to the tail wind $\Delta U = 2$ m/s.

(Δz) The advantage of altitude is also recognized. Determine first the relative reduction in air density ($\Delta \rho/\rho$) when the change is from sea level to $\Delta z = 2000$ m (e.g. Mexico City); next, determine the relative reduction in the drag force ($\Delta F/F$) overcome by the runner.

Compare the magnitudes of the three advantages estimate above, and comment on whether the warmth (ΔT) and altitude (Δz) advantages are negligible relative to the allowed tail wind advantage.

To learn more about physics by observing and questioning athletics, read Refs. [40, 41].

8

Convection with Change of Phase

8.1 Condensation

8.1.1 Laminar Film on Vertical Surface

The common thread of the convection phenomena treated in Chapters 5–7 is that the fluid – the convective medium – remains in its original single-phase state, in spite of the heating or cooling that it experiences. In the present chapter we consider convection phenomena where the fluid undergoes a change of phase. Early cases of phase change under the influence of heat transfer were the melting and solidification processes discussed in Section 4.7, as examples of time-dependent unidirectional conduction. The phase-change phenomena detailed in the present chapter are all caused by convection.

One of the simplest convection phase-change process is the condensation of a vapor on a cold vertical surface, Figure 8.1. The film of condensate that forms on the vertical surface can have three distinct regions. The *laminar* section is near the top, where the film is the thinnest. The film thickness increases in the downward direction, as more and more of the surrounding vapor condenses on the exposed surface of the film. There comes a point where the film becomes thick enough to show the first signs of transition to a nonlaminar flow regime. In this *wavy* flow region, the visible surface of the film shows a sequence of regular ripples. Finally, if the wall extends sufficiently far downward, the film enters and remains in the *turbulent* region, where the ripples appear irregular in both space and time.

The transitions from one flow regime to the next will be described in more precise terms in Figure 8.5. At this stage it is instructive to see the similarity between the three-regime sequence of the vertical film (Figure 8.1, left) and the flow regimes of a single-phase natural convection boundary layer (rotate Figure 7.5 by 180°). The laminar film of condensate is a laminar boundary layer flow where the vertical surface is so cold that it causes the change of phase (condensation) of the descending boundary layer flow.

On the right side of Figure 8.1, we see that even in the laminar film region, which is the simplest of the three regions, the flow of the liquid film generally interacts with the descending boundary layer of cooled vapor. The temperature of the liquid–vapor interface is the saturation temperature that corresponds to the local pressure along the wall, T_{sat}. The saturation temperature is sandwiched between the temperature of the isothermal vapor reservoir, T_∞, and the wall temperature, T_w. Through the shear stress at the liquid–vapor interface, the descending jet of vapor aids the downward flow of the liquid film. The vapor in the descending jet is colder than the vapor reservoir and warmer than the liquid in the film attached to the wall.

This two-phase flow is considerably more complicated in the wavy and turbulent sections of the wall. In applications where the film is sufficiently long to exhibit all three regimes, the overall heat transfer rate from the vapor reservoir to the wall is dominated by the contributions made by the wavy and turbulent

Heat Transfer: Evolution, Design and Performance, First Edition. Adrian Bejan.
© 2022 John Wiley & Sons, Inc. Published 2022 by John Wiley & Sons, Inc.
Companion website: www.wiley.com/go/bejan/heattransfer

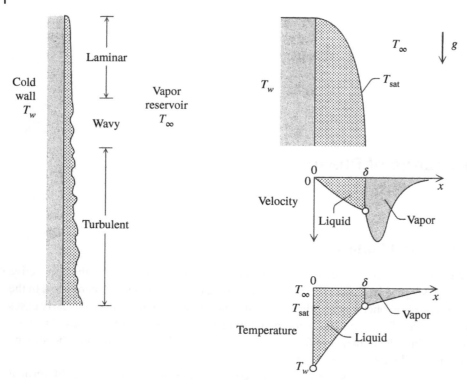

Figure 8.1 Flow regimes of the film of condensate on a cooled vertical surface.

sections. The same can be said about the total rate of condensation, which, as we shall learn in Eq. (8.21), is proportional to the overall heat transfer rate from the vapor to the vertical wall.

Consider the two-dimensional laminar film sketched in Figure 8.2, in which the distance y measures the downward length of the film. This flow is considerably simpler than the one seen in Figure 8.1, because this time the entire reservoir of vapor is isothermal at the saturation pressure, T_{sat}. This simplification allows us to focus on the flow of the liquid film and to neglect the movement of the nearest layers of vapor.

The analysis of the flow of liquid begins with the steady-state version of the momentum equations (M_x) and (M_y) of Table 5.1, which in the case of a *slender* film (i.e. in boundary layer-type flow) reduce to one equation:

$$\rho_l \left(u \frac{\partial u}{\partial x} + v \frac{\partial v}{\partial y} \right) = -\frac{dP}{dy} + \mu_l \frac{\partial^2 v}{\partial x^2} + \rho_l g \tag{8.1}$$

The last term on the right side represents the body force experienced by each small packet of liquid. Because of the slenderness of the film, the vertical pressure gradient in the liquid is the same as the hydrostatic pressure gradient in the outside vapor [1], $dP/dy = \rho_v g$. Equation (8.1) shows that the sinking force felt by the liquid is, in general, resisted by a combination of the effects of friction and inertia:

$$\rho_l \underbrace{\left(u \frac{\partial v}{\partial x} + v \frac{\partial v}{\partial y} \right)}_{\text{Inertia}} = \underbrace{\mu_l \frac{\partial^2 v}{\partial x^2}}_{\text{Friction}} + \underbrace{g(\rho_l - \rho_v)}_{\text{Sinking effect}} \tag{8.2}$$

We continue the analysis by assuming that the inertia effect is small when compared with the effect of friction, and we set the left side of Eq. (8.2) equal to zero. The resulting equation can be integrated twice in x and subjected to the conditions of no slip at the wall ($v = 0$ at $x = 0$) and zero shear at the liquid–vapor interface ($\partial v/\partial x = 0$ at $x = \delta$). The solution for the vertical liquid velocity profile is

$$v(x, y) = \frac{g}{\mu_l}(\rho_l - \rho_v)\delta^2 \left[\frac{x}{\delta} - \frac{1}{2}\left(\frac{x}{\delta}\right)^2\right] \tag{8.3}$$

in which the film thickness is an unknown function of longitudinal position, $\delta(y)$. One consequence of Eq. (8.3) is the local mass flowrate through a cross section of the film:

$$\Gamma(y) = \int_0^\delta \rho_l v \, dx = \frac{g\rho_l}{3\mu_l}(\rho_l - \rho_v)\delta^3 \tag{8.4}$$

The mass flowrate is $\Gamma(\text{kg/s} \cdot \text{m})$ and is expressed per unit length in the direction normal to the plane of Figure 8.2. Another symbol for this quantity would be $\dot{m}'$. The downward velocity and the flowrate are proportional to the sinking effect $g(\rho_l - \rho_v)$ and inversely proportional to the liquid viscosity.

The film thickness $\delta(y)$ can be determined by invoking the first law of thermodynamics for the control volume $\delta \times dy$ shown on the lower-left side of Figure 8.2. Entering this control volume from the right is the saturated vapor stream $d\Gamma$, with the enthalpy flowrate $h_g d\Gamma$. The vertical enthalpy inflow (W/m) associated with the mass flowrate Γ is

$$H = \int_0^\delta \rho_l v[h_f - c_{P,l}(T_{\text{sat}} - T)] \, dx \tag{8.5}$$

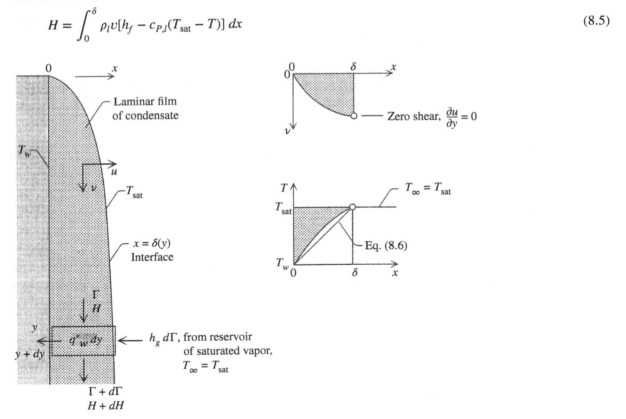

Figure 8.2 Laminar film of condensate in a reservoir of stationary saturated vapor.

where the quantity in the square brackets is the local specific enthalpy (kJ/kg) of the liquid at the point (x, y). Since the liquid is slightly subcooled $(T < T_{sat})$, its specific enthalpy is smaller than the specific enthalpy of saturated liquid, h_f. As was done originally by Nusselt [2], we assume that the local temperature T is distributed linearly across the film:

$$\frac{T_{sat} - T}{T_{sat} - T_w} \cong 1 - \frac{x}{\delta} \tag{8.6}$$

and, after using Eqs. (8.3) and (8.6) in the integral (8.5), we obtain

$$H = \left[h_f - \frac{3}{8} c_{P,l}(T_{sat} - T_w) \right] \Gamma \tag{8.7}$$

Finally, in accordance with the linear temperature profile assumption (8.6), the heat flux absorbed by the solid boundary of the control volume (i.e. the wall) is

$$q''_w \cong k_l \frac{T_{sat} - T_w}{\delta} \tag{8.8}$$

The steady-state form of the first law of thermodynamics for the $\delta \times dy$ system is

$$0 = H - (H + dH) + h_g d\Gamma - q''_w dy \tag{8.9}$$

or, after using Eqs. (8.7) and (8.8),

$$\frac{k_l}{\delta}(T_{sat} - T_w)dy = \underbrace{\left[h_{fg} + \frac{3}{8} c_{P,l}(T_{sat} - T_w) \right]}_{h'_{fg}} d\Gamma \tag{8.10}$$

Starting with this equation, we use h'_{fg} as shorthand for the *augmented* latent heat of condensation, which includes the proper latent heat, h_{fg}, and a contribution accounting for the cooling of the fresh condensate to temperatures below T_{sat}. Combined with the Γ expression (8.4), this equation becomes

$$\frac{k_l v_l (T_{sat} - T_w)}{h'_{fg} g(\rho_l - \rho_v)} dy = \delta^3 d\delta \tag{8.11}$$

and, after integrating from $y = 0$ where $\delta = 0$,

$$\delta(y) = \left[y \frac{4 k_l v_l (T_{sat} - T_w)}{h'_{fg} g(\rho_l - \rho_v)} \right]^{1/4} \tag{8.12}$$

In conclusion, the thickness of the laminar film increases as the longitudinal length raised to the power 1/4, that is, in the same way as the thickness (thermal, or velocity) of the vertical laminar boundary layer in a single-phase fluid, Eqs. (7.25c) and (7.31). Knowing $\delta(y)$, we can calculate, in order, the local heat transfer coefficient,

$$h_y = \frac{q''_w}{T_{sat} - T_w} = \frac{k_l}{\delta} = \left[\frac{k_l^3 h'_{fg} g(\rho_l - \rho_v)}{4 y v_l (T_{sat} - T_w)} \right]^{1/4} \tag{8.13}$$

the average heat transfer coefficient for a film of height L,

$$\bar{h}_L = \frac{h_{y=L}}{1 + (-1/4)} = \frac{4}{3} h_{y=L} \tag{8.14}$$

and the overall Nusselt number based on the L-averaged heat transfer coefficient,

$$\overline{Nu}_L = \frac{\overline{h}_L L}{k_l} = 0.943 \left[\frac{L^3 h'_{fg} g(\rho_l - \rho_v)}{k_l \nu_l (T_{sat} - T_w)} \right]^{1/4} \tag{8.15}$$

The dimensionless group formed on the right side of Eq. (8.15) is nearly equal to the slenderness ratio of the liquid film (cf. Eq. (8.12)):

$$\frac{L}{\delta(L)} = 0.707 \left[\frac{L^3 h'_{fg} g(\rho_l - \rho_v)}{k_l \nu_l (T_{sat} - T_w)} \right]^{1/4} \tag{8.16}$$

In numerical applications, the liquid properties that appear in these formulas are best evaluated at the average film temperature $(T_w + T_{sat})/2$. The latent heat of condensation h_{fg} is found in thermodynamic tables of saturated-state properties and takes the value that corresponds to the phase-change temperature T_{sat}. Rohsenow [3] refined the preceding analysis by not assuming the linear profile assumption (8.6) and by performing an integral analysis of the temperature distribution across the film. He found a temperature profile whose curvature increases with the degree of liquid subcooling, $c_{p,l}(T_{sat} - T_w)$. In place of the modified latent heat h'_{fg}, Eq. (8.10), Rohsenow recommended

$$h'_{fg} = h_{fg} + 0.68 c_{P,l}(T_{sat} - T_w) \tag{8.17}$$

This expression is also recommended for calculations involving the wavy and turbulent flow regimes. It can be rewritten as

$$h'_{fg} = h_{fg}(1 + 0.68\, Ja) \tag{8.18}$$

in which the *Jakob[1] number Ja* is a relative measure of the degree of subcooling experienced by the liquid film:

$$Ja = \frac{c_{P,l}(T_{sat} - T_w)}{h_{fg}} \tag{8.19}$$

This group is analogous to the Stefan number, Eq. (4.119), which in melting and solidification accounts for the degree of liquid superheating or solid subcooling.

To summarize, the total heat transfer rate absorbed by the wall, per unit length in the direction normal to the plane of Figure 8.2, is

$$q' = \overline{h}_L L(T_{sat} - T_w) \tag{8.20}$$
$$= k_l(T_{sat} - T_w)\overline{Nu}_L$$

The total (bottom-end) flowrate $\Gamma(L)$ can be calculated by substituting $y = L$ in what results from combining Eqs. (8.4) and (8.12). The total condensation rate $\Gamma(L)$ is proportional to the total cooling rate provided by the vertical wall:

$$\Gamma(L) = \frac{q'}{h'_{fg}} = \frac{k_l}{h'_{fg}}(T_{sat} - T_w)\overline{Nu}_L \tag{8.21}$$

---

1 Max Jakob (1879–1955) was a leading figure in German heat transfer engineering. He emigrated to the United States in 1936 and became a professor at the Illinois Institute of Technology. Through his research on boiling and condensation and his famous treatise on *Heat Transfer* (Wiley, 1949), he played a major role in the shaping of heat transfer as a core discipline in modern mechanical engineering.

The global Eqs. (8.20) and (8.21) hold true for the entire film, not only for the laminar section. Rewritten as $q' = \Gamma(L)h_{fg}(1 + 0.68\,Ja)$, Eq. (8.21) shows also that the cooling rate q' increases with both the latent heat h_{fg} and the degree of liquid subcooling, Ja. This trend should have been expected, because the cooling provided by the wall causes both the condensation of vapor at the $x = \delta$ interface and the cooling of the newly formed liquid to temperatures below T_{sat}.

The physical properties needed for evaluating the Nusselt number of Eq. (8.15) are listed in Table 8.2 and in Appendix C. In many instances, ρ_l is much greater than ρ_v, so that $\rho_l - \rho_v$ can safely be replaced with ρ_l.

The laminar film results discussed until now were derived by Nusselt [2] based on the assumption that the effect of inertia is negligible in the momentum balance (8.2). The complete momentum equation was used by Sparrow and Gregg [6] in a similarity formulation of the same problem. Their solution for $\overline{Nu}_L$ falls below Nusselt's equation (8.15) when the Prandtl number is smaller than 0.03 and the Jakob number is greater than 0.01.

In a subsequent analysis, Chen [7] abandoned the assumption of zero shear at the liquid–vapor interface (Figure 8.2, right) while retaining the effect of inertia in the momentum equation. The vapor was assumed saturated and stagnant *sufficiently far* from the interface. Next to the interface, the vapor is dragged downward by the falling film of condensate and forms a velocity boundary layer that bridges the gap between the downward velocity of the interface and the zero velocity of the outer vapor (see the small detail above Figure 8.3).

Chen's chart [7] for calculating the overall Nusselt number $\overline{Nu}_L$ is redrawn in Figure 8.3. Especially at low Prandtl numbers, the $\overline{Nu}_L$ values read off Figure 8.3 are smaller than those furnished by Sparrow and Gregg's solution [6] and agree better with experimental data. The lower $\overline{Nu}_L$ values are due to the additional restraining effect that the vapor drag has on the downward acceleration of the liquid film.

The scale analysis of laminar film condensation [1] showed that the fall of the liquid film is restrained by friction when $Pr_l < Ja$ and by inertia when $Pr_l > Ja$. The group that marks the transition from one type of

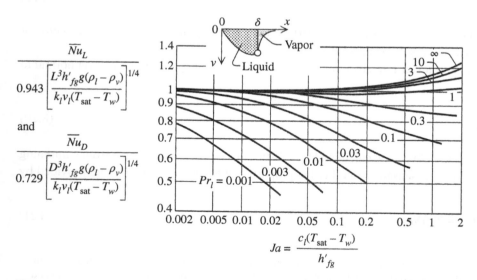

Figure 8.3 Effect of Prandtl number on heat transfer from a laminar film of condensate on a vertical wall or on a single horizontal cylinder.

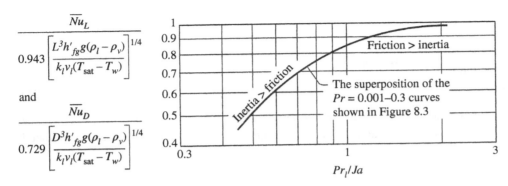

Figure 8.4 Inertia-restrained and friction-restrained film condensation on a vertical wall or on a single horizontal cylinder. Source: Bejan [1].

flow to the other is the ratio Pr_l/Ja. Indeed, if we use the group Pr_l/Ja on the abscissa of Figure 8.4, the low-Pr_l information of Figure 8.3 is represented well by the single curve shown in Figure 8.4.

Example 8.1 *Laminar Film Condensation: Vertical Surface*

A plane vertical wall of temperature 60 °C faces a space filled with stagnant saturated steam at atmospheric pressure. The height of the wall is $l = 2$ m. Assume that the film is laminar, and calculate the rate at which steam condenses on the vertical surface.

Solution

The relevant properties of water at atmospheric pressure and film temperature $(100 °C + 60 °C)/2 = 80 °C$ are (see Appendix C)

$$c_{P,l} = 4.196 \text{ kJ/kg} \cdot \text{K} \qquad \rho_l = 972 \text{ kg/m}^3$$
$$k_l = 0.67 \text{ W/m} \cdot \text{K} \qquad \nu_l = 3.66 \times 10^{-7} \text{ m}^2/\text{s}$$

The latent heat of condensation at atmospheric pressure (or at 100 °C) is $h_{fg} = 2257$ kJ/kg. The Jakob number is small:

$$Ja = \frac{c_{P,l}(T_{sat} - T_w)}{h_{fg}} = 0.074$$

and this means that h'_{fg} is nearly the same as h_{fg}:

$$h'_{fg} = h_{fg}(1 + 0.68Ja) = 1.05h_{fg}$$
$$= 2370.6 \text{ kJ/kg}$$

The total (bottom-end) condensation rate $\Gamma(L)$ is given by Eq. (8.21). The overall Nusselt number $\overline{Nu}_L$ can be calculated using Eq. (8.15) by assuming that the film is laminar over its entire height:

$$\frac{L^3 h'_{fg} g(\rho_l - \rho_v)}{k_l \nu_l (T_{sat} - T_w)} = \frac{(8 \text{ m}^3)(2370.6 \text{ kJ/kg})(9.81 \text{ m/s}^2)(972 \text{ kg/m}^3)}{(0.67 \text{ W/m} \cdot \text{k})(3.66 \times 10^{-7} \text{ m}^2/\text{s})(100 - 60)\text{K}} = 1.844 \times 10^{16}$$

$$\overline{Nu}_L = 0.943(1.844 \times 10^{16})^{1/4} = 10\,988 \tag{8.15}$$

$$\Gamma(L) = \frac{k_l}{h'_{fg}}(T_{\text{sat}} - T_w)\overline{Nu}_L$$

$$= \frac{0.67 \text{W}}{\text{m} \cdot \text{K}} \frac{\text{kg}}{2370.6 \text{ kJ}}(40 \text{ K})(10\,988)$$

$$= 0.124 \text{ kg}/\text{m} \cdot \text{s} \tag{8.21}$$

In Example 8.2 we learn that the laminar film assumption is inappropriate and that most of the film is wavy and turbulent. For the same reason, the actual condensation rate will be higher than the value calculated earlier.

8.1.2 Turbulent Film on Vertical Surface

The liquid film becomes wavy and, farther downstream, turbulent when the order of magnitude of its *local* Reynolds number is greater than 10^2 (see Appendix F). The local Reynolds number of the liquid film can be constructed in the form of the group $\rho_l \bar{u} \delta / \mu_l$, in which δ is the local thickness and $\bar{u}$ is the scale representative of the local downward velocity. And since the product $\rho_l \bar{u} \delta$ is of the same order of magnitude as the local liquid mass flowrate Γ, the *local* Reynolds number can be expressed as the ratio Γ/μ_l:

$$Re_y = \frac{4}{\mu_l}\Gamma(y) \tag{8.22}$$

The flowrate $\Gamma(y)$ and the Reynolds number Re_y increase in the downstream direction. Measurements indicate that the laminar section of the film expires in the vicinity of $Re_y \sim 30$. The film can be described as

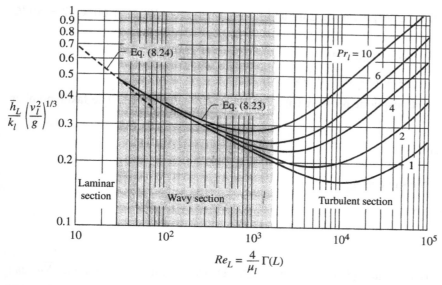

Figure 8.5 The *L*-averaged heat transfer coefficient for laminar, wavy, and turbulent film condensation on a vertical surface.

wavy in the segment corresponding approximately to $30 \leq Re_y \leq 1800$. Farther downstream the film appears turbulent. The succession of flow regimes is shown on the abscissa of Figure 8.5.

Experiments revealed that the heat transfer rate in the wavy and turbulent sections is considerably larger than the estimate based on the laminar film analysis, Eq. (8.15) and Figure 8.3. The sizeable record of experimental data and correlations on condensation heat transfer in the wavy and turbulent regimes was reviewed in Ref. [8], which developed a correlation for the *average* heat transfer coefficient for an L-tall film that may have wavy and turbulent regions:

$$\frac{\overline{h}_L}{k_l}\left(\frac{v_l^2}{g}\right)^{1/3} = [Re_L^{-0.44} + (5.82 \times 10^{-6})Re_L^{0.8}Pr_l^{1/3}]^{1/2} \tag{8.23}$$

Figure 8.5 shows that this correlation applies only above $Re_L \sim 30$. Equation (8.23) agrees within $\pm 10\%$ with measurements in experiments where the vapor was stagnant (or slow enough) so that the effect of shear at the interface was negligible. Below $Re_L \sim 30$, the recommended average heat transfer formula is Eq. (8.15), which, when $\rho_l \gg \rho_v$, can be projected on Figure 8.5 as the line:

$$\frac{\overline{h}_L}{k_l}\left(\frac{v_l^2}{g}\right)^{1/3} = 1.468 Re_L^{-1/3} \tag{8.24}$$

The usual unknown in a problem of vertical film condensation is the total condensation rate $\Gamma(L)$ or, alternatively, Re_L. Unfortunately, this unknown influences both sides of Eq. (8.23), or both the ordinate and abscissa parameters of Figure 8.5. Instead of the trial-and-error procedure that goes with using Eq. (8.23) or Figure 8.5, it is more convenient to rewrite the ordinate parameter of Figure 8.5 as

$$\frac{\overline{h}_L}{k_l}\left(\frac{v_l^2}{g}\right)^{1/3} = \frac{Re_L}{B} \tag{8.25}$$

where *the* dimensionless group B is proportional to the physical quantities that, when increasing, tend to augment the condensation rate, namely, L and $T_{sat} - T_w$:

$$B = L(T_{sat} - T_w)\frac{4k_l}{\mu_l h'_{fg}}\left(\frac{g}{v_l^2}\right)^{1/3} \tag{8.26}$$

The B group is a dimensionless temperature difference that serves as *driving parameter for film condensation*. Equation (8.25) is a consequence of the global statements (8.20)–(8.21), and it allows us to rewrite Eqs. (8.23) and (8.24) as

$$B = Re_L[Re_L^{-0.44} + (5.82 \times 10^{-6})Re_L^{0.8}Pr_l^{1.3}]^{-1/2} \tag{8.27}$$

$$B = 0.681 Re_L^{4/3} \tag{8.28}$$

Figure 8.6 displays this information by using the unknown Re_L on the abscissa and the driving parameter B on the ordinate. We now see the one-to-one relation between the condensation rate (the bottom-end Reynolds number) and the driving parameter B. The condensation rate increases faster when the film length is dominated by the turbulent regime.

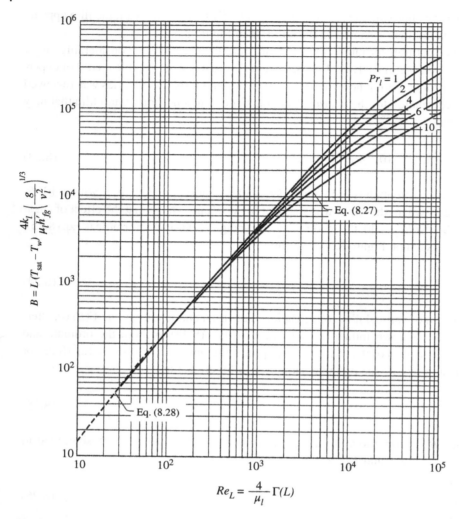

Figure 8.6 Total condensation rate (or Re_L) versus condensation driving parameter B.

Example 8.2 *Turbulent Film Condensation: Vertical Surface*

A plane vertical wall of temperature 60 °C faces a space filled with stagnant saturated steam at atmospheric pressure. The wall height is $L = 2$ m. Calculate the rate at which steam condenses on the vertical surface.

Solution

The properties of water at atmospheric pressure and film temperature $(100\,°C + 60\,°C)/2 = 80\,°C$ are (see Example 8.1 and Appendix C)

$$c_{Pl} = 4.196 \text{ kJ/kg} \cdot \text{K} \quad \mu_l = 3.55 \times 10^{-4} \text{ kg/m} \cdot \text{s}$$

$$k_l = 0.67 \text{ W/m} \cdot \text{K} \quad \nu_l = 3.66 \times 10^{-7} \text{ m}^2/\text{s}$$

$$h'_{fg} = 2370.6 \text{ kJ/kg} \quad Pr_l = 2.23$$

In the figure, the vertical axis is labeled $B = L(T_{sat} - T_w)\dfrac{4k_l}{\mu_l h'_{fg}}\left(\dfrac{g}{\nu_l^2}\right)^{1/3}$ and the horizontal axis is labeled $Re_L = \dfrac{4}{\mu_l}\Gamma(L)$. The curves are labeled $Pr_l = 1$, 2, 4, 6, 10, with Eq. (8.27) and Eq. (8.28) indicated.

For calculating the condensation rate $\Gamma(L)$, we have two options: Figure 8.6 and, depending on the Re_L range, Eqs. (8.27)–(8.28). In either case we must begin by calculating the driving parameter

$$B = L(T_{sat} - T_w)\frac{4k_l}{\mu_l h'_{fg}}\left(\frac{g}{v_l^2}\right)^{1/3}$$

$$= 2 \text{ m } (100 - 60)\text{K}\left(\frac{4 \times 0.67 \text{ W/m} \cdot \text{K}}{3.55 \times 10^{-4} \text{ kg/m} \cdot \text{s } 2370.6 \text{ kJ/kg}}\right)\left(\frac{9.81 \text{ m/s}^2}{(3.66)^2 \ 10^{-14} \text{ m}^4/\text{s}^2}\right)^{1/3}$$

$$= 10\ 659$$

On the graphic route, we enter $B = 10\,659$ and $Pr_l = 2.23$ in Figure 8.6 and read $Re_L \cong 2200$, which shows that the film is dominated by its turbulent section. The corresponding condensation rate is

$$\Gamma(L) = \frac{\mu_l}{4}Re_L = \frac{1}{4}(3.55 \times 10^{-4} \text{ kg/m} \cdot \text{s})2200$$

$$= 0.195 \text{ kg/m} \cdot \text{s}$$

Numerically, we can enter the same B and Pr_l values in Eq. (8.27) to obtain more precisely

$$Re_L = 2177 \text{ and } \Gamma(L) = 0.193 \text{ kg/m} \cdot \text{s}$$

This condensation rate is 56% greater than the estimate obtained in Example 8.1, by assuming that the film is laminar. The turbulence of the film augments the condensation rate that would have occurred had the flow remained laminar.

8.1.3 Film Condensation in Other Configurations

The vertical wall results described until now hold true not only for flat surfaces (Figure 8.7a) but also for curved vertical surfaces on which the condensate film is sufficiently thin. For the vertical cylindrical surface (internal or external) shown on the right side of Figure 8.7, the film is "thin" when its thickness scale is smaller than the diameter.

In the case of a plane wall inclined at an angle θ with respect to the vertical direction (Figure 8.8a), the gravitational acceleration component that acts along the surface is $g \cos \theta$. Condensation heat transfer results

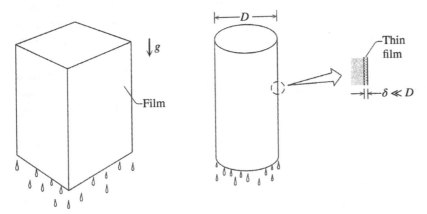

Figure 8.7 Vertical surfaces with thin films of condensate that can be regarded as plane.

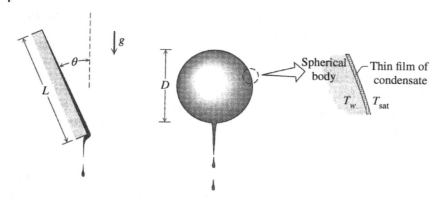

Figure 8.8 Film condensation on inclined plane and spherical surfaces.

for the inclined wall can be obtained by replacing g with $g \cos \theta$ in the results reported until now for a vertical plane wall.

More complicated surface shapes are those that are curved in such a way that the tangential component of gravity varies along the flowing film of condensate. One example is the *spherical surface* shown on the right side of Figure 8.8. If the film is laminar all around, then the diameter-averaged heat transfer coefficient is given by [9]

$$\overline{Nu}_D = \frac{\overline{h}_D D}{k_l} = 0.815 \left[\frac{D^3 h'_{fg} g (\rho_l - \rho_v)}{k_l \nu_l (T_{sat} - T_w)} \right]^{1/4} \tag{8.29}$$

The laminar film condensation on the surface of a *single horizontal cylinder* of diameter D (Figure 8.9, upper left) was first analyzed by Nusselt [2], who relied on the same simplifying assumptions as in the vertical wall analysis detailed in Section 8.1.1. The average heat transfer coefficient under a condensate film that is laminar all around the cylinder is [9]

$$\overline{Nu}_D = \frac{\overline{h}_D D}{k_l} = 0.729 \left[\frac{D^3 h'_{fg} g (\rho_l - \rho_v)}{k_l \nu_l (T_{sat} - T_w)} \right]^{1/4} \tag{8.30}$$

This expression is similar to the laminar film relationships for the sphere, Eq. (8.29), and the vertical wall, Eq. (8.15). For the horizontal cylinder and the sphere, the diameter D plays the role of vertical dimension in the same sense that L measures the height of the vertical plane wall in Eq. (8.15). The physical properties needed in Eqs. (8.29) and (8.30) and other formulas presented in this section are to be evaluated at the film temperature $(T_w + T_{sat})/2$.

The Prandtl number effect on laminar film condensation on a single horizontal cylinder was documented by Sparrow and Gregg [10] and Chen [11]. The latter considered also the effect of interfacial shear and found that the Pr_l effect is described fairly well by the curves plotted in Figure 8.3. Note the alternative $\overline{Nu}_D$ meaning of the ordinates of Figures 8.3 and 8.4.

Analogous to the analysis that yields the single-cylinder formula (8.30), the laminar film analysis of a *vertical column* of n horizontal cylinders (Figure 8.9, upper right) leads to

$$\overline{Nu}_{D,n} = \frac{\overline{h}_{D,n} D}{k_l} = 0.729 \left[\frac{D^3 h'_{fg} g (\rho_l - \rho_v)}{n k_l \nu_l (T_{sat} - T_w)} \right]^{1/4} \tag{8.31}$$

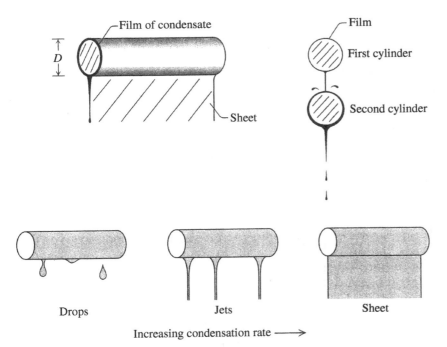

Figure 8.9 Film condensation on a single horizontal cylinder and on a vertical column ($n = 2$) of horizontal cylinders of the same size. The lower drawing shows the flow that impinges on the next cylinder.

The heat transfer coefficient $\overline{h}_{D,n}$ has been averaged over *all* the cylindrical surfaces, so that the total heat transfer rate per unit of cylinder length is $q' = \overline{h}_{D,n} n \pi D (T_{sat} - T_w)$. By comparing the right sides of Eqs. (8.30) and (8.31), we note that the average heat transfer coefficient of the n-tall column is generally smaller than that of the single cylinder:

$$\overline{h}_{D,n} = \frac{\overline{h}_D}{n^{1/4}} \tag{8.32}$$

The $\overline{h}_{D,n}$ values that are found experimentally are usually greater than the values calculated based on Eq. (8.31). This augmentation effect can be attributed to the splashing caused by the sheet or droplets of condensate as they impinge on the next cylinder. The lower part of Figure 8.9 shows that the condensation rate influences the type of flow that falls on the next cylinder, namely, drops, jets, or sheet, as the condensation rate increases. An additional factor can be the condensation that takes place on the sheet (or droplets) between two consecutive cylinders [11], because the falling sheet is at an average temperature below T_{sat} (i.e. it is subcooled; review the discussion under Eq. (8.5)). Furthermore, if each tube is slightly tilted or bowed (due to its weight or to a defect in the assembly), the condensate runs longitudinally along the tube and drips only from its lowest region. In such cases most of the length of the next cylinder is not affected by the condensate generated by the preceding cylinder.

When the cooled surface is perfectly horizontal and faces upward, Figure 8.10a, the condensate flows away from the central region and spills over the edges [12]. For a long *horizontal strip* of width L, the average heat transfer coefficient is given by a formula similar to Eq. (8.15), except that the exponent of the dimensionless

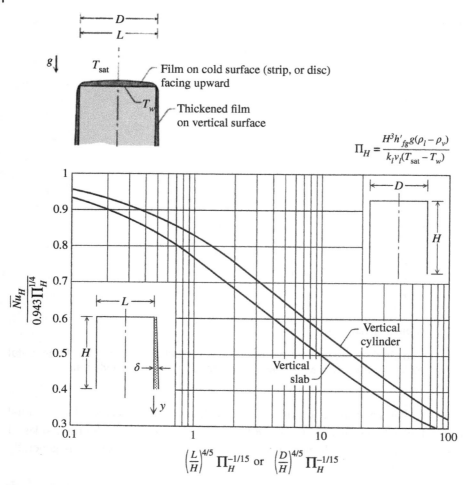

Figure 8.10 Film of condensate on a horizontal strip of width L or a horizontal disc of diameter D. Source: Drawn after Bejan [12].

group on the right side is 1/5:

$$\overline{Nu}_L = \frac{\overline{h}_L L}{k_l} = 1.079 \left[\frac{L^3 h'_{fg} g(\rho_l - \rho_v)}{k_l v_l (T_{sat} - T_w)} \right]^{1/5} \tag{8.33}$$

The average heat transfer coefficient for an upward facing *disc* with free edges is [12]

$$\overline{Nu}_D = \frac{\overline{h}_D D}{k_l} = 1.368 \left[\frac{D^3 h'_{fg} g(\rho_l - \rho_v)}{k_l v_l (T_{sat} - T_w)} \right]^{1/5} \tag{8.34}$$

Corresponding formulas for other surfaces whose shapes are somewhere between the "very long" shape of the strip and the "round" shape of the disc can be deduced from Eqs. (8.33) and (8.34) by using the concept of *characteristic length* of Eq. (7.76). Equations (8.33) and (8.34) are based on an analysis and set of assumptions of the same type as the ones outlined in Section 8.1.1.

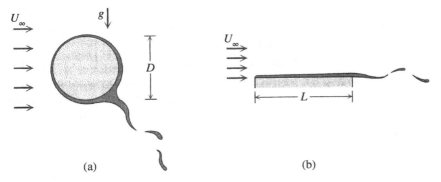

Figure 8.11 Film condensation on a horizontal cylinder in cross-flow (a) and on a flat plate parallel to the flow (b).

These results may be used to estimate the contribution made by the horizontal "roof" surface to the total condensation rate on a three-dimensional body. The horizontal surface contributes to the total condensation rate in two ways: directly, through the flowrate estimated based on Eqs. (8.33) and (8.34), and indirectly, by thickening the film that coats the vertical lateral surface [12]. In other words, when the condensate collected on the top surface spills over the edge, the vertical surface heat transfer coefficient is smaller than the value that would be calculated based on Eq. (8.15). This effect is documented in Figure 8.10b, in which, unlike in Eq. (8.15), the height of the vertical surfaces is labeled H. The symbol Π_H is defined on the figure and is used as shorthand for the dimensionless group that emerged on the right side of Eq. (8.15). Figure 8.10 shows that the roof condensate inhibits (i.e. the ordinate values in Figure 8.10 are less than 1) the condensation on the vertical surfaces when the abscissa parameter exceeds the order of magnitude of 1.

The film condensation processes described until now are examples of *natural convection*, because the flow is being driven by gravity. Considerably more complicated is condensation where the vapor is forced to flow over the cooled surface: the vapor and the condensate film interact across their mutual interface. The forced flow of vapor tends to drag the liquid in its direction, and the overall condensation heat transfer process is one of natural convection mixed with *forced convection*. One example is the film condensation on the outside of a *horizontal cylinder in cross-flow*, Figure 8.11a. The surface heat transfer coefficient depends on the free-stream velocity of the vapor, U_∞, as well as on gravity [13]:

$$\frac{\bar{h}_D D}{k_l} = 0.64 Re_D^{1/2} \left[1 + \left(1 + 1.69 \frac{g h'_{fg} \mu_l D}{U_\infty^2 k_l (T_{sat} - T_w)} \right)^{1/2} \right]^{1/2} \tag{8.35}$$

The Reynolds number is based on the kinematic viscosity of the liquid, $Re_D = U_\infty D / v_l$. Equation (8.35) holds for Reynolds numbers up to 10^6. In the limit of negligible gravitational effect, the right side of Eq. (8.35) approaches $0.64 Re_D^{1/2}$. In the opposite extreme, when the vapor stream slows to a halt, Eq. (8.35) becomes identical to Eq. (8.30), which represents the pure natural convection limit of the flow of Figure 8.11.

The results for laminar film condensation on a *flat plate in a parallel stream* of saturated vapor (Figure 8.11b) are represented well by [14]

$$\frac{\bar{h}_L L}{k_l} = 0.872 Re_L^{1/2} \left[\frac{1.508}{(1 + Ja/Pr_l)^{3/2}} + \frac{Pr_l}{Ja} \left(\frac{\rho_v \mu_v}{\rho_l \mu_l} \right)^{1/2} \right]^{1/3} \tag{8.36}$$

Here, the Reynolds number is again based on the liquid viscosity, $Re_L = U_\infty L/\nu_l$, and the Jakob number is the same as the one defined in Eq. (8.19). Equation (8.36) has the proper asymptotic behavior and has been tested in the range $(\rho_l \mu_l / \rho_v \mu_v)^{1/2} \sim 10 - 500$ and $Ja/Pr_l \sim 0.01 - 1$.

Inside a *vertical cylinder* with cocurrent vapor flow, Figure 8.12, the downward progress of the liquid is aided by the vapor that flows through the core of the cross section. The liquid film is therefore thinner than in the absence of downward vapor flow, and the L-averaged heat transfer coefficient $\bar{h}_L$ and the condensation rate are greater. Chen et al. [8] reviewed the experimental information available on this configuration and proposed a correlation that can be rearranged this way:

$$
\frac{\bar{h}_L}{k_l} \left(\frac{\nu_l^2}{g} \right)^{1/3} = \left[\begin{array}{l} Re_L^{-0.44} + (5.82 \times 10^{-6}) Re_L^{0.8} Pr_l^{1.3} \\ + (3.27 \times 10^{-4}) \dfrac{Pr_l^{1.3}}{D^2} \left(\dfrac{\nu_l^2}{g} \right)^{2/3} \left(\dfrac{\mu_v}{\mu_l} \right)^{0.156} \left(\dfrac{\rho_l}{\rho_v} \right)^{0.78} \dfrac{Re_L^{0.4} Re_t^{1.4}}{(1.25 + 0.39\, Re_L/Re_t)^2} \end{array} \right]^{1/2}
$$

(8.37)

This form is similar to Eq. (8.23), because the only difference between the present configuration (Figure 8.12) and that of Figure 8.7 is the presence of the core flow of vapor. The third group on the right side of Eq. (8.37) accounts for the increase in $\bar{h}_L$ that is due to the interfacial shear between the vapor and the liquid film. The Reynolds number Re_L is defined according to Eq. (8.22). The "terminal" Reynolds number Re_t is based not on the actual flowrate $\Gamma(L)$ but on $\dot{m}_v/\pi D$, in which $\dot{m}_v$ is the total flowrate of the vapor that enters through the top of the tube. The terminal Reynolds number Re_t is the maximum value approached by Re_L as a greater fraction of the original vapor stream is converted into liquid at the bottom of the tube (note: $Re_L < Re_t$).

The tube orientation is no longer a factor when *the vapor stream is fast* enough so that the last term overwhelms the others on the right side of Eq. (8.37). In this limit the gravitational effect is negligible, and the average heat transfer coefficient can be calculated with the simpler formula

$$
\frac{\bar{h}_L D}{k_l} = 0.0181 Pr_l^{0.65} \left(\frac{\mu_v}{\mu_l} \right)^{0.078} \left(\frac{\rho_l}{\rho_v} \right)^{0.39} \frac{Re_L^{0.2} Re_t^{0.7}}{1.25 + 0.39\, Re_L/Re_t}
$$

(8.38)

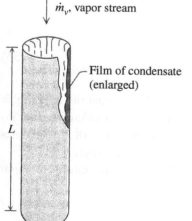

$\dot{m}_v$, vapor stream

Film of condensate (enlarged)

L

$\leftarrow D \rightarrow$

Figure 8.12 Condensation in a vertical tube with cocurrent flow of vapor.

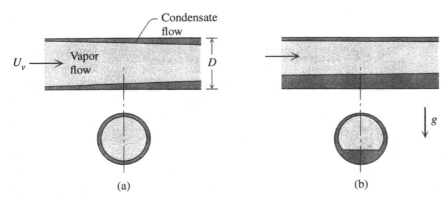

Figure 8.13 Annular film condensation in a tube with *fast* vapor flow (a) and accumulation of condensate at the bottom of a horizontal tube with *slow* vapor flow (b).

The rate of condensation inside a *horizontal tube* with fast vapor flow (Figure 8.13a) can be calculated based on Eq. (8.38). In this limit the liquid film coats uniformly the perimeter of the cross section. When the vapor flow is slow, the liquid flow favors the lower region of the tube cross section (Figure 8.13b). Chato [15] found that when the vapor flow Reynolds number is small,

$$\frac{\rho_v U_v D}{\mu_v} < 3.5 \times 10^4 \tag{8.39}$$

the condensation is dominated by natural convection, and $\overline{h}_D$ is given by

$$\frac{\overline{h}_D D}{k_l} = 0.555 \left[\frac{D^3 h'_{fg} g (\rho_l - \rho_v)}{k_l \nu_l (T_{\text{sat}} - T_w)} \right]^{1/4} \tag{8.40}$$

In Eq. (8.40) the effect of the buildup of condensate in the longitudinal direction has not been taken into account.

In cases where the vapor is *superheated*, $T_\infty > T_{\text{sat}}$, Rohsenow [16] recommends replacing h'_{fg} with a slightly larger quantity, h''_{fg}, which accounts also for the cooling experienced by the vapor en route to its saturation temperature at the interface:

$$h''_{fg} = h'_{fg} + c_{p,v} (T_\infty - T_{\text{sat}}) \tag{8.41}$$

A common feature of all the configurations discussed until now is that the vapor is pure, that is, it contains nothing but the substance that eventually condenses into the liquid film. When the vapor is a mixture containing not only the condensing species but also one or more *noncondensable* gases, the heat transfer coefficient is significantly lower than when the noncondensable gases are absent. The condensation rate is lower because the condensing species must first diffuse through the concentration boundary layer that coats the gas side of the interface. The condensing species must first overcome the mass transfer resistance posed by the concentration boundary layer [17]. The film condensation process can be complicated further by transient effects, such as a sudden change in wall temperature, vapor flow, or wall orientation.

8.1.4 Dropwise and Direct-Contact Condensation

The condensate distributes itself as a continuous thin film on the cooled surface only when the liquid wets the solid. This happens when the surface tension between the liquid and the solid material is sufficiently small, for example, when the solid surface is clean (grease-free), as in the condensation of steam on a clean metallic surface.

When the surface tension is large, the condensate coalesces into a multitude of droplets of many sizes. In time, each droplet grows as more vapor condenses on its exposed surface. The formation of each droplet is initiated at a point of surface imperfection (pit, scratch) called *nucleation site*. The droplet grows. There comes a time when the tangential pull of gravity dislodges the droplet and carries it downstream. The moving droplet devours the smaller droplets found in its path, creating in this way a clean trail ready for the initiation of a new generation of droplets of the smallest size.

Since the condensation rate is the highest in the absence of condensate on the surface (film or droplets), the periodic cleaning performed by the large drops renews finite-size regions of the surface for the restart of the time-dependent condensation process, Figure 8.14. This *surface renewal process* is the reason why dropwise condensation is a highly effective mechanism. The heat transfer coefficient for dropwise condensation is approximately 10 times greater than the corresponding coefficient estimated based on the assumption that the condensate forms a continuous film. When the droplet impacts a surface, it spreads as a splat or dendritic splash and "renews" the surface [18].

In the design of condensers, whose function is to cool a vapor stream and to convert it into liquid, there is a great incentive to promote the breakup of the condensate film into drops. This can be accomplished by (i) coating the solid surface with an organic substance (e.g. oil, wax, kerosene, oleic acid), (ii) injecting

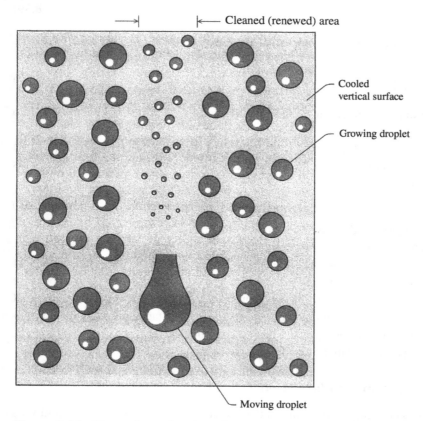

Figure 8.14 The surface cleaning effect due to the departure of one large drop.

nonwetting chemicals into the vapor so that they will be deposited on the condenser surface, and (iii) coating the surface with Teflon, silicone, or a noble metal (e.g. gold, silver). In methods (i) and (ii), unfortunately, the promoter material wears off, as it is gradually removed by the scraping action of the droplet movement. In method (iii), fluorocarbon coatings such as Teflon have good surface characteristics but a relatively low thermal conductivity. If the coating is thicker than about 20 μm, its conduction resistance tends to offset the heat transfer augmentation effect owing to dropwise condensation on the vapor side of the coating.

The phenomenon of dropwise condensation is complicated by its intermittent time-dependent character, the dominant effect of surface tension (drop size and shape), and the uncertainty associated with the location of nucleation sites and the time when the largest droplet will start its movement downstream. The film and drop condensation mechanisms are two examples of what is generally referred to as *surface condensation*. In both cases the condensate adheres to a solid surface that is being cooled by an external entity (from the back side). The mechanism of *direct-contact condensation* is conceptually different because the solid surface is absent and the cooling effect is provided by the large pool of subcooled liquid through which bubbles of condensing vapor rise. Surface tension plays an important role in determining the size, shape, and life of each vapor bubble. The progress on condensation has been reviewed in Ref. [19].

8.2 Boiling

8.2.1 Pool Boiling

In this section we turn our attention to the mechanism of boiling heat transfer, which occurs when the temperature of a solid surface is sufficiently higher than the saturation temperature of the liquid with which it comes in contact. The solid–liquid heat transfer is accompanied by the transformation of some of the heated liquid into vapor and by the formation of distinct vapor bubbles, jets, and films. The vapor and the surrounding packets of heated liquid are carried away by the effect of buoyancy (natural convection or pool boiling) or by a combination of buoyancy and the forced flow of liquid that may be sweeping the solid heater (mixed convection or flow boiling).

Boiling is therefore the short name for convective heat transfer with change of phase (liquid → vapor) when a *liquid is being heated* by a sufficiently hot surface. Boiling can be seen as the reverse of the condensation phenomenon discussed in the first part of this chapter, where the change of phase (vapor → liquid) was caused by the *cooling of a vapor* by contact with a sufficiently cold surface.

We begin with the case of *pool boiling*, Figure 8.15, in which the heater surface (T_w) is immersed in a pool of initially stagnant liquid (T_l). The basic question consists again of pinpointing the relationship between surface heat flux q''_w and temperature difference $T_w - T_{sat}$, where T_{sat} is the saturation temperature of the liquid. When, as shown on the left side of Figure 8.15, the liquid is at a temperature below saturation (i.e. subcooled, $T_l < T_{sat}$), boiling is confined to a layer in the immediate vicinity of the heater surface. The vapor bubbles collapse (recondense) as they rise through the subcooled liquid. When the liquid pool is at the saturation temperature (Figure 8.16, right), the vapor generated at the heater surface reaches the free surface of the pool. In what follows it is assumed that the liquid in the pool is all saturated ($T_l = T_{sat}$).

Figure 8.16 shows the main features of the relationship between q''_w and the excess temperature ($T_w - T_{sat}$). This particular curve corresponds to the pool boiling of water at atmospheric pressure; however, its "roller coaster" shape is a characteristic of the curves describing the pool boiling of other liquids. The nonmonotonic relationship between heat flux and excess temperature is due to the various forms (bubbles, film) that

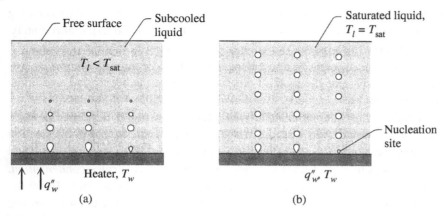

Figure 8.15 Nucleate pool boiling of a subcooled liquid (a) and a saturated liquid (b).

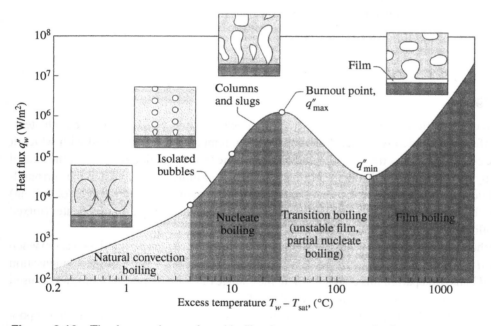

Figure 8.16 The four regimes of pool boiling in water at atmospheric pressure.

the generated vapor takes near the heater surface. The peculiar shape of the boiling curve is the basis for distinguishing between several pool boiling *regimes*.

The transition from one regime to the next can be seen by reading Figure 8.17 from left to right. This corresponds to a boiling experiment in which the heater surface temperature is increased monotonically and the resulting heat flux is measured. The experimental setups that can be used to trace the boiling curve are taken up at the end of this section.

One interesting aspect of the curve is that at low excess temperatures (in water, at $T_w - T_{sat} \lesssim 4\,°C$), the heat transfer occurs without the appearance of bubbles on the heater surface. In this regime the near-surface liquid becomes superheated and rises in the form of *natural convection* currents to the free surface of the

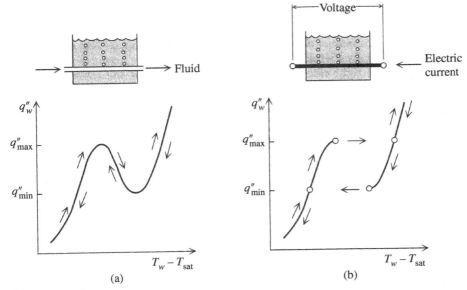

Figure 8.17 The pool boiling curve in a temperature-controlled experiment (a) and in a power-controlled experiment (b).

pool. If the heater surface is horizontal and large, and if the liquid pool is shallow, the convection currents are similar to the cellular (Bénard) flow shown on the right side of Figure 7.16. The relationship between q''_w and $T_w - T_{sat}$ depends on the shape and orientation of the immersed heater and can be determined by employing the formulas assembled in Chapter 7.

Proceeding toward larger excess temperatures, the next regime is *nucleate boiling*. This is characterized by the generation of vapor at a number of favored spots on the surface, which are called *nucleation sites*. A probable nucleation site is a tiny crack in the surface in which the trapped liquid is surrounded by a relatively large heater area per unit of liquid volume. At the low end of the nucleate boiling curve, the boiling process consists of isolated bubbles. At higher temperatures, the bubble frequency increases, the nucleation sites multiply, and the isolated bubbles interact and are replaced by slugs and columns of vapor.

The formation of more and more vapor in the vicinity of the surface has the effect of gradually insulating the surface against the T_{sat}-cold liquid. This effect is responsible for the gradual decrease of the slope of the nucleate boiling part of the curve and for its expiration at the point of maximum (peak) heat flux q''_{max}. The latter is also called *critical heat flux*, and in water it is of the order of 10^6 W/m^2. The excess temperature at this point is approximately 30 °C.

The next regime is the most peculiar, because the heat flux actually decreases as the excess temperature, $T_w - T_{sat}$, continues to increase. This trend reflects the fact that increasingly greater portions of the heater surface become coated with a continuous film of vapor. The vapor is unstable and is intermittently replaced by nucleate boiling. This regime is called *transition boiling*: it expires at the point[2] of minimum heat flux q''_{min}, where the excess temperature has become just large enough to sustain a stable vapor film on the heater

---

2 The corresponding temperature of the surface is called the Leidenfrost temperature, or Leidenfrost point, in honor of Johann Gottlob Leidenfrost, a German medical doctor who studied the evaporation of liquid droplets on hot surfaces. A portion of his 1756 treatise *De Aquae Communis Nonnullis Qualitatibus Tractatus* was translated from Latin [20].

surface. In water at 1 atm, the minimum heat flux is in the 10^4–10^5 W/m² range and occurs at an excess temperature in the 100–200 °C range.

At even larger excess temperatures, the vapor film covers the entire surface and the heat flux q''_w resumes its monotonic increase with $T_w - T_{sat}$. Radiation heat transfer across the film plays a progressively greater role as the excess temperature increases. This high-temperature mode is called the *film boiling* regime. It persists until T_w reaches the melting point of the surface material, that is, until the meltdown (burnout) of the heater surface. If, as in the case of a platinum surface, the melting point is very high (namely, 2042 K), the film boiling portion of the boiling curve can extend to heat fluxes above the critical q''_{max} of the nucleate boiling regime.

The tracing of the boiling curve from left to right in Figure 8.16 was based on the assumption that the excess temperature $T_w - T_{sat}$ can be controlled and increased monotonically. An experiment in which heater temperature control is possible is shown on the left side of Figure 8.17. The heater is a horizontal tube immersed in a pool of liquid. The heater surface temperature is controlled by a preheated stream that flows through the tube. In this temperature-controlled experiment, the boiling curve can be traced in either direction by gradually increasing or decreasing the excess temperature.

An alternative setup is the power-controlled experiment shown on the right side of Figure 8.17. The heater is a horizontal cylinder (a wire) stretched in a pool of liquid. The heat flux q''_w is controlled by the experimentalist who measures the power dissipated in the electrical resistance posed by the wire. In this experiment the shape of the emerging boiling curve depends on whether the power (q''_w) is increased or decreased.

When the power increases monotonically, the transition from the natural convection regime to the several forms of the nucleate boiling regime can be observed. As the imposed heat flux increases slightly above the critical value q''_{max}, the wire temperature increases abruptly (and dramatically) to the value associated with the film boiling portion of the boiling curve. In most cases this new temperature would be above the melting point of the surface material, and the wire burns up. This is why the peak of the nucleate boiling portion of the curve is often called the *burnout point*. The catastrophic event that can occur at heat fluxes comparable with and greater than q''_{max} is why in power-controlled applications of boiling heat transfer (e.g. nuclear reactors, electrical resistance heaters) it is advisable to operate at heat fluxes safely smaller than q''_{max}.

The power-controlled pool boiling experiment can be run in reverse by decreasing the heat flux. In that case the excess temperature decreases along the film boiling portion of the curve all the way down to the Leidenfrost temperature. As the heat flux is lowered slightly below the minimum heat flux of film boiling, q''_{min}, the vapor film collapses, the isolated bubbles form, and the wire temperature drops to the low level associated with the nucleate boiling regime.

To summarize, during heat flux-controlled boiling, the transition boiling regime is inaccessible, and certain portions of the boiling curve can be reached while varying q''_w in only one direction. For example, the nucleate boiling regime in the vicinity of the point of maximum heat flux can be established only by increasing the heat flux, starting from a sufficiently low level. Although much less important in practice, the film boiling regime in the vicinity of q''_{min} can be achieved by decreasing the heat flux, starting from a sufficiently high level. It is said that q''_w-controlled boiling is an example of *hysteresis*,[3] a phenomenon that depends not only on the imposed condition (q''_w) but also on its previous history, in this instance, the previous value of q''_w.

The boiling curve was first determined by Professor S. Nukiyama [21] of Tohoku University (Sendai, Japan), who employed the q''_w-controlled method illustrated on the right side of Figure 8.17. The existence

---

3 From the Greek words *hysteresis* (a deficiency) and *hysterein* (to lag behind, to fall short).

of the missing transition boiling portion of the curve was demonstrated based on T_w-controlled experiments by Drew and Mueller [22].

8.2.2 Nucleate Boiling and Peak Heat Flux

The most practical regime of the entire curve displayed in Figure 8.16 is the regime of nucleate boiling, because here the boiling heat transfer coefficient

$$h = \frac{q_w''}{T_w - T_{\text{sat}}} \tag{8.42}$$

reaches characteristically large values. These cover the range 10^3–10^5 W/m^2 K, as we saw in the beginning of this course at the top of Figure 1.11. Considerable research has been devoted to the measurement and correlation of the nucleate boiling heat transfer coefficient. One of the earliest and most successful correlations is [23]

$$T_w - T_{\text{sat}} = \frac{h_{fg}}{c_{p,l}} Pr_l^s C_{sf} \left[\frac{q_w''}{\mu_l h_{fg}} \left(\frac{\sigma}{g(\rho_l - \rho_v)} \right)^{1/2} \right]^{1/3} \tag{8.43}$$

This correlation applies to clean surfaces and, as an approximation, is insensitive to the shape and orientation of the surface. It depends on two empirical constants, C_{sf} and s, which are listed in Table 8.1. The dimensionless factor C_{sf} accounts for the particular combination of liquid and surface material, whereas the Prandtl number exponent s differentiates only between water and other liquids. The subscripts l and v denote saturated liquid and saturated vapor and indicate the temperature (T_{sat}) at which the properties are evaluated.

The symbol σ (N/m) denotes the surface tension of the liquid in contact with its own vapor. Representative values of this physical property have been collected in Table 8.2 along with other data needed for boiling heat transfer calculations. The surface tension plays an important role in the growth of each vapor bubble, and this role is being recognized in the theories aimed at predicting the nucleate boiling curve [25]. Written as Eq. (8.43), Rohsenow's nucleate boiling correlation can be used to calculate the excess temperature $T_w - T_{\text{sat}}$ when the heat flux q_w'' is known. The calculated excess temperature agrees within ±25% with experimental data. In the reverse case in which the excess temperature is specified, Eq. (8.43) can be rewritten as

$$q_w'' = \mu_l h_{fg} \left[\frac{g(\rho_l - \rho_v)}{\sigma} \right]^{1/2} \left[\frac{c_{P,l}(T_w - T_{\text{sat}})}{Pr_l^s C_{sf} h_{fg}} \right]^3 \tag{8.44}$$

to calculate the unknown heat flux. In this instance the calculated q_w'' agrees within a factor of 2 with actual heat flux measurements.

In conclusion, Eqs. (8.43) and (8.44) provide only an approximate estimate of the true position of the nucleate boiling curve. One reason for this is the S shape taken by the nucleate boiling curve on the logarithmic grid of Figure 8.16: this shape departs from the straight line that would correspond to Eq. (8.44). Another reason is the potential effect of surface roughness, which tends to increase the number of active nucleation sites. In artificially roughened surfaces, for example, the heat flux can be 1 order of magnitude greater than the q_w'' value furnished by Eq. (8.44).

For calculations involving the critical or peak heat flux on a large horizontal surface, the recommended relation is [26]

$$q_{\text{max}}'' = 0.149 h_{fg} \rho_v^{1/2} [\sigma g(\rho_l - \rho_v)]^{1/4} \tag{8.45}$$

Table 8.1 Empirical constants for the nucleate pool boiling correlations (8.43) and (8.44).

Liquid–surface combination	C_{sf}	S
Water–copper		
Polished	0.013	1.0
Scored	0.068	1.0
Emery polished, paraffin treated	0.015	1.0
Water–stainless steel		
Ground and polished	0.008	1.0
Chemically etched	0.013	1.0
Mechanically polished	0.013	1.0
Teflon pitted	0.0058	1.0
Water–brass	0.006	1.0
Water–nickel	0.006	1.0
Water–platinum	0.013	1.0
CCl_4–copper	0.013	1.7
Benzene–chromium	0.010	1.7
n-Pentane–chromium	0.015	1.7
n-Pentane–copper		
Emery polished	0.0154	1.7
Emery rubbed	0.0074	1.7
Lapped	0.0049	1.7
n-Pentane–nickel		
Emery polished	0.013	1.7
Ethyl alcohol–chromium	0.0027	1.7
Isopropyl alcohol–copper	0.0025	1.7
35% K_2CO_3–copper	0.0054	1.7
50% K_2CO_3–copper	0.0027	1.7
n-Butyl alcohol–copper	0.0030	1.7

Source: Rohsenow [23] and Vachon et al. [24].

The analytical form of this expression has a theoretical foundation, having been first proposed based on dimensional analysis by Kutateladze [27] and on the hydrodynamic stability of vapor columns by Zuber [28].

The peak heat flux formula (8.45) is clearly independent of the surface material. It applies to a sufficiently large surface whose linear length is considerably greater than the characteristic size of the vapor bubble. Equation (8.45) can also be used for a sufficiently large horizontal cylinder by replacing the 0.149 factor

Table 8.2 Surface tension and other physical properties needed for calculating boiling and condensation heat transfer rates.

Fluid	T_{sat} (K)	T_{sat} (°C)	P^a (10^5 N/m²)	ρ_l (kg/m³)	ρ_v (kg/m³)	h_{fg} (kJ/kg)	σ (N/m)
Ammonia	223	−50	0.409	702	0.38	1417	0.038
	300	27	10.66	600	8.39	1158	0.020
Ethanol	351	78	1.013	757	1.44	846	0.018
Helium	4.2	−269	1.013	125	16.9	20.42	10^{-4}
Hydrogen	20.3	−253	1.013	70.8		442	0.002
Lithium	600	327	4.2×10^{-9}	503		22 340	0.375
	800	527	9.6×10^{-6}	483	10^{-6}	21 988	0.348
Mercury	630	357	1.013	12 740	3.90	301	0.417
Nitrogen	77.3	−196	1.013	809	4.61	198.4	0.0089
Oxygen	90.2	−183	1.013	1134		213.1	0.013
Potassium	400	127	1.84×10^{-7}	814	2.2×10^{-7}	2196	0.110
	800	527	0.0612	720	0.037	2042	0.083
Refrigerant 12	243	−30	1.004	1488	6.27	165.3	0.016
Refrigerant 22	200	−73	0.166	1497	0.87	252.8	0.024
	250	−23	2.174	1360	9.64	221.9	0.016
	300	27	10.96	1187	46.55	180.1	0.007
Sodium	500	227	7.64×10^{-7}	898	4.3×10^{-7}	4438	0.175
	1000	727	0.1955	776	0.059	4022	0.130
Water	323	50	0.1235	988	0.08	2383	0.068
	373	100	1.0133	958	0.60	2257	0.059
	423	150	4.758	917	2.55	2114	0.048
	473	200	15.54	865	7.85	1941	0.037
	523	250	39.73	799	19.95	1716	0.026
	573	300	85.81	712	46.15	1405	0.014

a) The standard atmospheric pressure is nearly the same as the pressure of 10^5 N/m² (i.e. 1 bar). 1 atm = 1.0133×10^5 N/m².

Source: Adapted from Refs. Liley [4] and ASHRAE [5].

with 0.116 [29]. When the size of the heater is comparable with, or smaller than, the bubble size, the peak heat flux depends on the size and geometry of the heater. The peak heat flux can be calculated with a formula similar to Eq. (8.45), which contains an additional geometric correction factor [26]. Overall, the peak heat flux is *relatively insensitive to the shape and orientation of the heater surface*; therefore, Eq. (8.45) provides an adequate order-of-magnitude estimate of q''_{max} when more specific correlations are not available.

The maximum heat flux of Eq. (8.45) depends strongly on the pressure that prevails in the liquid pool. The pressure effect is brought into this relation through both h_{fg} and σ. Near the critical pressure, for example, the peak heat flux approaches zero because both h_{fg} and $\rho_l - \rho_v$ approach zero. The relationship between q''_{max} and pressure is not monotonic: in the case of water, q''_{max} increases as the pressure rises to about 70 atm and then decreases to zero as the pressure approaches the critical point pressure of 218.2 atm.

Regarding the effect of the gravitational acceleration, the nucleate boiling correlations presented in this and Section 8.2.3 are valid in the range 10–1 m/s². Despite the proportionality between q''_w and $g^{1/2}$ suggested by Eq. (8.44), it has been found that the gravitational acceleration has a considerably weaker effect on the nucleate boiling heat flux.

Example 8.3 *Nucleate Boiling: Critical Heat Flux*
A cylindrical heating element with a diameter of 1 cm and length of 30 cm is immersed horizontally in a pool of saturated water at atmospheric pressure. The cylindrical surface is plated with nickel. Calculate the heat flux q''_w and total heat transfer rate from the cylinder to the water pool, q_w, when the surface temperature is $T_w = 108\,°C$. Calculate also the critical heat flux q''_{max} and compare this value with q''_w and the approximate layout of the boiling curve shown in Figure 8.16.

Solution

Using Figure 8.16 as a guide, we see that the excess temperature of $108 - 100\,°C = 8\,°C$ is associated with the nucleate boiling regime. The wall-averaged heat flux can be evaluated using Eq. (8.44), for which the pertinent properties of water at 100 °C are (Table 8.2 and Appendix C)

$$\rho_l = 958 \text{ kg/m}^3 \qquad \sigma = 0.059 \text{ N/m}$$
$$\rho_v = 0.6 \text{ kg/m}^3 \qquad h_{fg} = 2257 \text{ kJ/kg}$$
$$c_{P,l} = 4.216 \text{ kJ/kg} \cdot \text{k} \qquad Pr_l = 1.78$$
$$\mu_l = 2.83 \times 10^{-4} \text{ kg/s} \cdot \text{m}$$

From Table 8.1 we collect the values of the empirical constants $C_{sf} = 0.006$ and $s = 1$. The right side of Eq. (8.44) can be calculated in modules (Figure E8.3):

$$\frac{g(\rho_l - \rho_v)}{\sigma} \cong \frac{g\rho_l}{\sigma} = \frac{\left(9.81\dfrac{\text{m}}{\text{s}^2}\right)\left(958\dfrac{\text{kg}}{\text{m}^3}\right)}{0.059\dfrac{1}{\text{m}}\dfrac{\text{kg}\cdot\text{m}}{\text{s}^2}}$$

$$= 1.593 \times 10^5 \text{ 1/m}^2$$

$$\frac{c_{P,l}(T_w - T_{sat})}{Pr_l^s C_{sf} h_{fg}} = \frac{4.216\dfrac{\text{kJ}}{\text{kg}\cdot\text{K}}(108-100)\text{K}}{1.78 \times 0.006 \times \left(2257\dfrac{\text{kJ}}{\text{kg}}\right)}$$

$$= 1.40$$

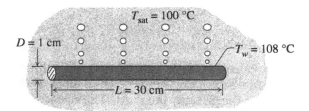

Figure E8.3

and the nucleate boiling heat flux becomes

$$q_w'' = \left(2.83 \times 10^{-4} \frac{\text{kg}}{\text{s} \cdot \text{m}}\right)\left(2257 \frac{\text{kJ}}{\text{kg}}\right)\left(1.593 \times 10^5 \frac{1}{\text{m}^2}\right)^{1/2}(1.40)^3$$

$$\cong 7 \times 10^5 \text{W/m}^2 \tag{8.44}$$

The total heat transfer rate q_w is proportional to the exposed cylindrical area:

$$q_w = \pi D L q_w'' = \pi(0.01 \text{ m})(0.3 \text{ m})(7 \times 10^5 \text{ W/m}^2)$$

$$\cong 6600 \text{ W}$$

The critical (peak) heat flux can be estimated using Eq. (8.45), in which $\rho_l - \rho_v \cong \rho_l$:

$$q_{max}'' \cong 0.149 h_{fg} \rho_v^{1/2}(\sigma g \rho_l)^{1/4}$$

$$= \left(0.149 \times 2257 \frac{\text{kJ}}{\text{kg}}\right)(0.6)^{1/2} \frac{\text{kg}^{1/2}}{\text{m}^{3/2}}\left[\left(0.059 \frac{\text{N}}{\text{m}}\right)\left(9.81 \frac{\text{m}}{\text{s}^2}\right)\left(958 \frac{\text{kg}}{\text{m}^3}\right)\right]^{1/4}$$

$$= 1.26 \times 10^6 \text{ W/m}^2 \tag{8.45}$$

$$q_{max} = \pi D L q_{max}'' = \pi(0.01 \text{ m})(0.3 \text{ m})\left(1.26 \times 10^6 \frac{\text{W}}{\text{m}^2}\right)$$

$$\cong 11900 \text{ W}$$

The critical heat flux is 80% greater than the nucleate boiling heat flux calculated in the first part of this example. The behavior of the same cylindrical heating element in the film boiling regime will be analyzed in Example 8.4.

8.2.3 Film Boiling and Minimum Heat Flux

The outstanding feature of the film boiling regime is the continuous layer of vapor (typically, 0.2–0.5 mm thick) that separates the heat surface from the rest of the liquid pool. The minimum heat flux q_{max}'' is registered at the lowest heater temperature where the film is still continuous and stable, Figure 8.16. The recommended correlation for the minimum heat flux on a sufficiently large horizontal plane surface is

$$q_{min}'' = 0.09 h_{fg} \rho_v \left[\frac{\sigma g(\rho_l - \rho_v)}{(\rho_l + \rho_v)^2}\right]^{1/4} \tag{8.46}$$

An interesting feature of this correlation is that q_{min}'' does not depend on the excess temperature $T_w - T_{sat}$. The analytical form of Eq. (8.46) was discovered by Zuber [28], who analyzed the stability of the horizontal vapor–liquid interface of the film. In the field of fluid mechanics, the unstable wavy shape that can be assumed by the horizontal interface between a heavy fluid (above) and lighter fluid (below) is called *Taylor instability*.

The minimum heat flux calculated with Eq. (8.46) agrees within 50% with laboratory measurements at low and moderate pressures. The accuracy of this correlation deteriorates as the pressure increases. The surface roughness has only a negligible effect on the minimum heat flux (of the order of 10%), because the asperities are cushioned by the film against the liquid.

For the rising portion of the film boiling curve (Figure 8.16), the correlations that have been developed have the same analytical form as the formulas encountered in our study of film condensation. For example, the formula for the average heat transfer coefficient on a *horizontal cylinder* [30]

$$\frac{\bar{h}_D D}{k_v} = 0.62 \left[\frac{D^3 h'_{fg} g (\rho_l - \rho_v)}{k_v v_v (T_w - T_{sat})} \right]^{1/4} \tag{8.47}$$

is similar to the film condensation formula (8.30). The difference is that in Eq. (8.47) the transport properties are those of vapor (k_v, v_v), because this time the film is occupied by vapor. The similarity between film boiling and film condensation is also geometric, as can be seen by comparing Figure 8.18 with the upper-left side of Figure 8.9. The corresponding formula for film boiling on a *sphere* is [26]

$$\frac{\bar{h}_D D}{k_v} = 0.67 \left[\frac{D^3 h'_{fg} g (\rho_l - \rho_v)}{k_v v_v (T_w - T_{sat})} \right]^{1/4} \tag{8.48}$$

In Eqs. (8.47) and (8.48), the augmented latent heat of vaporization h'_{fg} accounts for the superheating of the fresh vapor to temperatures above the saturation temperature [30]:

$$h'_{fg} = h_{fg} + 0.4 c_{P,v} (T_w - T_{sat}) \tag{8.49}$$

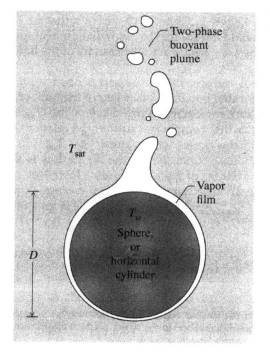

Figure 8.18 The film boiling regime on a sphere or horizontal cylinder.

The vapor properties k_v, v_v, ρ_v, and $c_{P,v}$ are best evaluated at the average film temperature $(T_w + T_{sat})/2$. Equations (8.47) and (8.48) further state that the heat transfer coefficient $\bar{h}_D$ is proportional to $(T_w - T_{sat})^{-1/4}$, which means that during film boiling the heat flux q''_w is proportional to $(T_w - T_{sat})^{3/4}$.

As the heater temperature increases, the effect of direct thermal radiation across the film contributes more and more to the overall heat transfer rate from the heater to the liquid pool. Bromley [30] showed that the thermal radiation effect can be incorporated into an effective average heat transfer coefficient $\bar{h}$,

$$\bar{h} \cong \bar{h}_D + \frac{3}{4}\bar{h}_{rad} \quad (\text{when } \bar{h}_D > \bar{h}_{rad}) \tag{8.50}$$

for which $\bar{h}_D$ is furnished by Eqs. (8.47) and (8.48) and $\bar{h}_{rad}$ is the radiation heat transfer coefficient:

$$\bar{h}_{rad} = \frac{\sigma \varepsilon_w (T_w^4 - T_{sat}^4)}{T_w - T_{sat}} \tag{8.51}$$

In water, the effect of thermal radiation begins to be felt as $T_w - T_{sat}$ increases above the 550–660 °C range. When $\bar{h}_{rad}$ is comparable with or greater than $\bar{h}_D$, Bromley's [30] recommended rule for the effective heat transfer coefficient $\bar{h}$ is

$$\bar{h} = \bar{h}_D \left(\frac{\bar{h}_D}{\bar{h}}\right)^{1/3} + \bar{h}_{rad} \quad (\text{when } \bar{h}_D \lesssim \bar{h}_{rad}) \tag{8.52}$$

Sparrow [31] showed that a rigorous analysis of combined convection and radiation in the vapor film leads to results that match within a few percentage points the values calculated based on Bromley's rule (8.52). It can be shown that the simpler Eq. (8.50) follows from Eq. (8.52) as $\bar{h}_{rad}/\bar{h}_D \to 0$.

The σ factor in Eq. (8.51) is the Stefan–Boltzmann constant, $\sigma = 5.669 \times 10^{-8}$ W/m^2 K^4, which should not be confused with the symbol used for surface tension. In the same equation, ε_w is the emissivity of the heater surface; and, numerically, the temperatures (T_w, T_{sat}) *must be expressed in degrees Kelvin*. In Chapter 10 we will learn that Eq. (8.51) can be derived from the more general formula for the net radiation heat transfer across a narrow gap, Eq. (10.84), by assuming that the emissivity of the liquid surface is equal to 1.

All the pool-boiling heat transfer correlations described until now apply when the pool contains saturated liquid. In cases where the bulk of the liquid is subcooled (e.g. Figure 8.16, left), the degree of *liquid subcooling* $(T_{sat} - T_l)$ constitutes an additional parameter that complicates the relationship between the actual heat flux q''_w and the excess temperature $T_w - T_{sat}$. On the natural convection portion of the boiling curve, where the liquid flow is single phase, the heat flux increases if the degree of liquid subcooling increases. For example, when the heater is small enough so that the natural convection flow is laminar, the heat flux increases as $(T_w - T_l)^{5/4}$, or, in terms of the degree of subcooling, as $[(T_w - T_{sat}) + (T_{sat} - T_l)]^{5/4}$. The subcooling parameter $T_{sat} - T_l$ has a relatively negligible effect on q''_{min} in the nucleate boiling regime, while both q''_{max} and q''_{min} increase linearly with $T_{sat} - T_l$. The effect of liquid subcooling is the most pronounced in the film boiling regime.

Example 8.4 *Film Boiling: Contribution due to Radiation*

The surface temperature of the cylindrical heating element described in Example 8.3 is raised to $T_w = 300\,°C$. The water pool is saturated at 1 atm, and the dimensions of the horizontal cylinder are $D = 1$ cm and $L = 30$ cm. Calculate the average heat transfer coefficient, the heat flux, and the total heat transfer rate from the heater to the water pool. The emissivity of the heater surface is $\varepsilon_w = 0.8$. Compare these heat transfer results with those obtained for nucleate boiling (Example 8.3) and with the boiling curve outlined in Figure 8.16.

Solution

According to the approximate position of the boiling curve shown in Figure 8.17, the present excess temperature $(300 - 100\,°C = 200\,°C)$ corresponds to the low-flux end of the film boiling portion of the curve. To calculate the heat flux,

$$q''_w = \bar{h}(T_w - T_{sat})$$

we use Eqs. (8.48)–(8.51), for which the physical properties of steam are evaluated at the average temperature of the film, $(300 + 100\,°C)/2 = 200\,°C = 473\,K$ (Appendix D):

$$\rho_v = 0.46\ \text{kg/m}^3 \qquad c_{P,v} = 1.982\ \text{kJ/kg} \cdot \text{K}$$
$$\nu_v = 3.55 \times 10^{-5}\ \text{m}^2/\text{s} \qquad k_v = 0.0334\ \text{W/m} \cdot \text{K}$$

From Table 8.2 we collect

$$\rho_l = 958\ \text{kg/m}^3 \quad h_{fg} = 2257\ \text{kJ/kg}$$

and, using Eq. (8.49), we calculate the augmented latent heat of vaporization:

$$
\begin{aligned}
h'_{fg} &= h_{fg} + 0.4\, c_{P,v}(T_w - T_{sat}) \\
&= 2257\ \text{kJ/kg} + (0.4 \times 1.982\ \text{kJ/kg} \cdot \text{K})(300 - 100)\text{K} \\
&= 2416\ \text{kJ/kg}
\end{aligned}
$$

The heat transfer coefficient for pure convection, $\bar{h}_D$, follows in two steps from Eq. (8.48), in which $\rho_l - \rho_v \cong \rho_l$:

$$
\frac{\bar{h}_D D}{k_v} \cong 0.62 \left[\frac{D^3 h'_{fg} g \rho_l}{k_v \nu_v (T_w - T_{sat})} \right]^{1/4}
$$

$$
= 0.62 \left[\frac{10^{-6}\text{m}^3 \, 2416\frac{\text{kJ}}{\text{kg}} 9.81\frac{\text{m}}{\text{s}^2} 958\frac{\text{kg}}{\text{m}^3}}{0.0334\frac{\text{W}}{\text{m} \cdot \text{K}} 3.55 \times 10^{-5}\frac{\text{m}^2}{\text{s}}(300 - 100)\text{K}} \right]^{1/4}
$$

$$
= 0.62(9.57 \times 10^7)^{1/4} = 61.3
$$

$$
\begin{aligned}
\bar{h}_D &= 61.3\frac{k_v}{D} = 61.3\frac{0.0334\ \text{W}}{\text{m} \cdot \text{K}}\frac{1}{0.01\ \text{m}} \\
&= 204.8\ \text{W/m}^2 \cdot \text{K}
\end{aligned}
$$

The correction due to radiation can be evaluated based on Eq. (8.51), in which $\varepsilon_w = 0.8$, and where T_w and T_{sat} are *thermodynamic* temperatures:

$$
\begin{aligned}
\bar{h}_{rad} &= \frac{\sigma \varepsilon_w (T_w^4 - T_{sat}^4)}{T_w - T_{sat}} = 5.669 \times 10^{-8}\frac{\text{W}}{\text{m}^2 \cdot \text{K}^4} 0.8\frac{(573.15^4 - 373.15^4)\text{K}^4}{(573.15 - 373.15)\text{K}} \\
&= 20.1\ \text{W/m}^2 \cdot \text{K}
\end{aligned}
$$

Since $h_{rad} < h_D$, we can use Eq. (8.50) to calculate, in order,

$$\bar{h} = \bar{h}_D + \frac{3}{4}\bar{h}_{rad} = 220\ \text{W/m}^2 \cdot \text{K}$$

$$q_w'' = \bar{h}(T_w - T_{\text{sat}}) = 220\frac{\text{W}}{\text{m}^2 \cdot \text{K}}200 \text{ K}$$

$$\cong 4.4 \times 10^4 \text{ W/m}^2$$

Note that this heat flux level agrees in an order-of-magnitude sense with the lower end of the film boiling curve in Figure 8.16. The total heat transfer rate from the cylindrical element to the pool of water is, finally,

$$q_w = \pi D L q_w'' = \pi(0.01 \text{ m})(0.3 \text{ m})\left(4.4 \times 10^4 \frac{\text{W}}{\text{m}^2}\right)$$

$$\cong 414 \text{ W}$$

8.2.4 Flow Boiling

The preceding material referred to pool boiling, in a stationary volume of liquid (Figure 8.15). The boiling heat transfer process is considerably more complicated in situations where the liquid is forced to flow past the heater. In nucleate flow boiling, the heat transfer rate is due to a combination of two closely interrelated effects: (i) the bubble formation and motion near the surface and (ii) the direct sweeping of the heater surface by the liquid itself. The heat transfer mechanism is a combination of the nucleate pool boiling of Section 8.2.2 and a forced convection phenomenon of Chapters 5 and 6.

There is no general, definitive method of correlating flow boiling data [32, 33]. Rohsenow showed that the experimental data on nucleate boiling with convection are represented adequately by the additive formula

$$q'' = q_w'' + q_c'' \tag{8.53}$$

In this expression, q_w'' is the nucleate pool boiling heat flux calculated based on Eq. (8.44) and the assumption that the bulk of the liquid is stationary. The second term, q_c'', is the single-phase convection heat flux to the liquid, $q_c'' = h_c(T_w - T_l)$, for which the convection heat transfer coefficient h_c can be estimated by employing the results listed in Chapters 5 and 6 or, in the case of a significant natural convection effect, Chapter 7. In particular, for nucleate boiling in duct flow, Rohsenow recommends calculating h_c by replacing the coefficient 0.023 with 0.019 in the Dittus–Boelter correlation (6.90). The superposition formula (8.53) works best when the flowing liquid is subcooled and the generation of vapor near the heater surface is not excessive.

Correlations for the peak heat flux q_{max}'' on a cylinder in cross-flow are in Refs. [34, 35]. Film boiling in the presence of forced convection was documented in Ref. [36].

8.3 Evolutionary Design

Heat transfer in the presence of phase change is a vast territory. In this chapter, we learned the fundamentals by considering just two phase-change phenomena: condensation and boiling. Examples of melting and solidification were considered in Chapter 4. The territory is vast because of the applications that rely on these heat transfer mechanisms. In addition to fundamentals, the applications signal the presence and importance of performance, which trace the direction for the evolution of designs.

Performance evolution (toward the better) is where the action is. This is amply illustrated by the evolution of condensers for stationary power plants [37] and vehicles that are driven by machine power [38, 39]. The phenomenon of performance evolution has its own fundamentals [40]. In this section we use melting and solidification to illustrate the fundamental content of performance evolution.

8.3.1 Latent Heat Storage

Figure 8.19 shows a hot stream of initial temperature T_∞ that comes in contact with a phase-change material through a finite thermal conductance UA, assumed known, where A is the heat transfer area between the melting material and the stream and U is the overall heat transfer coefficient based on A. The phase-change material (solid or liquid) is at the melting point T_m. The stream is well mixed at the temperature T_{out}, which is also the temperature of the stream discharged into the atmosphere (T_0).

The assumed steady operation of the installation in Figure 8.19 accounts for the cyclic operation in which every short storage (melting) stroke is followed by a short energy retrieval (solidification) stroke. During the solidification stroke, the flow $\dot{m}$ is stopped, and the recently melted phase-change material is solidified to its original state by the cooling effect provided by the heat engine positioned between T_m and T_0. The steady-state model represents the complete cycle: storage followed by retrieval. The cooling effect due to the power plant can be expressed in two ways:

$$q_m = UA(T_{out} - T_m) \tag{8.54}$$

$$q_m = \dot{m}c_P(T_\infty - T_{out}) \tag{8.55}$$

By eliminating T_{out} between these two equations, we obtain

$$q_m = \dot{m}c_P\frac{N_{tu}}{1 + N_{tu}}(T_\infty - T_m) \tag{8.56}$$

in which N_{tu} is the number of heat transfer units of the heat exchanger surface (cf. Section 9.4):

$$N_{tu} = \frac{UA}{\dot{m}c_P} \tag{8.57}$$

Of interest is the power w that can be extracted from the phase-change material. We model as reversible the cycle executed by the working fluid between T_m and T_0:

$$w = q_m\left(1 - \frac{T_0}{T_m}\right) \tag{8.58}$$

and after combining with Eq. (8.56),

$$w = \dot{m}c_P\frac{N_{tu}}{1 + N_{tu}}(T_\infty - T_m)\left(1 - \frac{T_0}{T_m}\right) \tag{8.59}$$

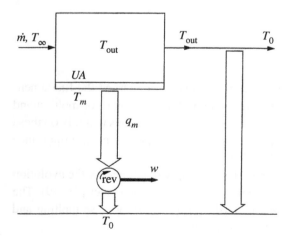

Figure 8.19 Steady production of power using a one phase-change material and one mixed stream.

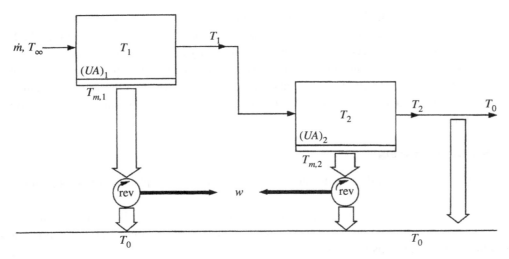

Figure 8.20 Cascade of melting and solidification in two materials placed in series. Source: Drawn after Lim et al. [42].

The type of phase-change material is free to be charged, and it is represented by the melting temperature T_m. The power w is maximum when

$$T_m = (T_0 T_\infty)^{1/2} \tag{8.60}$$

The power output that corresponds to this choice of phase-change material is [41]

$$w = \dot{m} c_p T_\infty \frac{N_{tu}}{1 + N_{tu}} \left[1 - \left(\frac{T_0}{T_m} \right)^{1/2} \right]^2 \tag{8.61}$$

One way to increase the power output of the single-element installation of Figure 8.19 is by placing the exhaust in contact with a second phase-change element of a lower temperature [42]. This cascade design proposed in 1992 [42] is shown in Figure 8.20, where individually, each phase-change element has the features of the element described in Figure 8.19. The two elements contain different phase-change materials $(T_{m,1}, T_{m,2})$, and their heat exchanger surfaces are not identical $[(UA)_1, (UA)_2]$. The well-mixed gas temperatures above each heat exchanger surface are T_1 and T_2. An analysis equivalent to Eqs. (8.54)–(8.57) yields the melting temperatures of the recommended materials and the manner in which to divide the total heat exchanger size $[UA = (UA)_1 + (UA)_2]$ between the two stages of the installation.

8.3.2 Shaping Inserts for Faster Melting

In the models of Figures 8.19 and 8.20, the phase-change material has a flat body heated by convection over its exposed surface. These simple models served their purpose, which was to identify the materials that should be used in order to increase the performance of the greater installations that employ periodic energy storage.

Rewards do come from more freedom, and in this case freedom means to abandon the simple model and to allow the shape of the heated surface to vary. In other words, the shape becomes a new degree of freedom (in addition to the freedom to select materials), and the number of eligible shapes is infinite.

This fresh opportunity was demonstrated recently with tree-shaped (constructal) heaters embedded in a block of phase-change material [43]. The amount of melted material increases in time as an S curve: the S becomes steeper, and the melting process becomes faster as the complexity of the tree design increases.

Another class of designs of inserts for faster melting are vertical tubes with a hot stream on the inside and natural convection melting on the outside [44], and also spiral heaters embedded in the phase-change material [45]. The shaping of inserted heaters is also effective in increasing the performance of conduction for single-phase energy storage [46].

8.3.3 Rhythmic Surface Renewal

Even more freedom can be found by abandoning the steady-state operation and replacing it with a periodic one. The rhythm is the payoff to be discovered.

This concept was first demonstrated in laminar boundary layer natural convection [47] and forced convection [48]. The idea behind it is that in steady state, the most intense thermal contact and heat transfer between the wall and the fluid that sweeps it is at the leading edge of the boundary layer, where the layer is the thinnest. Farther downstream the layer is progressively thicker and the heat transfer is less intense. On the other hand, during transient heat transfer across the contact between warm body and colder body, the most intense heat transfer occurs right after the two bodies make contact.

These two images of intense heat transfer justify the use of discrete time intervals of boundary layer heating, which are interspaced with periods of no heating, i.e. periods of surface renewal. The concept was subsequently demonstrated for rhythmic condensation [49], pool boiling [50], wiping a liquid film [51], and solidification, as in the design of ice-making machines [52]. New directions of designs for convection with two-phase flow are traced in Refs. [53–56].

References

1 Bejan, A. (2013). *Convection Heat Transfer*, 4e. Hoboken, NJ: Wiley.

2 Nusselt, W. (1916). Die Oberflächenkondensation des Wasserdampfes. *Z. Ver. Deutsch. Ing.* 60: 541–569.

3 Rohsenow, W.M. (1956). Heat transfer and temperature distribution in laminar-film condensation. *Trans. ASME* 78: 1645–1648.

4 Liley, P.E. (1987). Thermophysical properties. In: *Handbook of Single-Phase Convective Heat Transfer*, Chapter 22 (eds. S. Kakac, R.K. Shah and W. Aung). New York: Wiley.

5 ASHRAE (1981). *Handbook of Fundamentals*. New York: ASHRAE.

6 Sparrow, E.M. and Gregg, J.L. (1959). A boundary-layer treatment of laminar-film condensation. *J. Heat Transfer* 81: 13–18.

7 Chen, M.M. (1961). An analytical study of laminar film condensation: part 1—flat plates. *J. Heat Transfer* 83: 48–54.

8 Chen, S.L., Gerner, F.M., and Tien, C.L. (1987). General film condensation correlations. *Exp. Heat Transfer* 1: 93–107.

9 Dhir, V.K. and Lienhard, J.H. (1971). Laminar film condensation on plane and axisymmetric bodies in non-uniform gravity. *J. Heat Transfer* 93: 97–100.

10 Sparrow, E.M. and Gregg, J.L. (1959). Laminar condensation heat transfer on a horizontal cylinder. *J. Heat Transfer* 81: 291–296.

11 Chen, M.M. (1961). An analytical study of laminar film condensation: part 2—single and multiple horizontal tubes. *J. Heat Transfer* 83: 55–60.

12 Bejan, A. (1991). Film condensation on an upward facing plate with free edges. *Int. J. Heat Mass Transfer* 34: 578–582.

13 Shekriladze, I.G. and Gomelauri, V.I. (1966). Theoretical study of laminar film condensation of flowing vapour. *Int. J. Heat Mass Transfer* 9: 581–591.

14 Rose, J.W. (1989). A new interpolation formula for forced-convection condensation on a horizontal surface. *J. Heat Transfer* 111: 818–819.

15 Chato, J.C. (1962). Laminar condensation inside horizontal and inclined tubes. *J. ASHRAE* 4: 52–60.

16 Rohsenow, W.M. (1973). Film condensation. In: *Handbook of Heat Transfer*, Section 12A (eds. W.M. Rohsenow and J.P. Hartnett). New York: McGraw-Hill.

17 Minkowycz, W.J. and Sparrow, E.M. (1966). Condensation heat transfer in the presence of non-condensibles, interfacial resistance, superheating, variable properties, and diffusion. *Int. J. Heat Mass Transfer* 9: 1125–1144.

18 Bejan, A. and Gobin, D. (2006). Constructal theory of droplet impact geometry. *Int. J. Heat Mass Transfer* 49: 2412–2419.

19 Bejan, A. and Kraus, A.D. (2003). *Handbook of Heat Transfer*. Hoboken, NJ: Wiley.

20 Wares, C. (1966). On the fixation of water in diverse fire. *Int. J. Heat Mass Transfer* 9: 1153–1166.

21 Nukiyama, S. (1934). The maximum and minimum values of the heat Q transmitted from metal to boiling water under atmospheric pressure. *J. Jpn. Soc. Mech. Eng.* 37: 367–374; English translation in (1966). *Int. J. Heat Mass Transfer* 9: 1419–1433.

22 Drew, T.B. and Mueller, C. (1937). Boiling. *Trans. AIChE* 33: 449–473.

23 Rohsenow, W.M. (1952). A method for correlating heat transfer data for surface boiling of liquids. *Trans. ASME* 74: 969–976.

24 Vachon, R.I., Nix, G.H., and Tanger, G.E. (1968). Evaluation of constants for the Rohsenow pool-boiling correlation. *J. Heat Transfer* 90: 239–247.

25 Rohsenow, W.M. (1988). What we don't know and do know about nucleate pool boiling heat transfer. *ASME HTD-Vol. 104* 2: 169–172.

26 Lienhard, J.H. and Dhir, V.K. (1973). Extended hydrodynamic theory of the peak and minimum pool boiling heat fluxes. *NASA CR–2270*, July 1973.

27 Kutateladze, S.S. (1948). On the transition to film boiling under natural convection. *Kotloturbostroenie* 3: 10.

28 Zuber, N. (1958). On the stability of boiling heat transfer. *Trans. ASME* 80: 711–720.

29 Sun, K.H. and Lienhard, J.H. (1970). The peak pool boiling heat flux on horizontal cylinders. *Int. J. Heat Mass Transfer* 13: 1425–1439.

30 Bromley, A.L. (1950). Heat transfer in stable film boiling. *Chem. Eng. Prog.* 46: 221–227.

31 Sparrow, E.M. (1964). The effect of radiation on film-boiling heat transfer. *Int. J. Heat Mass Transfer* 7: 229–238.

32 Rohsenow, W.M. (1973). Boiling. In: *Handbook of Heat Transfer*, Section 13 (eds. W.M. Rohsenow and J.P. Hartnett). New York: McGraw-Hill.

33 Whalley, P.B. (1987). *Boiling, Condensation and Gas-Liquid Flow*, Chapters 16, 17, and 20. Oxford: Clarendon Press.

34 Lienhard, J.H. and Eichhorn, R. (1976). Peak boiling heat flux on cylinders in a cross flow. *Int. J. Heat Mass Transfer* 19: 1135–1142.

35 Kheyrandish, K. and Lienhard, J.H. (1985). Mechanisms of burnout in saturated and subcooled flow boiling over a horizontal cylinder. *ASME-AIChE National Heat Transfer Conference*, Denver, CO (4–7 August 1985).

36 Bromley, A.L., LeRoy, N.R., and Robbers, J.A. (1953). Heat transfer in forced convection film boiling. *Ind. Eng. Chem.* 45: 2639–2646.

37 Bejan, A., Lee, J., Lorente, S., and Kim, Y. (2015). The evolutionary design of condensers. *J. Appl. Phys.* 117: 125101.

38 Bejan, A., Charles, J.D., and Lorente, S. (2014). The evolution of airplanes. *J. Appl. Phys.* 116: 044901.

39 Chen, R., Wen, C.Y., Lorente, S., and Bejan, A. (2016). The evolution of helicopters. *J. Appl. Phys.* 120: 014901.

40 Bejan, A. (2016). *The Physics of Life: The Evolution of Everything*. New York: St. Martin's Press.

41 Bejan, A. (2016). *Advanced Engineering Thermodynamics*, 4e. Hoboken, NJ: Wiley.

42 Lim, J.S., Bejan, A., and Kim, J.H. (1992). Thermodynamic optimization of phase-change energy storage using two or more materials. *J. Energy Resour. Technol.* 114: 84–90.

43 Bejan, A., Ziaei, S., and Lorente, S. (2014). The S curve of energy storage by melting. *J. Appl. Phys.* 116: 114902.

44 Lorente, S., Bejan, A., and Niu, J.L. (2014). Phase change heat storage in an enclosure with vertical pipe in the center. *Int. J. Heat Mass Transfer* 72: 329–335.

45 Lorente, S., Bejan, A., and Niu, J.L. (2015). Constructal design of latent heat energy storage with vertical spiral heaters. *Int. J. Heat Mass Transfer* 81: 283–288.

46 Alalaimi, M., Lorente, S., and Bejan, A. (2015). Thermal coupling between a helical pipe and a conducting volume. *Int. J. Heat Mass Transfer* 83: 762–767.

47 Vargas, J.V.C. and Bejan, A. (1993). The resonance of natural convection in an enclosure heated periodically from the side. *Int. J. Heat Mass Transfer* 36: 2027–2038.

48 Morega, A.M., Vargas, J.V.C., and Bejan, A. (1995). Optimization of pulsating heaters in forced convection. *Int. J. Heat Mass Transfer* 38 (16): 2925–2934.

49 Vargas, J.V.C. and Bejan, A. (1999). Optimisation of film condensation with periodic wall cleaning. *Int. J. Therm. Sci. (Revue Générale de Thermique)* 38: 113–120.

50 Vargas, J.V.C. and Bejan, A. (1997). Optimization of pulsating heating in pool boiling. *J. Heat Transfer* 119: 298–304.

51 Bejan, A., Vargas, J.V.C., and Lim, J.S. (1994). When to defrost a refrigerator and when to remove the scale from the heat exchanger of a power plant. *Int. J. Heat Mass Transfer* 37: 523–532.

52 Vargas, J.V.C. and Bejan, A. (1995). Fundamentals of ice making by convection cooling followed by contact melting. *Int. J. Heat Mass Transfer* 38: 2833–2841.

53 Stafford, J. (2016). Principle-based design of distributed multiphase segmented flow. *Int. J. Heat Mass Transfer* 100: 508–521.

54 Norouzi, E. and Amidpour, M. (2012). Optimal thermodynamic and economic volume of a heat recovery steam generator by constructal design. *Int. Comm. Heat Mass Transfer* 39: 1286–1292.

55 Norouzi, E., Mehrgoo, M., and Amidpour, M. (2012). Geometric and thermodynamic optimization of a heat recovery steam generator: a constructal design. *J. Heat Transfer* 134: 111801.

56 Wei, P.S., Lin, C.L., Liu, H.J., and DebRoy, T. (2012). Scaling weld or melt pool shape affected by thermocapillary convection with high Prandtl numbers. *J. Heat Transfer* 134: 042101.

Problems

Condensation on Vertical Surfaces

8.1 Rely on the formulas developed for a vertical laminar film, and demonstrate that the total cooling rate provided by the wall is proportional to the total rate of condensation, $q' = h'_{fg}\Gamma(L)$.

8.2 Consider the control volume drawn around the entire film of height L shown in Figure B8.2, and make no assumption concerning the flow regimes that may be present inside the control volume. The vertical wall is isothermal, T_w. To the right of the film of condensate, the vapor is stagnant and saturated. Show that in this general configuration the total wall cooling rate is proportional to the condensate mass flowrate, $q' = h'_{fg}\Gamma(L)$.

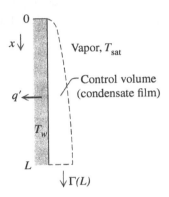

0

$x\downarrow$

Vapor, T_{sat}

Control volume
(condensate film)

$q' \leftarrow$

T_w

L

$\downarrow \Gamma(L)$

Figure P8.2

8.3 Show that regardless of the flow regime, the length L of a vertical film of condensate is related to the average heat transfer coefficient $\overline{h}_L$ and the total condensation rate $\Gamma(L)$ by the general formula

$$L = \frac{h'_{fg}\Gamma(L)}{(T_{sat} - T_w)\overline{h}_L}$$

Use this formula to derive Eq. (8.24), which holds only for a *laminar* vertical film, when $\rho_l > > \rho_v$.

8.4 Demonstrate that the local Reynolds number Re_x along a laminar vertical film of condensate is equal to

$$Re_x = \frac{8}{3}\frac{u_{max}(x)\,\delta(x)}{\nu_l}$$

where $u_{max}(x)$ is the downward velocity of the liquid–vapor interface. [Review the zero-shear boundary condition discussed in Eq. (8.3).]

8.5 Saturated vapor condenses on a cold vertical slab of height L. Both sides of the slab are covered by laminar films of condensate. A single horizontal cylinder of diameter D, and at the same temperature as the slab, is immersed in the same saturated vapor. For what special diameter D will the total condensation rate on the cylinder equal the total condensation rate produced by the slab?

8.6 a) Saturated steam at 1 atm condenses on a vertical wall of temperature 80 °C and height 1 m. Assume that the condensate forms a laminar film, and calculate the average heat transfer coefficient, the condensation rate, and the film Reynolds number at the bottom of the wall.

b) Is the film laminar over its entire height? If not, recalculate the quantities of part (a) by relying on the chart given in Figure 8.6. Compare the new condensation rate estimate with the value calculated in part (a), and you will see the condensation augmentation effect of the film waviness.

Condensation in Other Configurations

8.7 The horizontal thin-walled tube shown in Figure P8.7 is cooled by an internal fluid of temperature T_w. The tube is immersed in a stagnant atmosphere of saturated vapor, which condenses in laminar film fashion on the outer cylindrical surface.

It is proposed to increase the total condensation rate by flattening the tube cross section into the shape shown on the right side of the figure. Calculate the percent increase in condensation flowrate associated with this design change.

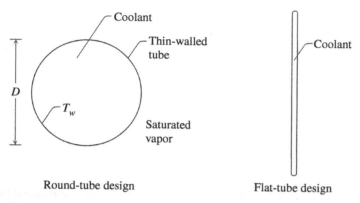

Round-tube design   Flat-tube design

Figure P8.7

8.8 A plane rectangular surface of width $L = 1$ m, length $Z \gg L$, and temperature $T_w = 80\,°\text{C}$ is suspended in saturated steam of temperature $100\,°\text{C}$. When this surface is oriented in such a way that L is aligned with the vertical (i.e. as in Problem 8.6), the steam condenses on it at the rate $0.063\,\text{kg/s·m}$. The purpose of this exercise is to show how the condensation rate decreases when the surface becomes tilted relative to the vertical direction (Figure P8.8).

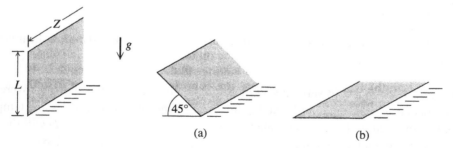

(a)   (b)

Figure P8.8

a) Calculate the condensation rate when the L width makes a 45° degree angle with the vertical and the Z length is aligned with the horizontal. Determine the film Reynolds number and the flow regime.

b) Assume that the surface is perfectly horizontal facing upward (as the top surface in Figure 8.10) and that the film of condensate is laminar. Calculate the condensation rate and the Reynolds number of the liquid film spilling over one edge, and verify the validity of the laminar film assumption.

8.9 Atmospheric-pressure saturated steam condenses on the outside of a horizontal tube of wall temperature $T_w = 60\,°C$ and outer diameter $D = 2$ cm. Assume that the condensate forms a laminar film, and calculate the mass flowrate of condensate dripping from the bottom of the tube. Calculate also the film Reynolds number, and in this way prove the validity of the laminar flow assumption.

8.10 a) The bank of horizontal tubes shown on the left side of Figure P8.10 is surrounded by $100\,°C$ saturated steam, which condenses on the outside of each tube. The tube surface is maintained at $60\,°C$ by a cold fluid that flows through each tube in the direction perpendicular to the plane of the figure. Assuming that the condensate film is laminar, calculate the total mass flowrate of condensate per unit length of tube bank.

   b) In a competing design, the same bundle of tubes appears rotated by 90°, as shown on the right side of the figure. Calculate the total condensate mass flowrate in this new design, and comment on the effect of the 90° rotation.

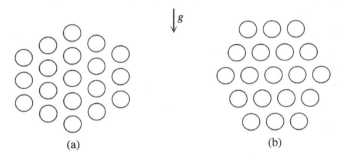

(a)   (b)

Figure P8.10

8.11 The average heat transfer coefficients for film condensation on an upward facing strip and disc (Figure 8.10) are listed in Eqs. (8.33) and (8.34). Rewrite each of these formulas by using as length scale the characteristic length of the surface, Eq. (7.76), $L_c = A/p$, where A and p are the area and perimeter of the surface, respectively. Show that the average heat transfer coefficient of any other surface whose shape is somewhere between the "very long" limit (the strip) and the "round" limit (the disc) is given by the approximate formula

$$\overline{Nu}_{L_c} = \frac{\overline{h}L_c}{k_l} \cong 0.8 \left[\frac{L_c^3 h'_{fg} g(\rho_l - \rho_v)}{k_l \nu_l (T_{sat} - T_w)} \right]^{1/5}$$

8.12 a) Saturated steam at 1 atm condenses as a laminar film on a metallic horizontal tube at $80\,°C$. The tube outside diameter is 4 cm. Calculate the condensation rate and verify that the laminar film assumption is adequate.

   b) It is proposed to coat the tube with a 0.5-mm layer of Teflon to achieve dropwise condensation on the exposed surface (Figure P8.12). Assume that the average heat transfer coefficient for dropwise condensation is 10 times greater than the value calculated in part (a). Calculate the new condensation rate, and explain why it is not greater than when the film was laminar.

   c) How thin must the Teflon coating be if it is to increase the condensation rate?

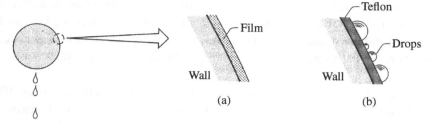

Figure P8.12

Nucleate Boiling

8.13 Consider the spherical vapor bubble of radius r shown in Figure P8.13. The pressure and temperature inside the bubble (P_v, T_v) are slightly above the pressure and temperature in the liquid (P_l, T_l). The liquid is saturated, $T_l = T_{sat}$.

a) Invoke the mechanical equilibrium of one hemispherical control volume, and show that the bubble radius varies inversely with the pressure difference: $r = 2\sigma/(P_v - P_l)$.

b) Rely on the Clausius–Clapeyron relation $dP/dT = h_{fg}/(Tv_{fg})$ to show that the bubble radius also varies inversely with the temperature difference: $r = 2\sigma T_{sat}/[h_{fg}\rho_v(T_v - T_{sat})]$.

c) Calculate the radius of a steam bubble with $T_v - T_{sat} = 2\,\mathrm{K}$ in water at $T_{sat} = 100\,°\mathrm{C}$.

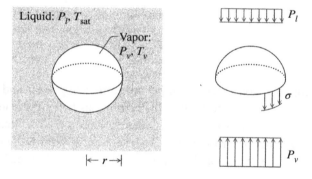

Liquid: P_l, T_{sat}

Vapor: P_v, T_v

$|\!\leftarrow r \rightarrow\!|$

P_l

σ

P_v

Figure P8.13

8.14 Consider the water on nickel nucleate boiling calculations outlined in Example 8.3. Obtain an estimate for the excess temperature $T_w - T_{sat}$ at critical heat flux conditions by equating the nucleate boiling heat flux q''_w with the calculated peak heat flux q''_{max}. Compare your estimate with the actual excess temperature at peak heat flux (Figure 8.16), and explain why your ($T_w - T_{sat}$) value is smaller.

8.15 Estimate the C_{sf} constant that corresponds to the nucleate boiling portion of the curve shown in Figure 8.16. Use Rohsenow's correlation (8.43) and a point (q''_w, $T_w - T_{sat}$) in the vicinity of the transition from the isolated bubbles regime to the regime of columns and slugs.

8.16 The vacuum insulation around a spherical liquid helium vessel breaks down (develops an air leak) and allows the heat flux $q''_w = 10^3\,\mathrm{W/m^2}$ to land on the external surface of the vessel. An amount of saturated liquid helium at atmospheric pressure boils at the bottom of the vessel (Figure P8.16). Calculate the excess temperature $T_w - T_{sat}$ by assuming nucleate boiling with $C_{sf} = 0.02$ and $s = 1.7$. Compare the heat leak q''_w with the peak heat flux for nucleate boiling in the pool of liquid helium.

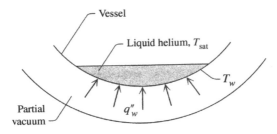

Figure P8.16

8.17 Water boils in the pressurized cylindrical vessel shown in Figure P8.17. The steam relief valve is set in such a way that the pressure inside the vessel is $4.76 \times 10^5 \text{N/m}^2$. The bottom surface is made out of copper (polished), and its temperature is maintained at $T_w = 160\,°C$. Assume nucleate boiling, and calculate the total heat transfer rate from the bottom surface to the boiling water. Later, verify the correctness of the nucleate boiling assumption. The calculation of the time when there is no liquid left in the vessel is the subject of the next problem.

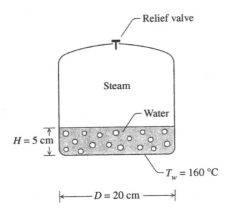

Figure P8.17

8.18 The cylindrical vessel described in the preceding problem has an inner diameter of 20 cm. The depth of the original amount of liquid is 5 cm, and the pressure is maintained at $4.76 \times 10^5 \text{ N/m}^2$. The nucleate boiling heat transfer rate to the liquid (calculated in the preceding problem) is $q_w = 12.44 \text{ kW}$.

a) Estimate the time needed to evaporate all the liquid. Base your estimate on the simple relation $q_w = \dot{m} h_{fg}$, which is routinely recommended.

b) The $q_w = \dot{m} h_{fg}$ relation is valid only approximately and is incorrect from a thermodynamics standpoint. With reference to the control volume defined by the pressurized vessel (Figure P8.18), show that the correct proportionality between q_w and $\dot{m}$ is

$$q_w = \dot{m}\left(h_g - \frac{u_f v_g - u_g v_f}{v_g - v_f}\right)$$

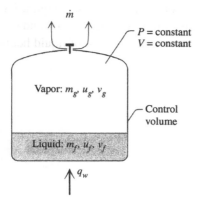

Figure P8.18

where u, v, $()_g$, and $()_f$ are the thermodynamics symbols for specific internal energy, specific volume, saturated vapor, and saturated liquid, respectively. Note that the pressure (or temperature) and the total volume V remain constant. Numerically, show that the quantity arrived at in the round brackets deviates from h_{fg}, assumed in part (a), as the saturated liquid–vapor mixture approaches the critical point.

8.19 The bottom of a shallow pan is a 20-cm-diameter disc made out of mechanically polished stainless steel. It is used to boil water at atmospheric pressure while the temperature of the bottom surface is 108 °C.

a) Calculate the heat flux supplied by the bottom surface to the water pool. Consult Figure 8.16 to anticipate the boiling regime, and later compare the calculated q''_w value with the one plotted on the figure.

b) Estimate also the heat transfer coefficient and the total heat transfer rate to the water pool. Compare the calculated h value with the range of values indicated in Figure 1.11.

Film Boiling

8.20 Let δ be the average thickness of the vapor film in the film boiling regime depicted in Figure 8.18. When the heat transfer across this film is dominated by convection, the heat flux is roughly the same as the conduction heat flux across a vapor layer of thickness δ:

$$q''_w \sim k_v \frac{T_w - T_{sat}}{\delta}$$

Use this idea and the numerical data of Example 8.4 to evaluate the film thickness δ.

8.21 In a power-controlled pool boiling experiment, a horizontal cylindrical heater is immersed in saturated water at atmospheric pressure. The peak heat flux is 10^6 W/m². The power is increased slightly above this level, and the nucleate boiling regime is replaced abruptly by film boiling (Figure 8.17a). Estimate the excess temperature in this new regime by assuming that radiation is the dominant (i.e. the only) mode of heat transfer across the film. Assume also $\varepsilon_w = 1$. Compare your estimate with the value read off Figure 8.16. Will the actual excess temperature be larger or smaller than this pure-radiation estimate?

8.22 Water boils in the film boiling regime on a horizontal surface with an excess temperature of $200\,^\circ C$. Calculate the minimum heat flux when:

(a) $T_{sat} = 100\,^\circ C$

(b) $T_{sat} = 300\,^\circ C$

and comment on the behavior of q''_{min} as the saturated water pool approaches the critical point.

Evolutionary Design

8.23 How can we freeze water faster? Consider three competing configurations:

(p) Plane layer of water of depth H, in a tray of length L, and width W

(c) Cylinders of diameter D_c and length L

(s) Spherical capsules of diameter D_s

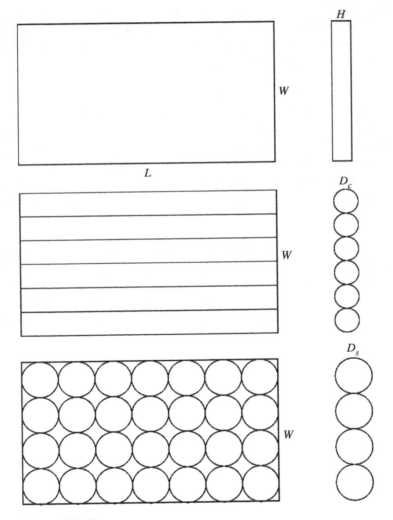

Figure P8.23

Figure P8.23 shows these configurations viewed from above and from the side. The water volume (V) is the same in all three configurations. The water body is cooled uniformly from all sides by the sub-zero surrounding air flow.

The scale of the freezing time is $t \sim d^2/\alpha$, where d is the depth to which thermal diffusion must penetrate the water in order for solidification to be complete. In the three configurations, the value of d is, respectively, $H/2$, $D_c/2$, and $D_s/2$.

Determine the scales of the three freezing times (t_p, t_c, t_s), and report the ratios t_c/t_p and t_s/t_p. In which configurations does the water freeze the fastest, plane, cylindrical, or spherical?

8.24 Ice is being produced cyclically in a horizontal tray, as in the first configuration of the preceding problem. During the first time interval of each cycle (t_1), the thickness of the ice layer grows to $H_1 \sim (\alpha t_1)^{1/2}$. During the second time interval (t_2), the ice is removed, a new amount of water is poured into the tray, and the ($t_1 + t_2$) cycle is repeated.

The time t_2 is a known constant of the ice-making machine. The freezing time (t_1 or H_1) is a degree of freedom. Show how the freezing time interval affects the rate of ice production averaged over time, or the amount of ice produced during one complete cycle. Report the optimal freezing time (or the optimal rhythm t_1/t_2) and the corresponding maximum rate of ice production.

8.25 Rain drops come in many sizes, big and small. Show that the big drops fall from high altitude (H), and the small drops form near ground. Construct your proof based on scale analysis. Model the drop as a roundish form with one length scale (D), and recognize that every drop falls at terminal velocity, such that the weight of the drop is balanced by the air drag. Consider separately the high Re regime where the drag coefficient is a constant of order 1, and the low Re regime where the drag coefficient is proportional to Re^{-1}. Condensation on the drop surface (during its time of travel from H to ground) accounts for the mass of the drop, ρD^3. For both Re regimes, determine the scaling relation between D and H.

8.26 You may have noticed that after washing the dishes and exposing them to air to dry, some dishes dry a lot more slowly than the rest. How their surfaces are oriented with respect to gravity has an important effect (cf. Problem 7.31). Another effect is due to the condition of the surface of the dish. Consider two plane surfaces, both wet and facing upward. One surface is wetted by a thin film of water, and the other is hydrophobic and is covered by isolated drops. Which surface will become dry faster?

9

Heat Exchangers

9.1 Classification of Heat Exchangers

In the convection phenomena discussed thus far (Chapters 5–8), the primary goal was to determine the relationship between the heat transfer rate and the driving temperature difference. This operation was reduced to calculating the heat transfer coefficient. In the present chapter, we have the opportunity to use the heat transfer coefficient in the greater effort of designing an actual apparatus called a *heat exchanger*.

A heat exchanger is a device, or piece of hardware, which has the function to promote the transfer of heat between two or more entities at different temperatures. In most cases, and especially in the examples considered in this chapter, the heat exchanging entities are two streams of fluid. To prevent stream-to-stream mixing, the two fluids are separated by solid walls that, together, constitute the *heat transfer surface*, or *heat exchanger surface*. In some heat exchangers, a solid heat transfer surface is not necessary, because of the natural immiscibility of the two streams or the separation (stratification) of the two fluids in the gravitational field. In such cases the heat transfer between the two fluids occurs through their mutual interface, and the apparatus is called a *direct contact heat exchanger*.

The heat exchanger is a system whose design involves the calculation not only of the heat transfer rate across the heat exchanger surface but also the pumping power needed to circulate the two streams through the various flow passages, the geometric layout of the flow pattern (the wrapping and weaving of one stream around and through the other), the construction of the actual hardware, and the ability to disassemble the apparatus for periodic cleaning. Over the past 100 years, the analysis, design, and manufacturing of heat exchangers have grown into a separate discipline in thermal engineering. The current status is described in heat exchanger monographs [1, 2]. In the present chapter, we examine the scientific basis of this discipline.

The other aspect that makes the study of heat exchangers challenging is their diversity. There are many types of heat exchangers, and there is even more than one way to categorize these types. One way to distinguish between the various types is by looking at the flow arrangement. Figures 9.1 and 9.2 show the three main flow configurations known as *parallel flow* (*or* cocurrent flow), *counterflow* (or countercurrent flow), and *cross-flow*. In parallel flow, the two inlet ports are positioned at the same end of the heat exchanger, where the stream-to-stream temperature difference is the greatest. In the counterflow scheme, the stream-to-stream temperature difference is more evenly distributed along the heat exchanger.

The cross-flow arrangements can also vary with respect to the degree of lateral mixing that is experienced by each stream inside the channel, cf. Figure 9.2. The lateral mixing can be inhibited by installing longitudinal corrugations that divide the "unmixed" stream into many ministreams that flow in parallel. In the absence of longitudinal partitions, the stream can mix transversely in each channel cross section. Whether the stream

Heat Transfer: Evolution, Design and Performance, First Edition. Adrian Bejan.
© 2022 John Wiley & Sons, Inc. Published 2022 by John Wiley & Sons, Inc.
Companion website: www.wiley.com/go/bejan/heattransfer

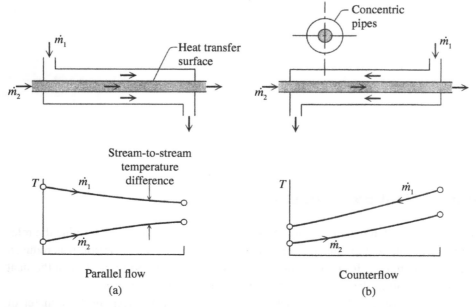

Figure 9.1 (a) Double-pipe parallel flow and (b) counterflow heat exchangers.

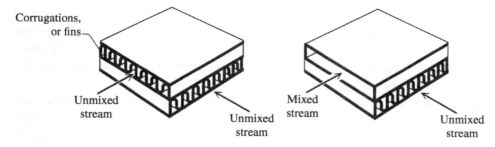

Figure 9.2 Plate fin cross-flow heat exchangers and the use of longitudinal corrugations to prevent the transversal mixing of the stream.

can be regarded as truly "mixed" depends on the design of the channel (width, length), the presence of turbulence, and the degree to which the mixing effect can propagate downstream.

The flow arrangements of Figures 9.1 and 9.2 are *single-pass* schemes, because in each case the stream passes only once through the heat exchanger volume. Examples of *multipass* arrangements are given in the second and third cases illustrated in Figure 9.3.

Another way to differentiate between various heat exchanger designs is to consider their construction. Perhaps the simplest design is the *double-pipe* arrangement used in the two examples shown in Figure 9.1. In the concentric pipe arrangement, the streams are separated by the wall of the inner pipe: the outer or inner surface plays the role of "heat transfer surface."

More complicated is the *shell-and-tube* heat exchanger illustrated in Figure 9.4, in which the inner stream flows through not one but several tubes. The outer stream is confined by a large-diameter vessel (the "shell"). Transversal baffles force the outer stream to flow across the tubes, augmenting in this way the overall heat

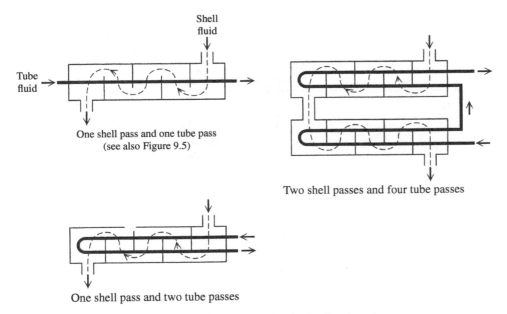

Figure 9.3 Single-pass and multipass shell-and-tube heat exchangers.

transfer coefficient between the two fluids. Three of the more common baffle designs are also illustrated in Figure 9.4. Worth noting is that the shell-and-tube heat exchanger is of the same type as the single-pass heat exchanger shown as the first case in Figure 9.3, where, for simplicity, only one of the tubes was drawn. Note also that the direction of gravity was not taken into account when the flow directions were indicated with arrows in Figures 9.1, 9.3, and 9.4.

Another construction type is the *plate fin* heat exchanger, in which each channel is defined by two parallel plates separated by fins or spacers. The fins are connected to the parallel plates by tight mechanical fit, gluing, soldering, brazing, welding, or extrusion. Alternating passages are connected in parallel to end chambers (called headers) and form one side (i.e. one stream) of the heat exchanger. Fins are employed on both sides in gas-to-gas applications, whereas in gas-to-liquid applications, fins are needed only on the gas side because the heat transfer coefficient there is lower. If present on the liquid side, the fins play a structural (stiffening) role. In bar-and-plate heat exchangers, solid parallel bars are used to seal the edges of each passage.

The fins can be continuous, as in the case of the unmixed streams of Figure 9.2, or interrupted, as on the left side of Figure 9.5. Many fin shapes and orientations have been developed. In the cross-flow heat exchanger shown on the right side of Figure 9.5, the outer surface of each tube is fitted with equidistant annular fins of constant thickness. In this arrangement the finned side of the tube wall usually faces a gaseous stream (mixed, or partially mixed), while the inner surface of each tube (unfinned) faces a liquid stream. Once again, the fins are needed more on the gas side of the tube wall, because the bare-wall heat transfer coefficient on the gas side is smaller than on the liquid side.

Heat exchangers can also be categorized with respect to their degree of compactness. The plate fin and tube fin designs illustrated in Figures 9.2–9.5 have more heat transfer area per unit volume than the corresponding designs without fins. Figure 9.6 shows that a high heat transfer area density corresponds to a small hydraulic diameter for the passages wetted by the streams. Compact heat exchangers have heat transfer area densities

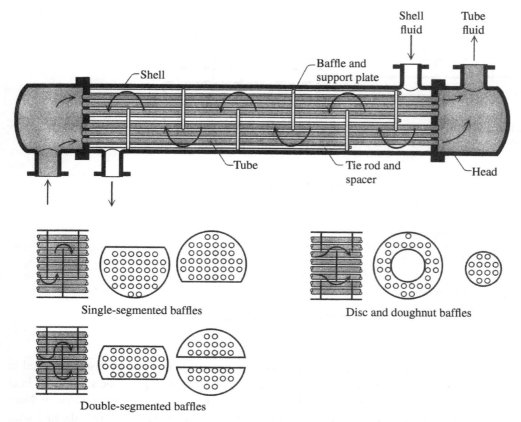

Figure 9.4 Shell-and-tube heat exchanger and three examples of baffle design.

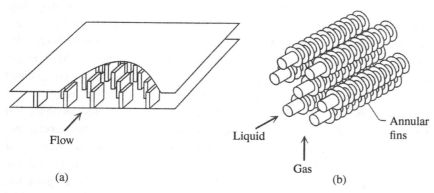

Figure 9.5 Parallel-plate channel with interrupted fins (a) and bank of finned tubes in cross-flow (b).

in excess of $700 \, m^2/m^3$ and are essential in applications where the size and weight of the heat exchanger is an important design constraint (e.g. automobiles, naval and airborne power plants, and air conditioning systems).

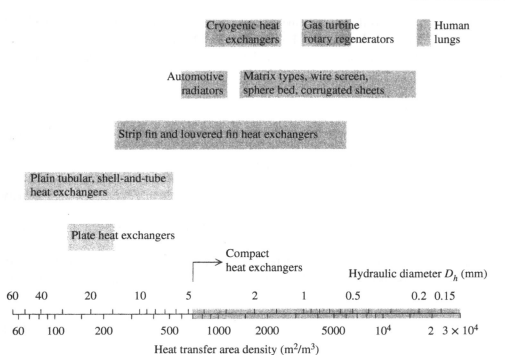

Figure 9.6 Ordering of heat exchangers according to their degree of compactness. Source: Drawn after Bejan and Kraus [2].

9.2 Overall Heat Transfer Coefficient

The heat exchanger surface is generally more complicated than the three-resistance sandwich analyzed earlier in Section 2.1.3 and Figure 2.3. In addition to the conduction resistance of the wall and the convective film resistances on the two sides of the wall, the heat exchanger wall may be complicated by fins and the gradual accumulation of a layer of oxides and other deposits on surfaces exposed to fluids. This configuration is illustrated in Figure 9.7, where the deposited layer is called *scale*. The time-dependent process by which this layer is formed is called *scaling* or *fouling*.

To derive the overall thermal resistance formula for the heat exchanger surface, consider only one side of the wall. The total area of this side, A, is the sum of the area contributed by the exposed surfaces of the fins, A_f, plus the area of the unfinned portions of the wall, A_u:

$$A = A_f + A_u \tag{9.1}$$

Assuming that the heat transfer coefficient has the same value h on both A_f and A_u, the total heat transfer rate through the surface A is

$$q = \eta h A_f (T_w - T_\infty) + h A_u (T_w - T_\infty) \tag{9.2}$$

The temperature T_w refers to the unfinned portions of the wall, or the base of each fin, while T_∞ is the bulk temperature of the fluid. The first term on the right side of Eq. (9.2) represents the heat transfer contribution made by the fins, the factor η being the fin efficiency defined in Eq. (2.98). The fin efficiency can be calculated based on formulas such as Eq. (2.99) and the charts of Figures 2.14 and 2.16.

The second term accounts for the direct heat transfer through the bare portions of the wall. The total heat transfer rate (9.2) can be rearranged as

$$q = \left(\eta \frac{A_f}{A} + \frac{A_u}{A} \right) hA(T_w - T_\infty)$$
$$= \varepsilon hA(T_w - T_\infty) \tag{9.3}$$

leading to the definition of the *overall surface efficiency* factor:

$$\varepsilon = \eta \frac{A_f}{A} + \frac{A_u}{A} = 1 - \frac{A_f}{A}(1 - \eta) \tag{9.4}$$

This factor should not be confused with the overall projected surface effectiveness ε_0, which was defined in Eq. (2.62).

When the surface A is covered with a layer of solid debris, the total heat transfer rate q must overcome not only the convective thermal resistance determined above, $(\varepsilon hA)^{-1}$, but also the thermal resistance of the deposited layer:

$$r_s = \frac{T_w - T_s}{q''} \tag{9.5}$$

The r_s resistance refers to the scale that covers only one unit of the total area A, while T_s is the temperature of the fluid side of the scale, Figure 9.7. When the scale is present, the total heat transfer rate is given by an expression similar to Eq. (9.3):

$$q = \varepsilon h_e A(T_w - T_\infty) \tag{9.6}$$

where the *effective heat transfer coefficient* h_e accounts for two effects – the conduction thermal resistance across the scale (r_s) and the convective film resistance ($1/h$) on the fluid side of the scale:

$$\frac{1}{h_e} = r_s + \frac{1}{h} \tag{9.7}$$

The r_s resistance is called *fouling factor* and has the units m²·K/W. Table 9.1 shows a set of the most up-to-date recommendations concerning the r_s values to be used in overall heat transfer coefficient calculations [3]. The order of magnitude of the convective heat transfer coefficient (h in Eq. (9.7)) depends on the flow arrangement and can be read off Figure 1.12.

In Eq. (9.7) was assumed that the size of the area crossed by the heat flux does not change as the flux travels the thickness of the scale. This assumption holds true as long as the scale is thin in comparison with the body that it covers (e.g. heat exchanger tube). Otherwise, the area variation must be taken into account, for example, by modeling the scale as a shell around the tube (cf. Eq. (2.33)).

In summary, according to Eq. (9.6), the total thermal resistance for one side of the heat exchanger surface is $(\varepsilon h_e A)^{-1}$. Let the subscripts h and c represent the two sides (hot and cold) of the heat exchanger surface shown in Figure 9.7. If $R_{t,w}$ is the conduction resistance of the wall itself, the total thermal resistance posed by the heat exchanger surface is

$$\frac{1}{U_c A_c} = \frac{1}{\varepsilon_h h_{e,h} A_h} + R_{t,w} + \frac{1}{\varepsilon_c h_{e,c} A_c} \tag{9.8}$$

Table 9.1 Representative values of the fouling factor r_s (m$^2 \cdot$K/W).

Temperature of heating medium water temperature	Up to 115°C 50°C or less		115–205°C Above 50°C	
Water velocity	1 m/s and less	Over 1 m/s	1 m/s and less	Over 1 m/s
Water types				
Distilled water	0.0001	0.0001	0.0001	0.0001
Sea water	0.0001	0.0001	0.0002	0.0002
Brackish water	0.0004	0.0002	0.0005	0.0004
City or well water	0.0002	0.0002	0.0004	0.0004
River water, average	0.0005	0.0004	0.0007	0.0005
Hard water	0.0005	0.0005	0.0009	0.0009
Treated boiler feed water	0.0002	0.0001	0.0002	0.0002
Liquids				
Liquid gasoline, oil, and liquefied petroleum gases			0.0002–0.0004	
Vegetable oils			0.0005	
Caustic solutions			0.0004	
Refrigerants, ammonia			0.0002	
Methanol, ethanol, and ethylene glycol solutions			0.0004	
Gases				
Natural gas			0.0002–0.0004	
Acid gas			0.0004–0.0005	
Solvent vapors			0.0002	
Steam (non-oil bearing)			0.0001	
Steam (oil bearing)			0.0003–0.0004	
Compressed air			0.0002	
Ammonia			0.0002	

Source: Based on the data compiled in Chenoweth [3].

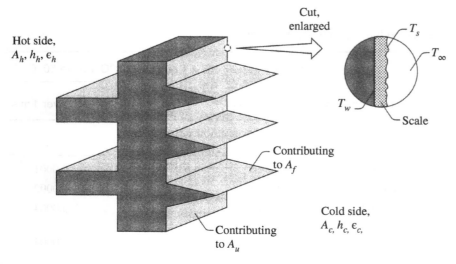

Figure 9.7 Heat exchanger surface with fins and scale on both sides.

On the left side, the overall heat transfer coefficient U_c is said to be based on the cold-side area A_c. Alternatively, the left side of Eq. (9.8) can be labeled $1/(U_h \cdot A_h)$, in which U_h is the overall heat transfer coefficient based on the hot-side surface A_h:

$$\frac{1}{U_c A_c} \equiv \frac{1}{U_h A_h} \tag{9.9}$$

It is a good idea to verify that all the resistance terms are significant (large enough) on the right side of Eq. (9.8). The representative order of magnitude of the overall heat transfer coefficient in various two-fluid heat exchangers is listed in Table 9.2.

The wall resistance $R_{t,w}$ can be calculated using the methods of Chapter 2, for example, based on Eq. (2.7) for a plane wall of thickness L, area A, and conductivity k and Eq. (2.33) for a tube wall of radii r_o and r_i, length l, and conductivity k. This method of calculating $R_{t,w}$ is based on the assumption that the conduction through the wall is unidirectional. (This conflicts somewhat with the wall sketched on the left side of Figure 9.7, because the size and density of the fins there were exaggerated intentionally.) The overall heat transfer coefficient formula for a plane wall with convective heat transfer on both sides, Eq. (2.17), is the special form taken by the more general Eq. (9.8) in the case where the fins are absent and both surfaces are clean.

Example 9.1 *Plane Wall: Pin Fins on One Side*

The heat exchanger surface shown in Figure E9.1 separates a stream of hot liquid ($T_h = 100\,°C$, $h_h = 200\,W/m^2 \cdot K$) from a stream of cold gas ($T_c = 30\,°C$, $h_c = 10\,W/m^2 \cdot K$). This surface is made of a slab of thickness $t = 0.8\,cm$, which has the frontal area of a square with side $H = 0.5\,m$. To offset the effect of the small heat transfer coefficient h_c, the area of the cold side was increased by adding a number of pin fins arranged in a square pattern. The dimensions of the wall, individual fin, and square array are indicated directly on the figure. The wall and fins are made of a metal with the conductivity $k = 40\,W/m \cdot K$. The effect of fouling is negligible on both sides of the wall; in other words,

$$h_{e,h} \cong h_h \qquad h_{e,c} \cong h_c$$

Table 9.2 Representative orders of magnitude of the overall heat transfer coefficient.

Hot fluid	Cold fluid	U (W/m²·K)
Water	Water	1000–2500
Ammonia	Water	1000–2500
Gases	Water	10–250
Light organics[a]	Water	370–730
Heavy organics[b]	Water	25–370
Steam	Water	1000–3500
Steam	Ammonia	1000–3500
Steam	Gases	25–250
Steam	Light organics	500–1000
Steam	Heavy organics	30–300
Light organics	Light organics	200–400
Heavy organics	Heavy organics	50–200
Light organics	Heavy organics	50–200
Heavy organics	Light organics	150–300

a) Organic liquids with viscosities below 0.0005 kg/s·m.
b) Organic liquids with viscosities greater than 0.001 kg/s·m.
Source: After Shah and Mueller [4]; Kern [5].

Calculate the overall thermal resistance of the heat exchanger surface ($1/U_hA_h$ or $1/U_cA_c$) and the total heat transfer rate q.

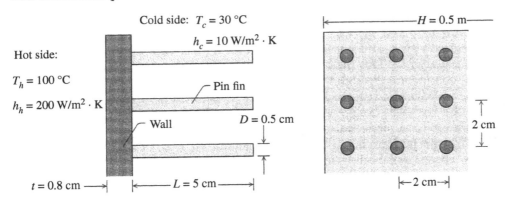

Figure E9.1

Solution

We begin with the heat exchanger surface. The hot side is unfinned; therefore its total area is

$$A_h = (0.5 \text{ m})^2 = 0.25 \text{ m}^2$$

On the cold side, we see one pin fin on each square with side 2 cm, in other words, one fin per 4 cm². The total number of fins is therefore

$$n = \frac{(0.5\ \mathrm{m})^2}{4\ \mathrm{cm}^2/\mathrm{fin}} = \frac{0.25\ \mathrm{m}^2}{0.004\ \mathrm{m}^2/\mathrm{fin}} = 625\ \text{fins}$$

The total unfinned area on the right side is the wall area left between the roots of the fins:

$$A_u = H^2 - n\pi\left(\frac{D}{2}\right)^2$$

$$= (0.5\ \mathrm{m})^2 - 625\pi\left(\frac{0.005\ \mathrm{m}}{2}\right)^2 = 0.238\ \mathrm{m}^2$$

To calculate the finned area A_f, we first model each fin (length L) as a slightly longer one with insulated tip (length L_c). This larger length is given by Harper and Brown's approximation, Eq. (2.97):

$$L_c = L + \frac{D}{4} = 5.125\ \mathrm{cm}$$

The finned area is the sum of the lateral (cylindrical) areas contributed by each fin:

$$A_f = n\pi D L_c$$

$$= 625\pi(0.005\ \mathrm{m})(0.05125\ \mathrm{m}) = 0.503\ \mathrm{m}^2$$

Therefore, the total area of the cold side is

$$A_c = A_f + A_u = 0.741\ \mathrm{m}^2$$

Next, to calculate the cold surface efficiency ε_c, we first estimate the efficiency of each pin fin (cf. Eq. (2.99)):

$$\eta = \frac{\tanh(mL_c)}{mL_c}$$

in which the dimensionless group mL_c has the value

$$mL_c = \left(\frac{h_c p}{kA}\right)^{1/2} L_c = \left(\frac{h_c \pi D}{k(\pi/4)}D^2\right)^{1/2} L_c$$

$$= 2\left(\frac{h_c}{kD}\right)^{1/2} L_c$$

$$= 2\left(\frac{10\ \mathrm{W}}{\mathrm{m}^2\cdot\mathrm{K}}\frac{\mathrm{m}\cdot\mathrm{K}}{40\ \mathrm{W}}\frac{1}{0.005\ \mathrm{m}}\right)^{1/2}0.05125\ \mathrm{m} = 0.724$$

The fin efficiency is then

$$\eta = \frac{\tanh(0.724)}{0.724} = 0.85$$

and this translates into an overall cold-surface efficiency of

$$\varepsilon_c = \eta\frac{A_f}{A_c} + \frac{A_u}{A_c}$$

$$= 0.85\frac{0.515\ \mathrm{m}^2}{0.753\ \mathrm{m}^2} + \frac{0.238\ \mathrm{m}^2}{0.753\ \mathrm{m}^2} = 0.856$$

The overall thermal resistance posed by the heat exchanger surface is

$$\frac{1}{U_h A_h} = \frac{1}{h_h A_h} + \frac{t}{k A_h} + \frac{1}{\varepsilon_c h_c A_c}$$

$$= \left(\frac{1}{200 \times 0.25} + \frac{0.008}{40 \times 0.25} + \frac{1}{0.856 \times 10 \times 0.753} \right) \frac{K}{W}$$

$$= (0.02 + 0.008 + 0.155) \frac{K}{W} = 0.183 \text{ K/W}$$

This calculation shows that the cold-side resistance (0.148 K/W) accounts for 88% of the overall resistance (0.169 K/W) and that the resistance of the plane wall of thickness t is negligible. Finally, the total heat transfer rate through the entire arrangement is

$$q = U_h A_h (T_h - T_c) = \frac{(100 - 30) \text{ K}}{0.169 \text{ K/W}} = 383 \text{ W}$$

9.3 Log-Mean Temperature Difference Method

9.3.1 Parallel Flow

In this and the next two sections, we analyze the relationship between the total heat transfer rate q, the heat transfer area A, and the inlet and outlet temperatures of the two streams.

Consider first the *parallel flow* heat exchanger shown in Figure 9.8. The abscissa shows the "length" of the heat exchanger or the heat transfer area that is sandwiched in between and swept by the two streams. Control volumes of width dA are drawn around each of the streams. The local stream-to-stream temperature difference is $T_h - T_c$, where T_h and T_c represent the bulk temperatures of the "hot" and "cold" streams, respectively. The heat transfer rate that crosses the area element dA is

$$dq = (T_h - T_c) U_A \, dA \tag{9.10}$$

where U_A is the overall heat transfer coefficient at this particular position along the heat exchanger surface. With reference to each of the control volumes, the first law of thermodynamics for steady flow requires that

$$dq = -C_h dT_h \tag{9.11}$$

$$dq = C_c dT_c \tag{9.12}$$

The symbol C (W/K) is shorthand for the product $\pm \dot{m} c_P$ (i.e. the capacity rate) of each stream. Note that by writing the first-law statements (9.11) and (9.12), we are assuming either that the two fluids are ideal gases or that the pressure drop along each stream is negligible so that the specific enthalpy change experienced by each fluid can be approximated by $di \cong c_P dT$.

An equation that describes the variation of $T_h - T_c$ along the heat transfer surface A can be obtained by using Eqs. (9.11) and (9.12):

$$d(T_h - T_c) = dT_h - dT_c = -\frac{dq}{C_h} - \frac{dq}{C_c}$$

$$= -\left(\frac{1}{C_h} + \frac{1}{C_c} \right) dq \tag{9.13}$$

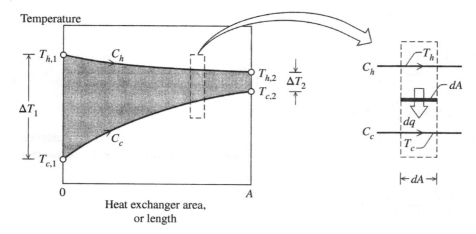

Temperature

Heat exchanger area,
or length

Figure 9.8 The temperature distribution in a parallel flow heat exchanger.

and by eliminating dq based on Eq. (9.10):

$$d(T_h - T)_c = -\left(\frac{1}{C_h} + \frac{1}{C_c}\right)(T_h - T_c)\, U_A\, dA \tag{9.14}$$

Separation of variables is achieved by dividing both sides by $T_h - T_c$; after integrating from $A = 0$ to any A, we obtain

$$\ln\frac{\Delta T_2}{\Delta T_1} = -\left(\frac{1}{C_h} + \frac{1}{C_c}\right) UA \tag{9.15}$$

where ΔT_1 and ΔT_2 are the stream-to-stream temperature differences in the front section and the back section of the parallel-flow heat exchanger (see Figure 9.8):

$$\Delta T_1 = T_{h,1} - T_{c,1} \qquad \Delta T_2 = T_{h,2} - T_{c,2} \tag{9.16}$$

The overall heat transfer coefficient U in Eq. (9.15) is the result of *averaging* U_A over the entire area A. As a special case, $U = U_A$ when U_A is constant along the heat transfer area.

What we have achieved in Eq. (9.15) is a relationship between the heat exchanger "size" (UA), the inlet and outlet temperatures (or ΔT_1 and ΔT_2), and the two capacity rates (C_h, C_c). The total stream-to-stream heat transfer rate q can be made to appear in this relationship by first integrating the first-law statements (9.11) and (9.12) along each stream:

$$q = -C_h(T_{h,2} - T_{h,1}) \tag{9.17}$$

$$q = C_c(T_{c,2} - T_{c,1}) \tag{9.18}$$

These integrations are based on the assumption that each C is a constant; in other words, the c_P of each fluid does not vary appreciably as the stream travels the entire length of the heat exchanger. Together, Eqs. (9.17) and (9.18) show that

$$\frac{1}{C_h} + \frac{1}{C_c} = \frac{1}{q}(-T_{h,2} + T_{h,1} + T_{c,2} - T_{c,1})$$

$$= \frac{1}{q}(\Delta T_1 - \Delta T_2) \tag{9.19}$$

Therefore, Eq. (9.15) can also be written as

$$\ln \frac{\Delta T_2}{\Delta T_1} = -\frac{1}{q}(\Delta T_1 - \Delta T_2)UA \tag{9.20}$$

In conclusion, the proportionality between the total heat transfer rate q and the overall thermal conductance of the heat exchanger surface is

$$q = UA\Delta T_{lm} \tag{9.21}$$

where ΔT_{lm} is the *log-mean temperature difference* notation encountered in Eq. (6.105):

$$\Delta T_{lm} = \frac{\Delta T_1 - \Delta T_2}{\ln \dfrac{\Delta T_1}{\Delta T_2}} = \frac{\Delta T_2 - \Delta T_1}{\ln \dfrac{\Delta T_2}{\Delta T_1}} \tag{9.22}$$

9.3.2 Counterflow

Equation (9.21) holds true not only for the parallel flow heat exchanger analyzed above but also for a *counterflow heat exchanger*, for which ΔT_{lm} is given by Eq. (9.22) with

$$\Delta T_2 = T_{h,in} - T_{c,out} \quad \text{and} \quad \Delta T_1 = T_{h,out} - T_{c,in} \tag{9.23}$$

All the flow arrangements to which the ΔT_{lm} heat transfer relation (9.21) applies are summarized in Figure 9.9. The single-stream examples treated earlier in Figure 6.11 are special cases of the more general result derived here in the form of Eq. (9.21).

The upper-right frame of Figure 9.9 shows that in the counterflow arrangement the stream-to-stream temperature difference is more *uniform* along the heat exchanger, certainly more uniform than in the other three arrangements. The temperature difference is also *smaller* than in the parallel flow arrangement if the only

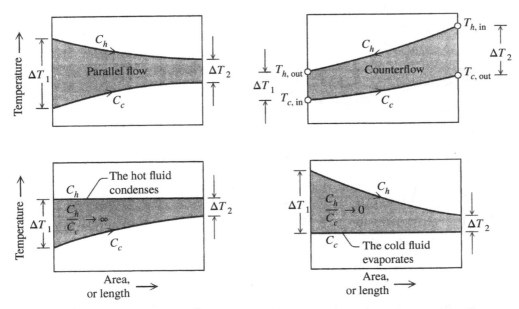

Figure 9.9 Flow arrangements to which the heat transfer relationship (9.21) applies.

difference between the two upper frames in Figure 9.9 is the direction of the C_h stream (i.e. if the two heat exchangers have the same size, capacity rates, and fluid inlet temperatures). For this reason, the counter-flow heat exchanger is *thermodynamically more efficient* (it generates less entropy) than the corresponding parallel flow heat exchanger.

9.3.3 Other Flow Arrangements

In *cross-flow* and *multipass* arrangements, the ratio q/UA is a more complicated function of $T_{h,1}$, $T_{h,2}$, $T_{c,1}$, and $T_{c,2}$, that is, more complicated than the ΔT_{lm} group identified on the right side of Eq. (9.21). For such arrangements, the heat transfer rate formula can be shaped in a way that mimics Eq. (9.21):

$$q = UA\Delta T_{lm}F \tag{9.24}$$

in which the new *correction factor F* is a function of two dimensionless parameters P and R. The definitions of these parameters are indicated directly on Figures 9.10–9.14, which show a sample of Bowman et al.'s [6] charts for the calculation of the correction factor. The ΔT_{lm} value that is to be substituted in Eq. (9.24) must be estimated by imagining that the flow arrangement is that of *counterflow*, that is, by using Eq. (9.23).

Note that F is generally less than 1 and that F approaches 1 as either P or R approaches the value 0. In either of these limits, Eq. (9.24) approaches the limit represented by Eq. (9.21). The limit $P \to 0$ corresponds to a heat exchanger in which the stream represented by the temperatures t_1 and t_2 experiences a change of phase (i.e. it condenses or evaporates; therefore, $t_1 = t_2$ when the pressure drop is sufficiently small). In the other limit, $R \to 0$, the condensing or evaporating stream is the one represented by T_1 and T_2. Since the temperature of a condensing or evaporating stream does not change along the heat exchanger surface, that stream acts

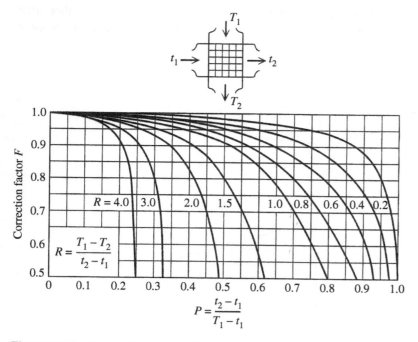

Figure 9.10 Correction factor F for cross-flow (single-pass) heat exchangers in which both streams remain unmixed.

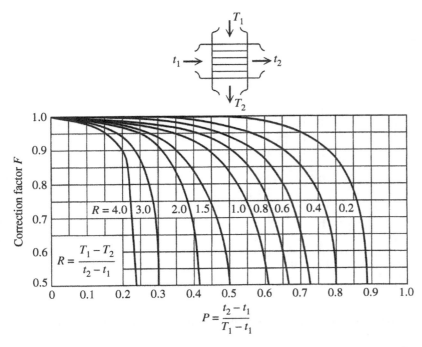

Figure 9.11 Correction factor F for cross-flow (single-pass) heat exchangers in which one stream is mixed and the other unmixed.

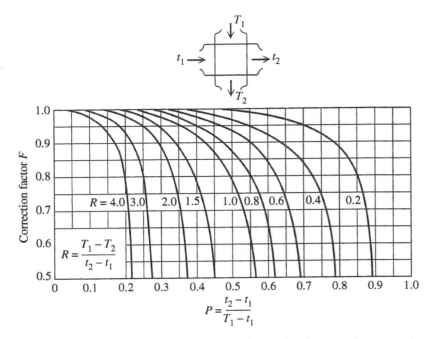

Figure 9.12 Correction factor F for cross-flow (single-pass) heat exchangers in which both streams are mixed.

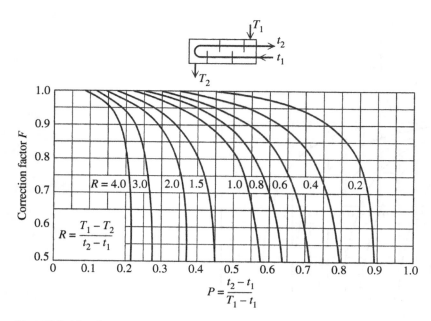

Figure 9.13 Correction factor F for shell-and-tube heat exchangers with one shell pass and any multiple of two tube passes (2, 4, 6, etc. tube passes).

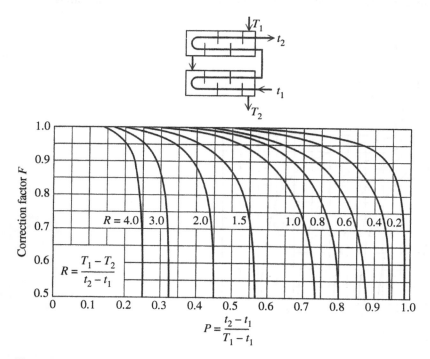

Figure 9.14 Correction factor F for shell-and-tube heat exchangers with two shell passes and any multiple of four tube passes (4, 8, 12, etc. tube passes).

as if its assumed (single-phase) capacity rate C is infinite. This observation is made also graphically in the lower half of Figure 9.9.

The temperature difference ratio R is, in fact, the ratio of the capacity rates of the two streams. To see this, assign the label C_t to the capacity rate of the $t_1 \rightarrow t_2$ stream and the label C_T to the capacity rate of the second stream, $T_1 \rightarrow T_2$. Under the assumption that the exterior of the heat exchanger enclosure is sufficiently well insulated with respect to the environment, the first law of thermodynamics for the entire heat exchanger reduces to

$$C_t(t_2 - t_1) = C_T(T_1 - T_2) \tag{9.25}$$

which means that

$$\frac{C_t}{C_T} = \frac{T_1 - T_2}{t_2 - t_1} = R \tag{9.26}$$

To summarize, the relationship between the total heat transfer rate q and the size of the heat exchanger area UA is given by Eq. (9.21), or more generally by Eq. (9.24). This relation must always be used in conjunction with the first-law statements of type (9.17) and (9.18), because the four inlet and outlet temperatures (which make up ΔT_{lm}) are interrelated and depend on the ratio of the two capacity rates, C_h/C_c. Indeed, one drawback of the ΔT_{lm}-type formula for the total heat transfer rate q is that, at first reading, it gives the impression that q does not depend on the capacity rates of the two streams.

Example 9.2 *Calculating the Needed Heat Exchanger Area by the ΔT_{lm} Method*

A hot water stream of flowrate $\dot{m}_h = 1$ kg/s is to be cooled from 90 to 60 °C in a heat exchanger, by contact with a larger stream of cold water, $\dot{m}_c = 2$ kg/s. The inlet temperature of the cold stream is 40 °C.

Calculate the heat exchanger area A needed for accomplishing this task. The overall heat transfer coefficient (based on A) is known, $U = 1000$ W/m²·K. Consider first the counterflow arrangement and later the cross-flow scheme in which both fluids are mixed (Figure E9.2).

Solution

Regardless of the flow arrangement, the outlet temperature of the cold stream is dictated by the continuity of the total heat transfer rate between the two streams:

$$q = \dot{m}_h c_{P,h}(T_{h,\text{in}} - T_{h,\text{out}}) = \dot{m}_c c_{P,c}(T_{c,\text{out}} - T_{c,\text{in}})$$

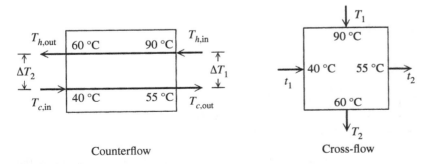

Counterflow   Cross-flow

Figure E9.2

Recognizing that for water $c_{P,h} \cong c_{P,c}$, we obtain

$$T_{c,\text{out}} = T_{c,\text{in}} + \frac{\dot{m}_h}{\dot{m}_c}(T_{h,\text{in}} - T_{h,\text{out}})$$

$$= 40\,°C + \frac{1}{2}(90 - 60\,°C) = 55\,°C$$

Consider first the counterflow arrangement, in which the end temperature differences are

$$\Delta T_1 = 35\,°C \quad \text{and} \quad \Delta T_2 = 20\,°C$$

The log-mean temperature difference has a value somewhere between 35 and 20 °C:

$$\Delta T_{\text{lm}} = \frac{\Delta T_1 - \Delta T_2}{\ln \dfrac{\Delta T_1}{\Delta T_2}} = \frac{(35 - 20)\,°C}{\ln \dfrac{35}{20}} = 26.8\,°C$$

The needed heat exchanger area can be calculated by using Eq. (9.21):

$$A_{\text{counterflow}} = \frac{q}{U \Delta T_{\text{lm}}} = \frac{\dot{m}_h c_{P,h}(T_{h,\text{in}} - T_{h,\text{out}})}{U \Delta T_{\text{lm}}}$$

$$= \frac{\left(1\,\dfrac{\text{kg}}{\text{s}}\right)\left(4.19\dfrac{10^3 \text{J}}{\text{kg} \cdot \text{K}}\right)(90 - 60)\,°C}{\left(1000\,\dfrac{\text{W}}{\text{m}^2 \cdot \text{K}}\right)(26.8\,°C)} = 4.69\,\text{m}^2$$

As an alternative, consider the cross-flow arrangement in which both fluids are mixed. In the attached sketch, the inlet and outlet temperatures have been relabeled in accordance with the notation employed in Figure 9.12. Next, we calculate the dimensionless parameters employed in Figure 9.12:

$$P = \frac{t_2 - t_1}{T_1 - t_1} = \frac{55 - 40}{90 - 40} = 0.3$$

$$R = \frac{T_1 - T_2}{t_2 - t_1} = \frac{90 - 60}{55 - 40} = 2 \quad \left(\text{note, again, that } R = \frac{C_c}{C_h}\right)$$

and to read on the ordinate of the graph,

$$F \cong 0.885$$

The log-mean temperature difference is the value calculated for the counterflow arrangement (see the comment under Eq. (9.24)):

$$\Delta T_{\text{lm}} = 26.8\,°C$$

Therefore, the new area requirement is (cf. Eq. (9.24))

$$A_{\text{cross-flow}} = \frac{q}{U \Delta T_{\text{lm}} F} = \frac{\dot{m}_h c_{P,h}(T_{h,\text{in}} - T_{h,\text{out}})}{U \Delta T_{\text{lm}} F}$$

$$= \frac{\left(1\,\dfrac{\text{kg}}{\text{s}}\right)\left(4.19\dfrac{10^3 \text{J}}{\text{kg} \cdot \text{K}}\right)(90 - 60)\,°C}{\left(1000\,\dfrac{\text{W}}{\text{m}^2 \cdot \text{K}}\right)(26.8\,°C)(0.885)} = 5.30\,\text{m}^2$$

In conclusion, the area required by the cross-flow arrangement is 13% larger than the area required by the counterflow heat exchanger.

Summary

The calculation of the area needed for the cross-flow arrangement was more complicated than in the case of the counterflow, because of the graphic procedure needed for estimating the correction factor F. It is as if in counterflow we began the calculation by setting $F = 1$ and then continued with the same steps as those employed for the cross-flow scheme. In summary, the numerical calculation of the area requirement involved the following steps:

1. The parameters P and R were calculated based on the known inlet and outlet temperatures.
2. The correction factor F was read off the appropriate chart.
3. The log-mean temperature difference ΔT_{lm} was calculated based on Eq. (9.22).
4. The area A was determined finally by using Eq. (9.24) and by invoking the first law to replace q in that equation.

Example 9.3 *Calculating the Outlet Temperatures and the Heat Transfer Rate by the ΔT_{lm} Method*

The counterflow heat exchanger shown in the sketch has a heat transfer area $A = 10\,\text{m}^2$ and a corresponding overall heat transfer coefficient $U = 500\,\text{W/m}^2\,\text{K}$. It is used to cool 1.5 kg/s of hot oil initially at 110 °C, by contact with a 0.5 kg/s stream of cold water with inlet temperature of 15 °C. The respective c_P values of oil and water are 2.25 and 4.18 kJ/kg·K, respectively.

Known in this problem are A, U, C_h, C_c, $T_{h,\text{in}}$, and $T_{c,\text{in}}$. Calculate the two outlet temperatures and the total heat transfer rate q.

Solution

The capacity rates of the two streams are

$$C_h = \dot{m}_h c_{P,h} = 1.5\,\frac{\text{kg}}{\text{s}}2.25\,\frac{\text{kJ}}{\text{kg}\cdot\text{K}} = 3.38\,\text{kW/K}$$

$$C_c = \dot{m}_c c_{P,c} = 0.5\,\frac{\text{kg}}{\text{s}}4.18\,\frac{\text{kJ}}{\text{kg}\cdot\text{K}} = 2.09\,\text{kW/K}$$

To calculate the total heat transfer rate q, we need ΔT_{lm}. Here, we run into difficulty because, with only the two inlet temperatures being specified, we cannot calculate ΔT_{lm} directly. We are forced to assume (i.e. guess) the value of one of the outlet temperatures. Later, after calculating q, we will check the correctness of the initial guess, and, if necessary, we shall correct it. We follow these steps:

1. Assume $T_{h,\text{out}} = 80\,°\text{C}$, and by recognizing the first law statement

$$C_h(T_{h,\text{in}} - T_{h,\text{out}}) = C_c(T_{c,\text{out}} - T_{c,\text{in}})$$

calculate the remaining outlet temperature:

$$T_{c,\text{out}} = T_{c,\text{in}} + \frac{C_h}{C_c}(T_{h,\text{in}} - T_{h,\text{out}})$$

$$= 15\,°\text{C} + \frac{3.38}{2.09}(110 - 80)\,°\text{C}$$

$$= 63.52\,°\text{C}$$

2. Calculate ΔT_{lm}:

$$\Delta T_1 = T_{h,\text{out}} - T_{c,\text{in}} = 80\,°C - 15\,°C = 65\,°C$$

$$\Delta T_2 = T_{h,\text{in}} - T_{c,\text{out}} = 110\,°C - 63.52\,°C = 46.48\,°C$$

$$\Delta T_{\text{lm}} = \frac{(65 - 46.48)\,°C}{\ln \dfrac{65}{46.48}} = 55.22\,°C$$

3. Determine q from the ΔT_{lm} expression (9.21):

$$q = UA\Delta T_{\text{lm}}$$

$$= \left(500\,\frac{W}{m^2 \cdot K}\right)(10\,m^2)(55.22\,°C) = 2.76 \times 10^5\,W$$

4. Calculate $T_{h,\text{out}}$ and compare it with the value assumed in the first step:

$$q = C_h(T_{h,\text{in}} - T_{h,\text{out}})$$

$$T_{h,\text{out}} = T_{h,\text{in}} - \frac{q}{C_h}$$

$$= 110\,°C - \frac{2.76 \times 10^5\,W}{3.39 \times 10^3\,\dfrac{W}{K}}$$

$$= 28.31\,°C$$

The $T_{h,\text{out}}$ value calculated above is considerably smaller than the assumed value (80 °C). The next move is to go back to step 1 and repeat the procedure by starting with a better guess for $T_{h,\text{out}}$. Will the new guess be a $T_{h,\text{out}}$ value greater or smaller than 80 °C?

To see the correct choice, it pays to review the preceding calculation in the reverse sequence in which it was made. In this way, the conclusion that the calculated $T_{h,\text{out}}$ is too low can be attributed to the fact that q is too high. In turn, q is too high because ΔT_{lm} is too large, and that means that the assumed value $T_{h,\text{out}}$ was also too large. In conclusion, for the second try, we begin with a lower guess for the $T_{h,\text{out}}$ value, and here are the results:

1. $T_{h,\text{out}} = 50\,°C$ (assumed)
   $T_{c,\text{out}} = 95.86\,°C$

2. $\Delta T_1 = 35\,°C$
   $\Delta T_2 = 14.14\,°C$
   $$\Delta T_{\text{lm}} = \frac{(35 - 14.14)\,°C}{\ln \dfrac{35}{14.14}} = 23.02\,°C$$

3. $q = 1.15 \times 10^5\,W$

4. $T_{h,\text{out}} = 110\,°C - \dfrac{1.15 \times 10^5\,W}{3.38 \times 10^3\,W/K}$
   $= 76\,°C$ (calculated)

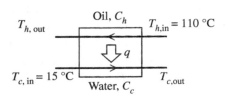

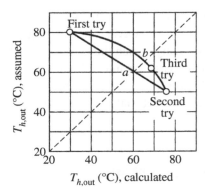

Figure E9.3

This time the calculated $T_{h,\text{out}}$ is greater than the assumed value; therefore, for the third try, we'll assume a $T_{h,\text{out}}$ value greater than the preceding guess. The accuracy of this third guess can be improved considerably by drawing the sketch shown on the right side of Figure E9.3 (or by performing the equivalent calculations on the computer). Each of the two tries executed until now is represented by one point. Connecting these points with a straight line and intersecting this line with the rising diagonal,

$$T_{h,\text{out}} \text{ (assumed)} = T_{h,\text{out}} \text{ (calculated)}$$

we obtain the point labeled a. In conclusion, $T_{h,\text{out}} \cong 61\,°C$ is a good candidate for the third guess, and the corresponding results are

1. $T_{h,\text{out}} = 61\,°C$ (assumed)
   $T_{c,\text{out}} = 94.24\,°C$
   $\Delta T_1 = 46\,°C$

2. $\Delta T_2 = 15.76\,°C$
   $\Delta T_{\text{lm}} = 28.23\,°C$

3. $q = 1.41 \times 10^5\,\text{W}$

4. $T_{h,\text{out}} = 68.24\,°C$ (calculated)

The agreement between the calculated and assumed $T_{h,\text{out}}$ values has improved. For an even better estimate, we connect with an arc of parabola the three points that now appear on the graph. This arc intersects the rising diagonal at point b, or $T_{h,\text{out}} \cong 63.5\,°C$. The last (the fourth) try yields, in order,

1. $T_{h,\text{out}} = 63.5\,°C$ (assumed)
   $T_{c,\text{out}} = 90.2\,°C$
   $\Delta T_1 = 48.5\,°C$

2. $\Delta T_2 = 19.8\,°C$
   $\Delta T_{\text{lm}} = 32.04\,°C$

3. $q = 1.6 \times 10^5\,\text{W}$

4. $T_{h,\text{out}} = 62.6\,°C$ (calculated)

What we have discovered in this example is that when the area (A, U), capacity rates, and inlet temperatures are specified, the calculation of q by the ΔT_{lm} method requires a *trial-and-error procedure*. This procedure is even lengthier for a heat exchanger such those of Figures 9.10–9.14, for which the correction factor F is generally less than 1. In such a case, an integral part of step 1 is also the calculation of R, P, and F, and the correct ΔT_{lm} formula for use in step 2 is Eq. (9.24). Fortunately, this type of heat exchanger problem can be solved in one shot (i.e. without trial and error) by applying the effectiveness–*NTU* method, which is described next.

9.4 Effectiveness–*NTU* Method

9.4.1 Effectiveness and Limitations Posed by the Second Law

An alternative to calculating the total heat transfer rate q – a method that brings out the individual effects of not only the total thermal conductance UA but also that of the capacity rates C_h and C_c – begins with defining two new dimensionless groups [1]. The first is the *number* of heat exchanger (heat *transfer*) *units*, *NTU*:

$$NTU = \frac{UA}{C_{min}} \tag{9.27}$$

in which C_{min} is the smaller of the two capacity rates:

$$C_{min} = \min(C_h, C_c) \tag{9.28}$$

The second dimensionless group is the heat exchanger *effectiveness* ε, which is defined as the ratio between the actual heat transfer rate q and the "ceiling value" of the heat transfer rate that could take place between the two streams, q_{max}:

$$\varepsilon = \frac{\text{actual heat transfer rate}}{\text{maximum heat transfer rate}} = \frac{q}{q_{max}} \tag{9.29}$$

To see what is meant by q_{max}, consider the counterflow heat exchanger temperature distribution shown in Figure 9.15. The actual heat transfer rate q can be written either as $C_h(T_{h,in} - T_{h,out})$, or as $C_c(T_{c,out} - T_{c,in})$. Imagine that we increase the size of the heat exchanger UA, in an attempt to increase q while holding the two inlet temperatures $(T_{h,in}, T_{c,in})$ fixed. As we do this, the stream-to-stream temperature differences decrease everywhere along the heat exchanger, and this means that the two temperature distribution curves migrate toward one another (i.e. the curves become increasingly steeper). In other words, $T_{c,out}$ migrates toward the stationary $T_{h,in}$ and $T_{h,out}$ migrates toward the fixed temperature level $T_{c,in}$. Note that in Figure 9.15, the steeper *of* the two temperature distribution curves is the one that corresponds to the smaller of the two capacity rates $(C_{min} = C_c)$.

As these changes take place, the first outlet temperature to equal its facing inlet temperature is that of the stream with the lower capacity rate, namely, C_c and $T_{c,out} = T_{h,in}$ in Figure 9.15. Beyond this stage, the outlet temperature of the cold stream, $T_{c,out}$, remains at the $T_{h,in}$ temperature level no matter how large a UA we build into the heat exchanger. Otherwise, with $T_{c,out}$ going above $T_{h,in}$, we would be looking at a picture in which the two temperature distributions would cross somewhere inside the heat exchanger. A heat exchanger with crossing temperature distribution curves is physically impossible (a violation of the second law of thermodynamics), because in the heat exchanger zone where the T_h curve happens to be below the T_c curve, the local heat transfer rate dq would be required to flow "uphill," from T_h to T_c (cf. Ref. [7]).

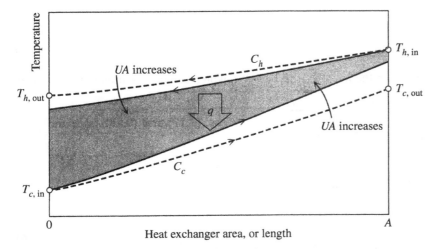

Figure 9.15 The effect of the heat transfer conductance *UA* on the temperature distributions inside a counterflow heat exchanger.

In summary, since the actual heat transfer rate is $C_{\min}(T_{c,\text{out}} - T_{c,\text{in}})$, and since $T_{c,\text{in}}$ is fixed while $T_{c,\text{out}}$ cannot exceed $T_{h,\text{in}}$, the ceiling value for q corresponds to the limit where the temperature excursion experienced by the $C_{\min}$ stream reaches its largest value, in this example, $T_{h,\text{in}} - T_{c,\text{in}}$,

$$q_{\max} = C_{\min}(T_{h,\text{in}} - T_{c,\text{in}}) \tag{9.30}$$

Although we derived this formula by analyzing the behavior of a counterflow heat exchanger, its validity as a definition for $q_{\max}$ is general. The temperature difference $T_{h,\text{in}} - T_{c,\text{in}}$ is the difference between the two inlet ports, that is, the largest difference between two states that belong to the two streams that make up the heat exchanger.

9.4.2 Parallel Flow

The total heat transfer rate relation developed for a parallel flow heat exchanger in Section 9.3.1 can be restated in terms of effectiveness and *NTU*. Consider again the parallel flow temperature distributions shown in Figure 9.8, and note that in this case the effectiveness ε is equal to

$$\varepsilon = \frac{C_h(T_{h,1} - T_{h,2})}{C_{\min}(T_{h,1} - T_{c,1})} = \frac{C_c(T_{c,2} - T_{c,1})}{C_{\min}(T_{h,1} - T_{c,1})} \tag{9.31}$$

Assume that the smaller of the two capacity rates is C_c,

$$C_c = C_{\min} \quad \text{and} \quad C_h = C_{\max} \tag{9.32}$$

The *NTU* definition (9.27) and Eq. (9.18) allow us to rewrite Eq. (9.20) as

$$\ln \frac{\Delta T_2}{\Delta T_1} = -NTU \left(1 - \frac{T_{h,2} - T_{h,1}}{T_{c,2} - T_{c,1}} \right) \tag{9.33}$$

Next, by eliminating q between Eqs. (9.17) and (9.18), we conclude that

$$\frac{T_{h,2} - T_{h,1}}{T_{c,2} - T_{c,1}} = -\frac{C_c}{C_h} = -\frac{C_{\min}}{C_{\max}} \tag{9.34}$$

or that Eq. (9.33) is the same as

$$\ln \frac{\Delta T_2}{\Delta T_1} = -NTU \left(1 + \frac{C_{min}}{C_{max}} \right) \tag{9.35}$$

The second part of this analysis focuses on the left side of Eq. (9.35), which can be rewritten sequentially as

$$\frac{\Delta T_2}{\Delta T_1} = \frac{T_{h,2} - T_{c,2}}{T_{h,1} - T_{c,1}} = \frac{T_{h,2} - T_{c,2} + T_{h,1} - T_{c,1} - (T_{h,1} - T_{c,1})}{T_{h,1} - T_{c,1}}$$

$$= 1 + \frac{T_{h,2} - T_{c,2} - T_{h,1} + T_{c,1}}{T_{h,1} - T_{c,1}}$$

$$= 1 + \frac{-(T_{c,2} - T_{c,1}) + T_{h,2} - T_{h,1}}{\frac{1}{\varepsilon}(T_{c,2} - T_{c,1})}$$

$$= 1 + \varepsilon \left(-1 + \frac{T_{h,2} - T_{h,1}}{T_{c,2} - T_{c,1}} \right) = 1 + \varepsilon \left(-1 - \frac{C_{min}}{C_{max}} \right) \tag{9.36}$$

Substituting this last expression on the left side of Eq. (9.35), we arrive at the expression for the effectiveness of a parallel flow heat exchanger:

$$\varepsilon_{\text{parallel flow}} = \frac{1 - \exp[-NTU(1 + C_{min}/C_{max})]}{1 + C_{min}/C_{max}} \tag{9.37}$$

This expression can be turned inside out to produce a formula for calculating the required NTU for an application in which the effectiveness and the capacity rate ratio are specified:

$$NTU_{\text{parallel flow}} = -\frac{\ln[1 - \varepsilon(1 + C_{min}/C_{max})]}{1 + C_{min}/C_{max}} \tag{9.38}$$

Equations (9.37) and (9.38) are also obtained at the end of an analysis in which the reverse of the assumption (9.32) is made, that is, when $C_h = C_{min}$ and $C_c = C_{max}$.

In conclusion, Eqs. (9.37) and (9.38) represent the effectiveness–NTU relationship of a parallel flow heat exchanger, regardless of which stream (hot or cold) has the smaller capacity rate. The same relationship is shown graphically in Figure 9.16. Two limiting cases of this relationship are worth recording. When $C_{min} = C_{max}$, the capacity rates are said to be *balanced* and Eq. (9.37) reduces to

$$\varepsilon_{\text{parallel flow}} = \frac{1}{2}[1 - \exp(-2NTU)] \quad (C_{min} = C_{max}) \tag{9.39}$$

The opposite limit, $C_{min}/C_{max} = 0$, represents a heat exchanger in which one of the streams (C_{max}) undergoes a phase change at nearly constant pressure:

$$\varepsilon = 1 - \exp(-NTU) \left(\frac{C_{min}}{C_{max}} = 0 \right) \tag{9.40}$$

9.4.3 Counterflow

Based on a completely analogous analysis, which this time would be focused on the counterflow arrangement of Figure 9.15, we can write the effectiveness definition

$$\varepsilon = \frac{C_h(T_{h,in} - T_{h,out})}{C_{min}(T_{h,in} - T_{c,in})} = \frac{C_c(T_{c,out} - T_{c,in})}{C_{min}(T_{h,in} - T_{c,in})} \tag{9.41}$$

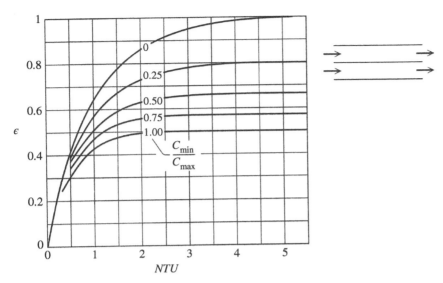

Figure 9.16 The effectiveness of a parallel flow heat exchanger. Source: Kays and London [1].

and derive the effectiveness–*NTU* relation

$$\varepsilon_{\text{counterflow}} = \frac{1 - \exp[-NTU(1 - C_{\min}/C_{\max})]}{1 - (C_{\min}/C_{\max})\exp[-NTU(1 - C_{\min}/C_{\max})]} \tag{9.42}$$

or, conversely, at the *NTU*–effectiveness relation

$$NTU_{\text{counterflow}} = \frac{\ln\left(\dfrac{1 - \varepsilon C_{\min}/C_{\max}}{1 - \varepsilon}\right)}{1 - C_{\min}/C_{\max}} \tag{9.43}$$

This relationship is presented graphically in Figure 9.17. The two extremes of this relationship are represented by the *balanced* counterflow heat exchanger

$$\varepsilon_{\text{counterflow}} = \frac{NTU}{1 + NTU} \quad (C_{\min} = C_{\max}) \tag{9.44}$$

and by the heat exchanger in which the $C_{\max}$ stream experiences a change of phase at nearly constant pressure:

$$\varepsilon = 1 - \exp(-NTU) \quad \left(\frac{C_{\min}}{C_{\max}} = 0\right) \tag{9.45}$$

Note that the effectiveness formula in this second extreme, Eq. (9.45), is the same as that for parallel flow heat exchangers, Eq. (9.40). This is caused by the fact that when the temperature of the condensing or evaporating stream ($C_{\max}$) does not vary along the heat exchanger surface, the direction in which this stream flows loses its significance. This is why in the lower part of Figure 9.9, the flow direction is not indicated on the temperature distribution curves of the streams that condense or evaporate.

9.4.4 Other Flow Arrangements

Figures 9.18–9.22 display the effectiveness–*NTU* curves for five other flow arrangements. This selection of cross-flow and multi-pass arrangements is the same as one covered by the ΔT_{lm} correction factor charts of

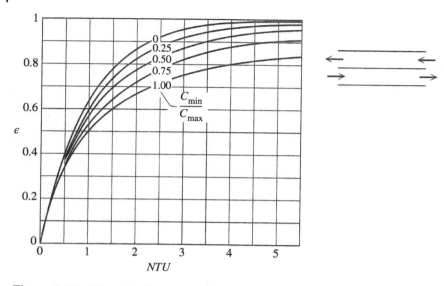

Figure 9.17 The effectiveness of a counterflow heat exchanger. Source: Shah and Mueller [4].

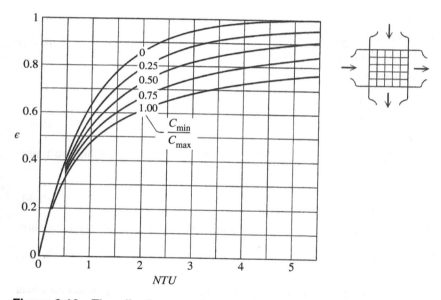

Figure 9.18 The effectiveness of a cross-flow (single-pass) heat exchanger in which both streams remain unmixed. Source: Kays and London [1].

Figures 9.10–9.14. The analytical expressions that stand behind the curves plotted in Figures 9.18–9.22 can be found in Ref. [1].

The effectiveness–*NTU* relationships and graphs presented in this entire section remind us that, in general, the effectiveness depends not only on *NTU* and the capacity rate ratio but also on the flow arrangement:

$$\varepsilon = \text{function}\left(NTU, \frac{C_{min}}{C_{max}}, \text{flow arrangement}\right) \tag{9.46}$$

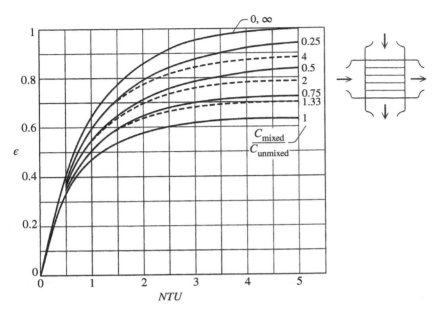

Figure 9.19 The effectiveness of a cross-flow (single-pass) heat exchanger in which one stream is mixed and the other is unmixed. Source: Kays and London [1].

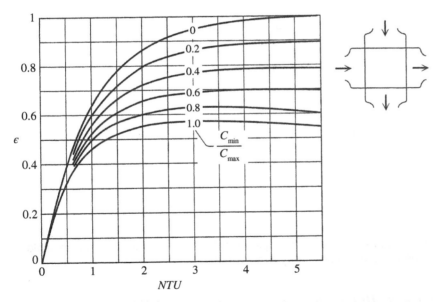

Figure 9.20 The effectiveness of a cross-flow (single-pass) heat exchanger in which both streams are mixed. Source: Kays and London [1].

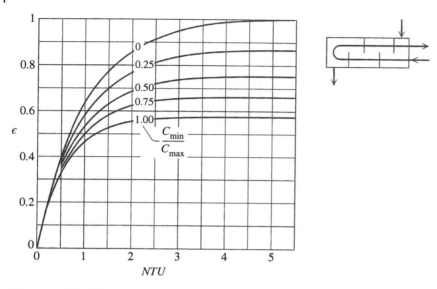

Figure 9.21 The effectiveness of shell-and-tube heat exchangers with one shell pass and any multiple of two tube passes (2, 4, 6, etc. tube passes). Source: Kays and London [1].

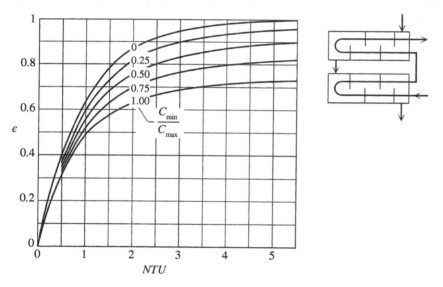

Figure 9.22 The effectiveness of shell-and-tube heat exchangers with two shell passes and any multiple of four tube passes (4, 8, 12, etc. tube passes). Source: Kays and London [1].

Example 9.4 *Calculating the Necessary Heat Exchanger Area by the ε–NTU Method*

To see the details and merits of the ε–NTU method relative to those of the ΔT_{lm} method, consider again the problem that we solved in Example 9.2. It dealt with the design of a heat exchanger in which a hot water stream had to be cooled to a certain final temperature by contact with a stream of cold water. The overall heat transfer coefficient, the two flowrates, the two inlet temperatures, and the outlet temperature of the hot stream were specified (Figure E9.4):

$$U = 1000 \ \text{W/m}^2 \cdot \text{K} \qquad T_{h,\text{in}} = 90\,°\text{C}$$
$$\dot{m}_h = 1 \ \text{kg/s} \qquad T_{h,\text{out}} = 60\,°\text{C}$$
$$\dot{m}_c = 2 \ \text{kg/s} \qquad T_{c,\text{in}} = 40\,°\text{C}$$

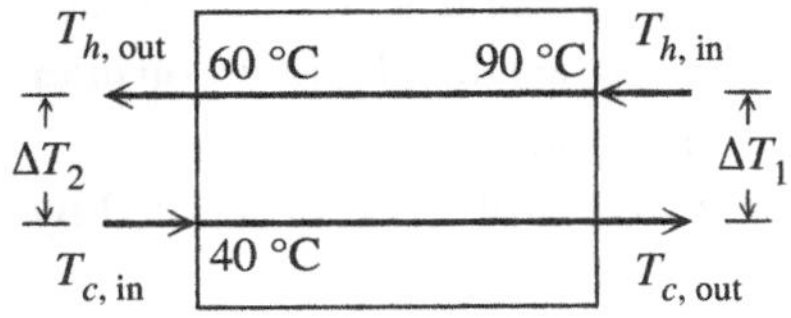

Figure E9.4

Consider only the counterflow arrangement. Calculate the heat exchanger area that is necessary to accomplish the task specified in the problem statement.

Solution

The unknown A appears in the *NTU* definition (9.27); therefore, we also need the value of *NTU* and, even before that, the value of the effectiveness ε. It is more instructive, however, if we construct the calculation of A as a sequence of three steps:

1. In the first step, we calculate the effectiveness by using one of the two formulas listed in Eq. (9.41), where

$$C_{\text{min}} = C_h \quad \text{and} \quad C_{\text{max}} = C_c$$

The first of these formulas is more convenient, because we know all the needed temperatures (in the second formula, we would first have to calculate $T_{c,\text{out}}$):

$$\varepsilon = \frac{C_h}{C_{\text{min}}} \frac{T_{h,\text{in}} - T_{h,\text{out}}}{T_{h,\text{in}} - T_{c,\text{in}}} = \frac{90 - 60}{90 - 40} = 0.6$$

2. The next objective is to calculate *NTU*. For this we need the capacity rate ratio

$$\frac{C_{\text{min}}}{C_{\text{max}}} = \frac{C_h}{C_c} = \frac{(\dot{m}c_{\text{P}})_h}{(\dot{m}c_{\text{P}})_c} \cong \frac{\dot{m}_h}{\dot{m}_c} = \frac{1}{2}$$

in which we assumed that $c_P \cong 4.19 \ \text{kJ/kg·K}$ applies to both water streams. The number of heat transfer units follows from Eq. (9.43):

$$NTU = \frac{\ln\left[\left(1 - 0.6\frac{1}{2}\right) / (1 - 0.6)\right]}{1 - \frac{1}{2}} = 1.12$$

3. In the last step, we recognize the *NTU* definition (9.27), from which follows A

$$A = NTU \frac{C_{\text{min}}}{U} = NTU \frac{(\dot{m}c_P)_h}{U}$$

$$= 1.12 \frac{(1 \ \text{kg/s})(4.19 \times 10^3 \ \text{J/kg·K})}{10^3 \ \text{W/m}^2 \cdot \text{K}} = 4.69 \ \text{m}^2$$

This area is, of course, the same as the answer obtained by the ΔT_{lm} method in Example 9.2. The ε–NTU method was more direct, because, unlike the ΔT_{lm} method, it did not require the calculation of three auxiliary quantities – ΔT_{lm}, the cold-stream outlet temperature $T_{c,out}$, and the stream-to-stream heat transfer rate q.

Example 9.5 *Calculating the Outlet Temperatures and Heat Transfer Rate by the ε–NTU Method*
It is much more advantageous to use the ε–NTU method instead of the ΔT_{lm} method when the surface (A, U), capacity rates (C_c, C_h), and inlet temperatures $(T_{c,in}, T_{h,in})$ are specified. To see this, let us reconsider the problem described in Example 9.3, which was solved by a trial-and-error method.

Known in this problem are the flow arrangement – counterflow – and that a stream of hot oil is cooled by a stream of cold water:

Oil	Water
$\dot{m}_h = 1.5\,\text{kg/s}$	$\dot{m}_c = 0.5\,\text{kg/s}$
$c_{P,h} = 2.25\,\text{kJ/kg}\cdot\text{K}$	$c_{P,c} = 4.18\,\text{kJ/kg}\cdot\text{K}$
$T_{h,in} = 15\,°\text{C}$	$T_{c,in} = 15\,°\text{C}$

The characteristics of the heat exchanger surface are also known:

$$A = 10\,\text{m}^2 \quad U = 500\,\text{W/m}^2\cdot\text{K}$$

Determine the two outlet temperatures and the total stream-to-stream heat transfer rate q.

Solution

The chief unknown is q, because if we know q we can calculate the outlet temperatures by writing the first law for each stream. The total heat transfer rate q is the "actual" rate in the definition of effectiveness ε, Eqs. (9.29) and (9.30). Therefore, working backward, we must calculate ε and, before that, the number of heat transfer units NTU. This reasoning suggests that the structure of the numerical solution consists of the following four steps:

1. In the first step, we identify the larger and smaller capacity rates, the capacity rate ratio, and the number of heat transfer units:

$$C_h = (\dot{m}c_P)_{\text{oil}} = \left(1.5\,\frac{\text{kg}}{\text{s}}\right)\left(2.25\,\frac{\text{kJ}}{\text{kg}\cdot\text{K}}\right) = 3.38\,\text{kW/K}$$

$$C_c = (\dot{m}c_P)_{\text{water}} = \left(0.5\,\frac{\text{kg}}{\text{s}}\right)\left(4.18\,\frac{\text{kJ}}{\text{kg}\cdot\text{K}}\right) = 2.09\,\text{kW/K}$$

$$C_{max} = C_h \quad \text{and} \quad C_{min} = C_c$$

$$\frac{C_{min}}{C_{max}} = \frac{2.09}{3.38} = 0.618$$

$$NTU = \frac{UA}{C_{min}} = \left(500\,\frac{\text{W}}{\text{m}^2\cdot\text{K}}\right)(10\,\text{m}^2)\left(\frac{\text{K}}{2.09\times 10^3\,\text{W}}\right) = 2.392$$

2. Now we have all the necessary ingredients for calculating ε using Eq. (9.42):

$$\varepsilon = \frac{1 - \exp\left[-2.392(1 - 0.618)\right]}{1 - 0.618 \exp\left[-2.392(1 - 0.618)\right]} = 0.796$$

3. The total heat transfer rate follows from Eqs. (9.29) and (9.30), which after eliminating q_{max} yields

$$q = \varepsilon C_{min}(T_{h,in} - T_{c,in})$$
$$= 0.796 \times 2.09 \frac{10^3 \text{ W}}{\text{K}} (110 - 15) \text{ K} = 1.58 \times 10^5 \text{ W}$$

4. In this final step, we draw a control volume around each stream, write the first law, and calculate the respective outlet temperatures:

$$\text{Water}: \quad q = C_c(T_{c,out} - T_{c,in})$$
$$T_{c,out} = T_{c,in} = \frac{q}{C_c}$$
$$= 15\,°\text{C} + \frac{1.58 \times 10^5 \text{ W}}{2.09 \times 10^3 \text{ W/K}} = 90.6\,°\text{C}$$

$$\text{Oil}: \quad q = C_h(T_{h,in} - T_{h,out})$$
$$T_{h,out} = T_{h,in} - \frac{q}{C_h}$$
$$= 110\,°\text{C} - \frac{1.58 \times 10^5 \text{ W}}{3.38 \times 10^3 \text{ W/K}} = 63.2\,°\text{C}$$

Here is a side-by-side comparison of the answers provided by the two methods:

Result	ε–NTU method	ΔT_{lm} method (four iterations, Example 9.3)
q	1.58×10^5 W	1.6×10^5 W
$T_{c,out}$	$90.6\,°\text{C}$	$90.2\,°\text{C}$
$T_{h,out}$	$63.2\,°\text{C}$	$62.6\,°\text{C}$

9.5 Pressure Drop

9.5.1 Pumping Power

The heat transfer characteristics of heat exchangers represent only half of the picture. Recall that, in general, the area-averaged overall heat transfer coefficient depends on the flow velocities on the two sides of the heat transfer surface. The second feature in the description of a heat exchanger deals with the power that is required for the purpose of forcing each stream to flow through its heat exchanger passage. This consideration leads to the calculation of the total pressure drop ΔP that is experienced by the stream across the passage. If the stream $\dot{m}$ carries a liquid that may be modeled thermodynamically as incompressible, the power required by an adiabatic pump is [7]

$$\dot{W}_p = \frac{1}{\eta_p}\frac{\dot{m}}{\rho}\Delta P \tag{9.47}$$

where ΔP is the pressure rise through the pump, ρ is the liquid density, and η_p is the isentropic efficiency of the pump. The group $\dot{m}\,\Delta P/\rho$ represents the minimum (or isentropic) power requirement.

Likewise, when the stream carries a gas that may be modeled as ideal (with constants R and c_P), a monotonic relationship also exists between the power required by a compressor and the pressure rise experienced by $\dot{m}$ through the compressor [7]:

$$\dot{W}_c = \frac{1}{\eta_c}\dot{m}c_P T_{in}\left[\left(\frac{P_{out}}{P_{in}}\right)^{R/c_P} - 1\right] \tag{9.48}$$

The subscripts in and out represent the compressor inlet and outlet conditions, and the temperature T_{in} is expressed in degrees Kelvin. Equation (9.48) refers to a single-stage adiabatic compressor with isentropic efficiency η_c.

The pressure drop through the heat exchanger passage is generally a complicated function of flow parameters and passage geometry. The pressure drop over a sufficiently straight portion of the passage ΔP_s may be evaluated with the method discussed in Section 6.1.3, provided that the density does not vary appreciably along the passage:

$$\Delta P_s = f\frac{4L}{D_h}\frac{1}{2}\rho V^2 \tag{9.49}$$

We recognize in this expression the passage length L, the hydraulic diameter D_h, the mean velocity V, and the friction factor f. Unlike in Chapter 6, where we wrote U for the mean longitudinal velocity through the duct, here we write V to avoid any confusion with the overall heat transfer coefficient. When ρ is not constant, the ΔP_s expression includes a correction for the pressure difference required to accelerate the stream while inside the heat exchanger, Eq. (9.61).

The total pressure drop ΔP consists of ΔP_s plus contributions made by other geometric features (sudden contractions, enlargements, bends, protuberances) that are encountered in the flow passage. The pressure drop contribution made by each feature of this kind is taken into account as the product between the dynamic pressure $\frac{1}{2}\rho V^2$ and a particular *pressure loss coefficient*, which can be found in the heat exchanger literature [1]. This procedure is illustrated in the following text.

9.5.2 Abrupt Contraction and Enlargement

Consider first the contraction experienced by the stream in Figure 9.23, as the passage cross section changes from A_a to A_b. The sharpness of this change in cross section may cause the separation of the flow upstream as well as downstream of the step change. The eddies visible in Figure 9.23 are a clear sign of the *irreversibility* of the flow in the direction $a \rightarrow b$.

In the absence of irreversibility, the total pressure $P + \frac{1}{2}\rho V^2$ would be conserved from a to b. In the present case, this quantity decreases in proportion with the obstructive role of the abrupt contraction:

$$\left(P_a + \frac{1}{2}\rho_a V_a^2\right) - \left(P_b + \frac{1}{2}\rho_b V_b^2\right) = K_c\frac{1}{2}\rho_b V_b^2 \tag{9.50}$$

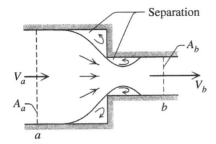

Figure 9.23 Abrupt contraction in a two-dimensional flow passage.

The *contraction loss coefficient* K_c is dimensionless. We can rewrite Eq. (9.50) by noting that at constant ρ, mass conservation requires $V_a A_a = V_b A_b$:

$$P_a - P_b = (1 - \sigma_{a-b}^2)\frac{1}{2}\rho_b V_b^2 + K_c\frac{1}{2}\rho_b V_b^2 \qquad (9.51)$$

In this equation, σ_{a-b} represents the passage contraction ratio

$$\sigma_{a-b} = \frac{A_b}{A_a} \qquad (9.52)$$

The dash lines in Figures 9.24 and 9.25 show the contraction loss coefficient for the flow entering a bundle of parallel tubes or parallel-plate channels. In each case the sudden contraction occurs as the stream crosses the plane of the entrance to the channels. In other words, the original cross section of the stream is the same as the frontal area of the bundle, whereas the final stream cross section is the sum of all the channel cross sections. The abscissa parameter in these figures is the contraction ratio defined in Eq. (9.52), which, in general, is defined by

$$\sigma = \frac{\text{flow cross-sectional area}}{\text{frontal area}} \qquad (9.53)$$

The Reynolds number Re is based on the hydraulic diameter of one channel and on the mean flow velocity through the *narrower* channel (e.g. V_b in Figure 9.23).

The irreversible flow through an abrupt enlargement of the passage is characterized similarly by introducing the *enlargement loss coefficient* K_e. With reference to the left side of Figure 9.26, this coefficient is defined by writing the pressure *rise* experienced by the stream through the enlargement of the passage cross section:

$$P_d - P_c = (1 - \sigma_{c-d}^2)\frac{1}{2}\rho_c V_c^2 - K_e\frac{1}{2}\rho_c V_c^2 \qquad (9.54)$$

The passage enlargement ratio σ_{c-d} is defined as in Eq. (9.53):

$$\sigma_{c-d} = \frac{A_c}{A_d} \qquad (9.55)$$

while V_c represents the mean velocity in the narrower portion of the channel. Figure 9.26 shows that the flow separates downstream from the step increase in flow cross-sectional area: In actual flows, one separation region is almost always larger than the other, that is, unlike in the axially symmetric sketch of Figure 9.26. Figures 9.24 and 9.25 provide the K_e values for the sudden enlargement from a heat exchanger core (tubes or flat channels) to a flow with cross section equal to the frontal or backward facing area of the core.

The pressure losses calculated on the basis of K_c and K_e add up in passages where both effects – contraction and enlargement – are present. An example in which contraction is followed by enlargement is presented

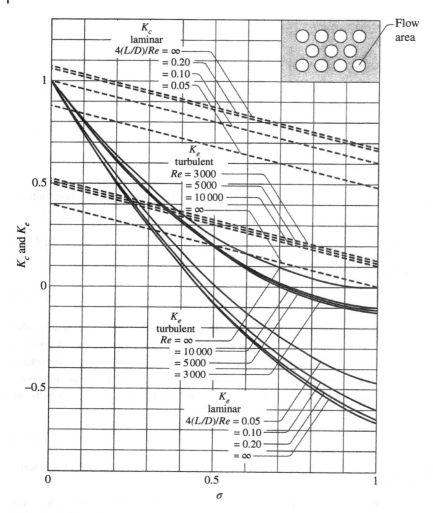

Figure 9.24 Abrupt contraction and enlargement loss coefficients for a heat exchanger core with multiple circular-tube passages. Source: Drawn after Kays and London [1].

in Figure 9.27, which shows the flow in and out of a heat exchanger core such as the multiple tubes and multiple parallel-plate channels sketched in the upper-right comers of Figures 9.24 and 9.25. An even earlier example was the tube side of the shell-and-tube heat exchanger presented in Figure 9.4. The pressure drop during contraction ($a \to b$) and the pressure rise during enlargement ($c \to d$) are illustrated in the lower part of Figure 9.27. These can be calculated with Eqs. (9.51) and (9.54). The pressure drop along the straight passage itself is $\Delta P_s = P_b - P_c$. The *total* pressure drop experienced by the stream is

$$
\begin{aligned}
\Delta P &= P_a - P_d \\
&= \underbrace{(P_a - P_b)}_{\text{Eq. (9.51)}} + \underbrace{(P_b - P_c)}_{\substack{\Delta P_s \\ \text{Eq. (9.61)}}} - \underbrace{(P_d - P_c)}_{\text{Eq. (9.54)}}
\end{aligned}
\tag{9.56}
$$

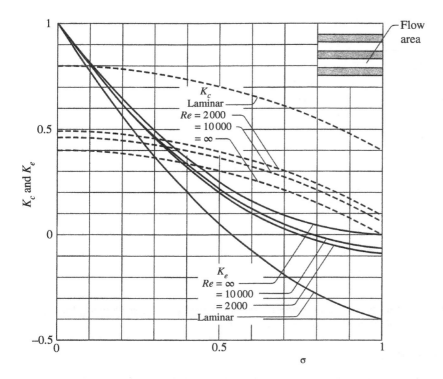

Figure 9.25 Abrupt contraction and enlargement loss coefficients for a heat exchanger core with multiple parallel-plate passages. Source: Drawn after Kays [8].

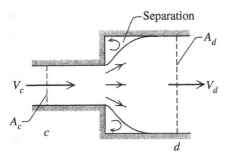

Figure 9.26 Abrupt enlargement in a two-dimensional flow passage and the separated flow downstream from an orifice.

In special cases where the density ρ is essentially constant from a to d (i.e. when $V_b = V_c = V$) and where the contraction ratio equals the enlargement ratio ($\sigma_{a-b} = \sigma_{c-d}$), Eq. (9.56) reduces to

$$\Delta P = K_c \frac{1}{2}\rho V^2 + \Delta P_s + K_e \frac{1}{2}\rho V^2 \tag{9.57}$$

The three components of the total ΔP represent, in order, the contraction loss, the straight-duct pressure drop, and the enlargement loss.

One way to decrease the pressure losses associated with contractions, enlargements, and bends is to *smooth* these geometric features, that is, to make them less abrupt. Separation and large swirls are likely to occur in bends of ducts, because the naturally curved stream is pinched as it tries to make the corner. The flow is considerably smoother and the separation regions are smaller when the elbow is rounded.

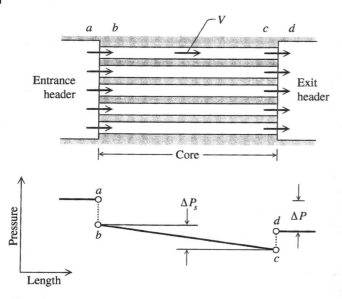

Figure 9.27 The cross-sectional contraction and enlargement experienced by a stream flowing through the core of a heat exchanger, and the associated pressure distribution.

9.5.3 Acceleration and Deceleration

While developing the special-case formula (9.57), we assumed that the fluid density does not vary appreciably between inlet and outlet. This is a good approximation for liquids such as water and oil. It is not always adequate for gases. For example, when the gas is heated while flowing through the heat exchanger passage, its outlet density ρ_{out} is smaller than the density at the inlet, ρ_{in}. If the inlet and outlet have the same cross-sectional area A_c, the inlet velocity is smaller than the outlet velocity, $V_{in} < V_{out}$. Conserved at every position along the flow passage is the mass flowrate $\dot{m} = \rho V A$, or the *mass velocity*

$$G = \frac{\dot{m}}{A} = \rho_{in} V_{in} = \rho_{out} V_{out} \tag{9.58}$$

When the flow is turbulent, the velocity distribution across the passage is relatively uniform (slug profile), so that the longitudinal momentum of the stream is approximately $\dot{m}V$. This increases from $\dot{m}V_{in}$ to $\dot{m}V_{out}$ because the gas accelerates from V_{in} to V_{out}. The change in longitudinal momentum $\dot{m}(V_{out} - V_{in})$ must be balanced by the pressure difference applied between the two ends of the passage:

$$\dot{m}(V_{out} - V_{in}) = \Delta P_{acc} A_c \tag{9.59}$$

In view of Eq. (9.58), we conclude that the pressure difference required to accelerate the stream from inlet to outlet is

$$\Delta P_{acc} = G^2 \left(\frac{1}{\rho_{out}} - \frac{1}{\rho_{in}} \right) \tag{9.60}$$

In conclusion, when the density is not constant, the pressure drop across the straight portion of the passage is equal to the wall friction effect (9.49) plus the acceleration effect (9.60):

$$\Delta P_s = f \frac{4L}{D_h} \frac{1}{2} \rho V^2 + G^2 \left(\frac{1}{\rho_{out}} - \frac{1}{\rho_{in}} \right) \tag{9.61}$$

In the frictional term on the right side, the density is averaged between inlet and outlet, $\rho = (\rho_{in} + \rho_{out})/2$, which means that $V = G/\rho$. The second term on the right side holds when the velocity does not vary appreciably across the passage. In the opposite case, for example, in laminar flow through a tube, the second term must be multiplied by 4/3 to account for the parabolic velocity distribution.

When the gas is being cooled, ρ_{out} is greater than ρ_{in}, and ΔP_{acc} is negative. In this case the deceleration of the stream works toward decreasing the overall pressure difference that must be maintained across the passage, ΔP. It is important to keep in mind that when the gas density varies, the contraction–enlargement relationships to use are Eqs. (9.51) and (9.54), not Eq. (9.57).

9.5.4 Tube Bundles in Cross-Flow

The flow of a stream perpendicular to an array of cylinders formed the subject of Section 5.5.4. There we learned how to calculate the heat transfer rate between the stream and a given number of cylinders. Arrays of aligned cylinders and staggered cylinders were described in Figure 5.16. In the present context, the cross-flow and the array of cylinders (tubes with internal flow) constitute a cross-flow heat exchanger. The pressure drop experienced by the cross-flow is proportional to the number of tube rows counted in the flow direction, n_l:

$$\Delta P = n_l\, f\chi \frac{1}{2}\rho V_{max}^2 \tag{9.62}$$

The dimensionless factors f and χ are presented in Figure 9.28 for *aligned arrays*. Each array of this kind is described by the longitudinal pitch X_l and the transversal pitch X_t or by their dimensionless counterparts

$$X_l^* = \frac{X_l}{D} \quad X_t^* = \frac{X_t}{D} \tag{9.63}$$

where D is the tube outside diameter. The f curves in Figure 9.28 correspond to a square array ($X_l = X_t$), while the χ correction factor accounts for arrangements in which $X_l \neq X_t$. Note that $\chi = 1$ for the square arrangement. The Reynolds number $Re_D = V_{max} D/\nu$ is based on the maximum mean velocity V_{max}, which occurs in the smallest spacing between two adjacent tubes in a row perpendicular to the flow direction.

Figure 9.29 contains the corresponding f and χ information for *staggered arrays*. The f curves have been drawn for the equilateral triangle arrangement [$X_t = X_d$ or $X_t = (3^{1/2}/2)X_t$]. The upper-right inset shows that the χ value plays the role of correction factor in instances where the tube centers do not form equilateral triangles. The Reynolds number is again based on the mean velocity through the narrowest tube-to-tube spacing, V_{max}. Equation (9.62) and Figures 9.28 and 9.29 are valid when $n_l > 9$. They were constructed based on extensive experiments in which the test fluids were air, water, and several oils [9].

9.5.5 Compact Heat Exchanger Surfaces

The pressure drop and heat transfer in passages with even more complicated geometries have also been measured and catalogued in dimensionless form. A representative sample is shown in Figure 9.30, which refers to the passage external to a bundle of finned tubes. The fluid (usually a gas) flows perpendicular to the tubes. The heat transfer surface area is increased by the fins, which occupy much of the space that would have been left open between the tubes. Configurations of this kind [1] have been designed to increase the heat transfer area per unit volume (heat transfer area density):

$$\alpha = \frac{A}{\mathcal{V}} \tag{9.64}$$

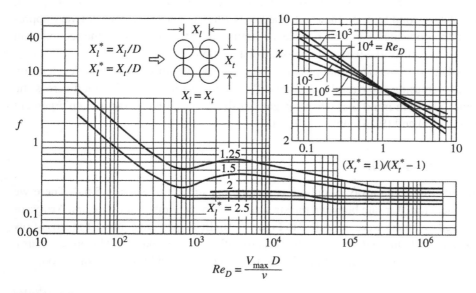

Figure 9.28 Arrays of aligned tubes: the coefficients f and χ for the pressure drop formula (9.62). Source: Drawn after Zukauskas [9].

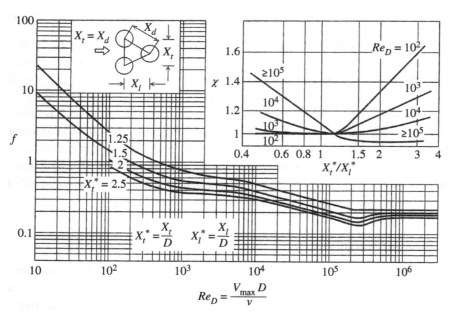

Figure 9.29 Arrays of staggered tubes: the coefficients f and χ for the pressure drop formula (1). Source: Drawn after Zukauskas [9].

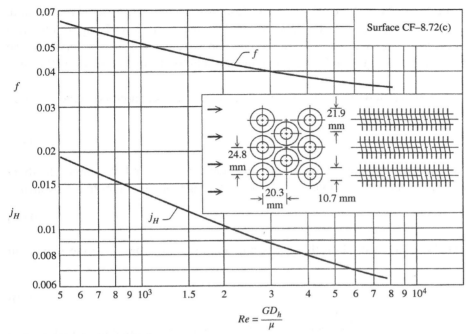

Tube outside diameter = 10.7 mm
Fin pitch = 343 m^{-1}
Flow passage hydraulic diameter D_h = 4.43 mm
Fin thickness, average (fins are tapered sligthtly) = 0.48 mm, copper
Free-flow area/frontal area, σ = 0.494
Heat transfer area/total volume, α = 446 m^2/m^3
Fin area/total area, A_f/A = 0.876

Figure 9.30 Pressure drop and heat transfer data for the flow passage through a bundle of staggered finned tubes Source: Kays and London [1].

A heat exchanger passage with a large α value exhibits a high degree of compactness. With $\alpha = 446\,\text{m}^2/\text{m}^3$, the example illustrated in Figure 9.30 almost qualifies as a compact heat exchanger, which according to Figure 9.6 requires $\alpha > 700\,\text{m}^2/\text{m}^3$. The symbol for area density α should not be confused with the symbol used for thermal diffusivity.

In addition to α, the other key parameters that describe the geometry of the flow passage are

$$\sigma = \frac{A_c}{A_{\text{fr}}} = \frac{\text{minimum free-flow area}}{\text{frontal area}} \tag{9.65}$$

$$\frac{A_f}{A} = \frac{\text{fin area}}{\text{total heat transfer area}} \tag{9.66}$$

and the hydraulic diameter

$$D_h = 4\frac{A_c L}{A} \tag{9.67}$$

where L is the length of the heat exchanger in the direction of flow. It is useful to take another look at the D_h definition for a duct with constant cross-sectional geometry, Eq. (6.28), and to recognize it as a special case of Eq. (9.67). The minimum free-flow area A_c refers to the smallest cross section encountered by the fluid. In Figure 9.30 this area occurs between two adjacent tubes aligned vertically: A_c is coplanar with the two centerlines. The dimensionless parameter σ accounts for the contraction and enlargement experienced by the stream. The (fin area)/(total area) ratio A_f/A is an ingredient in the formula for the overall surface efficiency ε, Eq. (9.4).

The mass flowrate of the stream that flows through this complicated passage is conserved in the direction of flow:

$$\dot{m} = \rho A_{\text{fr}} V \tag{9.68}$$

By this definition, V is the mean velocity based on the frontal area. On the other hand, the mass velocity G is defined based on the *maximum* mean velocity, that is, the velocity through the minimum free-flow area:

$$G = \rho V_{\max} = \frac{\dot{m}}{A_c} \tag{9.69}$$

where $A_c = \sigma A_{\text{fr}}$. The Reynolds number used on the abscissa of Figure 9.30 is based on the maximum velocity:

$$Re = \frac{V_{\max} D_h}{\nu} = \frac{G D_h}{\mu} \tag{9.70}$$

The total pressure drop across a heat exchanger core constructed as shown in the example of Figure 9.30, namely, the difference between the pressure on the left (the inlet) and the pressure on the right (the outlet), is given by [1]

$$\Delta P = \frac{G^2}{2\rho_{\text{in}}} \left[f \frac{A}{A_c} \frac{\rho_{\text{in}}}{\rho} + (1 + \sigma^2) \left(\frac{\rho_{\text{in}}}{\rho_{\text{out}}} - 1 \right) \right] \tag{9.71}$$

In this equation ρ is the average density evaluated at the temperature averaged between the inlet and the outlet, $(T_{\text{in}} + T_{\text{out}})/2$. The average density can also be estimated by averaging the fluid specific volume $(1/\rho)$ between the inlet and outlet values:

$$\frac{1}{\rho} = \frac{1}{2} \left(\frac{1}{\rho_{\text{in}}} + \frac{1}{\rho_{\text{out}}} \right) \tag{9.72}$$

or

$$\rho = \frac{2\rho_{\text{in}}\rho_{\text{out}}}{\rho_{\text{in}} + \rho_{\text{out}}} \tag{9.73}$$

The friction factor f that multiplies the first term in the square brackets of Eq. (9.71) has been found experimentally and plotted in Figure 9.30. The f factor accounts for fluid friction against solid walls *and* for the entrance and exit losses. The second term in the square brackets accounts for the acceleration or deceleration of the stream. This contribution is negligible when the density is essentially constant along the passage.

The term multiplied by f in the ΔP formula (9.71) is, as expected, proportional to the length of the flow passage, as $A/A_c = 4\,L/D_h$ according to Eq. (9.67). When the volume $\mathcal{V}$ is specified, the ratio A/A_c needed in Eq. (9.71) can be estimated by writing

$$\frac{A}{A_c} = \frac{\alpha \mathcal{V}}{\sigma A_{\text{fr}}} \tag{9.74}$$

Charts such as shown in Figure 9.30 also contain the information necessary for calculating the average heat transfer coefficient h for the particular flow passage. This is expressed in nondimensional terms as the Colburn j_H factor:

$$j_H = St\, Pr^{2/3} \tag{9.75}$$

in which the Stanton number is based on G (i.e. on maximum velocity):

$$St = \frac{h}{Gc_P} = \frac{h}{\rho c_P V_{max}} \tag{9.76}$$

In Figure 9.30 and many like it for other compact surfaces, factors j_H and f exhibit almost the same dependence on Reynolds number. Here, the ratio $j_H/(f/2)$ decreases only by one third (from 0.6 to 0.4) as Re increases from 500 to 800. The observation that even in a complicated passage the ratio $j_H/(f/2)$ is relatively constant and of order of magnitude 1 is important. It agrees qualitatively with the Colburn analogy for a duct with constant cross-sectional geometry, Eq. (6.88).

The calculation of the heat transfer between the two fluids that interact in cross-flow in Figure 9.30 also requires an estimate of the average heat transfer coefficient for the internal surface of the tubes. This estimate can be obtained by following the method outlined in Chapter 6. The complex procedure of selecting a certain heat exchanger (surface type, size) subject to various constraints (e.g. volume, pressure drop) is known as *heat exchanger design*. Several books have been devoted to this important area of thermal engineering practice [1, 2].

The newer work on the design and optimization of heat exchangers is rooted in thermodynamics and seeks to minimize the destruction of useful energy (exergy, or availability) in the heat exchanger [7, 10]. The destruction of useful energy is due to thermodynamic irreversibility, or the generation of entropy through one-way processes such as heat transfer (always from hot to cold) and fluid flow (always against friction). For this reason, the thermodynamic approach to heat exchanger design is also known as irreversibility minimization, or entropy generation minimization. This methodology has shown that certain dimensions, aspect ratios and operating conditions can be selected optimally, often on the basis of simple formulas, so that the overall irreversibility of the heat exchanger is minimum.

Example 9.6 *Compact Heat Exchanger, Pressure Drop, and Heat Transfer Coefficient*
Air enters at 1 atm and 300 °C in the core of a finned tube heat exchanger of the type shown in Figure 9.30. The air flows at the rate of 1500 kg/h perpendicular to the tubes and exits with a mean temperature of 100 °C. The core is 0.5 m long, with a 0.25-m² frontal area. Calculate the total pressure drop between the air inlet and outlet and the average heat transfer coefficient on the air side.

Solution

The air density at the inlet and outlet is listed in Appendix D:

$$\rho_{in} = 0.616\,\text{kg/m}^3 \quad \rho_{out} = 0.946\,\text{kg/m}^3$$

The other relevant properties of air at the average temperature (300 °C + 100 °C)/2 = 200 °C are

$$\rho = 0.746\,\text{kg/m}^3 \qquad Pr = 0.68$$
$$\mu = 2.58 \times 10^{-5}\,\text{kg/s·m} \quad c_P = 1.025\,\text{kJ/kg·K}$$

The pressure drop is furnished by Eq. (9.71), for which we calculate, in order,

$$\frac{A}{A_c} = \frac{\alpha \mathcal{V}}{\alpha A_{fr}} = \frac{\alpha}{\sigma} L = \frac{446 \text{ m}^2/\text{m}^3}{0.494}(0.5 \text{ m}) = 451.4$$

$$A_c = \sigma A_{fr} = 0.494 \times 0.25 \text{ m}^2 = 0.124 \text{ m}^2$$

$$G = \frac{\dot{m}}{A_c} = \frac{1500 \text{ kg}}{3600 \text{ s}} \frac{1}{0.124 \text{ m}^2} = 3.36 \text{ kg/m}^2 \cdot \text{s}$$

$$D_h = 4\frac{L}{A/A_c} = 4\frac{0.5 \text{ m}}{451.4} = 4.43 \text{ mm (listed also in Figure 9.30)}$$

$$Re = \frac{GD_h}{\mu} = 3.36 \frac{\text{kg}}{\text{m}^2 \cdot \text{s}} \frac{0.00443 \text{ m}}{2.58 \times 10^{-5} \text{ kg/s} \cdot \text{m}} = 577$$

$$f \cong 0.061 \quad (\text{at } Re = 577)$$

$$\Delta P = \left(3.36 \frac{\text{kg}}{\text{m}^2 \cdot \text{s}}\right)^2 \frac{1}{2 \times 0.616 \text{ kg/m}^3} \left[(0.061 \times 451.4)\frac{0.616}{0.746} + (1 + 0.494)^2 \left(\frac{0.616}{0.946} - 1\right)\right]$$

$$= 204.4 \text{ N/m}^2$$

The heat transfer coefficient emerges out of the j_H value read off Figure 9.30:

$$j_H \cong 0.018$$

$$St = j_H Pr^{-2/3} \cong 0.018(0.68)^{-2/3} = 0.0233$$

$$h = StGc_P = \left(0.0233 \times 3.36 \frac{\text{kg}}{\text{m}^2 \cdot \text{s}}\right)\left(1.025 \frac{10^3 \text{J}}{\text{kg} \cdot \text{K}}\right) \cong 80 \text{ W/m}^2 \cdot \text{K}$$

9.6 Evolutionary Design

9.6.1 Entrance-Length Heat Exchangers

Compactness and lightweight are the main attractions in heat exchanger design. The physics basis of this tendency was stated in Sections 5.6.1 and 6.5.1. Size is important because the object must be built and carried natural, and the required power expenditure is proportional to the size. This facet of design physics is also visible through the lens of miniaturization (Section 7.5.2), because the relentless evolution toward smaller size is synonymous with the evolution toward higher density of heat transfer in the volume occupied by the object.

These thoughts bring us to the selection of spacings as degrees of freedom (Sections 5.6.3 and 7.5.1), where we learned that the most dense packing of heat transfer rate in a fixed volume filled with flow channels occurs when the boundary layers touch at the channel exit. This holds true for forced and natural convection, in both laminar and turbulent flow. The conclusion is that the channel length for high heat transfer density should match the thermal entrance length. This principle was unveiled in 2002 [11] and was emphasized in textbooks [12, 13] and applications (Figures 9.31–9.33).

9.6.2 Dendritic Heat Exchangers

The first applications of the entrance-length concept where in the invention of dendritic heat exchangers, that is, in volumes where the two mating streams followed tree-shaped paths. Dendritic flow architectures too are features of high-density packing. This was predicted based on the constructal law [12–15]. Dendritic flow

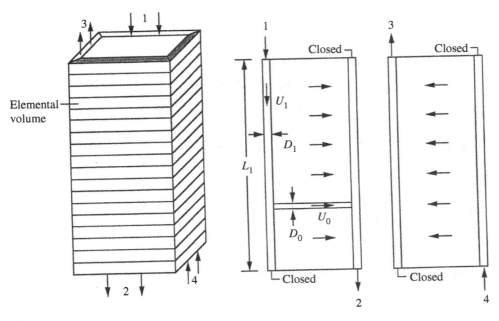

Figure 9.31 First construct containing a large number of stacked elemental volumes with dendritic streams in cross flow.

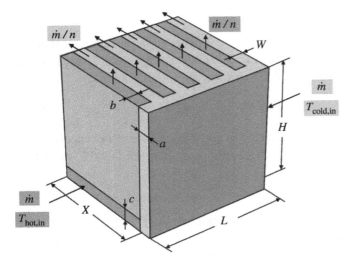

Figure 9.32 Three-dimensional flow construct with two dendritic streams in cross flow.

paths are also visible in the design of animal organs with volumetric heat transfer and mass transfer. The muscle, for example, is bathed by two streams of the blood (arterial, venous); each stream enters the volume through one point and exits through another point. Throughout the volume, each stream flares out as a delta and then recollects itself as a river basin. Each stream is configured like two trees sharing the same canopy. The heat or mass transfer between the two streams is between their superimposed tree canopies, which fill the same volume.

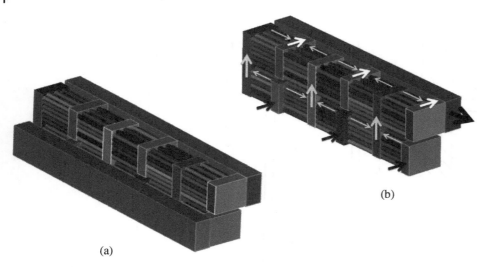

Figure 9.33 Tube-based heat exchanger-architecture illustrating the use of channels of different length scale levels: (a) Entire heat exchanger and (b) heat exchanger with the first-level inlet channel removed.

To illustrate the impact that the entrance-length concept has on the tree-shaped architecture of one stream in a dendritic heat exchanger, recall that branches (bifurcations, pairings) are places where step changes occur in channel sizes. For example, if the channels are very slender (with negligible junction losses, Section 9.5.2) and have round cross sections, the recommended step change for minimum power loss in laminar flow (Section 6.5.2) is $D_1/D_2 = 2^{1/3}$, independently of channel lengths.

If, in addition, all the channels (the trunk and the two branches) have lengths that match their thermal entrance lengths, then the ratio (trunk length)/(branch length) is $L_1/L_2 = 2$, and what changes stepwise (in addition to D_1/D_2) is the aspect ratio of the channel profile (L/D); this way (Ref. [13], Example 4.1, p. 118):

$$\frac{(L/D)_1}{(L/D)_2} = 2^{2/3} \tag{9.77}$$

The conclusion is that the channel slenderness ratio should decrease in going toward smaller channels. In the same direction, the channels become more numerous as they fill the canopy of their tree [11–13, 16–18].

9.6.3 Heat Exchanger Size

As their name indicates, heat exchangers are devices that facilitate the transfer of heat between two or more fluid streams, which are usually separated by thin solid walls (heat transfer surfaces). The wall acquires a temperature distribution that results from the thermal interaction between the mating streams.

The fundamental question in heat exchanger design is this: How does the flow configuration affect the performance of the flow component? The performance has a thermal aspect (the degree of thermal contact, e.g. the effectiveness) and a fluid-mechanical aspect (the pressure drop or the pumping power). The two aspects of performance are pursued separately or together, as in the search for reduced entropy generation rate or exergy destruction rate. Inside a specified size (volume, weight, or cost), the search is for heat and fluid flow configuration – the aspect ratio, the shape of the profile of each flow passage – that offers improved thermal and mechanical performance. This traditional activity concerns the designs that occupy the "shape" plane in Figure 9.34.

A fundamental step beyond the traditional course is to question the size itself [19]. What size should the heat exchanger have in order to serve best the larger installation (a vehicle, for example) that must use the heat exchanger as a component?

The size of a flow component in a complex flow system can be determined from the trade-off between the losses associated with the flow component and the losses associated with moving the complex flow system on the landscape. This trade-off is illustrated in Figure 5.20 and its caption. Reference [19] demonstrated the existence of the trade-off by determining the diameter of a single tube with specified fluid mass flowrate and the surface of a heat transfer area with specified heat transfer rate.

The design domain occupied by heat exchangers is considerably wider than the single tube example. The common feature of these applications is that the heat exchangers belong to vehicles that move on the landscape with a certain speed. At least as prevalent are the stationary heat exchangers, such as in plants for power, refrigeration, and air conditioning. For these, the power required for transporting the heat exchanger does not apply. There is however an additional power requirement ($\dot{W}_3$) that is of economics origin, and it is greater when the heat exchanger is larger.

The contribution to $\dot{W}_3$ is a function of factors that vary from one installation to the next, stationary or not. Examples are the power spent on manufacturing and assembling the heat exchanger, the power spent on bringing the heat exchanger components to the site, and the power spent on maintaining and cleaning periodically [12, 20] the heat exchanger. All these features of stationary heat exchanger design and operation contribute to the $\dot{W}_3$ requirement such that each contribution increases monotonically with the size of the heat exchanger.

In conclusion, the $\dot{W}_3$ loss increases monotonically with the size M, and this means that the qualitative impact of the $\dot{W}_3(M)$ function for stationary heat exchangers has the same character as in the analysis of moving heat exchangers. That is, the trade-offs unveiled in Figure 9.34 are also present in stationary applications.

Figure 9.34 is a qualitative summary of the fundamental step made in this section. Any flow configuration has two basic characteristics, shape and size. The overall loss of the heat exchanger ($\dot{W}_1 + \dot{W}_2 + \dot{W}$)₃ can be decreased in two ways: (i) in the "shape" plane at fixed size, by balancing the heat transfer and fluid flow irreversibilities and (ii) by exploring the "size" dimension of the design, as was anticipated in Figure 9.34.

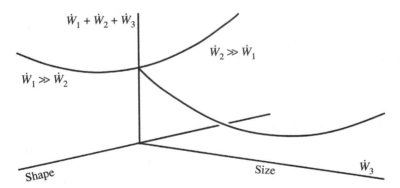

Figure 9.34 A flow system has shape and size. The shape is the trade-off between heat transfer irreversibility ($\dot{W}_2$) and fluid flow irreversibility ($\dot{W}_1$). The size is the trade-off between the sum of the first two irreversibilities and the irreversibility associated with carrying and constructing the flow system ($\dot{W}_3$).

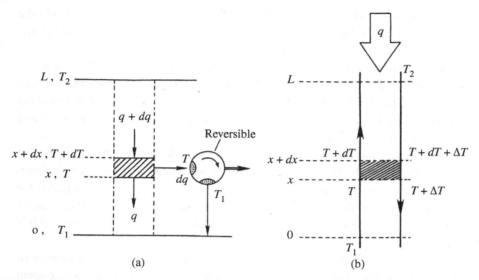

Figure 9.35 The space occupied by a general system with heat transfer across a finite temperature difference: (a) The useful power destroyed by heat transfer irreversibility. (b) Two streams in counterflow are a convection heat tube that fits inside the general system.

9.6.4 Heat Tubes with Convection

The heat tube is the slender space through which a current (q) of energy flows unidirectional from a high temperature (T_2) to a lower temperature (T_1), Figure 9.35. The space occupied by the heat tube has a known length (L) and volume (V). What happens along L is free to vary, depending on the nature of the energy current that flows through the tube, conduction, convection, or radiation [21]. Earlier in this book, we encountered this in the context of heat conduction (Figures 2.17 and 2.19), in the shaping of fins for high conductance or reduced weight.

The freedom to vary what happens along L is the physics basis of the phenomenon of evolution, which unites technological designs with natural occurrences of the phenomenon [14, 15]. What unites the conduction with convection and radiation to constitute the general heat tube phenomenon is the performance, which is ruled by thermodynamics. In all the heat tubes, the current flows one way, from high to low. All heat tubes are placed where thermodynamic irreversibility occurs. All heat tubes generate entropy, which is a measure of irreversibility and the destruction of useful power in the spaces that the heat tubes occupy [7, 10].

With reference to Figure 9.35a, the rate of entropy generation in the heat tube is

$$S_{gen} = \int_{T_1}^{T_2} \frac{q}{T^2} dT \qquad (9.78)$$

If the local heat current (q) is proportional to the local longitudinal temperature gradient,

$$q = K \frac{dT}{dx} \qquad (9.79)$$

where the conductance K is a known constant, and then by integrating Eq. (9.79) from $x = 0$ to L, we reach the integral constraint

$$\int_{T_1}^{T_2} \frac{dT}{q} = \frac{L}{K}, \text{ constant} \qquad (9.80)$$

The minimization of the s_{gen} integral (9.78) subject to the size constraint (9.80) reveals how the heat current (or the distributed cooling effect dq/dT) should be distributed along the heat tube [7, 10, 22]:

$$q = T\frac{K}{L} \ln \frac{T_2}{T_1} \tag{9.81}$$

$$\frac{dq}{dT} = \frac{K}{L} \ln \frac{T_2}{T_1}, \text{ constant} \tag{9.82}$$

The corresponding rate of entropy generation is

$$s_{gen} = \frac{K}{L} \left(\ln \frac{T_2}{T_1} \right)^2 \tag{9.83}$$

This solution is general, valid for any heat tube in which the heat current ($q(x)$ or $q(T)$) is proportional to the local temperature gradient dT/dx, cf., Eq. (9.79). The nonuniform distribution of heat current from high T to low T finds general applicability in the evolution of energy-system design. This was illustrated first in terms of heat tubes with heat conduction. Such tubes fill the cold zone (the insulation) of all refrigeration systems and they limit the performance most severely in cryogenic systems. In conduction heat tubes the proportionality constant is

$$K = kA \tag{9.84}$$

where k is the thermal conductivity of the material, and A is the cross-sectional area of the heat tube. The solution was generalized for materials with temperature-varying conductivity [22, 23].

In insulation applications at temperatures comparable with the temperature of the environment (T), where $T_2 - T_1$ is much smaller than the thermodynamic temperature, $T_2 - T_1 \ll T$, the solution reduces to the Fourier expression

$$q = \frac{kA}{L} (T_2 - T)_1, \text{ constant} \tag{9.85}$$

This is why throughout the evolution of insulation systems in technology and animal design (hair, fur, sub-skin) [24–26], the distributed cooling feature (Eq. (9.82)) is absent.

The solution for the limit $(T_2 - T_1) \ll T$ was generalized for nonuniform $k(T)$ and $A(x)$ in Section 2.7.6. With reference to Figure 2.19, the problem was to determine the shape $A(x)$ for which the overall thermal resistance is minimal subject to the volume constraint. The variational solution was extended to convection heat tubes [27] and then generalized for all modes of heat transfer in [28], Figure 9.35b. The purpose of proposing convection heat tubes was to explain the evolution of counterflow pairs of blood vessels in the long legs of wading birds and also the evolution of counterflow heat exchangers in low temperature installations.

In the convection tube, there are two streams in counterflow. The local temperature difference between them (at x, or T) is ΔT. The descending stream is warmer and has the capacity rate $(\dot{m}c_P)_h$. The second stream is colder, and its capacity rate $(\dot{m}c_P)_c$ is different.

The idea behind the convection tube concept is that any imaginary surface (any cut) at $x = $ constant is pierced by two enthalpy streams in counterflow. The net stream of energy (as enthalpy flow) is oriented the same way as the warmer stream, downhill,

$$q = (\dot{m}h)_h - (\dot{m}h)_c \tag{9.86}$$

If the streams carry fluids that behave as ideal gases or incompressible substances, then, along each stream, $dh_{h,c} = (\dot{m}c_P)_{h,c}dT_{h,c}$. In particular, if the capacity rates are balanced $(\dot{m}c_P)_h = (\dot{m}c_P)_c = \dot{m}c_P$, then the net

(a)

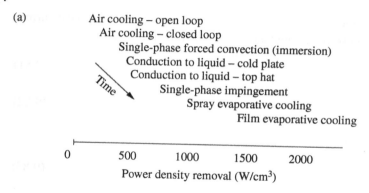

Air cooling – open loop
Air cooling – closed loop
Single-phase forced convection (immersion)
Conduction to liquid – cold plate
Conduction to liquid – top hat
Single-phase impingement
Spray evaporative cooling
Film evaporative cooling

Time

Power density removal (W/cm³)

(b)

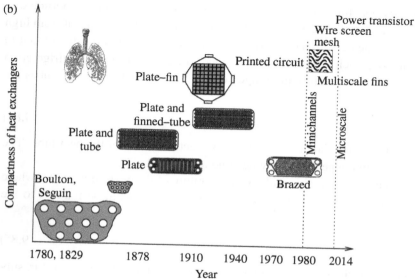

Figure 9.36 (a) The evolution of heat transfer mechanisms and (b) compact configurations of heat exchangers during the past two centuries. Source: Bejan and Errera [30].

enthalpy current is

$$q = \dot{m}c_P \left(T_h - T_c\right) \tag{9.87}$$

When there is no distributed cooling (unlike in Figure 9.35a), q is constant. The convective current (9.87) [10, 27, 28] was recognized later in bio heat transfer [26, 29]. The local rate of heat transfer between the two streams (perpendicular to x) is

$$dq = U(pdx)\left(T_h - T_c\right) \tag{9.88}$$

where U, p, and (pdx) are the overall heat transfer coefficient, the perimeter of thermal contact between the two streams, and the mutual contact area per unit length in the longitudinal direction, respectively. According to the first law, the temperature change along either stream is given by

$$dq = \dot{m}c_P \, dT \tag{9.89}$$

Combining Eqs. (9.87)–(9.89) yields

$$q = \frac{(\dot{m}c_P)^2}{UA} \frac{dT}{dx} \tag{9.90}$$

The group in front of dT/dx is the constant K of the general treatment, Eq. (9.79). This is why the general treatment [7, 10] applies to configurations with x-dependent q and with x-distributed cooling dq/dT, which for convection means different capacity rates, i.e. unbalanced counterflow. The main results are

$$q = \left[\frac{(\dot{m}c_P)^2}{UA} \ln \frac{T_2}{T_1} \right] T, \quad \frac{\Delta T}{T} = \frac{\dot{m}c_P}{UA} \ln \frac{T_2}{T_1} \tag{9.91}$$

where the local stream-to-stream temperature difference is $\Delta T = T_h - T_c$ and A is the total contact area between the streams. In the limit where variations in the absolute T are negligible over the heat exchanger volume (i.e. $\Delta T \cong$ constant), Eq. (9.91) reduces to $\Delta T \cong$ constant, as commented in Ref. [21]. That limit is a good approximation for heat exchangers near room temperature, but not for the entire T range of applications, certainly not in heat exchangers at very low and very high temperatures.

Progress on evolving the architecture and performance of heat exchangers continues (Figure 9.36) [30]. A clear trend is toward the use of entropy generation minimization [31, 32], dendritic flow architectures [33–36], novel flow paths [37], and evolutionary design (constructal method) [30, 38–43]. Along the way, it becomes increasingly necessary to be vigilant about publishing attempts that confuse the readers with false physics and false claims of novelty and paternity of ideas [21, 44–54].

References

1 Kays, W.M. and London, A.L. (1984). *Compact Heat Exchanges*, 3e. New York: McGraw-Hill.
2 Bejan, A. and Kraus, A.D. (2003). *Heat Transfer Handbook*. Hoboken, NJ: Wiley.
3 Chenoweth, J.M. (1990). Final report of the HTRI/TEMA joint committee to review the fouling section of the TEMA Standards. *Heat Transfer Eng.* 11 (1): 73–107.
4 Shah, R.K. and Mueller, A.C. (1985). Heat exchangers. In: *Handbook of Heat Transfer Applications*, Chapter 4,, 2e (eds. W.M. Rohsenow, J.P. Hartnett and E.N. Ganic). New York: McGraw-Hill.
5 Kern, D.Q. (1950). *Process Heat Transfer*. New York: McGraw-Hill.
6 Bowman, R.A., Mueller, A.C., and Nagle, W.M. (1940). Mean temperature difference in design. *Trans. ASME* 62: 283–294.
7 Bejan, A. (2016). *Advanced Engineering Thermodynamics*, 4e. Hoboken, NJ: Wiley.
8 Kays, W.M. (1950). Loss coefficients for abrupt changes in flow cross-section with low Reynolds number flow in single and multiple-tube systems. *Trans. ASME* 72: 1067–1074.
9 Zukauskas, A. (1987). Convective heat transfer in cross flow. In: *Handbook of Single-Phase Convective Heat Transfer*, Chapter 6 (eds. S. Kakac, R.K. Shah and W. Aung). New York: Wiley.
10 Bejan, A. (1982). *Entropy Generation Through Heat and Fluid Flow*. New York: Wiley.
11 Bejan, A. (2002). Dendritic constructal heat exchanger with small-scale cross flows and large-scale counterflows. *Int. J. Heat Mass Transfer* 45: 4607–4620.
12 Bejan, A. (2000). *Shape and Structure, from Engineering to Nature*. Cambridge: Cambridge University Press.
13 Bejan, A. and Lorente, S. (2008). *Design with Constructal Theory*. Hoboken, NJ: Wiley.
14 Bejan, A. (2016). *The Physics of Life*. New York: St. Martin's Press.
15 Bejan, A. and Zane, J.P. (2012). *Design in Nature*. New York: Doubleday.

16 Bejan, A., Alalaimi, M., Sabau, A.S., and Lorente, S. (2017). Entrance-length dendritic plate heat exchangers. *Int. J. Heat Mass Transfer* 114: 1350–1356.

17 Sabau, A.S., Nejad, A.H., Klett, J.W. et al. (2018). Novel evaporator architecture with entrance-length crossflow – paths for supercritical Organic Rankine Cycles. *Int. J. Heat Mass Transfer* 119: 208–222.

18 Raja, V.A.P., Basak, T., and Das, S.K. (2008). Thermal performance of a multi-block heat exchanger designed on the basis of Bejan's constructal theory. *Int. J. Heat Mass Transfer* 51: 3582–3594.

19 Bejan, A., Lorente, S., Martins, L., and Meyer, J.P. (2017). The constructal size of a heat exchanger. *J. Appl. Phys.* 122: 064902.

20 Bejan, A. and Vargas, J.V.C. (1994). When to defrost a refrigerator, and when to remove the scale from the heat exchanger of a power plant. *Int. J. Heat Mass Transfer* 37: 523–532.

21 Bejan, A. (2019). Heat tubes: conduction and convection. *Int. J. Heat Mass Transfer* 137: 1258–1262.

22 Bejan, A. and Smith, J.L. Jr., (1974). Thermodynamic optimization of mechanical supports for cryogenic apparatus. *Cryogenics* 14: 158–163.

23 Jany, P. and Bejan, A. (1988). Ernst Schmidt's approach to fin optimization: an extension to fins with variable conductivity and the design of ducts for fluid flow. *Int. J. Heat Mass Transfer* 31: 1635–1644.

24 Bejan, A. (1990). Theory of heat transfer from a surface covered with hair. *J. Heat Mass Transfer* 112: 662–667.

25 Lage, J.L. and Bejan, A. (1991). Natural convection from a vertical surface covered with hair. *Int. J. Heat Fluid Flow* 12: 46–53.

26 Bejan, A. (2001). The tree of convective heat streams: its thermal insulation function and the predicted 3/4-power relation between body heat loss and body size. *Int. J. Heat Mass Transfer* 44: 699–704.

27 Bejan, A. and Smith, J.L. Jr., (1976). Heat exchangers for vapor-cooled conducting supports of cryostats. *Adv. Cryog. Eng.* 21: 247–256.

28 Bejan, A. (1976). A general variational principle for thermal insulation system design. *Int. J. Heat Mass Transfer* 22: 219–228.

29 Weinbaum, S. and Jiji, L.M. (1985). A new simplified bioheat equation for the effect of blood flow on local average tissue temperature. *J. Biomech. Eng.* 107: 131–139.

30 Bejan, A. and Errera, M.R. (2015). Technology evolution, from the constructal law: heat transfer designs. *Int. J. Energy Res.* 39: 919–928. https://doi.org/10.1002/er.3262.

31 Manjunath, K. and Kaushik, S.C. (2014). Second law thermodynamic study of heat exchangers: a review. *Renewable Sustainable Energy Rev.* 40: 348–374.

32 Manjunath, K. and Kaushik, S.C. (2014). Entropy generation and thermo-economic analysis of constructal heat exchanger. *Heat Transfer Asian Res.* 43 (1): 39–60.

33 da Silva, A.K., Lorente, S., and Bejan, A. (2004). Constructal multi-scale tree-shaped heat exchangers. *J. Appl. Phys.* 96 (3): 1709–1718.

34 da Silva, A.K. and Bejan, A. (2006). Dendritic counterflow heat exchanger experiments. *Int. J. Therm. Sci.* 45: 860–869.

35 Zimparov, V.D., da Silva, A.K., and Bejan, A. (2006). Constructal tree-shaped parallel flow heat exchangers. *Int. J. Heat Mass Transfer* 49: 4558–4566.

36 Zhang, F., Sundén, B., Zhang, W., and Xie, G. (2015). Constructal parallel-flow and counterflow microchannel heat sinks with bifurcations. *Numer. Heat Transfer, Part A* 68: 1087–1105.

37 Sabau, A.S., Bejan, A., Brownell, D. et al. (2020). Design, additive manufacturing, and performance of heat exchanger with novel flow-path architecture. *Appl. Therm. Energy* 180: 115775.

38 Yang, J., Fan, A., Liu, W., and Jacobi, A.M. (2014). Optimization of shell-and-tube heat exchangers conforming to TEMA standards with designs motivated by constructal theory. *Energy Convers. Manage.* 78: 468–576.

39 Saha, S.K. and Baelmans, M. (2014). A design method for rectangular microchannel counter flow heat exchangers. *Int. J. Heat Mass Transfer* 74: 1–12.

40 Yang, S., Ordonez, J.C., and Vargas, J.V.C. (2017). Constructal vapor compression refrigeration (CVR) systems design. *Int. J. Heat Mass Transfer* 115: 754–768.

41 Klein, R.J., Lorenzini, G., Zinani, F.S.F., and Rocha, L.A.O. (2017). Dimensionless pressure drop number for non-Newtonian fluids applied to constructal design of heat exchangers. *Int. J. Heat Mass Transfer* 115: 910–914.

42 Azad, A.V. and Amidpour, M. (2011). Economic optimization of shell and tube heat exchanger based on constructal theory. *Energy* 36: 1087–1096.

43 Klein, R.J., Zinani, F.S.F., Rocha, L.A.O., and Biserni, C. (2018). Effect of Bejan and Prandtl numbers on the design of tube arrangements in forced convection of shear thinning fluids: a numerical approach motivated by constructal theory. *Int. Commun. Heat Mass Transfer* 93: 74–82.

44 Bejan, A. (2020). *Freedom and Evolution: Hierarchy in Nature, Society and Science*, Chapter 11. New York: Springer Nature.

45 Bejan, A. (2017). Evolution in thermodynamics. *Appl. Phys. Rev.* 4: 011305.

46 Bejan, A. (2018). Thermodynamics today. *Energy* 160: 1208–1219.

47 Bejan, A. (2019). Thermodynamics of heating. *Prof. R. Soc. A* 475: 20180820.

48 Bejan, A. (2018). Comment on "Study on the consistency between field synergy principle and entransy dissipation extremum principle". *Int. J. Heat Mass Transfer* 120: 1187–1188.

49 Bejan, A. (2018). Letter to the Editor on "Temperature-heat diagram analysis method for heat recovery physical adsorption refrigeration cycle—taking multi stage cycle as an example". *Int. J. Refrig* 90: 277–279.

50 Bejan, A. and Lorente, S. (2012). Letter to the Editor. *Chem. Eng. Process. Process Intensif.* 56: 34.

51 Bejan, A. (2014). "Entransy", and its lack of content in physics. *J. Heat Transfer* 136: 055501.

52 Bejan, A. (2014). Comment on "Application of entransy analysis in self-heat recuperation technology". *Ind. Eng. Chem. Res.* 53: 1274–1285.

53 Bejan, A. (2015). Heatlines (1983) versus synergy (1998). *Int. J. Heat Mass Transfer* 81: 654–658.

54 Bejan, A. (2016). Letter to the editor of renewable and sustainable energy reviews. *Renewable Sustainable Energy Rev.* 53: 1636–1637.

55 Lorente, S., Wechsatol, W., and Bejan, A. (2002). Tree-shaped flow structures designed by minimizing path lengths. *Int. J. Heat Mass Transfer* 45: 3299–3312.

56 Bejan, A. and Almahmoud, H. (2021). Tree flows through hierarchical slits and orifices. *Int. Comm. Heat and Mass Transfer* 128: 1055.

Problems

Overall Heat Transfer Coefficient

9.1 Assume that a scale layer of resistance r_s covers the entire area A of one side of the heat exchanger surface shown in Figure 9.7. Defining $h_s = 1/r_s$ as an equivalent scale heat transfer coefficient (across the temperature difference $T_w - T_s$), repeat the analysis contained in Eqs. (9.1)–(9.3) to arrive at the total heat transfer rate formula (9.6). In other words, prove that the effective heat transfer coefficient for one side of the heat exchanger surface accounts for the two effects shown on the right side of Eq. (9.7).

9.2 The heat exchanger surface shown in Figure P9.2 is exposed to a stream of city water on the hot side and a stream of compressed air on the cold side. The thermal conductivity of the wall and pin fin material is 50 W/m·K. The heat transfer coefficients (h_h, h_c) are distributed uniformly over their respective surfaces. The total area of the hot side of the wall is 1 m².

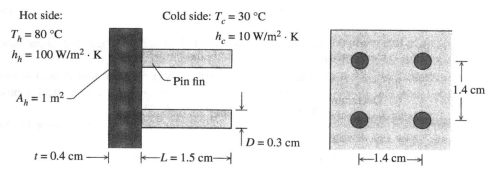

Figure P9.2

a) Calculate the total heat transfer rate through the heat exchanger surface, q_a, by taking into account the effect of fouling on both sides of the surface. On which side is the effect of fouling insignificant?
b) Repeat the calculation of the total heat transfer rate (q_b) by assuming that the pin fins are absent from the cold side of the surface. Compare q_b and q_a, and comment on the extent to which the finning of the cold side augments (enhances) the total heat transfer rate.

9.3 Consider the heat exchanger surface illustrated in Figure P9.3. The cold side is flat and of size $A_c = 0.4$ m². The hot side of the wall is covered with plate fins arranged in a staggered array. The geometric dimensions and thermal conditions are indicated directly on the drawing. The wall and the fins have the same thermal conductivity, $k = 20$ W/m·K. The effect of fouling is negligible on both sides of the heat exchanger surface.

Calculate the overall thermal resistance $(U_c A_c)^{-1}$ and the total heat transfer rate through the heat exchanger surface. Show that the heat transfer through each plate fin can be modeled as one-dimensional conduction.

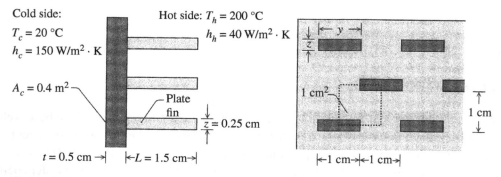

Figure P9.3

9.4 Hot air flows through the finned tubes dimensioned on Figure 9.30 in the text, while cold water flows inside the tubes. The heat transfer coefficients on the hot and cold sides of the surface are $h_h = 80\,\text{W/m}^2\cdot\text{K}$ and $h_c = 1000\,\text{W/m}^2\cdot\text{K}$, respectively. The tube wall and the fins are made of aluminum; their average temperature is $200\,°\text{C}$. The scale resistances are negligible on both sides of the surface.

a) Assume that the fin thickness is negligible, and prove that the area ratio is given by

$$\frac{A_c}{A_h} \cong \frac{D_i}{D_o}\left(1 - \frac{A_{f,h}}{A_h}\right)$$

In this relationship, D_i and D_o are the inner and outer diameters of the bare tube and $A_{f,h}$ is the contribution the fins make to A_h. Note that $(A_f/A)_h$ is listed under the chart of Figure 9.30.

b) Calculate the overall surface efficiency for the hot side, ε_h.

c) Determine the overall heat transfer coefficient based on the cold side, U_c, and then deduce the overall heat transfer coefficient based on the hot side, U_h.

Log-Mean Temperature Difference Method

9.5 A 1 kg/s stream of hot water is cooled from 90 to $60\,°\text{C}$ in a parallel flow heat exchanger in which the cooling agent is a 2 kg/s stream of water with the initial temperature of $40\,°\text{C}$. The overall heat transfer coefficient between the two streams is constant and known, $U = 1000\,\text{W/m}^2\cdot\text{K}$.

a) Calculate by the ΔT_{lm} method the size of the required heat exchanger area A.

b) Note that in the first part of Example 9.2, the same stream was cooled in a counterflow heat exchanger arrangement, requiring a total heat transfer area of $4.69\,\text{m}^2$. Compare this area with your answer to part (a) and comment on which of the two heat exchangers (parallel flow or counterflow) is preferable.

9.6 A cross-flow heat exchanger in which both streams are unmixed has the function to cool a 1 kg/s stream of hot water from 90 to $60\,°\text{C}$. The cold stream carries 2 kg/s of water with an initial (inlet) temperature of $40\,°\text{C}$. The overall heat transfer coefficient is $U = 1000\,\text{W/m}^2\cdot\text{K}$.

a) Calculate by the ΔT_{lm} method the required heat transfer area A.

b) Compare this result with the area needed by a cross-flow heat exchanger with both fluids mixed (see the second part of Example 9.2), and comment on the effect of stream mixing on the size of the required heat exchanger surface.

9.7 Show that the total heat transfer rate between the two streams of a counterflow heat exchanger is also given by Eq. (9.21) in which ΔT_{lm} assumes the form listed in Eqs. (9.22) and (9.23). In other words, derive Eq. (9.21) while analyzing the counterflow heat exchanger shown in the upper-right quadrant of Figure 9.9. As a guide, use the analysis listed between Eqs. (9.10) and (9.22) in the text.

9.8 Consider the parallel flow oil cooler in which an oil stream of 1.5 kg/s is cooled by a water stream of 0.5 kg/s. The oil inlet temperature is $110\,°\text{C}$, and the water inlet temperature is $15\,°\text{C}$. The respective specific heats c_p of oil and water are, respectively, 2.25 and 4.18 kJ/kg·K. The total heat transfer area is $A = 10\,\text{m}^2$, while the overall heat transfer coefficient is $U = 500\,\text{W/m}^2\cdot\text{K}$.

a) Calculate the two outlet temperatures and the total stream-to-stream heat transfer rate by using the ΔT_{lm} method. As a starting guess for the trial-and-error procedure, use $T_{h,\text{out}} = 70\,°\text{C}$ for the outlet temperature of the oil stream.

b) The counterflow alternative to this parallel flow design was analyzed in Example 9.3. Compare the heat transfer rates and oil outlet temperatures promised by the two designs, and comment on the relative attractiveness of the counterflow arrangement.

Effectiveness–*NTU* Method

9.9 The effectiveness–*NTU* relation for a parallel flow heat exchanger, Eq. (9.37), was derived in the text by assuming that $C_c = C_{\min}$ and $C_h = C_{\max}$. Rederive the effectiveness–*NTU* relationship by making the opposite assumption, $C_h = C_{\max}$ and $C_h = C_{\min}$. Prove in this way that Eqs. (9.37) and (9.38) are valid regardless of whether the hot stream or the cold stream has the smaller capacity rate.

9.10 The purpose of a counterflow oil cooler is to lower the temperature of a 2 kg/s stream of oil from a 110 °C inlet to an outlet temperature of 50 °C. The oil specific heat is 2.25 kJ/kg·K. The coolant is a 1 kg/s stream of cold water with an inlet temperature of 20 °C. The specific heat of water is 4.18 kJ/kg·K and the overall heat transfer coefficient has the value $U = 400\,\text{W/m}^2\text{·K}$.

a) Calculate the necessary heat exchanger area A by the ε–*NTU* method.

b) Later, calculate the total stream-to-stream heat transfer rate and the water outlet temperature.

9.11 The heat transfer surface of a counterflow heat exchanger is characterized by $U = 600\,\text{W/m}^2\text{·K}$ and $A = 10\,\text{m}^2$. The hot side is bathed by a 1 kg/s stream of hot water with an inlet temperature of 90 °C. On the cold side, the surface is cooled by a 4 kg/s stream of water with an inlet temperature of 20 °C. Calculate the stream-to-stream heat transfer rate and the two outlet temperatures by using the ε–NTU method.

9.12 The specified job of a heat exchanger is to lower the temperature of a 1 kg/s stream of hot water from 80 to 40 °C. The available coolant is a 1 kg/s stream of cold water, with an inlet temperature of 20 °C. The overall stream-to-stream heat transfer coefficient is $U = 800\,\text{W/m}^2\text{·K}$. Calculate the required heat transfer area A by assuming that the two streams are oriented in the following manner:

a) Parallel flow (provide a physical or thermodynamic explanation for why this flow arrangement will not work).

b) Counterflow.

9.13 The pores of the granular material shown on the left side of Figure P9.13 are filled by a fluid with thermal diffusivity α. Such a combination of solid and fluid is known as a saturated porous medium (see the end of Appendix B). The fluid flows through the spaces between the grains with an imposed average velocity V. A temperature difference is maintained across the saturated porous medium, and as a result heat is transferred through it.

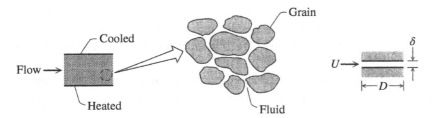

Figure P9.13

Consider now the sample enlarged on the right side of Figure P9.13. Generally speaking, there is always a difference between the local bulk temperature of the fluid and the temperature of the most immediate grain. When the grains and the interstitial spaces are sufficiently small, this local temperature difference can be neglected, and the saturated porous medium is described (approximately) by a single temperature at each point. A critical question, then, is this: How small the grains and fluid spaces should be for this local thermal equilibrium approximation to be valid?

a) Let D and δ represent the sizes of the grain and the fluid channel. As fluid channel in your analysis use the parallel-plate model illustrated in Figure P9.13. Review the ε–NTU charts exhibited in the text, and note that the notion of good thermal contact between fluid and solid walls is associated with the range $NTU \gg 1$. Rely on this observation to arrive at the following criterion for local fluid–grain thermal equilibrium:

$$\frac{U\delta}{\alpha} \ll \frac{D}{\delta}$$

b) In a certain laboratory apparatus, the grains are spheres of diameter 0.4 cm, the fluid is water at roughly 25 °C, and the average water velocity through the pores is 0.1 cm/s. The spheres are packed in such a way that $\delta \sim 0.1$ cm is a good estimate for the average fluid channel thickness. Determine numerically whether the water and the spheres are in local thermal equilibrium, that is, whether the criterion derived in part (a) is satisfied.

9.14 Figure P9.14 shows the low temperature (preheating) zone of a long reheating oven used in the manufacture of commercial steel plate. The plate enters the oven at 100 °C and is heated in counterflow by a stream of hot gas with an inlet temperature of 300 °C. The gas channel is 30 m long, with a ceiling-to-plate spacing of 15 cm. The gas mean velocity is 2 m/s, and the gas properties are very similar to those of air.

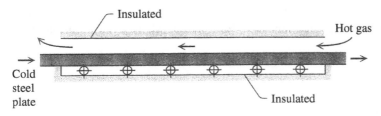

Figure P9.14

The steel plate is continuous and has a thickness of 10 cm. Its motion is slow: one spot on the plate spends a total of two hours inside the oven. This is called the residence time. The oven is sufficiently wide in the direction perpendicular to the figure. The top and bottom surfaces of the oven are well insulated.

a) Calculate the gas outlet temperature, and the final (outlet) temperature of the steel plate.

b) Estimate the needed residence time if the final temperature of the steel plate is to be 200 °C.

9.15   The heat exchanger analyses presented in the text are based on the assumption that all of the heat transfer occurs internally (between two streams) and that the heat exchanger enclosure is perfectly insulated with respect to its environment. When this assumption is not valid, the stream-to-stream heat transfer depends on the environmental temperature and on the properties of the external wall of the heat exchanger.

To examine this dependence, consider a counterflow heat exchanger of length L in which the cold stream is isothermal at T_c (e.g. a liquid boiling at nearly constant pressure). The hot stream is single-phase and enters the heat exchanger with the high temperature T_h. The hot stream engulfs the cold stream, so that the wall of the heat exchanger separates the hot stream of local bulk temperature $T(x)$, from the environment of temperature T_0. The external wall has the constant thickness t, thermal conductivity k, and a geometry that does not depend on the longitudinal coordinate x. Figure P9.15 shows a symmetric (round) geometry for simplicity. The outer wall is relatively "thin," that is, its thickness is smaller than the effective radius of the cross section. Assume that the temperatures of the external and internal surfaces of the outer wall are equal to T_0 and, respectively, $T(x)$. Note further that $T_h > T_c > T_0$. The overall stream-to-stream heat transfer coefficient is known and is based on the internal surface of the external wall.

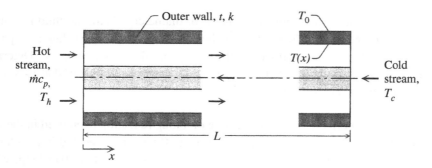

Figure P9.15

Derive expressions for the temperature distribution along the hot stream, the heat transfer rate received by the cold stream, and the heat transfer through the outer wall. Show that these results approach the more familiar forms (given in the text) for the limit where the wall insulation becomes perfect.

Pressure Drop

9.16   The core of a shell-and-tube heat exchanger contains 50 tubes, each with an inside diameter of 2 cm and length of 5 m. The header is shaped as a disc with a diameter of 30 cm. Air at 1 atm flows through the tubes with a total flowrate of 500 kg/h. The air inlet and outlet temperatures are 100 and 300 °C, respectively. Calculate the total pressure drop across the tube bundle by evaluating, in order, the pressure drops due to friction in the tube, the acceleration, the abrupt contraction, and the abrupt enlargement. In the end, compare the frictional pressure drop estimate with the total pressure drop.

9.17 Water flows through a two-dimensional (parallel-plates) channel that has a sudden contraction followed by sudden enlargement. The water has the mean velocity of 15 cm/s through the narrower channel. The width of the narrower portion channel is 0.4 m. The water temperature is 25 °C.
a) Determine the pressure drop due to the abrupt contraction.
b) How long must the narrower channel be if its pressure drop due to wall friction is to equal the pressure drop calculated in part (a)?

9.18 A 5000 kg/h stream of 500 °C air at atmospheric pressure is cooled in a shell-and-tube heat exchanger. The air flows through 300 parallel tubes. Each tube has an internal diameter of 1.5 cm and a length of 3 m. The header is built in such a way that the total cross-sectional area of the tubes is equal to 60% of the header (frontal) area. The outlet temperature of the air stream is 100 °C. Calculate the pressure differences associated with the abrupt contraction, friction and deceleration in the straight tubes, and, finally, with enlargement. Explain why the total pressure drop is smaller than the pressure drop due to friction in the straight tube.

9.19 The core of a cross-flow heat exchanger employs a bank of staggered *bare* tubes with a longitudinal pitch of 20.3 mm and transverse pitch of 24.8 mm. The outer diameter of each tube is 10.7 mm. Air flows perpendicular to the bare tubes. The frontal area seen by the air stream is a 0.5 m × 0.5 m square. The length of the heat exchanger core is 0.5 m. The air mass flowrate is 1500 kg/h, and the air properties may be evaluated at 200 °C and 1 atm.
a) Calculate the air pressure drop across the core of the heat exchanger.
b) The equivalent calculation was performed in Example 9.6 for the case where the tubes were finned. Compare the ΔP calculated in part (a) with the ΔP calculated in Example 9.6, and comment on the effect of finning on the total pressure drop (Figure P9.19).

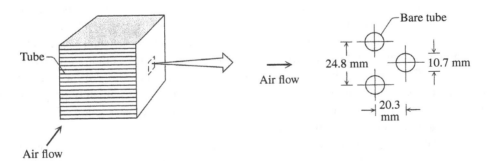

Figure P9.19

9.20 a) Show that the following relationship exists between hydraulic diameter, heat transfer area density, and contraction ratio:

$$D_h = 4\frac{\sigma}{\alpha}$$

b) The scale drawn along the baseline of Figure 9.6 was constructed by assuming a certain value for σ. Calculate that value.

c) Derive Eq. (9.74) and show that

$$\frac{A}{A_c} = \frac{4L}{D_h}$$

Evolutionary Design

9.21 In a cross-flow heat exchanger, air flows across a bundle of tubes with dimensions $D = 5$ cm and $X_t = X_t = 9$ cm. The air velocity averaged over the frontal area of the bundle is 3 m/s. There are 21 rows of tubes counted in the direction of air flow. The average air temperature inside the heat exchanger is 100 °C. Calculate the air pressure drop caused by the tubes, by assuming that the tubes are (a) aligned and (b) staggered. Compare the two pressure drops, and comment on the effect of staggering the tubes in the array.

9.22 The entrance-length concept for heat exchanger design can be illustrated with the Y-shaped design of three flow passages. For simplicity, consider a tube of length L_1 and diameter D_1 that branches into two identical tubes of length L_2 and diameter D_2. The flow is laminar, and each tube has an entrance length along with boundary layers that grow until they merge on the tube axis. To draw the Y-shaped design, we need two aspect ratios, D_1/D_2 and L_1/L_2. Assume that $D_1/D_2 = 2^{1/3}$. Determine L_1/L_2 from the requirement that L_1 and L_2 are the same as the entrance lengths of their respective flows. In other words, the annular boundary layer of thickness δ_1 that lines the inner surface of the tube reaches the scale $D_1/2$ at the longitudinal flow distance of order L_1. Compare your L_1/L_2 result with the L_1/L_2 ratio for a Y-shaped design on a fixed area, Eq. (6.113). Determine the ratio of the two slenderness ratios, $(L_1/D_1)/(L_2/D_2)$, and comment on how the slenderness ratio changes in going from the large tube (L_1, D_1) to the smaller branches.

9.23 Dendritic heat exchangers are packages of flow paths shaped as trees. The origin and promise of such architectures become clear even in their simplest setting, such as Figure P9.23. The flow is guided into tree-like branches (or tributaries) by partitions that have openings in the right places. Only two partitions are shown in Figure P9.23, one with a single opening (A_1) and the other with two identical openings ($2A_2$). The fluid flow can be oriented in either direction. According to measurements of the pressure drop across a plate with one orifice, the pressure drop is approximately proportional to $(\dot{m}/A)^2$, where $\dot{m}$ is the mass flowrate through the orifice and A is the cross section of the opening. The flow is steady, and the total size of the openings ($A_1 + 2A_2$) is fixed.

Determine the relative size of the openings (A_2/A_1) such that the total pressure drop across the two partitions is minimal. In such a design, what is the ratio of the individual pressure drops across each partition ($\Delta P_2/\Delta P_1$)?

Rely on the above solution as you contemplate a more complex flow configuration where the number of parallel partitions is large, and the number of orifices doubles from one partition to the next ($A_1, 2A_2, 4A_3, \ldots, 2^{n-1}A_n$). In this limit, the flow architecture is a tree that connects one "point" (A_1)

with a "line" of many small openings [55, 56]. Determine the relative sizes of the openings (A_1, A_2, A_3, ...) and the relation between the pressure drops across successive partitions. If the partitions are positioned equidistantly, how is the pressure distributed (approximately) between A_1 and the line of A_n openings?

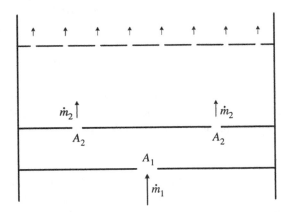

Figure P9.23

9.24 Dendritic heat exchangers are packages of tree-shaped flows guided by solid walls. In Figure P9.24 the guides are two-dimensional channels (slits) arranged in such a way that the flowing streams pair or bifurcate. The elemental configuration that gives birth to the tree consists of one channel (L_1, D_1) that is continued by two identical channels (L_2, D_2). The objective in this exercise is to determine the configuration that offers minimum pressure drop and maximum heat transfer rate. The configuration is represented by the ratios L_2/L_1 and D_2/D_1 that, if known, allow the thinker to imagine the drawing. The configuration is two-dimensional, as drawn.

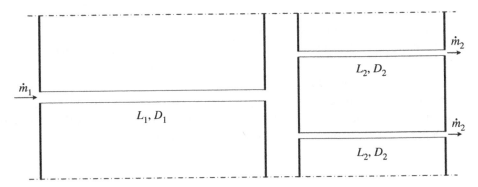

Figure P9.24

The direction of the flow does not matter. Assume that the flow is steady and fully developed turbulent (fully rough) in every channel. This means that the friction coefficient is a constant independent of flowrate. The pressure drop along each channel is due to friction along its surfaces. The total volume of the three channels is fixed. Determine the ratio D_2/D_1 such that the pressure drop along the entire assembly is minimum. Show that the result $D_2/D_1 = 2^{-3/4}$ is independent of the lengths of the channels.

Next, show that the minimized pressure drop is proportional to $(L_1 + 2^{1/4}L_2)^4$. If the total length of the assembly is fixed ($L_1 + L_2 =$ constant), what is the ratio L_2/L_1 for which the minimized pressure drop has the lowest value?

Finally, consider the heat transfer between the stream ($\dot{m}$) and the surfaces of the three channels. For simplicity, assume that the scale of the temperature difference (ΔT) between wetted surfaces and flowing fluid is known, and show that the total thermal resistance between surfaces and fluid varies as $[L_1^{-1} + (2L_2)^{-1}]$. Minimize the total thermal resistance subject to fixed ($L_1 + 2^{1/4}L_2$), which means fixed pressure drop according to the preceding paragraph. Show that the resulting ratio of lengths is $L_2/L_1 = 2^{-5/8}$.

The aspect ratios D_2/D_1 and L_2/L_1 derived this way serve as recurrence relations in the construction of a tree flow consisting of many stages (plenums) with bifurcation (or pairing), as in Figure P9.24. Show how the slenderness ratio of one channel (L_i/D_i) changes from one stage to the next. How does the pressure drop along one channel change from one stage to the next?

9.25 The biggest challenge in heat exchanger performance maximization is the parallel trends in heat transfer and fluid friction (pressure drop) associated with changes in flow configuration. When one change leads to an increase in heat transfer coefficient (desirable), it also causes an increase in fluid friction (not desirable). See the parallel trends in f and j_H in Figure 9.30 and the discussion under Eq. (9.75). In this problem you will discover that this behavior is well known and documented in much simpler terms earlier in this book. Proceed in these steps:

(i) Assume a single spherical body in forced convection, for which the drag coefficient is plotted in Figure 5.15. For the Re_D range 10–1000, assume that the drag coefficient varies as $C_D = c_1 Re_D^{-a}$. Determine the values of c_1 and a.

(ii) The Nusselt number for the sphere in forced convection is given by Eq. (5.139). For the Re_D range 10–1000, neglect the temperature-induced changes in viscosity (μ) and the first term (the constant 2) that accounts for the conduction (no flow) limit. In other words, use the simpler expression $\overline{Nu_D} \cong (0.4Re_D^{1/2} + 0.06Re_D^{2/3}) Pr^{0.4}$. Fit this expression to the simpler form $\overline{Nu_D} = c_2 Re_D^b$ by using the ($\overline{Nu_D}$, Re_D) data furnished by Eq. (5.139), and determine b and c_2.

(iii) Determine the St expression that corresponds to $\overline{Nu_D} = c_2 Re_D^b$, and compare it with the C_D expression obtained in part (i). You will see the opposing trends in $\overline{Nu_D}$ and C_D as Re_D increases. Finally, express St in terms of Re_D, C_D, and Pr.

(iv) Recall the definition of drag coefficient, Eq. (5.138), and approximate the drag force as $F_D \sim \overline{\tau}\pi D^2$, where $\overline{\tau}$ is the surface-averaged shear stress aligned with F_D and πD^2 is the surface of the sphere. Define the dimensionless drag friction coefficient that corresponds to $\overline{\tau}$:

$$C_{f,D} = \frac{\overline{\tau}}{\frac{1}{2}\rho U^2}$$

and discover that (in the Re_D range 10–1000) the St expression found in part (iii) takes the form

$$St Pr^{0.6} \sim \frac{1}{2}C_{f,D}$$

Compare this result with the Colburn relations (5.131) and (6.88), and comment on the physics basis for the coincidence.

10

Radiation

10.1 Introduction

In this chapter we consider the phenomenon of radiative heat transfer, which relative to conduction and convection represents a distinct mechanism of heat exchange between two entities at different temperatures. In Section 1.5 we noted that unlike conduction and convection, radiation is the mechanism by which bodies can exchange heat *from a distance*, without making direct contact. Heat transfer by radiation can take place even when the space between two surfaces is completely evacuated.

Thermal radiation is the stream of electromagnetic radiation emitted by a material entity (solid body, pool of liquid, cloud of reacting gaseous mixture) on account of its *finite* temperature T (K). The temperature and the emitted thermal radiation are reflections of the molecular agitation of the material.

Look at Figure 10.1, where the two bodies, (T_1) and (T_2), have arbitrary shapes. These bodies emit their respective streams of thermal radiation in all the directions to which they have access. Every point (e.g. area element) of each body emits radiation in all the directions in which one can look from that point. Only a fraction of the total stream emitted by (T_1) is intercepted and possibly absorbed by body (T_2). This fraction depends not only on the shapes and sizes of the two bodies but also on their relative position, on the condition (e.g. smoothness, cleanliness) of their surfaces, and on the nature of the surroundings. Similarly, only a fraction of the radiation emitted by (T_2) is intercepted and possibly absorbed by (T_1). With reference to one of the bodies, the heat transfer problem reduces to calculate the following:

1 The radiation heat transfer rate that leaves the body surface (e.g. the radiation directly emitted by the surface, plus the reflected portion of the radiation that strikes the surface).
2 The radiation heat transfer rate that enters the surface (e.g. the absorbed portion of the radiation that strikes the surface).

The difference between quantities 1 and 2 represents the *net* heat transfer rate that leaves the body that is being analyzed. A similar accounting can be made of the radiation heat transfer experienced by the other surfaces (bodies) that take part in the configuration.

It is already apparent that radiation heat transfer calculations involve several new complications relative to the conduction and convection configurations studied until now. Chief among these is the optical aspect, that is, the manner in which one body "sees" its neighboring entities. For this reason, space geometry and the ability to see in three dimensions play an important role in the analyses that emerge. The surface condition and the manner in which it affects the emission and absorption of radiation represents another complication. We begin with the simplest configuration of radiation heat exchange and continue with the complications that may alter the configuration.

Heat Transfer: Evolution, Design and Performance, First Edition. Adrian Bejan.
© 2022 John Wiley & Sons, Inc. Published 2022 by John Wiley & Sons, Inc.
Companion website: www.wiley.com/go/bejan/heattransfer

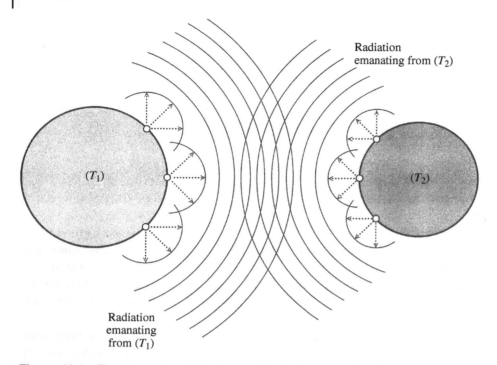

Radiation
emanating from (T_2)

(T_1)

(T_2)

Radiation
emanating
from (T_1)

Figure 10.1   The thermal radiation interaction between two bodies at different temperatures.

10.2   Blackbody Radiation

10.2.1   Definitions

The surface of a system (e.g. solid body) that participates in radiation heat transfer is classified according to its ability to absorb the radiation heat current that strikes it. Consider one unit area dA of such a surface, and let G represent the total radiation heat flux (W/m^2) that arrives on dA from all the other surfaces that can see dA. This quantity is called *total irradiation* and will be defined in Eq. (10.59). In general, three things can happen to the incident flux G:

1 A portion, αG, can be *absorbed* at the surface, that is, in the molecular layers situated immediately below the surface.
2 Another portion, ρG, can be *reflected* back toward the entire space that can be seen from dA.
3 Possibly, a third portion, τG, can pass right through the body and exit through the other side.

   Figure 10.2a shows that the conservation of energy in the coin-shaped control volume that encloses dA requires $\alpha G + \rho G + \tau G = G$; in other words,

$$\alpha + \rho + \tau = 1 \tag{10.1}$$

The nonnegative dimensionless numbers that appear on the left side are properties of the system that is being analyzed. They are known as the *total absorptivity* (α), *total reflectivity* (ρ), and *total transmissivity* (τ) of the material that resides under dA. These properties will be defined more rigorously in Section 10.4.2. Equation (10.1) implies that none of the numbers (α, ρ, τ) can be greater than 1.

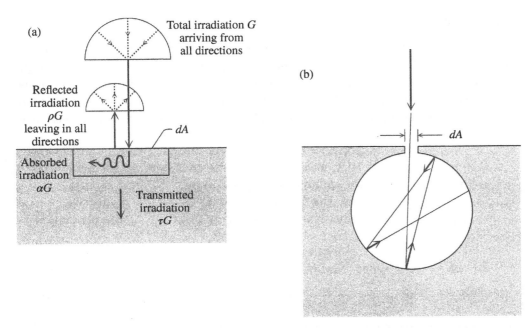

Figure 10.2 The definitions of total absorptivity, reflectivity, and transmissivity (a) and the black surface of the cavity aperture *dA* (b).

Opaque is the system (or the region associated with *dA*) that is characterized by zero total transmissivity. For this class, Eq. (10.1) reduces to

$$\alpha + \rho = 1 \quad \text{(opaque,} \quad \tau = 0) \tag{10.2}$$

If, in addition, the reflectivity is zero, the surface is said to be *black*, and the system covered by the black surface is referred to as *blackbody*. This class is represented by

$$\alpha = 1 \quad \text{(black,} \quad \rho = 0, \quad \tau = 0) \tag{10.3}$$

This terminology is not meant to imply that the color of the surface is actually black. Many surfaces that are not colored black have α values that come close to 1 and, for engineering purposes, can be modeled as black. The "black surface" terminology is a useful reminder that all the incident radiation G disappears into the surface, so that none of it "shines back" at the sources that produced G. A black surface that illustrates this interpretation is shown on Figure 10.2b. The surface that appears black to an outside observer is the small opening *dA* – the aperture – through which the incident flux G enters the cavity that has been machined into the solid body. The incident flux bounces many times around the cavity, and during each bounce it is partially absorbed and reflected by the cavity wall. When *dA* is small enough, there are so many bounces that the fraction of G that finally escapes (is "reflected") through the opening is negligible relative to the incident stream. The *dA* that the external observer sees is, therefore, a surface with $\alpha = 1$, namely, a black surface. The small-aperture cavity illustrated in Figure 10.2 is the basic design by which a black surface of known temperature is constructed in the laboratory. Note that the cavity wall itself is not black: this is why it can partially absorb and reflect the radiation that impinges on it. Only the aperture *dA* is black, that is, "black" from the point of view of an external entity that radiates toward *dA*.

10.2.2 Temperature and Energy

Two alternative descriptions are available for the manner in which the radiation emitted by a surface propagates through the surrounding medium. One is the continuum description, in which the shower of energy G discussed in the preceding paragraphs is an electromagnetic wave or a superposition of waves of many wavelengths. The alternative is the discrete-particle description provided by quantum theory. According to the latter, thermal radiation is a stream of particles (photons), each being characterized by zero rest mass and the energy quantum

$$e = h\nu \tag{10.4}$$

where $h = 6.6256 \times 10^{-34}\,\text{J}\cdot\text{s}$ is Planck's constant. In either description, thermal radiation is also characterized by its frequency $\nu\,(\text{s}^{-1})$ or range of frequencies. The frequency information can be conveyed also by means of the wavelength λ (m) or range of wavelengths that are present in a particular radiation stream. The relationship between frequency and wavelength is

$$\lambda = \frac{c}{\nu} \tag{10.5}$$

where c is the speed of propagation of the wave or of the photons. In an evacuated space, that speed is the speed of light, $c = 2.998 \times 10^8$ m/s.

The one-to-one relationship between frequency and wavelength is also illustrated in Figure 10.3, which shows the complete spectrum of electromagnetic radiation and the relatively narrow band occupied by thermal radiation. In wavelength terms, the thermal radiation domain is sandwiched between approximately 0.1 and 100 μm, where the length unit (the micron) is 1 μm = 10^{-6} m.

An even narrower band inside the thermal radiation domain is the *visible range*, that is, the thermal radiation that can be seen by the human eye. The visible range is sandwiched roughly between the wavelengths of 0.4 and 0.7 μm. Proceeding in order of increasing wavelengths, the thermal radiation domain is divided into three subdomains: the ultraviolet range, the visible range, and the infrared range.

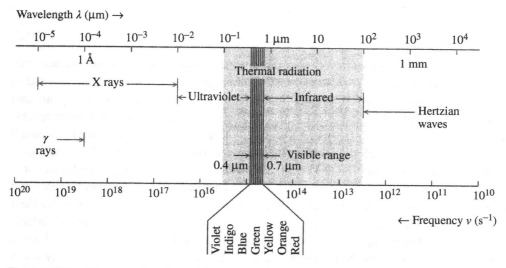

Figure 10.3 The wavelength and frequency domains of thermal radiation and their position on the electromagnetic spectrum.

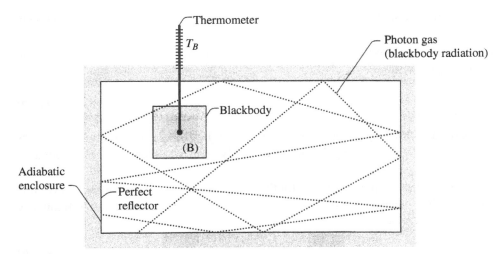

Figure 10.4 Enclosure with perfectly reflecting internal surfaces and blackbody radiation of temperature T_B.

The visible range contains the wavelengths that correspond to the principal colors discerned by the human eye. No stream of thermal radiation is truly of a single color (i.e. of a single wavelength λ). The radiation characterized by an infinitesimally narrow band of wavelengths, λ–$(\lambda + d\lambda)$, is called *monochromatic* (literally, single-color) radiation of wavelength λ. This terminology is used over the entire thermal radiation domain, not just in the visible range.

An important and rarely acknowledged property of thermal radiation is the temperature. To develop an understanding for this property, consider the evacuated enclosure illustrated in Figure 10.4. The inner surface of the wall of this enclosure is a perfect reflector: all the radiation that impinges on it is reflected fully toward the enclosure. Seen from the outside, the enclosure is therefore surrounded by an adiabatic surface, because the net heat transfer rate across the wall is zero. The enclosure and all of its contents constitute an isolated thermodynamic system.

Next, assume that a certain blackbody (B) is introduced in the enclosure. This body emits radiation of all wavelengths and absorbs fully the radiation streams that strike it. In time, the enclosure fills with thermal radiation (a photon "gas") of all wavelengths; Figure 10.4 shows the traces of only a few photons that travel through the enclosure. Simultaneously, the temperature of body (B) – a property that can be measured with a thermometer – approaches an equilibrium value T_B. At equilibrium, the body (B) can only have the same temperature as the photon gas that surrounds it, because the enclosure contents constitute an isolated system. The temperature of the photon gas – in this case, the temperature of blackbody radiation – is T_B.

The photons of the blackbody radiation trapped inside the enclosure cover all the frequencies (or wavelengths); however, some frequencies are represented by considerably more photons than other frequencies. The distribution of the blackbody photon population over the frequency spectrum can be deduced on the basis of quantum-statistical thermodynamics:

$$n_v = \frac{8\pi v^2 c^{-3}}{\exp(hv/kT) - 1} \tag{10.6}$$

where $k = 1.3805 \times 10^{-23}$ J/K is the Boltzmann's constant. The number n_v represents the number of photons per unit volume and frequency interval; in other words, the units of n_v are photons/($m^3 \cdot s^{-1}$). The temperature

T in the denominator of Eq. (10.6) is the thermodynamic (Kelvin) temperature of the particular blackbody radiation.

Recalling that the energy of one photon is $h\nu$, Eq. (10.4), the energy per unit volume and frequency interval u_ν ($J/m^3 \cdot s^{-1}$) can be calculated by writing

$$u_\nu = n_\nu h\nu = \frac{8\pi h\nu^3 c^{-3}}{\exp(h\nu/kT) - 1} \tag{10.7}$$

This formula was first proposed in 1901 by Planck [1], which is why Eq. (10.7) is commonly known as Planck's radiation equation. It is also the starting point in the thermodynamic treatment of thermal radiation as a photon gas system [2].

A more useful alternative to Eq. (10.7) is the expression for the blackbody radiation energy per unit volume and wavelength, u_λ ($J/m^3 \cdot m$). This quantity can be obtained from Eq. (10.7) by noting that the frequency unit $\Delta\nu$ and the wavelength unit $\Delta\lambda$ obey

$$|\Delta\nu| = \frac{c}{\lambda^2}|\Delta\lambda| \tag{10.8}$$

This relationship is obtained by differentiating Eq. (10.5). The energy per unit volume and wavelength is

$$u_\lambda = u_\nu \frac{\Delta\nu}{\Delta\lambda} = \frac{8\pi hc\lambda^{-5}}{\exp(hc/k\lambda T) - 1} \tag{10.9}$$

10.2.3 Intensity

These energy considerations allow us to calculate the energy transport that is associated with a certain stream of photons. Consider an infinitesimal area dA that is exposed to equilibrium blackbody radiation of temperature T. For example, dA can be a small patch on the surface of body (B) in Figure 10.4, or it can be the area bordered by a small wire loop suspended somewhere in the photon gas that fills the enclosure. This dA area has been magnified in Figure 10.5a. Drawn perpendicular to dA is the axis dA–n, which serves as centerline for the "pencil of rays" of infinitesimal solid angle $d\omega$. Figure 10.5b also shows the definition of the *solid angle*:

$$d\omega = \frac{dA_n}{r^2} \tag{10.10}$$

where dA_n is an infinitesimal patch on the sphere of radius r. Note that when $d\omega$ is fixed, dA_n increases proportionally with r^2; this means that the solid angle is a measure of that fraction of space that can be viewed from the center of the sphere through the window dA_n. The maximum solid angle is enjoyed by an observer who can look in all the directions (i.e. through all the points of the spherical surface $4\pi r^2$ drawn around him):

$$\omega_{sphere} = \frac{4\pi r^2}{r^2} = 4\pi \text{ sr} \tag{10.11}$$

The solid angle unit is the steradian (symbol sr); this particular word serves as a reminder that the solid angle is the solid-body geometry analog of the concept of angle, whose unit is the radian (symbol rad) used in plane geometry. The first part of the word "steradian" is based on the Greek word *stereos* (hard, firm, solid).

Returning to the original infinitesimal area dA shown in Figure 10.5a, we see that the product $u_\lambda c$ represents the energy flowrate per unit time, wavelength, and area normal to a particular ray located inside the thin

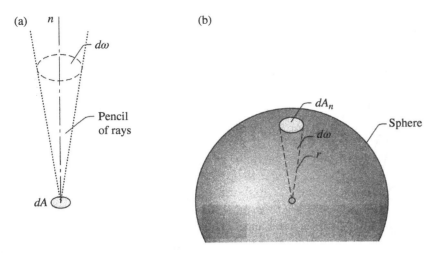

Figure 10.5 Pencil of rays aligned with the direction normal to dA (a) and the definition of infinitesimal solid angle (b).

pencil. Dividing $u_\lambda c$ by 4π steradians, we obtain the intensity of monochromatic blackbody radiation, $I_{b,\lambda}$ (W/m^3 · sr):

$$I_{b,\lambda} = \frac{u_\lambda c}{4\pi} = \frac{2hc^2 \lambda^{-5}}{\exp(hc/k\lambda T) - 1} \tag{10.12}$$

in other words, the number of watts per unit wavelength (m), unit area normal to the ray (m^2), and unit solid angle (sr). The monochromatic energy current that flows through the pencil of rays and through the opening of area dA is therefore equal to $I_{b,\lambda}\, dA\, d\omega$. The intensity of monochromatic blackbody radiation ($I_{b,\lambda}$) is a function of only wavelength and temperature; in Section 10.2.4 we examine this function by focusing on a quantity that is proportional to $I_{b,\lambda}$.

The total intensity of blackbody radiation, I_b (W/m^2 · sr), is obtained by integrating the monochromatic intensity $I_{b,\lambda}$ over the entire radiation spectrum:

$$I_b(T) = \int_0^\infty I_{b,\lambda}(\lambda, T)d\lambda \tag{10.13}$$

It turns out that I_b is proportional to the fourth power of the absolute temperature of the particular ray of blackbody radiation; this proportionality will be derived in Eq. (10.23). The total intensity I_b represents the energy conveyed by the ray (i.e. in a certain direction) per unit time, solid angle, and unit area normal to the direction of the ray.

10.2.4 Emissive Power

The actual direction of the photons that travel along the rays of the pencil of Figure 10.5 was not discussed until now. The quantity $I_{b,\lambda}\, dA\, d\omega$ can represent either the monochromatic energy current that arrives through the pencil at dA or the current that leaves dA through the pencil aligned with the normal direction dA–n. Of interest is the energy current that leaves dA in *all the directions* to which the dA observer has access. Those directions are all intercepted by the hemisphere that uses the plane of dA as base and dA itself as center. This new situation is illustrated in Figure 10.6.

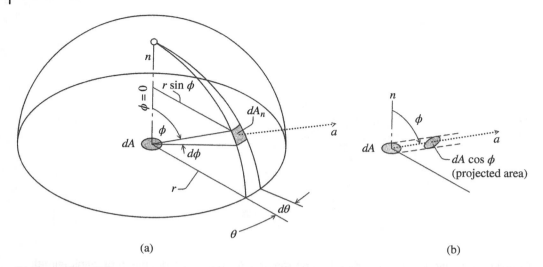

(a)   (b)

Figure 10.6 The infinitesimal solid angle dA_n/r^2 associated with the direction dA–a (a) and the projection of dA on a plane normal to the direction dA–a (b).

To calculate the per-unit-wavelength energy current "emitted" by dA in all the directions of the hemisphere, we recall the two angular coordinates (ϕ, θ) of the spherical system defined in Figure 1.8. Consider the arbitrary direction dA–a in Figure 10.6. The intensity of monochromatic blackbody radiation emitted along this ray is given by Eq. (10.12). The solid angle of the pencil of rays centered on dA–a is

$$d\omega = \frac{dA_n}{r^2} = \frac{(r\,\sin\phi\,d\theta)(r\,d\phi)}{r^2} = \sin\phi\,d\theta\,d\phi \qquad (10.14)$$

The area normal to the dA–a direction is not dA but the *projected* area $dA\cos\phi$. The energy current per unit wavelength that leaves dA through the pencil of rays aligned with dA–a is therefore $I_{b,\lambda}$ ($\sin\phi\,d\theta\,d\phi$)($dA\cos\phi$). Integrating this quantity over all the directions intercepted by the hemisphere, we obtain the per-unit-wavelength energy current emitted by dA in all directions:

$$\int_{\theta=0}^{2\pi}\int_{\phi=0}^{\pi/2} I_{b,\lambda}\,\sin\phi\,\cos\phi\,d\phi\,d\theta\,dA = \pi I_{b,\lambda}dA \qquad (10.15)$$

The product $\pi I_{b,\lambda}$ represents the monochromatic hemispherical emissive power of the black surface to which dA belongs:

$$E_{b,\lambda} = \pi I_{b,\lambda} = \frac{C_1\lambda^{-5}}{\exp(C_2/\lambda T) - 1} \qquad (10.16)$$

for which the values of the constants C_1 and C_2 can be deduced from Eq. (10.12):

$$C_1 = 2\pi hc^2 = 3.742 \times 10^{-16}\ \mathrm{W \cdot m^2}$$
$$C_2 = \frac{hc}{k} = 1.439 \times 10^{-2}\ \mathrm{m \cdot K} \qquad (10.17)$$

The units of $E_{b,\lambda}$ are $\mathrm{W/m \cdot m^2}$, in other words, energy per unit time, wavelength, and surface area. The factor of π steradians appearing in the product $\pi I_{b,\lambda}$ is the result of performing the special integral shown in

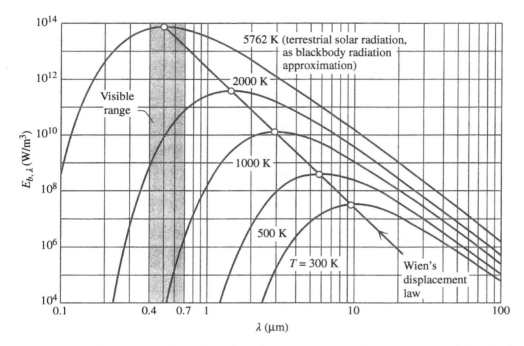

Figure 10.7 The effects of wavelength and temperature on the monochromatic hemispherical blackbody emissive power.

Eq. (10.15). The π factor should not be confused with the hemispherical solid angle, which is equal to 2π steradians (cf. Eq. (10.11)).

The main features of the $E_{b,\lambda}$ (λ,T) function are illustrated in Figure 10.7. The monochromatic emissive power increases dramatically with the absolute temperature, as we'll discover in Eq. (10.19). A maximum $E_{b,\lambda}$ value is registered at a characteristic wavelength for each temperature; solving $\partial E_{b,\lambda}/\partial\lambda = 0$ in conjunction with Eq. (10.16), we obtain

$$\lambda T = 2.898 \times 10^{-3} \text{ m} \cdot \text{K} \tag{10.18}$$

This simple relation is known as Wien's displacement law [3]. It describes the locus of the $E_{b,\lambda}$ maxima, which on the logarithmic field of Figure 10.7 is represented by a straight line. The wavelength of maximum $E_{b,\lambda}$ varies inversely with the absolute temperature. This means that as the temperature increases, the bulk of the energy emitted by a blackbody shifts to (is "displaced" toward) progressively shorter wavelengths. The maximum value of the monochromatic emissive power is obtained by substituting Eq. (10.18) into Eq. (10.16):

$$E_{b,\lambda,\text{max}} = (12.87 \times 10^{-6} \text{ W/m}^3 \cdot \text{K}^5)T^5 \tag{10.19}$$

It is possible now to replace all the $E_{b,\lambda}(\lambda, T)$ curves of Figure 10.7 with a unique curve, by dividing the $E_{b,\lambda}$ ordinate by T^5 (or $E_{b,\lambda,\text{max}}$) and by multiplying the λ abscissa by T. The end result of this construction is the bell-shaped curve labeled $E_{b,\lambda}/E_{b,\lambda,\text{max}}$ on Figure 10.8. The constant that appears on the right side of Wien's displacement law, Eq. (10.18), is now an important entry on the λT abscissa of Figure 10.8.

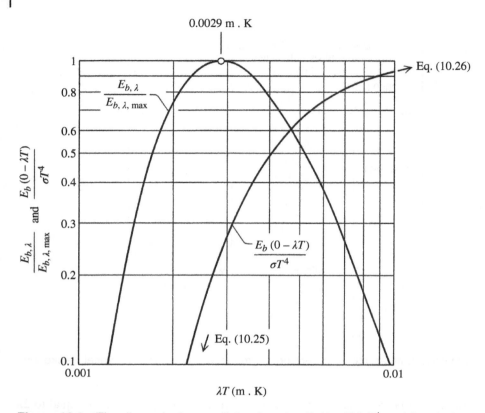

Figure 10.8 The dimensionless radiation function $E_b(0\text{--}\lambda T)/\sigma T^4$ and the single-curve substitute for the family of $E_{b,\lambda}(\lambda, T)$ curves shown in Figure 10.7.

To summarize, the monochromatic emissive power $E_{b,\lambda}$ represents the heat flux that leaves dA along all the rays intersected by the hemisphere, per unit of wavelength interval. The total hemispherical emissive power E_b is the heat flux integrated over all the wavelengths of the radiation spectrum:

$$E_b = \int_0^\infty E_{b,\lambda}\, d\lambda \tag{10.20}$$

In view of Eq. (10.16), the result of this integration is the Stefan–Boltzmann law [4, 5]:

$$E_b = \sigma T^4 \tag{10.21}$$

where the constant $\sigma = 5.67 \times 10^{-8}\ \text{W/m}^2 \cdot \text{K}^4$ is shorthand for the group

$$\sigma = \frac{C_1}{C_2^4} \int_0^\infty \frac{u^3\, du}{e^u - 1} = \frac{C_1}{C_2^4} \frac{\pi^4}{15} \tag{10.22}$$

Incidentally, Eqs. (10.13), (10.16), and (10.20)–(10.22) can be used to prove that the total intensity of black-body radiation I_b is also proportional to T^4:

$$I_b = \frac{\sigma}{\pi} T^4 = \frac{E_b}{\pi} \tag{10.23}$$

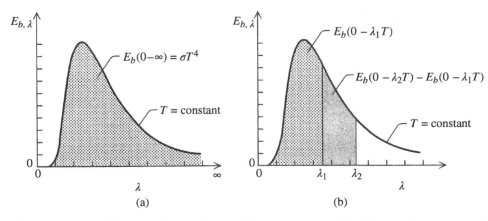

Figure 10.9 (a) The definition of the radiation function $E_b(0-\lambda T)$ and (b) the calculation of the total hemispherical emissive power associated with the band of wavelengths $\lambda_1-\lambda_2$.

The total hemispherical emissive power σT^4 is equal to the area under the curve $E_{b,\lambda}$ ($\lambda, T =$ constant) when this curve is plotted in a Cartesian set of linear coordinates (Figure 10.9a). The total number of watts per unit area emitted by the black surface in the finite band of wavelengths $\lambda_1-\lambda_2$ can be calculated by defining the radiation function:

$$E_b(0 - \lambda_1 T) = \int_0^{\lambda_1} E_{b,\lambda}(\lambda, T)\, d\lambda \tag{10.24}$$

This quantity represents the area trapped under the $E_{b,\lambda}$ isotherm between $\lambda = 0$ and $\lambda = \lambda_1$ (Figure 10.9b), which also means that $\sigma T^4 = E_b(0-\infty)$. The heat flux emitted hemispherically in the finite band $\lambda_1-\lambda_2$ is equal to the difference between the respective radiation functions:

$$\int_{\lambda_1}^{\lambda_2} E_{b,\lambda}(\lambda, T)d\lambda = E_b(0 - \lambda_2 T) - E_b(0 - \lambda_1 T) \tag{10.25}$$

The values of the radiation function $E_b(0-\lambda T)$ can be placed in a surprisingly compact form, by noting first that the dimensionless ratio $E_b(0-\lambda T)/\sigma T^4$ depends only on the value of the group λT. Furthermore, the ratio $E_b(0-\lambda T)/\sigma T^4$ is greater than 0 and less than 1, because it represents the fraction of σT^4 that is emitted in the interval $0-\lambda$. That fraction is the same as the result of dividing the dotted area of Figure 10.9b by the dotted area of Figure 10.9a.

The values of the dimensionless radiation function $E_b(0-\lambda T)/\sigma T^4$ have been tabulated by Dunkle [6]. Table 10.1 shows a set of representative values; the intermediate-λT portion of the table accounts for the rising curve illustrated in Figure 10.8. In the small-λT limit, the dimensionless radiation function is approximated within less than 1% by the asymptote

$$\frac{E_b(0 - \lambda T)}{\sigma T^4} \cong \frac{15}{\pi^4}(u^3 + 3u^2 + 6u + 6)e^{-u} \tag{10.26}$$

where $u = \dfrac{C_2}{\lambda T}$ if $\lambda T < 0.0042\ \text{m} \cdot \text{K}$

In the large-λT limit, the data of Table 10.1 approach within 1% the asymptote

$$\frac{E_b(0 - \lambda T)}{\sigma T^4} \cong 1 - \left(\frac{0.005\,35\ \text{m} \cdot \text{K}}{\lambda T}\right)^3 \quad \text{if } \lambda T > 0.016\ \text{m} \cdot \text{K} \tag{10.27}$$

Table 10.1 Principal values and the asymptotic behavior of the dimensionless radiation function.

λT (m·K)	$\dfrac{E_b(0-\lambda T)}{\sigma T^4}$	λT (m·K)	$\dfrac{E_b(0-\lambda T)}{\sigma T^4}$
The small-λT limit: Eq. (10.26)		0.0065	0.776
0.001	0.002 13	0.007	0.808
0.002	0.0667	0.0075	0.834
0.0025	0.162	0.008	0.856
0.003	0.273	0.009	0.890
0.0035	0.383	0.010	0.914
0.004	0.481	0.011	0.932
0.0045	0.564	0.012	0.945
0.005	0.634	0.015	0.970
0.0055	0.691	0.020	0.986
0.006	0.738	The large-λT limit: Eq. (10.27)	

Example 10.1   *Intensity, Projected Area, and Solid Angle*
The black surface $A_1 = 1\,\mathrm{cm}^2$ emits radiation with total intensity $I_{b,1} = 1400\,\mathrm{W/m}^2 \cdot \mathrm{sr}$. A portion of this radiation, $q_{1 \rightarrow 2}$ (W), strikes the surface $A_2 = 1\,\mathrm{cm}^2$ shown in Figure E10.1. The relative position of A_1 and A_2 is fixed by the distance $r = 1\,\mathrm{m}$ and the two angles (30° and 60°) indicated on the drawing. Calculate the heat current $q_{1 \rightarrow 2}$.

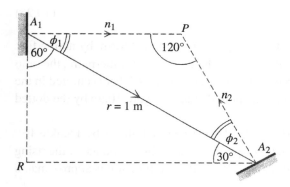

Figure E10.1

Solution

The angles ϕ_1 and ϕ_2 can be calculated by remembering the following lessons of plane geometry:

$$\phi_1 = 30° \tag{1}$$

$$\phi_1 + \phi_2 = 180° - 120° \tag{2}$$

Equation (1) refers to two angles formed in the "underarms" of the line $\overline{A_1 A_2}$ as it cuts the parallel lines $\overline{A_1 P}$ and $\overline{A_2 R}$. Equation (2) is the statement that in the triangle $\overline{A_1 P A_2}$ the angles add up to 180°. Combining Eqs. (1) and (2), we conclude that

$$\phi_2 = 30° \tag{3}$$

The radiation of intensity $I_{b,1}$ is emitted by the surface A_1 in all the directions of the hemisphere that has A_1 as base. The heat current emitted by A_1 and intercepted by A_2 is

$$q_{1 \to 2} = I_{b,1} A_{n,1} \omega_{1-2} \tag{4}$$

in which $A_{n,1}$ is the projection of A_1 on the plane normal to the direction A_1–A_2:

$$A_{n,1} = A_1 \cos \phi_1 = \frac{3^{1/2}}{2} \text{ cm}^2 \tag{5}$$

The solid angle ω_{1-2} is subtended by the projected area $A_{n,2}$:

$$\omega_{1-2} \cong \frac{A_{n,2}}{r^2} \tag{6}$$

where $A_{n,2}$ is the projection of A_2 on the plane perpendicular to the direction $A_1 - A_2$:

$$A_{n,2} = A_2 \cos \phi_2 = \frac{3^{1/2}}{2} \text{ cm}^2 \tag{7}$$

Note that the solid angle relationship (6) is approximately valid only when the target area A_2 is small relative to r^2. Numerically, Eq. (6) yields

$$\omega_{1-2} = \frac{3^{1/2}}{2} 10^{-4} \text{ sr} \tag{8}$$

so that the one-way heat current $q_{1 \to 2}$ amounts to

$$q_{1 \to 2} = \left(1400 \, \frac{\text{W}}{\text{m}^2 \cdot \text{sr}} \right) \left(\frac{3^{1/2}}{2} \text{ cm}^2 \right) \left(\frac{3^{1/2}}{2} 10^{-4} \text{ sr} \right) \tag{9}$$

$$= 1.05 \times 10^{-5} \text{ W}$$

Example 10.2 *Radiation Function and the Human Skin as a Selective Absorber*

The human skin is "selective" when it comes to the absorption of the solar radiation that strikes it perpendicularly. The skin absorbs only 50% of the incident radiation with wavelengths between $\lambda_1 = 0.52$ and $\lambda_2 = 1.55 \, \mu\text{m}$. The radiation with wavelengths shorter than λ_1 and longer than λ_2 is fully absorbed. The solar surface may be modeled as black with temperature $T = 5800 \text{ K}$. Calculate what fraction of the incident solar radiation is absorbed by the human skin.

Solution

Guided by how Table 10.1 was constructed, we calculate, in order,

$$\lambda_1 T = (0.52 \times 10^{-6} \text{ m})(5800 \text{ K}) = 0.003 \text{ m} \cdot \text{K}$$

$$\lambda_2 T = (1.55 \times 10^{-6} \text{ m})(5800 \text{ K}) = 0.009 \text{ m} \cdot \text{K}$$

Reading Table 10.1, we obtain

$$\frac{E_b(0 - \lambda_1 T)}{\sigma T^4} = 0.273$$

$$\frac{E_b(0 - \lambda_2 T)}{\sigma T^4} = 0.890$$

which means that the λ_1–λ_2 band contains 61.7% of the total emissive power of the solar surface:

$$\frac{E_b(\lambda_1 T - \lambda_2 T)}{\sigma T^4} = 0.890 - 0.273 = 0.617$$

while the λ_2–∞ band contains 11%:

$$\frac{E_b(\lambda_2 T - \infty)}{\sigma T^4} = 1 - 0.89 = 0.11$$

The absorbed fraction of the incident solar radiation is due to 100% absorption below λ_1, 50% absorption between λ_1 and λ_2, and 100% absorption above λ_2:

$$1 \times 0.273 + 0.5 \times 0.617 + 1 \times 0.11 = 0.692$$

In conclusion, the human skin absorbs roughly 70% of the incident solar radiation. It is important to note that the group σT^4 is not the solar radiation heat flux that strikes the skin. The actual heat flux is considerably smaller than σT^4 because it depends on the relative size and position of the two surfaces (sun, skin). On this aspect we focus in the next section.

10.3 Heat Transfer Between Black Surfaces

10.3.1 Geometric View Factor

Consider now the problem of estimating the net heat transfer rate q_{1-2} (W) between the two isothermal black surfaces (A_1, T_1) and (A_2, T_2) shown in Figure 10.10. The shapes and relative positions of the two surfaces are meant to be arbitrary. The q_{1-2} analysis consists of estimating, in order, the following:

1 The fraction of the radiation emitted by the area element dA_1 and intercepted (i.e. absorbed fully) by the area element dA_2.
2 The fraction of the radiation emitted by dA_2 and intercepted (absorbed) by dA_1.
3 The net heat transfer rate from dA_1 to dA_2, namely, the difference between the answers to parts (1) and (2).
4 Finally, the net heat transfer rate from A_1 to A_2, that is, between the two isothermal areas.

The thinking behind step (1) was presented already in the discussion of Figure 10.6. If this time the segment r is the distance between the area elements dA_1 and dA_2, then the solid angle through which dA_2 is seen by an observer stationed at dA_1 is equal to $dA_2 \cos \phi_2/r^2$. Note that $dA_2 \cos \phi_2$ is the size of dA_2 after it has been projected onto a plane perpendicular to the dA_1–dA_2 line.

Traveling from dA_1 toward dA_2 (and toward the rest of the space) is blackbody radiation of total intensity $I_{b,1} = I_b(T_1)$. According to the definition (10.13), $I_{b,1}$ is the number of watts of T_1 radiation that are emitted in the r direction, per unit solid angle and per unit of area normal to the r direction. The size of the emitting area that is normal to the r direction is the projected-dA_1 area, namely, $dA_1 \cos \phi_1$. All this means that if we multiply $I_{b,1}$ by the projected emitting area $dA_1 \cos \phi_1$ and by the solid angle subtended by the target,

Figure 10.10 The geometric parameters needed for calculating the view factor integral (10.33).

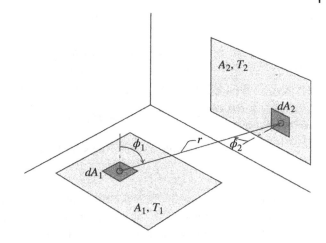

$dA_2 \cos \phi_2 / r^2$, we obtain the answer to step (1):

$$q_{dA_1 \to dA_2} = I_{b,1} dA_1 \cos \phi_1 \frac{dA_2 \cos \phi_2}{r^2} \tag{10.28}$$

The arrow used in the $dA_1 \to dA_2$ subscript is a reminder that $q_{dA_1 \to dA_2}$ represents a one-way transfer of energy per unit time, in this case, from dA_1 (emitter) to dA_2 (target). The argument of the preceding two paragraphs can be used one more time in step (2) of the analysis; however, the end result can be read in Eq. (10.28) by simply switching the positions of the subscripts 1 and 2. The fraction of T_2 radiation emitted by dA_2 and absorbed by dA_1 is (note again the one-way arrow in the $dA_2 \to dA_1$ subscript)

$$q_{dA_2 \to dA_1} = I_{b,2} dA_2 \cos \phi_2 \frac{dA_1 \cos \phi_1}{r^2} \tag{10.29}$$

The third step consists of subtracting Eq. (10.29) from Eq. (10.28) to evaluate the net heat transfer rate that leaves dA_1 and arrives at dA_2:

$$q_{dA_1 - dA_2} = q_{dA_1 \to dA_2} - q_{dA_2 \to dA_1}$$

$$= (I_{b,1} - I_{b,2}) \frac{\cos \phi_1 \cos \phi_2}{r^2} dA_1 dA_2 \tag{10.30}$$

This net heat transfer rate is distinguished from the previous one-way contributions ($q_{dA_1 \to dA_2}$ and $q_{dA_2 \to dA_1}$) by means of the new subscript $dA_1 - dA_2$: the first area in this subscript (dA_1) represents the origin of the net heat transfer rate, whereas the second area (dA_2) represents the destination. Equation (10.30) can also be written in terms of the total hemispherical emissive powers[1] associated with T_1 and T_2 (cf. Eq. (10.23)):

$$q_{dA_1 - dA_2} = \sigma(T_1^4 - T_2^4) \frac{\cos \phi_1 \cos \phi_2}{\pi r^2} dA_1 dA_2 \tag{10.31}$$

The macroscopic areas A_1 and A_2 communicate through a very large number of dA_1, dA_2 pairs of the kind analyzed until now. The net heat transfer rate from A_1 to A_2 is the sum of all the contributions of type (10.31) that are possible. This sum, q_{1-2}, can be obtained by integrating $q_{dA_1 - dA_2}$ over the two areas A_1 and A_2:

$$q_{1-2} = \sigma(T_1^4 - T_2^4) \int_{A_1} \int_{A_2} \frac{\cos \phi_1 \cos \phi_2}{\pi r^2} dA_1 dA_2 \tag{10.32}$$

---

1 Note the appearance of π in the denominator on the right side of Eq. (10.31).

On the left side, the subscript 1–2 states that the net heat transfer rate q_{1-2} leaves the surface A_1 and enters (crosses) the surface A_2.

The units of the double-area integral on the right side of Eq. (10.32) are the units of area (m²). It is permissible (and convenient) to replace this cumbersome integral expression with the product A_1F_{12}, in which the dimensionless factor F_{12} is the *geometric view factor*[2] based on A_1:

$$F_{12} = \frac{1}{A_1} \int_{A_1} \int_{A_2} \frac{\cos\phi_1 \, \cos\phi_2}{\pi r^2} \, dA_1 \, dA_2 \tag{10.33}$$

In the end, the q_{1-2} formula assumes the simpler form

$$q_{1-2} = \sigma(T_1^4 - T_2^4)A_1F_{12} \tag{10.34}$$

The geometric view factor is a purely geometric quantity, one that depends only on the sizes, orientations, and relative position of the two surfaces. The same definition holds when the surfaces are not black (Section 10.4.4). Here we derived the formula for F_{12} by considering two black surfaces because this model leads to the simplest analysis.

Alternatively, the double-area integral of Eq. (10.32) can be replaced with the product A_2F_{21}, in which F_{21} is the geometric view factor *based on* A_2:

$$F_{21} = \frac{1}{A_2} \int_{A_1} \int_{A_2} \frac{\cos\phi_1 \, \cos\phi_2}{\pi r^2} \, dA_1 \, dA_2 \tag{10.35}$$

The net heat transfer rate from A_1 to A_2 is now given by the expression

$$q_{1-2} = \sigma(T_1^4 - T_2^4)A_2F_{21} \tag{10.36}$$

Equations (10.34) and (10.36) show that the final calculation of q_{1-2} depends on being able to evaluate one geometric view factor (F_{12} or F_{21}). Before plunging into the manipulation of the double-area integral (10.33), it pays to review the physical meaning of the view factor F_{12}. For this we return to the expression for the radiation heat transfer rate emitted by dA_1 and absorbed by dA_2. If we integrate this expression over both finite-size areas, we obtain the radiation heat transfer rate emitted by A_1 and absorbed by A_2:

$$q_{1\to2} = I_{b,1} \int_{A_1} \int_{A_2} \frac{\cos\phi_1 \, \cos\phi_2}{r^2} \, dA_1 \, dA_2 = \cdots$$

$$= \sigma T_1^4 A_1 F_{12} \tag{10.37}$$

On the right side, the group $\sigma T_1^4 A_1$ is the same as $E_{b,1}A_1$, or the number of watts of blackbody radiation emitted by the surface A_1 in all the directions in which the points of A_1 can look. Only a portion of $E_{b,1}A_1$ is intercepted and absorbed by A_2 (because, in general, A_1 may be surrounded by more than just A_2): that portion[3] is $q_{1\to2}$ or (cf. Eq. (10.37)) $E_{b,1}A_1F_{12}$. In conclusion, the physical meaning of the view factor is this

$$F_{12} = \frac{q_{1\to2}(\text{radiation leaving } A_1 \text{ and being intercepted by } A_2)}{E_{b,1}A_1(\text{radiation leaving } A_1 \text{ in all directions})} \tag{10.38}$$

---

2 This factor is also known as geometric configuration factor, view factor, radiation shape factor, and angle factor.
3 The *one-way* heat transfer rate $q_{1\to2}$ should not be confused with the *net* heat transfer rate q_{1-2} of Eqs. (10.34) and (10.36). The relationship that unites these concepts is [see also the top line of Eq. (10.30)]

$$q_{1-2} = q_{1\to2} - q_{2\to1}$$

where $q_{2\to1}$, is the portion of the T_2 temperature radiation emitted by A_2 and intercepted by A_1. ($q_{2\to1} = E_{b,2}A_2F_{21}$).

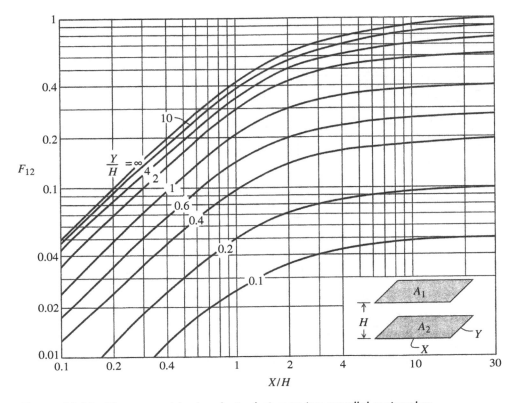

Figure 10.11 The geometric view factor between two parallel rectangles.

The definition formulated above indicates that the value of a geometric view factor falls between 0 and 1. This feature is evident on the ordinates of Figures 10.11 and 10.12. These figures and Table 10.2 contain the view factors of some of the two-surface configurations that are encountered frequently in radiation heat transfer calculations, most notably the parallel coaxial discs. Several other configurations are covered in Howell's catalog [7] as well as in Siegel and Howell's treatise [8].

10.3.2 Relations Between View Factors

The calculation of each shape factor can be traced to the purely geometric definition (10.33), as shown in Example 10.3. In some instances, however, it is possible to deduce the F_{12} value based on a shortcut, that is, by manipulating the known view factors of one or more related configurations. Essential to this simpler approach are the following relationships also known as "view factor algebra." These relationships are purely geometric, that is, independent of surface description (black versus nonblack).

10.3.2.1 Reciprocity

The first relation is obtained by comparing Eqs. (10.34) and (10.36):

$$A_1 F_{12} = A_2 F_{21} \tag{10.38}$$

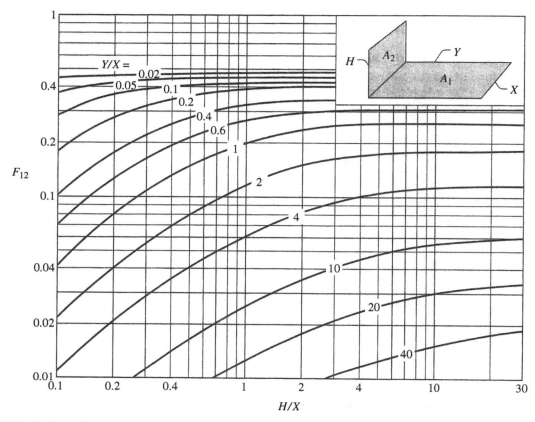

Figure 10.12 The geometric view factor between two perpendicular rectangles with a common edge.

If we know F_{21} from a chart, formula, or table, we can calculate F_{12} if we also know the areas involved in the current configuration (A_1, A_2). Equation (10.38) is known as the reciprocity rule or the reciprocity property of the two view factors associated with the same pair of surfaces.

10.3.2.2 Additivity

Let A_{2a} and A_{2b} be the two surfaces that, together, make up the A_2 surface discussed until now (Figure 10.13). If F_{12} represents the fraction of the A_1 radiation that is intercepted by A_2, then F_{12} can only be the sum of F_{12a} and F_{12b}, because F_{12a} is the fraction absorbed by A_{2a} and F_{12b} is the fraction absorbed by A_{2b}:

$$F_{12} = F_{12a} + F_{12b} \tag{10.39}$$

Of course, we can write as many terms on the right side of this equation as there are pieces in the A_2 mosaic, for example,

$$F_{12} = \sum_{i=1}^{n} F_{12_i} \tag{10.40}$$

when A_2 is made up of n pieces, $A_2 = A_{2_1} + A_{2_2} + \cdots + A_{2_n}$. Equation (10.40) is the additivity rule of the view factors between one surface (A_1) and all the pieces of a neighboring surface (A_2). This property can be

Table 10.2 Minicatalog of geometric view factors.

Configuration	Geometric view factor
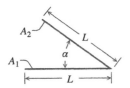	Two infinitely long plates of width L, joined along one of the long edges: $$F_{12} = F_{21} = 1 - \sin\frac{\alpha}{2}$$
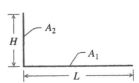	Two infinitely long plates of different widths (H, L), joined along one of the long edges and with a 90° angle between them: $$F_{12} = \frac{1}{2}[1 + x - (1 + x^2)^{1/2}]$$ where $x = H/L$
	Triangular cross-section enclosure formed by three infinitely long plates of different widths (L_1, L_2, L_3): $$F_{12} = \frac{L_1 + L_2 - L_3}{2L_1}$$
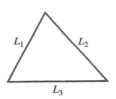	Disc and parallel infinitesimal area positioned on the disc centerline: $$F_{12} = \frac{R^2}{H^2 + R^2}$$
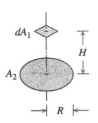	Parallel discs positioned on the same centerline: $$F_{12} = \frac{1}{2}\left\{ X - \left[X^2 - 4\left(\frac{x_2}{x_1}\right)^2 \right]^{1/2} \right\}$$ where $x_1 = \dfrac{R_1}{H}$, $x_2 = \dfrac{R_2}{H}$, and $X = 1 + \dfrac{1 + x_2^2}{x_1^2}$
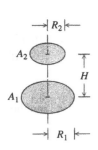	Infinite cylinder parallel to an infinite plate of finite width $(L_1 - L_2)$: $$F_{12} = \frac{R}{L_1 - L_2}\left(\tan^{-1}\frac{L_1}{H} - \tan^{-1}\frac{L_2}{H} \right)$$

(Continued)

Table 10.2 (Continued)

Configuration	Geometric view factor
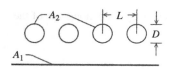	Two parallel and infinite cylinders: $$F_{12} = F_{21} = \frac{1}{\pi}\left[(X^2-1)^{1/2} + \sin^{-1}\left(\frac{1}{X}\right) - X\right]$$ where $X = 1 + \dfrac{L}{2R}$
(row of cylinders diagram with A_2, L, D, A_1)	Row of equidistant infinite cylinders parallel to an infinite plate: $$F_{12} = 1 - (1-x^2)^{1/2} + x\tan^{-1}\left(\frac{1-x^2}{x^2}\right)^{1/2}$$ where $x = \dfrac{D}{L}$
Elemental areas are presented in Example 10.3 and Problems 10.32 and 10.33 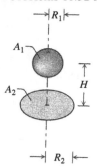	Sphere and disc positioned on the same centerline: $$F_{12} = \frac{1}{2}[1 - (1+x^2)^{-1/2}]$$ where $x = \dfrac{R_2}{H}$

Source: After Howell [7] and Siegel and Howell [8].

used to great advantage in the calculation of the view factor F_{12a} in the two cases shown in Figure 10.13. In the two-rectangle geometry on the left side, $F_{12a} = F_{12} - F_{12b}$, where both F_{12} and F_{12b} can be read off Figure 10.12 (note that the A_1–A_{2b} and A_1–A_2 pairs touch along one edge).

On Figure 10.13b, the view factor between the small disc and the annulus is simply $F_{12a} = F_{12} - F_{12b}$. Since A_2 and A_{2b} are both discs, F_{12} and F_{12b} can be calculated with the parallel discs' formula listed as the fifth entry in Table 10.2.

10.3.2.3 Enclosure

Returning to the (A_1, A_2) pair of surfaces in Figure 10.10, we recall that not all the radiation emitted by A_1 is intercepted by A_2. The remainder is intercepted by the other surfaces that surround A_1. The emitting surface (A_1) and all the surfaces that surround it ($A_2, A_3, ..., A_n$) form the enclosure shown in Figure 10.14.

Think now of the radiation emitted by A_1 in all the directions ($E_{b,1}A_1$): the portions intercepted by $A_2, A_3, ..., A_n$ are, respectively, $E_{b,1}A_1F_{12}, E_{b,1}A_1F_{13}, ..., E_{b,1}A_1F_{1n}$. If the A_1 surface is concave, some of the $E_{b,1}A_1$ current is intercepted by A_1 itself (that portion is $E_{b,1}A_1F_{11}$). The conservation of $E_{b,1}A_1$ inside

Figure 10.13 Two cases in which the additivity property can be used to calculate the view factor F_{12a}.

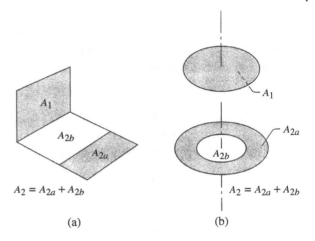

$A_2 = A_{2a} + A_{2b}$

(a)

$A_2 = A_{2a} + A_{2b}$

(b)

Figure 10.14 Enclosure formed by n surfaces and the dependence of the view factor F_{ii} on the curvature of each surface.

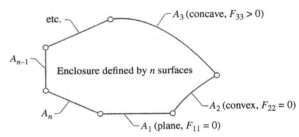

the enclosure requires that

$$E_{b,1}A_1 = E_{b,1}A_1 F_{11} + E_{b,1}A_1 F_{12} + \cdots + E_{b,1}A_1 F_{1n} \tag{10.41}$$

or, after dividing by $E_{b,1}A_1$,

$$1 = F_{11} + F_{12} + \cdots + F_{1n} \tag{10.42}$$

An equation of type (10.42) can be written for each of the surfaces that participates in the enclosure; therefore, the preceding energy conservation argument yields a system of n equations:

$$1 = \sum_{j=1}^{n} F_{ij} \quad (i = 1, 2, \ldots, n) \tag{10.43}$$

The subscript i indicates the surface that serves as emitter (i.e. the surface for which the equation is written), while the subscript j counts all the surfaces that make up the enclosure. The enclosure relationships (10.43) hold true regardless of whether the surfaces are black or not.

10.3.3 Two-Surface Enclosures

The simplest application of the preceding analysis is to calculate the net heat transfer rate q_{1-2} between two black surfaces that form an enclosure. Three examples of two-surface enclosures are illustrated in Figure 10.15. In each of these examples, the enclosed space is bounded only by A_1 and A_2. The two cylinders and the two spheres are not necessarily concentric.

The formula for calculating q_{1-2} is listed as Eq. (10.34) or Eq. (10.36). These relations apply to steady as well as unsteady (time-dependent) situations. The lower part of Figure 10.15 shows an electrical circuit analog of the net heat transfer rate across the enclosure. This circuit is based on reading Eq. (10.34) as a proportionality between the current q_{1-2} and the potential difference $E_{b,1} - E_{b,2}$, that is,

$$q_{1-2} = (E_{b,1} - E_{b,2})A_1F_{12} \tag{10.44}$$

where $E_{b,1} = E_b(T_1) = \sigma T_1^4$ and $E_{b,2} = E_b(T_2) = \sigma T_2^4$. The product A_1F_{12} plays the role of thermal conductance for the net heat transfer rate between the potential nodes represented by $E_{b,1}$ and $E_{b,2}$. The inverse of A_1F_{12} (or A_2F_{21}) is the *radiation thermal resistance*, R_r (m^{-2}):

$$R_r = \frac{1}{A_1F_{12}} = \frac{1}{A_2F_{21}} \tag{10.45}$$

One aspect that is brought to light by the resistance analog of the two-surface enclosure is that the net surface-to-surface heat transfer rate q_{1-2} must first enter A_1 through its back side and then exit through the back side of A_2. If the physical sense of q_{1-2} is indeed the sense assumed in Figure 10.15, the steady state prevails only if an external device heats the A_1 surface at the rate q_{1-2}, and if another external device cools the A_2 surface at the same rate. Relative to the environment that surrounds the enclosure, the back sides of both A_1 and A_2 are not adiabatic.

Listed under each configuration in Figure 10.15 are the respective geometric view factors. The F_{12} factors are equal to 1 because all the radiation emitted by the surface A_1 is intercepted by A_2. In the cylindrical and spherical geometries, A_1 is the inner (smaller) surface. The F_{21} factors are determined based on the reciprocity relationship $F_{21} = F_{12}A_1/A_2$. The row of F_{21} values demonstrates that the

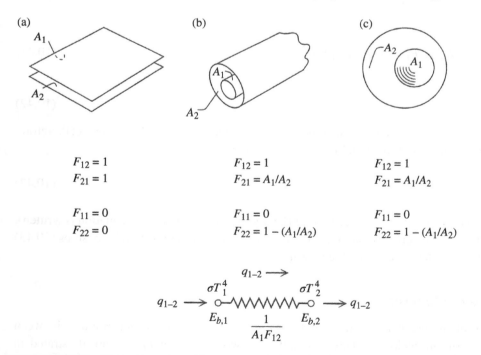

Figure 10.15 Examples of enclosures consisting of only two surfaces. (a) Two infinite parallel plates, (b) two infinite cylinders, and (c) two spheres.

infinite parallel-plate geometry is indeed the $A_1/A_2 \rightarrow 1$ limit of the two-cylinder and two-sphere geometries.

Although not essential to calculating q_{1-2}, the view factors F_{11} and F_{22} have also been listed to illustrate the observations written around the enclosure of Figure 10.14. These factors are zero in the case of plane and convex surfaces and finite in the case of concave surfaces.

Example 10.3   *View Factor Between Two Parallel Strips*

Calculate the geometric view factor between the infinitesimal area dA_1 and an infinitely long strip, A_2, of small width Δy situated in a plane parallel to dA_1. The width Δy is small relative to the distance R between the two strips.

Solution

With reference to Figure E10.3a and Eq. (10.33), we note that the angles ϕ_1 and ϕ_2 are equal and that $\cos \phi_1 = H/r$ and $r^2 = R^2 + x^2$. The segment of length R is the shortest distance between the dA_1 and the A_2 strip; therefore, according to a theorem of space geometry, the R segment is perpendicular to the x direction. The integral (10.33) becomes

$$F_{12} = \frac{1}{dA_1} \int_{dA_1} \int_{A_2} \frac{(H/r)^2}{\pi r^2} \, dA_1 \, dA_2 = \frac{H^2 \Delta y}{\pi} \int_{-\infty}^{\infty} \frac{dx}{(R^2 + x^2)^2}$$

$$= \frac{H^2 \Delta y}{\pi} \left| \frac{1}{2R^3} \tan^{-1} \frac{x}{R} + \frac{x}{2R^2(x^2 + R^2)} \right|_{-\infty}^{\infty}$$

$$= \frac{H^2 \Delta y}{2R^3} \tag{1}$$

in which we used $dA_2 = \Delta y \, dx$. Equation (1) can be rearranged using the terminology shown on the right side of the Figure E10.3a, namely,

$$\frac{H}{R} = \cos \phi \qquad \frac{\Delta y \cos \phi}{R} = d\phi \tag{2}$$

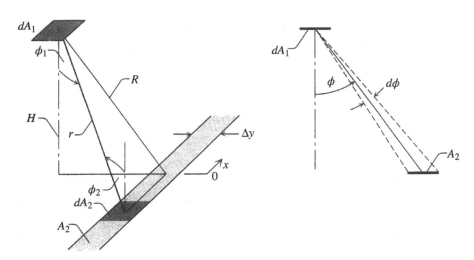

Figure E10.3a

That side of Figure E10.3a shows the projection of both dA_1 and A_2 on the plane defined by R and the normal to dA_1. Combining Eqs. (1) and (2), we obtain

$$F_{12} = \frac{\cos \phi}{2} d\phi = \frac{1}{2} d(\sin \phi) \tag{3}$$

Equation (3) is a powerful building block that can be used in the construction of the F_{12} values of considerably more complicated two-surface configurations. Worth noting is the fact that Eq. (3) is more general than the case defined in Figure E10.3a. The same view factor formula applies when dA_1 is not parallel to A_2, and/or dA_1 is itself a strip of infinitesimal width and finite length in the direction parallel to the A_2 strip. This general configuration is shown on the left side of Figure E10.3b.

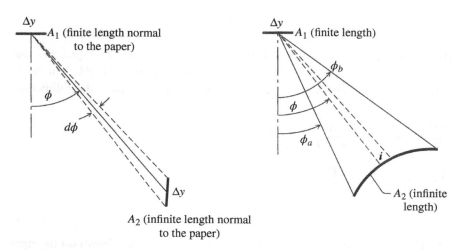

Figure E10.3b

The right side of Figure E10.3b shows the even more general case in which the width of the infinitely long strip A_2 is finite. This new strip can be reconstructed by gluing edge-to-edge a large number of Δy-narrow strips of the type indicated by i. The additivity property (10.40) can be invoked to write that the view factor from A_1 to the finite-width strip A_2 is equal to the sum of all the view factors F_{1i} given by Eq. (3):

$$F_{12} = \sum_i F_{1i} = \int_{\phi_a}^{\phi_b} \frac{1}{2} d(\sin \phi)$$

$$= \frac{1}{2}(\sin \phi_b - \sin \phi_a) \tag{4}$$

Example 10.4   *Enclosure Formed by Two Black Surfaces*
A stack of vertical circuit boards is shown in Figure E10.4. The arrangement is long in the direction perpendicular to the plane of Figure E10.4. Consider the two-surface enclosure formed in the space between two consecutive boards, and assume that both surfaces (1 = solid walls, 2 = ambient) are black. Assume that radiation is the only significant mode of heat transfer. Calculate the net heat transfer rate from $T_1 = 150\,°C$ to $T_2 = 25\,°C$ through the top "surface" of the enclosure.

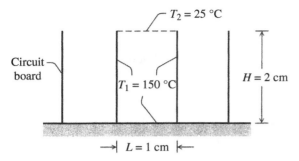

Figure E10.4

Solution

The solid walls represented by two boards and horizontal holder (T_1), and the ambient (T_2), form an enclosure in which the two surfaces are black. The net heat transfer rate from T_1 to T_2 through the enclosure is

$$q_{1-2} = A_1 F_{12}(\sigma T_1^4 - \sigma T_2^4)$$

The reciprocity property requires

$$A_1 F_{12} = A_2 F_{21}$$

in which $F_{21} = 1$ and $A_2 = LB$, where B is the long dimension of the boards, in the direction perpendicular to Figure E10.4. The net heat transfer rate per unit length is

$$\frac{1}{B}q_{1-2} = L\sigma(T_1^4 - T_2^4)$$

$$= (0.01 \text{ m}) \left(5.67 \times 10^{-8} \frac{\text{W}}{\text{m}^2 \cdot \text{K}^4} \right) [(150 + 273.15)^4 - (25 + 273.15)^4] \text{ K}^4$$

$$= 13.7 \text{ W/m}$$

An interesting aspect of this result is that it is independent of the board height H. Indeed, it is easy to obtain the same result when the boards are absent ($H = 0$ cm), that is, when the heat exchange is between two parallel and infinite black surfaces (the horizontal holder, 150 °C, and the ambient, 25 °C).

10.4 Diffuse-Gray Surfaces

10.4.1 Emissivity

The reason we devoted so much space to the analysis of heat transfer between black surfaces is that, under certain conditions, the same type of analysis – the same approach (structure) – can be used in the case of real or "nonblack" surfaces.[4] For example, the geometric view factors of Section 10.3.1 apply unchanged to nonblack surfaces. The challenge is to identify the special conditions that make the preceding analysis extendable to nonblack surfaces. Those conditions are condensed in the "diffuse-gray" surface model.

---

4 Surfaces that do not absorb all the incoming radiation.

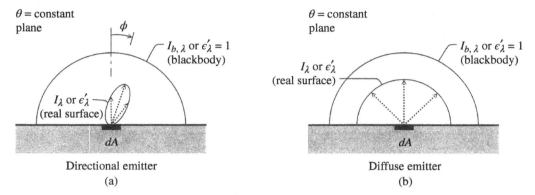

Figure 10.16 The emission characteristics of a directional emitter (a) and a diffuse emitter (b).

The intensity of the radiation emitted by a real surface of temperature T is only a fraction of the corresponding intensity of blackbody radiation. Relative to the intensity of monochromatic blackbody radiation considered in the preceding sections, $I_{b,\lambda}(\lambda, T)$, a real surface emits monochromatic radiation of intensity $I_\lambda(\lambda, T, \phi, \theta)$. Figure 10.16a and the arguments of $I_\lambda(\lambda, T, \phi, \theta)$ stress that, in general, the intensity of the radiation emitted by a real surface depends on the direction (ϕ, θ) in which each ray is pointed.

The inequality between the real surface and black surface intensities, $I_\lambda \leq I_{b,\lambda}$, is expressed alternatively as

$$\epsilon'_\lambda(\lambda, T, \phi, \theta) = \frac{I_\lambda(\lambda, T, \phi, \theta)}{I_{b,\lambda}} \leq 1 \tag{10.46}$$

where ϵ'_λ is the directional monochromatic emissivity of the real surface. The dimensionless number ϵ'_λ is a property of the emitting surface. The type of surface illustrated on Figure 10.16a, where ϵ'_λ and I_λ depend on the direction of the emitted ray, is recognized as a directional nonblack emitter. The prime notation (′) indicates that ϵ'_λ is a "directional" property, that is, a number associated with a certain direction (ϕ, θ).

Figure 10.16b shows a nonblack surface that emits uniformly in all the directions that point away from the infinitesimal surface area dA. The intensity of the emitted monochromatic radiation is smaller than in the blackbody limit, $I_{b,\lambda}$. This time, however, I_λ and the ratio $I_\lambda/I_{b,\lambda} = \epsilon'_\lambda$ are independent of the direction (ϕ, θ). The surface that has this feature is a diffuse nonblack emitter.

The comparison between the emission properties of real and black surfaces can be carried out also in terms of hemispherical quantities. Let $E_\lambda(\lambda, T)$ be the monochromatic emissive power of the real surface, that is, the number of watts emitted per unit of real surface area and wavelength, in all the directions intercepted (cut) by the hemisphere, in accordance with the integral (10.15):

$$E_\lambda(\lambda, T) = \int_{\theta=0}^{2\pi} \int_{\phi=0}^{\pi/2} I_\lambda(\lambda, T, \phi, \theta) \sin \phi \, \cos \phi \, d\phi \, d\theta \tag{10.47}$$

The ratio between $E_\lambda(\lambda, T)$ and the blackbody limit of the same quantity – the blackbody monochromatic emissive power $E_{b,\lambda}$ – is known as the monochromatic hemispherical emissivity of the real surface and is labeled ϵ_λ:

$$\epsilon_\lambda(\lambda, T) = \frac{E_\lambda(\lambda, T)}{E_{b,\lambda}(\lambda, T)} \leq 1 \tag{10.48}$$

A global measure of the difference between the emissions from a real surface and from a black surface of the same temperature is the total hemispherical emissivity ϵ:

$$\epsilon(T) = \frac{E(T)}{E_b(T)} \leq 1 \qquad (10.49)$$

The numerator in this definition is the emissive power (W/m²) of the real surface:

$$E(T) = \int_0^\infty E_\lambda(\lambda, T) \, d\lambda \qquad (10.50)$$

Combining Eqs. (10.48)–(10.50), we find that the total hemispherical emissivity ϵ is the $E_{b,\lambda}$-weighted average of the monochromatic hemispherical emissivity ϵ_λ:

$$\epsilon(T) = \frac{1}{\sigma T^4} \int_0^\infty E_\lambda(\lambda, T) \, d\lambda = \frac{1}{\sigma T^4} \int_0^\infty \epsilon_\lambda E_{b,\lambda} \, d\lambda \qquad (10.51)$$

Figure 10.17a shows the general outlook of the real-surface monochromatic hemispherical emissive power $E_\lambda(\lambda, T)$ next to the black surface-limiting curve $E_{b,\lambda}(\lambda, T)$. Figure 10.17b tells the same story in terms of monochromatic hemispherical emissivities (the curves shown on Figure 10.17a have been illustrated to scale based on the ϵ_λ curves assumed on the right side).

In most instances, the measured $\epsilon_\lambda(\lambda, T)$ value of a real surface is a complicated function of wavelength. The simplest approximation of this function is the constant-ϵ_λ dashed line shown on Figure 10.17b, namely, the statement

$$\epsilon_\lambda(\lambda, T) \cong \epsilon_\lambda(T) \quad \text{or} \quad \epsilon_\lambda \neq \text{function } (\lambda) \text{ (gray)} \qquad (10.52)$$

This approximation is appropriate in instances where the dominant values of the $\epsilon_\lambda(\lambda, T)$ function are located in that portion of the λ spectrum that is responsible for the bulk of the blackbody emissive power σT^4. The example illustrated in Figure 10.17 is one such case: note the position of the $\lambda_1-\lambda_2$ interval on the two abscissas of that figure. The $\epsilon_\lambda = 0.6$ value was chosen on Figure 10.17b in such a way that it approximates

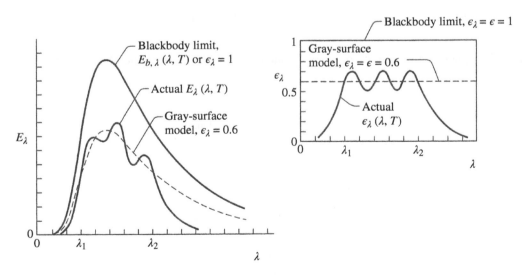

Figure 10.17 The gray-surface model: the monochromatic emissive power (a) and the corresponding monochromatic hemispherical emissivity (b).

the actual $\epsilon_\lambda(\lambda, T)$ curve over those wavelengths that contribute the most to the actual emissive power of the nonblack surface.

When the approximation (10.52) is justified, the "reduced" Planck curve $\epsilon_\lambda E_{b,\lambda}$ (in which $\epsilon_\lambda \neq \text{function}(\lambda)$) is an adequate approximation of the actual E_λ curve of the real surface. On Figure 10.17a, the reduced Planck curve was drawn with a dashed line using the λ-independent reduction factor $\epsilon_\lambda = 0.6$.

A gray surface or gray body of temperature T is the surface whose monochromatic hemispherical emissivity is independent of wavelength (i.e. a constant if T is fixed), Eq. (10.52). The weighted average indicated on the extreme right-hand side of Eq. (10.51) shows further that the total hemispherical emissivity of a gray surface is equal to its monochromatic hemispherical emissivity:

$$\epsilon = \epsilon_\lambda \quad (\text{gray}) \tag{10.53}$$

In the real-surface analyses presented in this chapter, it will be assumed that the surface is a diffuse emitter and that it can be modeled as gray. It will also be assumed that the surface is a diffuse absorber and a diffuse reflector, in accordance with the definitions that will be given in Figures 10.18 and 10.19 and the accompanying discussion.

The diffuse-gray model approximates fairly well the behavior of many surfaces encountered in practice, for example, copper, aluminum oxide, paints, and paper. Tables 10.3 and 10.4 show a compilation

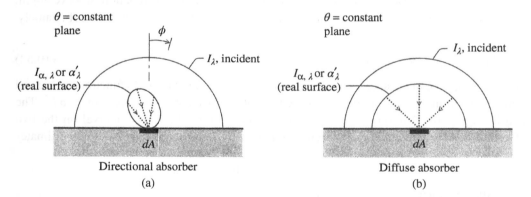

Figure 10.18 The absorption characteristics of a directional absorber (a) and a diffuse absorber (b).

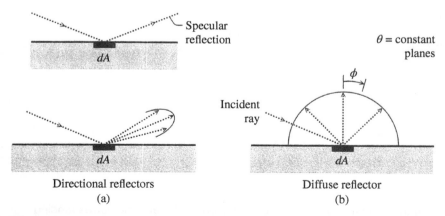

Figure 10.19 The reflection characteristics of a directional reflector (a) and a diffuse reflector (b).

of total hemispherical emissivity values (ϵ) of many real surfaces that have been modeled as diffuse and gray. Clean, well-polished metallic surfaces are characterized by small ϵ values. Nonmetallic surfaces, on the other hand, have high emissivities; in fact, some of these come close to satisfying the blackbody model $\epsilon = 1$ (e.g. soot, smooth glass, frost). Metallic surfaces that, in time, become coated with oxides and other impurities also acquire considerably higher emissivity values.

In general, then, the total hemispherical emissivities listed in Tables 10.3 and 10.4 are functions of the emitter temperature, type of material, and degree of surface smoothness and cleanliness. There are also difficulties that stem from the qualitative description of the surface condition in Tables 10.3 and 10.4, for example, "polished" versus "highly polished," or "oxidized" versus "black oxidized" (copper, Table 10.3). Furthermore, the temperature effect is itself a function of the type of material that makes up the surface. For example, the total hemispherical emissivity of metallic surfaces increases with the absolute temperature. At cryogenic temperatures (5–100 K), the ϵ value of a metallic surface is almost proportional to the temperature T. The ϵ values of nonmetallic surfaces can either increase or decrease as T increases.

10.4.2 Absorptivity and Reflectivity

A real opaque surface absorbs only a portion of the energy brought by the incoming (incident) radiation. The remainder – the portion that is not absorbed – is reflected back toward the surroundings. Let $I_\lambda(\lambda, T, \phi, \theta)$ represent the intensity of the general (nonblack) incident radiation that strikes the surface element from the direction (ϕ, θ). The relative size of the portion that is absorbed at the surface, $I_{a,\lambda}(\lambda, T, \phi, \theta)$, is indicated by the directional monochromatic absorptivity α'_λ:

$$\alpha'_\lambda(\lambda, T, \phi, \theta) = \frac{I_{a,\lambda}(\lambda, T, \phi, \theta)}{I_\lambda(\lambda, T, \phi, \theta)} \tag{10.54}$$

A surface is called a directional absorber when the directional monochromatic absorptivity α'_λ varies with the orientation of the incoming ray (Figure 10.18a). When α'_λ exhibits the same value regardless of the direction (ϕ, θ), the surface is called a diffuse absorber (Figure 10.18b). The incident intensity I_λ that strikes dA generally depends on the direction from which it comes, because its magnitude is determined by the various emitters that can look toward dA. In Figure 10.18 the size of I_λ was drawn constant (independent of ϕ and θ) only to facilitate the comparison between it and its absorbed fraction $I_{a,\lambda}$, i.e. $I_{a,\lambda}$ is generally smaller than the incident I_λ.

In a way that mimics the ($\epsilon'_\lambda, \epsilon_\lambda, \epsilon$) string of emissivity definitions reviewed in Section 10.4.1, we can define the monochromatic hemispherical absorptivity α_λ:

$$\alpha_\lambda(\lambda, T) = \frac{G_{a,\lambda}(\lambda, T)}{G_\lambda(\lambda, T)} \tag{10.55}$$

The denominator G_λ (W/m^2 · m) is the monochromatic irradiation, or the number of watts that strike the unit area from all the directions and per unit wavelength. G_λ is related to the intensity of the incident radiation (I_λ of Eq. (10.54)) through the double integral (see Eq. (10.47))

$$G_\lambda(\lambda, T) = \int_{\theta=0}^{2\pi} \int_{\phi=0}^{\pi/2} I_\lambda(\lambda, T, \phi, \theta) \sin\phi \, \cos\phi \, d\phi \, d\theta \tag{10.56}$$

The numerator $G_{a,\lambda}$ in the definition (10.55) is the absorbed portion of the monochromatic irradiation G_λ:

$$G_{a,\lambda}(\lambda, T) = \int_{\theta=0}^{2\pi} \int_{\phi=0}^{\pi/2} I_{a,\lambda}(\lambda, T, \phi, \theta) \sin\phi \, \cos\phi \, d\phi \, d\theta \tag{10.57}$$

Table 10.3 Metallic surfaces: representative values of total hemispherical emissivity.

Material		$\epsilon(T)$
Aluminum	Crude	0.07–0.08 (0–200 °C)
	Foil, bright	0.01 (−9 °C), 0.04 (1 °C), 0.087 (100 °C)
	Highly polished	0.04–0.05 (1 °C)
	Ordinarily rolled	0.035 (100 °C), 0.05 (500 °C)
	Oxidized	0.11 (200 °C), 0.19 (600 °C)
	Roughed with abrasives	0.044–0.066 (40 °C)
	Unoxidized	0.022 (25 °C), 0.06 (500 °C)
Bismuth	Unoxidized	0.048 (25 °C), 0.061 (100 °C)
Brass	After rolling	0.06 (30 °C)
	Browned	0.5 (20–300 °C)
	Polished	0.03 (300 °C)
Chromium	Polished	0.07 (150 °C)
	Unoxidized	0.08 (100 °C)
Cobalt	Unoxidized	0.13 (500 °C), 0.23 (1000 °C)
Copper	Black oxidized	0.78 (40 °C)
	Highly polished	0.03 (1 °C)
	Liquid	0.15
	Matte	0.22 (40 °C)
	New, very bright	0.07 (40–100 °C)
	Oxidized	0.56 (40–200 °C), 0.61 (260 °C), 0.88 (540 °C)
	Polished	0.04 (40 °C), 0.05 (260 °C), 0.17 (1100 °C), 0.26 (2800 °C)
	Rolled	0.64 (40 °C)
Gold	Polished or electrolytically deposited	0.02 (40 °C), 0.03 (1100 °C)
Inconel	Sandblasted	0.79 (800 °C), 0.91 (1150 °C)
	Stably oxidized	0.69 (300 °C), 0.82 (1000 °C)
	Untreated	0.3 (40–260 °C)
	Rolled	0.69 (800 °C), 0.88 (1150 °C)
Inconel X		0.74–0.81 (100–440 °C)
	Stably oxidized	0.89 (300 °C), 0.93 (1100 °C)

Table 10.3   (Continued)

Material		$\epsilon(T)$
Iron (see also steel)		
	Cast	0.21 (40 °C)
	Cast, freshly turned	0.44 (40 °C), 0.7 (1100 °C)
	Galvanized	0.22–0.28 (0–200 °C)
	Molten	0.02–0.05 (1100 °C)
	Plate, rusted red	0.61 (40 °C)
	Pure polished	0.06 (40 °C), 0.13 (540 °C), 0.35 (2800 °C)
	Red iron oxide	0.96 (40 °C), 0.67 (540 °C), 0.59 (2800 °C)
	Rough ingot	0.95 (1100 °C)
	Smooth sheet	0.6 (1100 °C)
	Wrought, polished	0.28 (40–260 °C)
Lead	Oxidized	0.28 (0–200 °C)
	Unoxidized	0.05 (100 °C)
Magnesium		0.13 (260 °C), 0.18 (310 °C)
Mercury		0.09 (0 °C), 0.12 (100 °C)
Molybdenum		0.071 (100 °C), 0.13 (1000 °C), 0.19 (1500 °C)
	Oxidized	0.78–0.81 (300–540 °C)
Monel	Oxidized	0.43 (20 °C)
	Polished	0.09 (20 °C)
Nichrome	Rolled	0.36 (800 °C), 0.8 (1150 °C)
	Sandblasted	0.81 (800 °C), 0.87 (1150 °C)
Nickel	Electrolytic	0.04 (40 °C), 0.1 (540 °C), 0.28 (2800 °C)
	Oxidized	0.31–0.39 (40 °C), 0.67 (540 °C)
	Wire	0.1 (260 °C), 0.19 (1100 °C)
Platinum	Oxidized	0.07 (260 °C), 0.11 (540 °C)
	Unoxidized	0.04 (25 °C), 0.05 (100 °C), 0.15 (1000 °C)
Silver	Polished	0.01 (40 °C), 0.02 (260 °C), 0.03 (540 °C)
Steel	Calorized	0.5–0.56 (40–540 °C)
	Cold rolled	0.08 (100 °C)
	Ground sheet	0.61 (1100 °C)
	Oxidized	0.79 (260–540 °C)
	Plate, rough	0.94–0.97 (40–540 °C)

(Continued)

Table 10.3 (Continued)

Material		$\epsilon(T)$
	Polished	0.07 (40 °C), 0.1 (260 °C), 0.14 (540 °C), 0.23 (1100 °C), 0.37 (2800 °C)
	Rolled sheet	0.66 (40 °C)
	Type 347, oxidized	0.87–0.91 (300–1100 °C)
	Type AISI 303, oxidized	0.74–0.87 (300–1100 °C)
	Type 310, oxidized and rolled	0.56 (800 °C), 0.81 (1150 °C)
	Sandblasted	0.82 (800 °C), 0.93 (1150 °C)
Stellite		0.18 (20 °C)
Tantalum		0.19 (1300 °C), 0.3 (2500 °C)
Tin	Unoxidized	0.04–0.05 (25–100 °C)
Tungsten	Filament	0.03 (40 °C), 0.11 (540 °C), 0.39 (2800 °C)
Zinc	Oxidized	0.11 (260 °C)
	Polished	0.02 (40 °C), 0.03 (260 °C)

Source: Data collected from Gubareff et al. [9] and Touloukian and Ho [10].

The total hemispherical absorptivity is defined as the ratio

$$\alpha(T) = \frac{G_a(T)}{G(T)} \tag{10.58}$$

where $G(T)$ is the total irradiation, which was noted already in Figure 10.2a. The $G_a(T)$ numerator is the absorbed portion of the total irradiation. Both G and G_a have units of heat flux (W/m²). The total irradiation is obtained by integrating the monochromatic irradiation over the entire spectrum:

$$G(T) = \int_0^\infty G_\lambda(\lambda, T)\, d\lambda \tag{10.59}$$

Similarly, the absorbed portion G_a is the same as the λ integral of the absorbed monochromatic irradiation $G_{a,\lambda}$:

$$G_a(T) = \int_0^\infty G_{a,\lambda}(\lambda, T)\, d\lambda \tag{10.60}$$

Combined, Eqs. (10.55) and (10.58) to (10.60) demonstrate that the total hemispherical absorptivity α is the G_λ-weighted average of the monochromatic hemispherical absorptivity α_λ:

$$\alpha(T) = \frac{1}{G(T)} \int_0^\infty \alpha_\lambda(\lambda, T)\, G_\lambda(\lambda, T)\, d\lambda \tag{10.61}$$

The difference between the total irradiation G and the absorbed portion G_a is the reflected portion G_r. We did this accounting once in Eq. (10.2), which now can be restricted to an opaque surface and written in terms

Table 10.4 Nonmetallic surfaces: representative values of total hemispherical emissivity.

Material	$\epsilon(T)$
Bricks	
Chrome refractory	0.94 (540 °C), 0.98 (1100 °C)
Fire clay	0.75 (1400 °C)
Light buff	0.8 (540 °C), 0.53 (1100 °C)
Magnesite refractory	0.38 (1000 °C)
Sand lime red	0.59 (1400 °C)
Silica	0.84 (1400 °C)
Various refractories	0.71–0.88 (1100 °C)
White refractory	0.89 (260 °C), 0.68 (540 °C)
Building materials	
Asbestos, board	0.96 (40 °C)
Asphalt pavement	0.85–0.93 (40 °C)
Clay	0.39 (20 °C)
Concrete, rough	0.94 (0–100 °C)
Granite	0.44 (40 °C)
Gravel	0.28 (40 °C)
Gypsum	0.9 (40 °C)
Marble, polished	0.93 (40 °C)
Mica	0.75 (40 °C)
Plaster	0.93 (40 °C)
Quartz	0.89 (40 °C), 0.58 (540 °C)
Sand	0.76 (40 °C)
Sandstone	0.83 (40 °C)
Slate	0.67 (40–260 °C)
Carbon	
Baked	0.52–0.79 (1000–2400 °C)
Filament	0.95 (260 °C)
Graphitized	0.76–0.71 (100–500 °C)
Rough	0.77 (100–320 °C)
Soot (candle)	0.95 (120 °C)
Soot (coal)	0.95 (20 °C)
Unoxidized	0.8 (25–500 °C)

(Continued)

Table 10.4 (Continued)

Material	$\epsilon(T)$
Ceramics	
Coatings	
Alumina on inconel	0.65 (430 °C), 0.45 (1100 °C)
Zirconia on inconel	0.62 (430 °C), 0.45 (1100 °C)
Earthenware, glazed	0.9 (1 °C)
Matte	0.93 (1 °C)
Porcelain	0.92 (40 °C)
Refractory, black	0.94 (100 °C)
Light buff	0.92 (100 °C)
White Al_2O_3	0.9 (100 °C)
Cloth	
Cotton	0.77 (20 °C)
Silk	0.78 (20 °C)
Glass	
Convex D	0.8–0.76 (100–500 °C)
Fused quartz	0.75–0.8 (100–500 °C)
Nonex	0.82–0.78 (100–500 °C)
Pyrex	0.8–0.9 (40 °C)
Smooth	0.92–0.95 (0–200 °C)
Waterglass	0.96 (20 °C)
Ice, smooth	0.92 (0 °C)
Oxides	
Al_2O_3	0.35–0.54 (850–1300 °C)
C_2O	0.27 (850–1300 °C)
Cr_2O_3	0.73–0.95 (850–1300 °C)
Fe_2O_3	0.57–0.78 (850–1300 °C)
MgO	0.29–0.5 (850–1300 °C)
NiO	0.52–0.86 (500–1200 °C)
ZnO	0.3–0.65 (850–1300 °C)
Paints	
Aluminum	0.27–0.7 (1–100 °C)
namel, snow white	0.91 (40 °C)
Lacquer	0.85–0.93 (40 °C)
Lampblack	0.94–0.97 (40 °C)
Oil	0.89–0.97 (0–200 °C)
White	0.89–0.97 (40 °C)

Table 10.4 (Continued)

Material	$\epsilon(T)$
Paper, white	0.95 (40 °C), 0.82 (540 °C)
Roofing materials	
Aluminum surfaces	0.22 (40 °C)
Asbestos cement	0.65 (1400 °C)
Bituminous felt	0.89 (1400–2800 °C)
Enameled steel, white	0.65 (1400 °C)
Galvanized iron, dirty	0.90 (1400–2800 T)
New	0.42 (1400 °C)
Roofing sheet, brown	0.8 (1400 °C)
Green	0.87 (1400 °C)
Tiles, uncolored	0.63 (1400–2800 °C)
Brown	0.87 (1400 °C)
Black	0.94 (1400 °C)
Asbestos cement	0.66 (1400–2800 °C)
Weathered asphalt	0.88 (1400–2800 °C)
Rubber	
Hard, black, glossy surface	0.95 (40 °C)
Soft, gray	0.86 (40 °C)
Snow	
Fine	0.82 (−10 T)
Frost	0.98 (0 °C)
Granular	0.89 (−10 °C)
Soils	0.92–0.96 (0–20 T)
Black loam	0.66 (20 °C)
Plowed field	0.38 (20 T)
Water	0.92–0.96 (0–40 T)
Wood	
Beech	0.91 (70 °C)
Oak, planed	0.91 (40 T)
Sawdust	0.75 (40 °C)
Spruce, sanded	0.82 (100 T)

Source: Data collected from Gubareff et al. [9] and Touloukian and Ho [10].

of the total hemispherical reflectivity $\rho = 1 - \alpha$:

$$G_r = G - G_a = (1 - \alpha)G = \rho G \tag{10.62}$$

The reflected total irradiation portion G_r is the hemispherical solid angle integral of all the reflected "offspring" generated by each incident ray. In most cases there are many reflected rays (directions) associated with a single incident ray. This feature is illustrated in Figure 10.19. The surface is a diffuse reflector when the intensity of the reflected radiation is uniform in all directions, that is, independent of ϕ and θ. Conversely, when the reflected rays prefer one or a limited bundle of directions, the surface is called a directional reflector. The extreme directional reflector shown in the upper-left corner of Figure 10.19 – the *specular*,[5] or mirrorlike, reflector – directs the reflected portion of each incident ray into a single direction.

To summarize, the *diffuse-gray model* that we adopt for the remainder of this chapter describes a surface that is simultaneously as follows:

1 Gray, Eq. (10.52) and Figure 10.17
2 Diffuse emitter, Figure 10.16b
3 Diffuse absorber, Figure 10.18b
4 Diffuse reflector, Figure 10.19b
5 Opaque

The total hemispherical emissivity of the diffuse-gray surface can be found in Tables 10.3 and 10.4. We now turn to the problem of estimating the total hemispherical absorptivity α, which is equivalent to determining the total hemispherical reflectivity ($\rho = 1 - \alpha$).

10.4.3 Kirchhoff's Law

Consider first the inner face (Σ) of the large surface of the enclosure shown in Figure 10.20. This face is nonblack and isothermal, at the temperature T_Σ. In the discussion associated with Figure 10.2b, we learned that a cavity bounded by a nonblack surface that is isothermal absorbs all the incident radiation that enters through a small orifice. Some of the radiation (photon gas) that fills the cavity of Figure 10.2b escapes through the orifice. Thermodynamically it can be argued that, from the viewpoint of an external observer, the radiation that streams out through the aperture can only be a function of the temperature of the cavity wall, that is, a blackbody radiation. This argument leads to the conclusion that a cavity surrounded by an isothermal wall fills with blackbody radiation of the same temperature as the wall, regardless of whether the wall is black or nonblack (good absorber or a good reflector). This means that the cavity surrounded by the surface Σ fills with blackbody radiation of temperature T_Σ.

Any elementary area dA suspended inside this cavity is struck by blackbody radiation of monochromatic intensity $I_{b,\lambda}(\lambda, T_\Sigma)$. Now consider the solid body of temperature T_A that is delineated by the nonblack surface A. If this body is introduced in the cavity bounded by the surface Σ, then the entire surface A will absorb heat transfer at the rate [2]:

$$q_{\text{absorbed by } A} = \int_{A=0}^{A} \int_{\omega=0}^{2\pi} \int_{\lambda=0}^{\infty} \alpha'_\lambda(\lambda, T_A, \phi, \theta) I_{b,\lambda}\left(\lambda, T_\Sigma\right) \cos\phi \, d\lambda \, d\omega \, dA \tag{10.63}$$

---

5 The term "specular" has its origin in the Latin adjective *specularis* (like a minor), which is related to the noun *speculum* (mirror, made of polished metal). The modern English word "speculate," which comes from the feminine Latin noun *specula* (a lookout, watchtower; or a little hope, ray of hope), is also related to these terms.

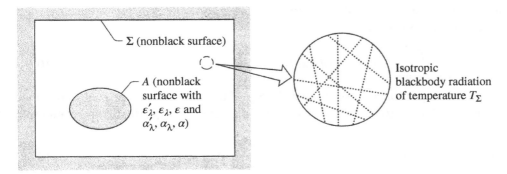

Figure 10.20 Two-surface enclosure for the derivation of Kirchhoff's law, $\alpha'_\lambda = \epsilon'_\lambda$.

The second of these nested integrals accounts for the arrival of blackbody radiation from all the directions of the hemisphere ($0 \leq \omega \leq 2\pi$) based on the area element dA; this particular integral and its notation were spelled out in Eqs. (10.14) and (10.15). The factor α'_λ in the integrand is the directional monochromatic absorptivity of A.

Let $\epsilon'_\lambda(\lambda, T_A, \phi, \theta)$ be the directional monochromatic emissivity of the surface A. The total heat current (watts) emitted by the entire area A is the result of another triple integral:

$$q_{\text{emitted by } A} = \int_{A=0}^{A} \int_{\omega=0}^{2\pi} \int_{\lambda=0}^{\infty} \epsilon'_\lambda(\lambda, T_A, \phi, \theta) I_{b,\lambda}(\lambda, T_A) \cos\phi \, d\lambda \, d\omega \, dA \tag{10.64}$$

In the integrand is the monochromatic intensity of blackbody radiation of temperature T_A, which is not the same as the $I_{b,\lambda}(\lambda, T_\Sigma)$ in the integrand of Eq. (10.63).

When the body (A) reaches thermal equilibrium with the enclosure wall (Σ), the temperature T_A becomes steady and equal to the outer wall temperature:

$$T_A = T_\Sigma \quad \text{(constant)} \tag{10.65}$$

The steadiness of T_A implies that the net heat transfer rate through the surface A is zero, meaning that the absorbed current, Eq. (10.63), matches (and cancels) the effect of the emitted current, Eq. (10.64). In addition, the thermal equilibrium condition $T_A = T_\Sigma$ guarantees that $I_{b,\lambda}(\lambda, T_A) = I_{b,\lambda}(\lambda, T_\Sigma)$.

All of these observations lead to the conclusion that, at equilibrium, the integrals of Eqs. (10.63) and (10.64) are equal and that their equality requires

$$\alpha'_\lambda(\lambda, T_A, \phi, \theta) = \epsilon'_\lambda(\lambda, T_A, \phi, \theta) \tag{10.66}$$

This result is known as Kirchhoff's law [11]. It states that the α'_λ value of a nonblack surface is always equal to the ϵ'_λ value of the same surface when the surface is in thermal equilibrium with the radiation that impinges on it. In problems of heat transfer, surfaces are almost never in equilibrium with their surroundings.[6] In spite of this, Eq. (10.66) is considered to be sufficiently accurate and is an integral part of the foundation of radiation heat transfer theory.

From the point of view of a user of the diffuse-gray model, an important question is whether the total absorptivity α can be calculated based on Kirchhoff's law, that is, based solely on emissivity information.

---

6 Recall that under the conditions of thermal equilibrium, the fundamental heat transfer problem becomes trivial, Eq. (1.14).

To answer this question, we note that the hemispherical monochromatic absorptivity α_λ is a weighted average of the directional monochromatic absorptivity α'_λ:

$$\alpha_\lambda(\lambda, T) = \frac{1}{G_\lambda(\lambda, T)} \int_{\theta=0}^{2\pi} \int_{\phi=0}^{\pi/2} \alpha'_\lambda(\lambda, T, \phi, \theta) I_\lambda(\lambda, T, \phi, \theta) \sin \phi \, \cos \phi \, d\phi \, d\theta \tag{10.67}$$

This relationship follows from Eqs. (10.54), (10.55), and (10.57). If the surface is a diffuse absorber, then α'_λ is not a function of the direction (ϕ, θ) (see Figure 10.18b); and, after using Eq. (10.56), the relationship (10.67) reduces to

$$\alpha_\lambda(\lambda, T) = \alpha'_\lambda(\lambda, T) \quad \text{(diffuse absorber)} \tag{10.68}$$

Similarly, we find that the hemispherical monochromatic emissivity ϵ_λ is related to the directional monochromatic emissivity ϵ'_λ through the hemispherical solid angle integral

$$\epsilon_\lambda(\lambda, T) = \frac{1}{E_{b,\lambda}(\lambda, T)} \int_{\theta=0}^{2\pi} \int_{\phi=0}^{\pi/2} \epsilon'_\lambda(\lambda, T, \phi, \theta) I_{b,\lambda}(\lambda, T) \sin \phi \, \cos \phi \, d\phi \, d\theta \tag{10.69}$$

This is the result of combining Eqs. (10.46)–(10.48). For the diffuse emitter illustrated in Figure 10.16b, the directional monochromatic emissivity is not a function of ϕ and θ. Therefore, pulling ϵ'_λ in front of the integral (10.69), and recognizing Eqs. (10.15) and (10.16), we obtain

$$\epsilon_\lambda(\lambda, T) = \epsilon'_\lambda(\lambda, T) \quad \text{(diffuse emitter)} \tag{10.70}$$

In conclusion, for a surface that is both a diffuse absorber and a diffuse emitter, Kirchhoff's law states that

$$\alpha_\lambda(\lambda, T) = \epsilon_\lambda(\lambda, T) \quad \text{(diffuse absorber and emitter)} \tag{10.71}$$

This statement follows from Eqs. (10.66), (10.68), and (10.70). If, in addition, the surface can be modeled as gray, ϵ_λ is independent of wavelength, and (cf. Eq. (10.53)) $\epsilon_\lambda = \epsilon(T)$. From Eq. (10.71) we conclude that α_λ is also λ independent and, more specifically, that

$$\alpha_\lambda = \epsilon(T) \tag{10.72}$$

This α_λ estimate can be substituted in the integrand on the right side of Eq. (10.61) to obtain

$$\alpha(T) = \epsilon(T) \quad \text{(diffuse-gray surface)} \tag{10.73}$$

Equation (10.73) represents Kirchhoff's law for a surface that obeys the diffuse-gray model. It states that the total hemispherical absorptivity α of a surface of temperature T is equal to the total hemispherical emissivity $\epsilon(T)$ of the same surface. To the user of the diffuse-gray model, Eq. (10.73) means that α can be estimated based on the ϵ information assembled in Tables 10.3 and 10.4, provided that the incident radiation has the same temperature as the surface temperature T.

In applications in which the temperature of the incident radiation does not differ substantially from that of the target surface, the thermal equilibrium restriction can be overlooked so that $\alpha(T)$ can be estimated (approximately) by using the $\epsilon(T)$ value listed in Tables 10.3 and 10.4. This approximation may not be appropriate when the incident radiation and the target surface have vastly different temperatures. In the case of granular snow, for example, Table 10.4 shows that $\epsilon = 0.89$ when the snow temperature is $T = -10\,°\text{C}$. Kirchhoff's law assures us that $\alpha = 0.89$ when the snow temperature is $-10\,°\text{C}$ and the temperature of the incident radiation is not much different than $-10\,°\text{C}$. In those situations, the snow approaches the black surface model. On the other hand, when the incident radiation has a much higher temperature (e.g. solar radiation, $T \cong 5800\,\text{K}$), the absorptivity of the snow is much lower, $\alpha \sim 0.4$, which is why the snow looks bright. The same observation holds true for the absorptivities exhibited by white paper.

To review the progress made in this section, the general form of Kirchhoff's law refers to directional monochromatic quantities, Eq. (10.66). The equality between the total hemispherical quantities α and ϵ, Eq. (10.73), was derived specifically for a diffuse-gray surface. It turns out that Eq. (10.73) holds not only for diffuse-gray surfaces but also for several other special sets of circumstances. These can be found in more advanced radiation texts [8].

10.4.4 Two-Surface Enclosures

Consider now the problem of estimating the net heat transfer rate between two diffuse-gray surfaces that form an enclosure. The areas (A_1, A_2), temperatures (T_1, T_2), and total hemispherical emissivities (ϵ_1, ϵ_2) are specified (Figure 10.21). Assume also that the smaller of the two surfaces, A_1, is not concave, so that $F_{11} = 0$. The object of the following analysis is the calculation of the net heat transfer rate q_{1-2}. This problem is related to the enclosure analyzed in Section 10.3.3, in which both surfaces were modeled as black. This time, both surfaces reflect some of the radiation that impinges on them.

Figure 10.22 shows an expanded view of the area element dA_1. Let G_1 (W/m^2) be the total irradiation that arrives at the dA_1 location. Proceeding in the opposite direction (i.e. departing from dA_1 toward the cavity) is the reflected portion $\rho_1 G_1$ plus the heat flux emitted by dA_1 itself, $\epsilon_1 E_{b,1}$. The total one-way heat flux that departs from dA_1 represents the surface radiosity and is labeled J_1 (W/m^2):

$$J_1 = \rho_1 G_1 + \epsilon_1 E_{b,1} \tag{10.74}$$

The difference between the heat flux that leaves, J_1, and the heat flux that arrives, G_1, is the net heat flux that leaves dA_1:

$$q_1'' = J_1 - G_1 \tag{10.75}$$

Eliminating G_1 between Eqs. (10.74) and (10.75), and recognizing that for a diffuse-gray surface $\rho_1 = 1 - \alpha_1 = 1 - \epsilon_1$, we obtain, in order,

$$
\begin{aligned}
q_1'' &= J_1 - \frac{J_1 - \epsilon_1 E_{b,1}}{\rho_1} \\
&= \frac{\epsilon_1}{1 - \epsilon_1}(E_{b,1} - J_1)
\end{aligned}
\tag{10.76}
$$

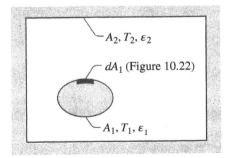

Figure 10.21 Enclosure defined by two diffuse-gray surfaces, and the thermal resistance analog of the net heat transfer rate from A_1 to A_2.

The net heat current that leaves the entire A_1 area is simply $q_1 = q_1'' A_1$, or [12]

$$q_1 = \frac{\epsilon_1 A_1}{1 - \epsilon_1}(E_{b,1} - J_1) \tag{10.77}$$

It pays to examine the structure of this analytical result. The net heat current that leaves the A_1 surface, q_1, must be provided by an external agent (a heater); this current must be "pumped" through the A_1 surface, that is, from its back side to the side that faces the enclosure. Equation (10.77) shows that what drives the q_1 current through the A_1 surface is the potential difference $E_{b,1} - J_1$. What impedes the passage of q_1 through A_1 is the "internal resistance" $(1 - \epsilon_1)/\epsilon_1 A_1$,

$$R_i = \frac{1 - \epsilon}{\epsilon A} \tag{10.78}$$

These observations are summarized in the electrical resistance analog shown in Figure 10.21. The diffuse-gray surface A_1 is represented by two nodes: by the radiosity node J_1 on the enclosure side of the surface A_1 and by the blackbody emissive power $E_{b,1} = E_b(T_1)$ on the back side of the same surface. An analogous internal resistance, $(1 - \epsilon_2)/\epsilon_2 A_2$, connects the J_2 and $E_{b,2}$ nodes that account for the surface A_2.

It remains to analyze what happens inside the enclosure, that is, in the space confined between A_1 and A_2. These two surfaces make themselves visible through their respective radiosities. The surface A_1, for example, sends out the total heat current $J_1 A_1$ in all the directions to which A_1 has access. Furthermore, the radiosity heat flux is distributed uniformly over the hemispherical solid angle based on each area element dA_1. The diffuse character of J_1 is due to the assumed diffuse-gray model: note the diffuse emission and diffuse reflection that contribute to J_1 in Figure 10.22.

The total heat current $J_1 A_1$ that originates from A_1 has all the features of the one-way heat current $E_{b,1} A_1$ discussed in Section 10.3.1. That discussion can be repeated word for word here, to conclude that the share of the one-way current $J_1 A_1$ that is intercepted by the surface A_2 is

$$q_{1 \to 2} = J_1 A_1 F_{12} \tag{10.79}$$

This one-way radiation current is also equal to $J_1 A_2 F_{21}$ because of the reciprocity relation $A_1 F_{12} = A_2 F_{21}$. In using the geometric view factors of Section 10.3.1, we are finally reaping the rewards promised by the diffuse-gray model. It is because of all the features of the diffuse-gray model that the view factor methodology can be applied to the present problem.

If we are on the surface A_2 and look toward the enclosure, we see that out of the entire one-way radiation that leaves A_2 (namely, $J_2 A_2$), only a certain portion is intercepted by the surface A_1:

$$q_{2 \to 1} = J_2 A_2 F_{21} = J_2 A_1 F_{12} \tag{10.80}$$

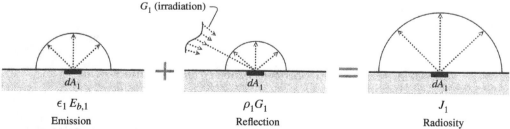

Figure 10.22 The surface radiosity as the superposition of diffuse emission and diffuse reflection.

The *net* heat current that proceeds in the direction $A_1 \rightarrow A_2$ is therefore

$$q_{1-2} = q_{1\rightarrow 2} - q_{2\rightarrow 1} = A_1 F_{12}(J_1 - J_2) \tag{10.81}$$

Comparing this last result with the resistance network started in Figure 10.21, we conclude that the net heat current from node J_1 to node J_2 is impeded by the purely geometric resistance $1/A_1 F_{12}$, which is also equal to $1/A_2 F_{21}$. This resistance is inserted between the internal resistances of the two diffuse-gray surfaces.

The series of three resistances makes it easy to express the net current q_{1-2} in terms of the overall blackbody emissive power difference $E_{b,1} - E_{b,1}$ or $\sigma(T_1^4 - T_2^4)$:

$$q_{1-2} = \frac{\sigma(T_1^4 - T_2^4)}{\dfrac{1 - \epsilon_1}{\epsilon_1 A_1} + \dfrac{1}{A_1 F_{12}} + \dfrac{1 - \epsilon_2}{\epsilon_2 A_2}} \tag{10.82}$$

Note that the continuity of energy through the surface A_1 requires that $q_{1-2} = q_1$, where q_1 is the heat transfer rate supplied to the back side of A_1, Eq. (10.77). If q_2 is the label of the heat transfer rate supplied by an external agent to the back side of the surface A_2, then the same energy continuity argument requires $q_{1-2} = -q_2$ and, in summary,

$$q_1 = q_{1-2} = -q_2 \tag{10.83}$$

Three important configurations that fall in the "two-surface enclosure" category were illustrated in Figure 10.15. Substituting into the general solution (10.82) the $F_{12} = 1$ value listed under each of those configurations, it is easy to derive the following special relationships:

Two infinite parallel plates $(A_1 = A_2 = A)$:

$$q_{1-2} = \frac{\sigma A(T_1^4 - T_2^4)}{\dfrac{1}{\epsilon_1} + \dfrac{1}{\epsilon_2} - 1} \tag{10.84}$$

The annular space between two infinite cylinders or between two spheres:

$$q_{1-2} = \frac{\sigma A_1(T_1^4 - T_2^4)}{\dfrac{1}{\epsilon_1} + \dfrac{A_1}{A_2}\left(\dfrac{1}{\epsilon_2} - 1\right)} \tag{10.85}$$

Extremely large surface (A_2) *surrounding a convex surface* $(A_1, F_{11} = 0)$:

$$q_{1-2} = \sigma A_1 \epsilon_1 (T_1^4 - T_2^4) \tag{10.86}$$

The two cylinders and two spheres referred to in Eq. (10.85) do not have to be positioned concentrically: note the eccentric positions in Figure 10.15.

Equations (10.84) and (10.85) are particularly useful in the analysis of thermal insulation systems in which the dominant mode of heat leakage is radiation. These systems employ "sandwiches" of two or more cavities of the kind defined by the surfaces A_1 and A_2 in the preceding analysis. In the sandwich, the cavities are separated by means of one or more opaque partitions called radiation shields.

Example 10.5 *Radiation Shielding*

Consider the heat transfer in the configuration defined in Figure E10.5, in which a third surface, A_s, has been inserted in the space between the two surfaces of Figure 10.21. The intermediate surface A_s shields A_2 from

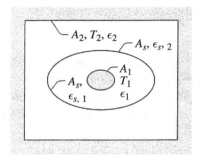

$$q_{1-2} \rightarrow \underset{\text{or}}{\underset{q_1}{\circ}} \overset{\sigma T_1^4}{\underset{\dfrac{1-\epsilon_1}{\epsilon_1 A_1}}{-\!\!\!\bigvee\!\!\!\bigvee\!\!\!-}} \overset{J_1}{\underset{\dfrac{1}{A_1 F_{1s}}}{-\!\!\!\bigvee\!\!\!\bigvee\!\!\!-}} \overset{J_{s,1}}{\underset{\dfrac{1-\epsilon_{s,1}}{\epsilon_{s,1} A_s}}{-\!\!\!\bigvee\!\!\!\bigvee\!\!\!-}} \overset{\sigma T_s^4}{\underset{\dfrac{1-\epsilon_{s,2}}{\epsilon_{s,2} A_s}}{-\!\!\!\bigvee\!\!\!\bigvee\!\!\!-}} \overset{J_{s,2}}{\underset{\dfrac{1}{A_s F_{s2}}}{-\!\!\!\bigvee\!\!\!\bigvee\!\!\!-}} \overset{J_2}{\underset{\dfrac{1-\epsilon_2}{\epsilon_2 A_2}}{-\!\!\!\bigvee\!\!\!\bigvee\!\!\!-}} \overset{\sigma T_2^4}{\circ} \rightarrow \underset{-q_2}{\underset{\text{or}}{q_{1-2}}}$$

Figure E10.5

the radiosity emanating from A_1; at the same time, A_s shields A_1 from that fraction of the A_2 radiosity that would otherwise be intercepted by A_1. Determine the net heat transfer rate from A_1 to A_2.

Solution

This configuration is a sandwich of two of the two-surface enclosures analyzed in Figure 10.21. The electrical network analog of the heat transfer path from A_1 all the way to A_2 consists of six resistances in series. Labeling with $\epsilon_{s,1}$ and $\epsilon_{s,2}$ the total hemispherical emissivities of the two sides of the radiation shield A_s, we can write immediately

$$q_{1-2} = \cfrac{\sigma(T_1^4 - T_2^4)}{\underbrace{\dfrac{1-\epsilon_1}{\epsilon_1 A_1} + \dfrac{1}{A_1 F_{1s}} + \dfrac{1-\epsilon_{s,1}}{\epsilon_{s,1} A_s}}_{\substack{\text{Total resistance} \\ \text{of the } A_1 - A_s \text{ gap}}} + \underbrace{\dfrac{1-\epsilon_{s,2}}{\epsilon_{s,2} A_s} + \dfrac{1}{A_s F_{s2}} + \dfrac{1-\epsilon_2}{\epsilon_2 A_2}}_{\substack{\text{Total resistance} \\ \text{of the } A_s - A_2 \text{ gap}}}} \tag{1}$$

The net heat current q_{1-2} leaves A_1, penetrates the shield A_s, and arrives finally at the outermost surface A_2. Equation (1) should be compared with the no-shield limit, Eq. (10.82), to observe the insulating effect of having inserted one shield in the space between A_1 and A_2; the denominator in Eq. (1) contains six terms, whereas the denominator of Eq. (10.82) contains only three.

Assume the special case where A_1, A_s, and A_2 are three infinite parallel plates of area A, and where all the emissivities have the same value, $\epsilon_1 = \epsilon_{s,1} = \epsilon_{s,2} = \epsilon_2 = \epsilon$. In this case Eq. (1) reduces to

$$q_{1-2} = \frac{\sigma A(T_1^4 - T_2^4)}{2\left(\dfrac{2}{\epsilon} - 1\right)} \quad \text{(one shield)} \tag{2}$$

In the same case the net heat transfer rate in the absence of the shield is (cf. Eq. (10.82))

$$q_{1-2} = \frac{\sigma A(T_1^4 - T_2^4)}{\frac{2}{\epsilon} - 1} \quad \text{(no shield)} \tag{3}$$

Equations (2) and (3) show that the shield reduces the net heat transfer rate to one half of the rate that prevails when the shield is absent.

Note that, unlike A_1 and A_2, the shield is not heated or cooled by an external entity. The shield is suspended between A_1 and A_2. The temperature T_s "floats" to that special level that guarantees the continuity of the net heat current through the shield node σT_s^4. In other words, the temperature T_s must be such that the heat current driven by the potential difference $\sigma T_1^4 - \sigma T_s^4$ is equal to the current driven by $\sigma T_s^4 - \sigma T_2^4$.

10.4.5 Enclosures with More than Two Surfaces

As a generalization of the two-surface analysis of the preceding section, consider the enclosure of Figure 10.23, in which all n surfaces are diffuse-gray. Equations (10.74)–(10.77) continue to apply to surface 1 in this new enclosure. What changes is the expression for the total irradiation current that strikes the surface A_1. In general, the observer seated on A_1 may see the radiosities of all n surfaces of the enclosure. For example, the irradiation current (watts) that emanates from the jth surface A_j and strikes the surface A_1 is $J_j A_j F_{j1}$. It follows that the irradiation current that impinges on A_1 is

$$
\begin{aligned}
A_1 G_1 &= J_1 A_1 F_{11} + J_2 A_2 F_{21} + \cdots + J_n A_n F_{n1} \\
&= \sum_{j=1}^{n} J_j A_j F_{j1} \\
&= \sum_{j=1}^{n} J_j A_1 F_{1j}
\end{aligned} \tag{10.87}
$$

Note that the last step in the above sequence is justified by the reciprocity relation $A_j F_{j1} = A_1 F_{1j}$.

From the point of view of surface A_1, the heat transfer problem is still the calculation of the net heat transfer rate q_1 that must be supplied to the back of A_1. This heat current can be evaluated using Eq. (10.77), provided

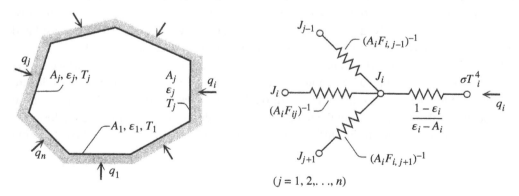

Figure 10.23 Enclosure formed by n diffuse-gray surfaces, and the resistance network portion associated with the surface A_j.

the radiosity J_1 is known. The problem reduces then to the calculation of J_1, which, as we will soon discover, depends on the geometry and radiative properties of all the surfaces of the enclosure. The needed relation between J_1 and the rest of the enclosure parameters follows from substituting into the J_1 definition (10.74) the general expression that was just derived for the irradiation G_1, namely, Eq. (10.87):

$$J_1 = (1 - \alpha_1) \sum_{j=1}^{n} J_j F_{1j} + \epsilon_1 \sigma T_1^4 \tag{10.88}$$

Starting with Eq. (10.74), we have also used $\rho_1 = 1 - \alpha_1$ and $E_{b,1} = \sigma T_1^4$.

Equation (10.88) states that the radiosity of the surface A_1 depends on the properties of the A_1 surface (α_1, ϵ_1, T_1), the radiosities of all the surfaces in the enclosure (J_j; $j = 1, 2, \ldots, n$), and the respective view factors through which all these surfaces are visible from A_1. A system of n equations for the n radiosities J_j is obtained by writing an equation of type (10.88) for each surface:

$$J_i = (1 - \alpha_i) \sum_{j=1}^{n} J_j F_{ij} + \epsilon_i \sigma T_i^4 \quad (i = 1, 2, \ldots, n) \tag{10.89}$$

If the configuration and all the surface properties are specified, then the system (10.89) pinpoints the values of the n radiosities; and, in the end, Eq. (10.77) delivers the value of the heat current q_1. An equation of type (10.77) holds for every single surface that participates in the enclosure:

$$q_i = \frac{\epsilon_i A_i}{1 - \epsilon_i} (\sigma T_i^4 - J_i) \quad (i = 1, 2, \ldots, n) \tag{10.90}$$

Therefore, the n radiosities delivered by system (10.89) can also be used to calculate the heat current q_i that must be supplied to the back of each surface. Equation (10.90) is restricted to gray surfaces, while Eq. (10.89) is not.

By convention, the heat transfer rate q_i has a positive value when it crosses the A_i surface and proceeds toward the enclosure. One important relationship among all the heat currents q_i follows from the steady-state first law of thermodynamics, applied to the thermodynamic system defined by the enclosure:

$$\sum_{i=1}^{n} q_i = 0 \tag{10.91}$$

The solution to the multisurface enclosure problem is summarized in Eqs. (10.89) and (10.90). When all of the surface temperatures T_i are specified, the system (10.89) can be solved first to determine all the J_is, leaving the q_is to be calculated from Eq. (10.90). When only some of the T_is and q_is are specified, the two systems (10.89) and (10.90) must be solved simultaneously. For example, when a surface A_k is insulated (adiabatic) with respect to the exterior of the enclosure, its net heat current is specified, $q_k = 0$.

The network analog [12] of this multisurface analysis is presented on Figure 10.23b. Each surface (A_i) is represented by two nodes, σT_i^4 and J_i, linked by an internal resistance of type $(1 - \epsilon_i)/\epsilon_i A_i$. On its enclosure side, the surface A_i is linked to each of the other surfaces through a resistance of type $1/A_i F_{ij}$. This feature can be illustrated also analytically by writing that the net heat current q_i deposited by A_i into the enclosure is the difference between what emanates from the surface (namely, $A_i J_i$) and what is intercepted by the same surface ($A_i G_i$):

$$q_i = A_i J_i - A_i G_i$$

$$= A_i J_i - \sum_{j=1}^{n} J_j A_i F_{ij} \tag{10.92}$$

$$= A_i J_i \sum_{j=1}^{n} F_{ij} - \sum_{j=1}^{n} J_j A_i F_{ij} \tag{10.93}$$

$$= \sum_{j=1}^{n} A_i F_{ij}(J_i - J_j) \tag{10.94}$$

The first step in this derivation, Eq. (10.92), is based on writing an equation of type (10.87) for the surface A_i. The second step, Eq. (10.93), consists of invoking the enclosure rule (10.43) for the surface A_i.

The end expression for q_i, Eq. (10.94), shows that on the enclosure side of the surface A_i, the current q_i breaks up into n minicurrents of type $A_i F_{ij}(J_i - J_j)$. Each minicurrent is driven by the respective radiosity difference between A_i and a neighboring surface A_j. The resistance overcome by each minicurrent is $1/A_i F_{ij}$. This resistance is shown in the network on Figure 10.6b.

A final observation concerns the geometric view factors F_{ij} of the enclosure. These can be arranged in the following matrix to show that the n-surface enclosure possesses n^2 view factors:

$$\begin{vmatrix} F_{11} & F_{12} & \cdots & F_{1n} \\ F_{21} & F_{22} & \cdots & F_{2n} \\ \vdots & \vdots & \ddots & \vdots \\ F_{n1} & F_{n2} & \cdots & F_{nn} \end{vmatrix}$$

Not all of these numbers can be specified independently; in other words, only some of the F_{ij}s constitute true "degrees of freedom" in the design (sizing, shaping, structuring) of the enclosure. Reciprocity relations of type (10.38) relate the view factors situated under and to the left of the $F_{11} - F_{nn}$ diagonal to their counterparts across the diagonal. There are $(n^2 - n)/2$ such relationships, because there are n view factors on the diagonal and $(n^2 - n)/2$ view factors on each side of the diagonal. Additionally, n enclosure relationships of type (10.43) can be written, one relationship for each surface. In conclusion, the number of independent geometric view factors is

$$n^2 - \frac{1}{2}(n^2 - n) - n = \frac{n}{2}(n - 1) \tag{10.95}$$

The number $(n/2)(n-1)$ is always an integer. For some of the simplest enclosures, which form the object of the problems proposed at the end of the chapter, this number is as follows:

Number of surfaces	2	3	4	5
Number of independent geometric view factors	1	3	6	10

This table shows that the number of independent view factors increases rapidly as the number of surfaces (isothermal patches) of the enclosure increases. The algebra associated with solving systems (10.89) and (10.90) becomes more challenging at the same time. Therefore, we have an incentive to model the heat transfer configuration (i.e. to define the enclosure) by means of the smallest number of isothermal surfaces possible. A modeling decision of this kind is illustrated in the next example.

Example 10.6 *Enclosure with Three Surfaces: One of Which Is Adiabatic ("Reradiating")*
Figure E10.6 shows the key features of an oven, namely, the source of heat (e.g. flame, electrical resistance) and the object that is to be heated (e.g. cut of meat, block of metal). The surfaces of both items can be modeled as diffuse-gray and can be characterized by A_1, ϵ_1, T_1, and A_2, ϵ_2, T_2, respectively.

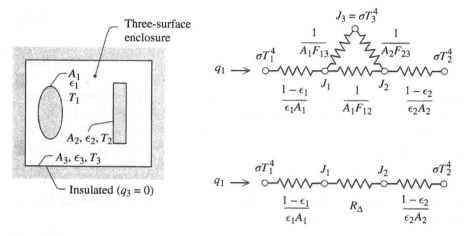

Figure E10.6

Surrounding the heater A_1 and the heat sink A_2 is a third surface that must be insulated (adiabatic) to direct all of the heating effect of A_1 into the heat sink A_2. In an industrial furnace, this third surface is built out of a high temperature-resistant (refractory) material. In real life, the temperature will vary along this surface, the highest temperatures being registered in the zone that is the closest to the heat source. The simplest model of the adiabatic surface is the isothermal surface (A_3, ϵ_3, T_3) identified in Figure E10.6. The temperature T_3 is nothing more than a first-cut (an average) estimate of the temperatures reached by the real-life refractory surface. This isothermal and refractory surface model simplifies the analysis considerably; in fact, it makes a fully analytical solution possible.

Solution

Of interest is the relationship between the heat source and the heat sink temperatures and the net heat transfer rate between the two. This can be read off the resistance network that corresponds to this problem. Starting with the upper version of the network, note that the three radiosity nodes (J_1, J_2, J_3) and that the internal resistance of the third surface, $(1 - \epsilon_3)/\epsilon_3 A_3$, are not shown. If this resistance had been drawn, say, above the node J_3, we would have discovered that the current through it is, in fact, zero because the surface A_3 is adiabatic $(q_3 = 0)$. In conclusion, the potentials (nodes) σT_3^4 and J_3 coincide on the diagram, and this means that the internal resistance $(1 - \epsilon_3)/\epsilon_3 A_3$ does not play any role in the functioning of the network. The heat transfer rate from A_1 to A_2 does not depend on ϵ_3.

Examining the lower version of the resistance network, we can write that

$$q_{1-2} = q_1 = \frac{\sigma(T_1^4 - T_2^4)}{\dfrac{1 - \epsilon_1}{\epsilon_1 A_1} + R_\Delta + \dfrac{1 - \epsilon_2}{\epsilon_2 A_2}} \tag{1}$$

The second term in the denominator, R_Δ, is shorthand for the resistance contributed by the triangular loop of the upper diagram:

$$R_\Delta = \frac{R_s(1/A_1 F_{12})}{R_s + (1/A_1 F_{12})} \tag{2}$$

where

$$R_s = \frac{1}{A_1 F_{13}} + \frac{1}{A_2 F_{23}} \tag{3}$$

A more accurate solution to the same problem can be obtained based on a model where the insulated surface A_3 is divided into two isothermal areas, A_{3a} and A_{3b}. This approach leads to a four-surface enclosure problem, for which an analytical solution of type (1)–(3) is not possible. Instead, the problem must be solved numerically by using Eqs. (10.89) and (10.90), assuming that the known parameters of the problem have been specified numerically.

10.5 Participating Media

10.5.1 Volumetric Absorption

The preceding analyses of radiation between two or more surfaces were based on the assumption that the medium that separates the surfaces is perfectly transparent and does not participate in the heat transfer process. The nonparticipating medium model is appropriate when the space between the radiating surfaces is occupied by vacuum or by a gas with monatomic or symmetrical diatomic molecule. Examples are the main components of air (N_2, O_2), H_2, Ne, Ar, and Xe.

The nonparticipating medium model is not appropriate when the medium absorbs and scatters the radiation that passes through it. Primary examples of such media are strongly polar gases: their molecules exhibit asymmetry (e.g. H_2O, CO_2, NH_3, O_3, CO, SO_2, NO, hydrocarbons, and alcohols). Gases of this type are called participating media, because at each point in their volume, they absorb and scatter the incident radiation while emitting their own.

In this section we consider the additional complication that is introduced in the previous model by a participating medium. The major difference is that, while until now we analyzed heat transfer processes that occurred strictly between *surfaces*, when the medium is a participating one, the analysis must account also for the *volumetric* absorption, transmission, and emission of radiation. This volumetric effect is usually associated with monochromatic radiation that falls within certain bands or wavelength intervals. Radiation with a wavelength outside these bands will pass unattenuated through the medium; with respect only to this radiation, the medium is nonparticipating.

Consider the one-dimensional layer of thickness L shown in Figure 10.24. This layer is being penetrated by a beam of monochromatic radiation. The intensity of this beam, $I_\lambda(x)$, is attenuated locally by the volumetric absorption in the sublayer of thickness dx. Experiments have shown that the local attenuation dI_λ is proportional to the local intensity of the beam:

$$dI_\lambda = -\kappa_\lambda I_\lambda \, dx \tag{10.96}$$

This proportionality serves as definition for the monochromatic extinction coefficient κ_λ, which has the units m^{-1}. The extinction coefficient is proportional to the concentration of absorbing molecules (i.e. the partial pressure of the absorbing gas) in the medium. Integrating Eq. (10.96) across the layer, we obtain

$$I_\lambda(x) = I_\lambda(0) \exp(-\kappa_\lambda x) \tag{10.97}$$

This exponential decay in beam intensity across the layer is known as Beer's law. In particular, when $x = L$, Eq. (10.97) shows the fraction of the incident intensity that escapes through the other side of the layer:

$$\tau_\lambda = \frac{I_\lambda(L)}{I_\lambda(0)} = \exp(-\kappa_\lambda L) \tag{10.98}$$

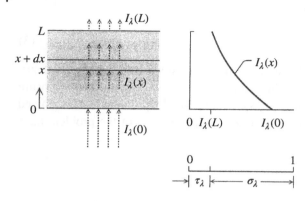

Figure 10.24 The attenuation of the monochromatic radiation that penetrates a volumetrically absorbing medium.

This fraction is less than 1, and it represents the monochromatic transmissivity (τ_λ) of the layer. The remaining fraction is absorbed in the L-thick layer:

$$\alpha_\lambda = 1 - \tau_\lambda = 1 - \exp(-\kappa_\lambda L) \tag{10.99}$$

as gases do not reflect the radiation that passes through them ($\alpha_\lambda + \tau_\lambda = 1$). The number α_λ is the monochromatic absorptivity of the L-thick medium.

If the gas temperature T_g is uniform and does not differ much from the temperature T_s of the surface that produced the beam $I_\lambda(0)$, we may invoke Kirchhoff's law:

$$\epsilon_\lambda = \alpha_\lambda = 1 - \exp(-\kappa_\lambda L) \tag{10.100}$$

In this equation ϵ_λ is the monochromatic emissivity of the layer.

Accurate analyses of gas absorption and emission must account for every band of wavelengths that contribute significantly to the process. In practice it is more convenient to work with total quantities, which are obtained by integrating Eqs. (10.98)–(10.100) over the entire spectrum:

$$\tau_g = \frac{\displaystyle\int_0^\infty I_\lambda(L)\,d\lambda}{\displaystyle\int_0^\infty I_\lambda(0)\,d\lambda} \tag{10.101}$$

$$\alpha_g = 1 - \tau_g \tag{10.102}$$

$$\epsilon_g \cong \alpha_g \quad (\text{if } T_g \cong T_s, \text{ i.e. if Kirchhoff's law applies}) \tag{10.103}$$

These coefficients are, in order, the gas transmissivity, absorptivity, and emissivity.

As a special case, the gray gas is defined by $\alpha_g = \alpha_\lambda$, $\epsilon_g = \epsilon_\lambda$, and $\alpha_g = \epsilon_g$. The gray gas represents a medium in which the monochromatic coefficients (τ_λ, α_λ, ϵ_λ) are independent of wavelength and where the gas emissivity is equal to the gas absorptivity regardless of the source of incident radiation.

10.5.2 Gas Emissivities and Absorptivities

The emission and absorption characteristics of participating gases are described quantitatively by the following procedure developed by Hottel [13]. Consider the hemispherical enclosure shown in Figure 10.25, in which the gas mixture contains only one participating component (CO_2, for example) and one or more nonparticipating (transparent) species. The gas mixture is at the uniform temperature T_g and total pressure P, while the partial pressure of carbon dioxide is P_c. The radius of the hemispherical gas volume is L.

Figure 10.25 Hemispherical space filled with gas radiating to a black surface element.

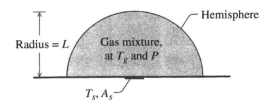

The gas mixture exchanges thermal radiation with the small area A_s situated in the center of the base.[7] The radiation heat current that is emitted by the gas and arrives at A_s is

$$q_{g \to s} = \epsilon_g A_s \sigma T_g^4 \tag{10.104}$$

where ϵ_g is the emissivity of the participating gas. Note that $q_{g \to s}$ is a one-way radiation current, and that the surface A_s absorbs this current entirely only if A_s is black. Note further that $\epsilon_g \sigma T_g^4$ (with $\epsilon_g < 1$) represents only a fraction of the emissive power associated with a blackbody of temperature T_g.

Measurements of the emitted radiation $q_{g \to s}$ have been used in conjunction with the ϵ_g definition (10.104) for the purpose of calculating and recording the gas emissivity ϵ_g. For carbon dioxide the emissivity is labeled ϵ_c; it is a function of the mixture temperature, the CO_2 partial pressure times the radius of the gas volume, and the total pressure of the mixture:

$$\epsilon_c = f_c(T_g, P_c L, P) \tag{10.105}$$

The function f_c is reported graphically by the following combination of Figures 10.26 and 10.27. When the mixture pressure is equal to 1 atm, the emissivity of carbon dioxide is simply the ϵ_c valued indicated on the ordinate of Figure 10.26. If the total pressure of the mixture happens to be different from 1 atm, the ϵ_c reading provided by Figure 10.26 must be multiplied by the correction factor C_c provided by Figure 10.27. In general, then, the f_c function indicated in Eq. (10.105) is reconstructed by multiplying the readings from the two charts:

$$\underset{\text{Figure 10.26}}{\epsilon_{c,P \neq 1\ \text{atm}}} = \underset{\text{Figure 10.26}}{\epsilon_{c,P=1\ \text{atm}}} \times \underset{\text{Figure 10.27}}{C_{c,P \neq 1\ \text{atm}}} \tag{10.106}$$

The carbon dioxide charts show that the emissivity increases monotonically as the product $P_c L$ increases (see Figure 10.26, $P = 1$ atm). This trend is understandable because when the partial pressure and the gas volume increase, the number of CO_2 molecules that emit radiation toward the target A_s also increases. The same explanation holds true for the trend made visible in Figure 10.27; ϵ_c also increases as the total pressure P increases.

Consider now the reverse of the interaction between the participating gas and the surface A_s in Figure 10.25. Let $q_{s \to g}$ be the radiation heat flux that is emitted by A_s in all directions and reaches the gas (note that if A_s is black, $q_{s \to g} = A_s \sigma T_s^4$). Only a fraction of this one-way heat current is absorbed by the gas:

$$q_a = \alpha_g q_{s \to g} \tag{10.107}$$

namely, the fraction represented by the gas absorptivity α_g. As noted previously in Eq. (10.103), α_g is equal to ϵ_g when T_s and T_g are equal, or almost equal. When the surface and gas temperatures differ greatly, the gas

---

7 The subscript used for "wall" in this book is w. Here we make an exception by using s (for "surface") and reserve w for quantities associated with water vapor as a radiating component in a mixture.

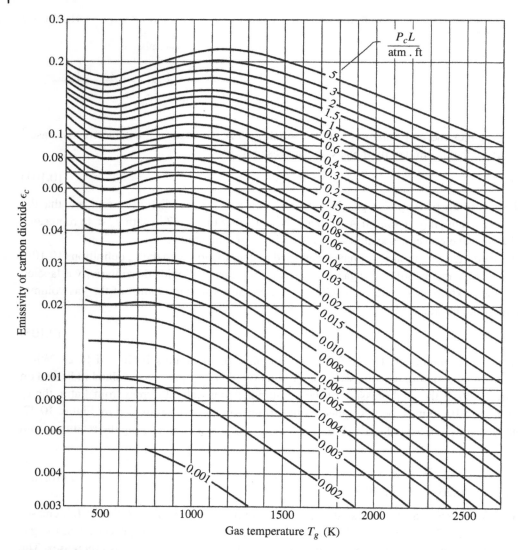

Figure 10.26 The emissivity of carbon dioxide in a mixture with nonparticipating gases at a mixture pressure of 1 atm. The length L is defined in Figure 10.25 or in Table 10.5. Source: Hottel [13].

absorptivity can be estimated based on the following empirical rule [13]. In the case of carbon dioxide as the lone participating gas in the mixture, the absorptivity α_c with respect to incident radiation of temperature T_s is

$$\alpha_c = f_c\left(T_s, P_c L \frac{T_s}{T_g}, P\right) \times \left(\frac{T_g}{T_s}\right)^{0.65} \tag{10.108}$$

The f_c function must be compared with Eq. (10.105) to see that the first factor on the right side of Eq. (10.108) is obtained from Figures 10.26 and 10.27, by replacing T_g with T_s and $P_c L$ with $P_c L(T_s/T_g)$. In the ratio T_s/T_g the temperatures are expressed in degrees Kelvin.

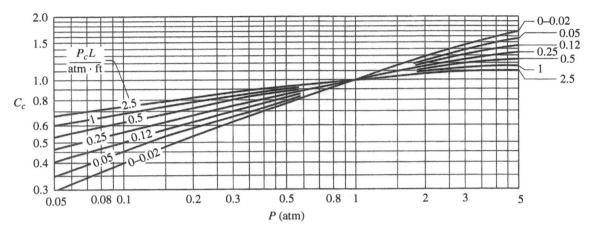

Figure 10.27 Correction factor for the emissivity of carbon dioxide in a mixture with nonparticipating gases at mixture pressures other than 1 atm. Source: Hottel [13].

A similar procedure is available for calculating the total emissivity and absorptivity of water vapor as the lone radiating component in a gas mixture. The charts shown in Figures 10.28 and 10.29 deliver the emissivity of water vapor as the function

$$\epsilon_w = f_w(T_g, P_w L, P + P_w) \tag{10.109}$$

where P_w is the partial pressure of water vapor. Figure 10.28 alone provides the value for ϵ_w when $P + P_w = 1$ atm. For other values of $P + P_w$, the ϵ_w reading obtained from Figure 10.28 must be multiplied by the correction factor C_w recommended by Figure 10.29:

$$\epsilon_{w,P+P_w \neq 1 \text{ atm}} = \epsilon_{w,P+P_w=1 \text{ atm}} \times C_{w,P+P_w \neq 1 \text{ atm}} \tag{10.110}$$
$$\text{Figure 10.28} \qquad \text{Figure 10.29}$$

The water vapor absorptivity with respect to incident radiation produced by a surface of temperature T_s can be evaluated with the empirical formula [13]

$$\alpha_w = f_w\left(T_s, P_w L \frac{T_s}{T_g}, P + P_w\right) \left(\frac{T_g}{T_s}\right)^{0.45} \tag{10.111}$$

By comparing it with Eq. (10.109), we note that the f_w factor on the right side of Eq. (10.111) is obtained from Figures 10.28 and 10.29 by replacing T_g with T_s and $P_w L$ with $P_w L(T_s/T_g)$.

The calculations described earlier referred to cases in which only one participating component was present in the mixture. That component was carbon dioxide (Figures 10.26 and 10.27) or water vapor (Figures 10.28 and 10.29). When CO_2 and H_2O are present together in the mixture, the emitted radiation is a bit smaller than the radiation emitted by CO_2 alone plus the radiation emitted by H_2O alone. This effect is expected, because each gas obstructs somewhat the radiation emitted by the other. The overall gas emissivity ϵ_g is, therefore, smaller than the sum of the ϵ_c and ϵ_g values discussed earlier:

$$\epsilon_g = \underset{\text{Figures 10.26–10.27}}{\epsilon_c} + \underset{\text{Figures 10.28–10.29}}{\epsilon_w} - \underset{\text{Figure 10.30}}{\Delta\epsilon} \tag{10.112}$$

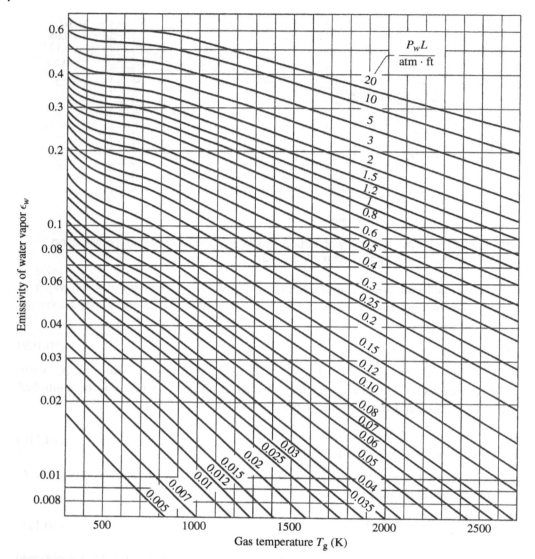

Figure 10.28 The emissivity of water vapor in a mixture with nonparticipating gases at a mixture pressure of 1 atm. The length L is defined in Figure 10.25 or in Table 10.5. Source: Hottel [13].

The $\Delta\epsilon$ correction is reported in Figure 10.30; it is primarily a function of the amount of radiating molecules $L(P_c + P_w)$ and combination (proportions) of CO_2 and H_2O in the mixture. The $\Delta\epsilon$ correction is largest (a few percentage points) when the partial pressures of carbon dioxide and water vapor are comparable. The emissivities of sulfur dioxide, carbon monoxide, and ammonia can be evaluated by consulting a set of similar charts compiled by Hottel [13].

The length L employed in the construction of Figures 10.26–10.30 refers strictly to the radius of the hemispherical gas volume (Figure 10.25). The same charts and formulas can be used for calculating ϵ_g and α_g in other geometric combinations of gas volumes and radiating surfaces. Listed in Table 10.5 are several configurations to which this procedure can be applied by replacing L with the equivalent length L_e shown in the right column.

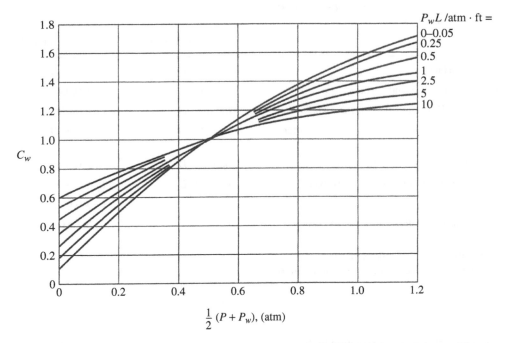

Figure 10.29 Correction factor for the emissivity of water vapor in a mixture with nonparticipating gases at mixture pressures other than 1 atm. Source: Hottel [13].

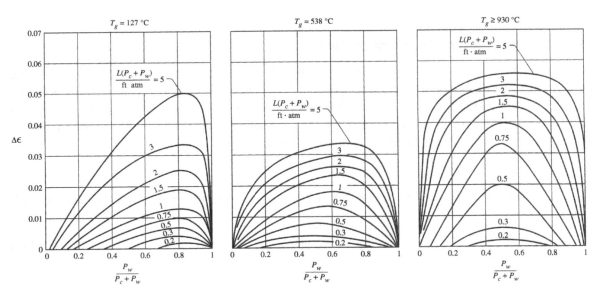

Figure 10.30 Correction for gas emissivity when carbon dioxide and water vapor are present simultaneously in a mixture with nonparticipating gases. Source: Hottel [13].

Table 10.5 The equivalent length L_e for several gas volume shapes.[a]

Shape of gas volume	Actual dimension	Equivalent length
Sphere, radiation to internal surface	Diameter D	$L_e \cong 0.60D$
Infinite cylinder, radiation to entire internal surface	Diameter D	$L_e \cong 0.95D$
Circular cylinder with height $= D$, radiation to entire surface	Diameter D	$L_e \cong 0.6D$
Circular cylinder with height $= D$, radiation to spot in the center of base	Diameter D	$L_e \cong 0.77D$
Semi-infinite circular cylinder, radiation to entire base	Diameter D	$L_e \cong 0.65D$
Semi-infinite cylinder, radiation to spot in the center of base	Diameter D	$L_e \cong 0.9D$
Cube, radiation to one face	Side a	$L_e \cong 0.67a$
Space between two infinite parallel planes, radiation to both planes	Spacing s	$L_e \cong 1.8s$
Space between tubes in an infinite tube bundle with tube diameter = clearance between two closest tube walls, radiation to a single tube, tube centers on		
Equilateral triangles	Tube diameter D	$L_e \cong 2.8D$
Squares	Tube diameter D	$L_e \cong 3.5D$
Arbitrary volume V surrounded by surface A, radiation to A	V, A	$L_e \cong 3.6\, V/A$

a) L_e replaces L in Figures 10.26–10.29.
Source: Hottel [13].

10.5.3 Gas Surrounded by Black Surface

As an application of the method described in Section 10.5.2, consider the heat transfer between a radiating gas and an enclosure of surface area A_s and temperature T_s, which contains the gas. The gas is characterized by T_g, P, L_e, and the partial pressure(s) of the radiating component(s) in the gas mixture. This gas information can be used to determine ϵ_g and the α_g value that refers to radiation arriving from T_s. If the internal surface A_s is black, it absorbs all the radiation that is emitted by the gas:

$$q_{g \to s} = \epsilon_g \sigma T_g^4 A_s \tag{10.113}$$

The radiation heat current emitted by A_s is $\sigma T_s^4 A_s$, while the portion that is absorbed by the entire gas volume is

$$q_{s \to g} = \alpha_g \sigma T_s^4 A_s \tag{10.114}$$

The instantaneous *net* rate of heat transfer from the gas to its black enclosure is

$$\begin{aligned} q_{g-s} &= q_{g \to s} - q_{s \to g} \\ &= \sigma A_s (\epsilon_g T_g^4 - \alpha_g T_s^4) \end{aligned} \tag{10.115}$$

The expression in round brackets shows that in this calculation the radiation emitted by the surface can be neglected if the gas temperature is sufficiently higher than the surface temperature.

10.5.4 Gray Medium Surrounded by Diffuse-Gray Surfaces

Consider now the two-surface enclosure shown in Figure 10.31. Each surface is diffuse-gray and isothermal, (A_1, T_1, ϵ_1 and A_2, T_2, ϵ_2), whereas the medium (e.g. gas) is isothermal and gray (T_g, $\epsilon_g = \alpha_g$). Of interest is the net heat transfer between A_1 and A_2, and how the gas "interferes" with this process.

An observer standing on A_1 sees not only the medium but also a surface of different temperature, A_2. The latter is visible only to the extent that the medium is transparent. The radiation that leaves A_1 is $J_1 A_1$. The portion that reaches A_2 is $J_1 A_1 F_{12} \tau_g$, where τ_g is the medium transmissivity ($\tau_g = 1 - \alpha_g$). In the reverse direction, the portion of the radiation leaving A_2 and arriving through the medium at A_1 is $J_2 A_2 F_{21} \tau_g$, where $A_2 F_{21} = A_1 F_{12}$. The net heat transfer rate from A_1 to A_2 directly (i.e. by penetrating the medium) is therefore $A_1 F_{12} \tau_g (J_1 - J_2)$. In conclusion, the radiosity nodes J_1 and J_2 are linked by a direct resistance of size $(A_1 F_{12} \tau_g)^{-1}$ in the network constructed in Figure 10.31.

The surfaces also communicate indirectly, by using the medium as an intermediary. Take the interaction between A_1 and the medium. The fraction of the radiation leaving A_1 and being absorbed volumetrically by the medium is $J_1 A_1 F_{1g} \alpha_g$, where $\alpha_g = \epsilon_g$. Traveling in the opposite direction is the medium emission that is intercepted by A_1, namely, $\epsilon_g \sigma T_g^4 A_1 F_{1g}$. The net heat current along this path is $A_1 F_{1g} \epsilon_g (J_1 - \sigma T_g^4)$. This shows that the J_1 node and the gas node σT_g^4 are linked by the resistance $(A_1 F_{1g} \epsilon_g)^{-1}$.

The radiation network is completed by a similar resistance between the gas volume (σT_g^4) and the second gray surface (J_2). Between each radiosity node and the respective blackbody emissive power node of the surface, we see the internal resistances defined in Eq. (10.78). The node that represents the gas volume is a floating one when the gas is not heated by an independent mechanism (e.g. electrically or by chemical reactions). Under these circumstances only, the net rate of heat transfer from A_1 to A_2 can be written by looking at the network:

$$q_{1-2} = \frac{\sigma(T_1^4 - T_2^4)}{\dfrac{1-\epsilon_1}{\epsilon_1 A_1} + R_\Delta + \dfrac{1-\epsilon_1}{\epsilon_2 A_2}} \tag{10.116}$$

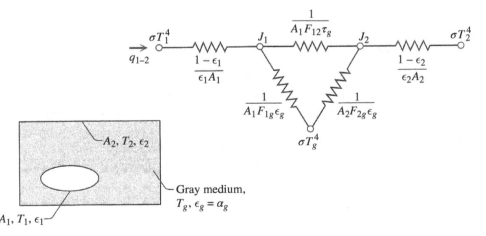

Figure 10.31 Gray medium enclosed by two diffuse-gray surfaces, and the radiation network for the case in which the temperature of the medium is floating.

where R_Δ is the equivalent series resistance of the triangular loop of the network:

$$R_\Delta = \frac{(A_1 F_{12} \tau_g)^{-1}[(A_1 F_{1g} \epsilon_g)^{-1} + (A_2 F_{2g} \epsilon_g)^{-1}]}{(A_1 F_{12} \tau_g)^{-1} + (A_1 F_{1g} \epsilon_g)^{-1} + (A_2 F_{2g} \epsilon_g)^{-1}} \qquad (10.117)$$

The radiative heat transfer between surfaces and participating media is discussed in greater depth in more advanced treatments [14, 15].

10.6 Evolutionary Design

Important applications have stimulated the interest in radiative heat transfer throughout its history. Most of the developments that led to the science taught in this chapter occurred during the first 100 years of industrial revolution. This happened for two main reasons.

First, in the practical domain, the rapid development and spreading of power plant technology (power from "fire") called for scientific principles of designing burners, heaters, boilers, and condensers. These are applications at higher than atmospheric temperature, where convective heat transfer is accompanied by significant radiative heat transfer.

Second, the rapid increase in the availability of power and the resulting rise in advancement (wealth, mobility, freedom, leisure) in Europe led to great advances in scientific research and higher education. The physics of radiative transport found its theory and principles. Naturally, this new physics attracted curious scientists who contemplated phenomena far removed from the power plants that gave them the freedom to contemplate. Geophysics was the chief beneficiary, with answers to questions about sun–Earth interaction, Earth temperature history, cloud cover, and similar questions about other celestial bodies.

Today, theory returns the favor to practice. With the physics of radiative heat transfer, we can address the design, performance, and improvement of power plants that are driven from great distances by radiation. Chief examples are the power plants driven by heating provided by collectors "fueled" by solar power. In this section we examine the essential features of solar power plants, which is a broad class that includes the Earth itself.

10.6.1 Terrestrial Solar Power

The simplest model [2] of a power plant driven by solar heating is shown in Figure 10.32. There are three systems, with different temperatures: the sun surface (T_s), the atmospheric temperature on Earth (T_0), and, sandwiched between them, the temperature of the solar collector (T). The power plant (not shown) is driven by the difference between the net radiative heat current collected from the sun (q_s) and the net convective heat current from the collector to the atmosphere (q_0). The design of the power plant is not the object of this treatment; therefore we account for the power plant by assuming that it is operating reversibly, i.e. as a Carnot engine the power output of which is

$$w = (q_s - q_0)\left(1 - \frac{T_0}{T}\right) \qquad (10.118)$$

The solar heat current intercepted by the collector (q_s) is proportional to the collector size, which in this model is represented by the collector area A. We assume that A is given, and consequently q_s is given. For the

Figure 10.32 Convective-cooling model for moderate temperature solar collectors.

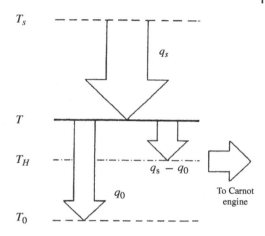

convective heat loss from collector to ambient, we assume the model (cf. Chapter 5)

$$q_0 = UA(T - T_0) \tag{10.119}$$

where U is the overall collector-ambient heat transfer coefficient based on A and the product UA is the overall thermal conductance between collector and ambient.

Free to change in this model is the collector temperature (T), or q_0, or the heat input to the power plant ($q_s - q_0$). Why is this degree of freedom useful? It is useful because by varying T we discover that with a collector of fixed size (A), we can extract more power (w) from the fixed current received from the sun (q_s).

A good problem to practice on is to eliminate q_0 between Eqs. (10.118) and (10.119) to derive w as a function of its lone variable (T). The value of w peaks at the collector temperature

$$T = (T_{max} T_0)^{1/2} \tag{10.120}$$

where T_{max} is the temperature ceiling that would be reached by the collector when all the solar heating (q_s) is lost undiminished directly to the atmosphere (q_0). In this limit the net heat transfer rate to the Carnot engine is zero, and so is the engine power output. In the opposite extreme, when the collector is the coldest ($T = T_0$), the heat input to the engine is the greatest, the Carnot efficiency of the engine is zero, and so is the engine power output. The design represented by the intermediary temperature ($T_{max} T_0)^{1/2}$ has the merit that it delivers the most power per unit of collector size:

$$\frac{w}{UAT_{max}} = \left[1 - \left(\frac{T_0}{T_{max}} \right)^{1/2} \right]^2 \tag{10.121}$$

The main message from this simple model is that there is freedom in how the collector is changed, and that rewards (higher w) are due to the freedom to change. This, the rewards of freedom, is an essential characteristic of all solar power plants models, simple or not so simple. They all offer the opportunity to place the collector at a temperature that makes the power plant most productive. More realistic models are collected in Ref. [2], and they illustrate this common feature.

10.6.2 Extraterrestrial Solar Power

In outer space things are a bit different but not much more complicated (Figure 10.33) [2]. The hot end of the power plant is heated by the sun (T_s) in the collector (T_H), and cooled by the cold background (T_∞) in the radiator (T_L). All the surfaces (sun, A_H, A_L, background) are modeled as black. The internal operation of the heat engine cycle (the system contained between A_H and A_L) is modeled as reversible. The mass of the power plant and the cost of constructing it and placing it in orbit are influenced strongly by the total surface (mass) of the collector and the radiator; consequently, the total surface is constrained:

$$A = A_H + A_L \quad \text{(constant)} \tag{10.122}$$

There are 2 degrees of freedom in this model. One is the freedom to divide A between A_H and A_L such that the power output w is maximum. Fixed are A, T_s, and T_∞. The other is the freedom to place the engine at a temperature level between T_s and T_∞. The analysis begins with the first and second laws for the reversible compartment (middle of Figure 10.33), which lead to $w = q_H(1 - T_L/T_H)$. To calculate q_H, consider the enclosure with three black surfaces (T_s, T_∞, T_H) formed above A_H:

$$q_H = A_H F_{Hs}\sigma(T_s^4 - T_H^4) - A_H F_{H\infty}\sigma(T_H^4 - T_\infty^4) \tag{10.123}$$

where F_{Hs} and $F_{H\infty}$ are the collector–sun and collector–background view factors. The relation between them is $F_{H\infty} = 1 - F_{Hs}$. Note further that T_∞^4 is negligible relative to T_H^4, because $T_\infty \cong 4\,\text{K}$, while the expected order of magnitude of T_H is 10^3 K. With these observations, Eq. (10.123) becomes

$$q_H = \sigma A_H(F_{Hs}T_s^4 - T_H^4) \tag{10.124}$$

The enclosure with two black surfaces formed under the radiator A_L shows that $q_L = \sigma A_L(T_L^4 - T_\infty^4)$, in which T_∞^4 is negligible relative to T_L^4:

$$q_L = \sigma A_L T_L^4 \tag{10.125}$$

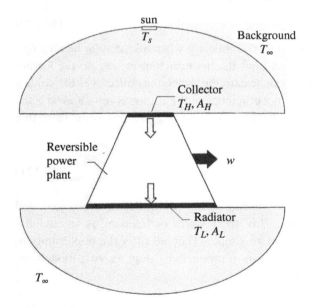

Figure 10.33 Extraterrestrial solar power plant with radiative heat transfer at the hot and cold ends.

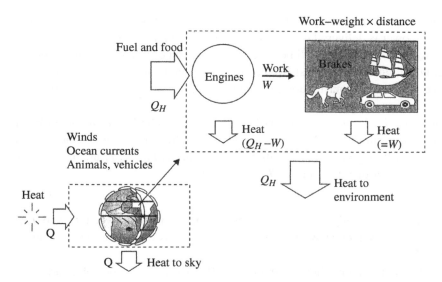

Figure 10.34 The whole Earth is an engine + brake system, containing innumerably smaller engine + brake systems (winds, ocean currents, animals, and human and machine species).

If the area constraint (10.122) is stated in terms of the area allocation ratio x, namely, $A_H = xA$, $A_L = (1 - x)A$, we find that the problem statement reduces to

$$\frac{T_L}{T_H} = \left[\frac{x}{1-x}(\zeta - 1) \right]^{1/3} \qquad q_H = \sigma A F_{Hs} T_s^4 x \left(1 - \frac{1}{\zeta} \right) \tag{10.126}$$

$$\tilde{w} = \frac{w}{\sigma A F_{Hs} T_s^4} = x \left(1 - \frac{1}{\zeta} \right) \left\{ 1 - \left[\frac{x}{1-x}(\zeta - 1) \right]^{1/3} \right\} \tag{10.127}$$

where $\zeta = F_{Hs}(T_s/T_H)^4$. The dimensionless power output $\tilde{w}$ emerges as a function of the 2 degrees of freedom, now represented by ζ and x. The maximum power ($\tilde{w} = 0.0414$) occurs at $\zeta = 1.538$ and $x = 0.35$. In this design the collector accounts for roughly one-third of the total surface available, which means that A_H is about half the size of A_L. This conclusion is independent of whether the collector is fitted with a concentrator.

Finally, by substituting $T_s = 5762 \, \text{K}$ in $\zeta = 1.538$ and the first of Eq. (10.126), we find that $T_H = (5174 F_{Hs}^{1/4}) \, \text{K}$ and $T_L = (3423 F_{Hs}^{1/4}) \, \text{K}$. Depending on the degree of concentration of solar radiation, F_{Hs} varies from 10^{-4} to 1; therefore T_H and T_L can be as low as roughly 520 and 340 K.

10.6.3 Climate

The extraterrestrial solar power plant model of Figure 10.33 is related to the conversion of solar heating into wind power (natural convection) on Earth, where the solar power is destroyed totally by friction and heat transfer across finite temperature differences (Figure 10.34). The heat engine that drives any natural convection process was first noted in the field of heat transfer in a drawing in my 1984 book [16], Figure 7.1. This is the basis of the constructal theory of global circulation and climate [17–20].

To start, notice that in the extraterrestrial model (Figure 10.33) the power w is delivered to another system (human) that uses the power. If the power is delivered to (and stored in) a battery, then the model is correct, but short-lived. Even the power stored in the battery will be used, i.e. destroyed eventually. More generally and permanently then, the power w in Figure 10.33 is destroyed by the user, which means that w is dissipated into heat transfer to the cold space. This added contribution to q_L, Eq. (10.125), makes the total heat transfer rate to T_∞ be the same as q_H.

Said another way, the power plant and its dissipative system (the user acting as a brake) form one node – a way station – in the flow of q_H from the sun (T_s) to the background (T_∞). If the temperature of the node is T_e, then setting Eqs. (10.124) and (10.125) equal, and noting that, approximately, T_H and T_L are the same as T_e because the ratio T_H/T_L is much smaller than the ratio T_s/T_∞, we obtain

$$A_H(F_{Hs}T_s^4 - T_e^4) = A_L T_e^4 \tag{10.128}$$

This equality proclaims the existence of a steady average temperature for the node between T_s and T_∞. The node could be the extraterrestrial power plant with its user (dissipator), as discovered at the end of Section 10.6.2, or a planet such as Earth.

This discovery was the start of the theory that predicts Earth climate. Beyond the existence of a particular climate (one T_e), more realistic questions to be answered concern the variations in T_e on the Earth's surface, for example, the heat transport effected by latitudinal circulation; the effective conductance and the variation of temperature in the north–south direction; the latitude of the boundaries between the Hadley, Ferrel, and Polar cells, wind speeds, and atmospheric circulation at the diurnal scale; and the average temperature between day and night [2, 17, 18].

The climate theory was extended to time-dependent conditions [2, 19], and it unveiled the close relationship between changes in the radiative properties of the atmosphere and changes in climate. This empowers theory to predict changes in climate over time and also in the distribution of temperature over the Earth's surface.

The fundamentals covered in this chapter are the basis for the expanding field of solar power collection and distribution. This work proceeds in many directions, for example, solar chimney power plants [21, 22], solar collector design [23–25], solar heat storage [26], urban climate [27], and animal niche construction [28].

References

1 Planck, M. (1901). Distribution of energy in the spectrum. *Ann. Phys.* 4 (3): 553–563.

2 Bejan, A. (2016). *Advanced Engineering Thermodynamics*, Chapter 9, 4e. Hoboken, NJ: Wiley.

3 Wien, W. (1894). Temperatur und Entropie der Strahlung. *Ann. Phys., Ser.* 2 52: 132–165.

4 Stefan, J. (1879). Über die Beziehung zwischen d\er Wärmestrahlung und der Temperatur. *Stiz. her. Akad. Wiss. Wien.* 79: 391–428.

5 Boltzmann, L. (1884). Ableitung des Stefan'schen Gesetzes, betreffend der Abhängigkeit der Wärmestrahlung von der Temperatur aus der electromagnetischen Lichtteorie. *Ann. Phys. (Leipzig), Ser.* 3 22: 291–294.

6 Dunkle, R.V. (1954). Thermal radiation tables and applications. *Trans. ASME* 65: 549–552.

7 Howell, J.R. (1982). *A Catalog of Radiation Configuration Factors*. New York: McGraw-Hill.

8 Siegel, R. and Howell, J.R. (1972). *Thermal Radiation Heat Transfer*, 783–791. New York: McGraw-Hill.

9 Gubareff, G.G., Janssen, J.E., and Torborg, R.H. (1960). *Thermal Radiation Properties Survey*, 2e. Minneapolis, MN: Honeywell Research Center.

10 Touloukian, Y.S. and Ho, C.Y. (eds.) (1972). *Thermophysical Properties of Matter*, vol. 7–9. New York: Plenum.

11 Kirchhoff, G. (1882). *Gesammelte Abhandlungen*, 574. Leipzig: Johann Ambrosius Barth.

12 Oppenheim, A.K. (1956). Radiation analysis by the network method. *Trans. ASME* 78: 725–735.

13 Hottel, H.C. (1954). Radiant-heat transmission. In: *Heat Transmission*, Chapter 4, 3e (ed. W.H. McAdams). New York: McGraw-Hill.

14 Sparrow, E.M. and Cess, R.D. (1978). *Radiation Heat Transfer*. Washington, DC: Hemisphere.

15 Hottel, H.C. and Sarofim, A.F. (1967). *Radiative Heat Transfer*, 31–39. New York: McGraw-Hill.

16 Bejan, A. (1984). *Convection Heat Transfer*, 1e, 110. New York: Wiley.

17 Bejan, A. and Reis, A.H. (2005). Thermodynamic optimization of global circulation and climate. *Int. J. Energy Res.* 29: 303–316.

18 Reis, A.H. and Bejan, A. (2006). Constructal theory of global circulation and climate. *Int. J. Heat Mass Transfer* 49: 1857–1875.

19 Clause, M., Meunier, F., Reis, A.H., and Bejan, A. (2012). Climate change, in the framework of the constructal law. *Int. J. Glob. Warm.* 4: 242–260.

20 Bejan, A. (2015). Sustainability: the water and energy problem, and the natural design solution. *Eur. Rev.* 23: 481–488.

21 Koonsrisuk, A., Lorente, S., and Bejan, A. (2010). Constructal solar chimney configuration. *Int. J. Heat Mass Transfer* 53: 327–333.

22 Lorente, S., Koonsrisuk, A., and Bejan, A. (2010). Constructal distribution of solar chimney power plants: few large and many small. *Int. J. Green Energy* 7: 577–592.

23 Ojeda, J.A. and Messina, S. (2017). Enhancing energy harvest in a constructal solar collector by using alumina-water as nanofluid. *Sol. Energy* 147: 381–389.

24 Zhou, X., Xu, Y., Yuan, S. et al. (2014). Pressure and power potential of sloped-collector solar updraft tower power plant. *Int. J. Heat Mass Transfer* 75: 450–461.

25 Botsaris, P.N. (2015). An approach of the spatial planning of a photovoltaic park using the constructal theory. *Sustain. Energy Technol. Assess.* 11: 11–16.

26 Solé, A., Falcoz, Q., Cabeza, L.F., and Neveu, P. (2018). Geometry optimization of a heat storage system for concentrated solar power plants (CSP). *Renew. Energy* 123: 227–235.

27 Mavromatidis, L.E., Mavromatidi, A., and Lequay, H. (2014). The unbearable lightness of expertness or space creation in the "climate change" era: a theoretical extension of the "constructal law" for building and urban design. *City Cult. Soc.* 5: 21–29.

28 Kasimova, R.G., Tishin, D., Obnosov, Y.V. et al. (2014). Ant mound as an optimal shape in constructal design. *J. Theor. Biol.* 355: 21–32.

29 Watts, R.G. (1990). Climate change due to greenhouse gases: change, impacts, and responses. In: *Heat Transfer 1990*, vol. 1 (ed. G. Hetsroni), 419–434. Washington, DC: Hemisphere.

Problems

Radiative Properties

10.1 Calculate the solid angle through which the sun can be seen from the Earth. With reference to Figure P10.1, the dimensions of the Earth–sun configuration are

$$d = 12\,756\,\text{km}, \quad \text{or} \quad 7926\,\text{miles}$$

$$r = 1.447 \times 10^8\,\text{km}, \quad \text{or} \quad 9.302 \times 10^7\,\text{miles}$$

$$D = 1.392 \times 10^8\,\text{km}, \quad \text{or} \quad 8.65 \times 10^5\,\text{miles}$$

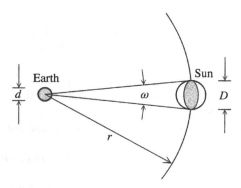

Figure P10.1

10.2 Determine analytically the wavelength at which the monochromatic hemispherical emissive power $E_{b,\lambda}$ is maximum, and verify that your result matches Wien's displacement law. Also determine the corresponding expression for the maximum monochromatic hemispherical emissive power $E_{b,\lambda,\text{max}}$.

10.3 Prove that the dimensionless radiation function $E_b(0 - \lambda T)/\sigma T^4$ depends only on the value of the group λT.

10.4 Fresh snow absorbs 80% of the incident solar radiation with wavelengths shorter than $\lambda_1 = 0.4\,\mu\text{m}$, 10% of the radiation between λ_1 and $\lambda_2 = 1\,\mu\text{m}$, and 100% of the radiation with wavelengths longer than λ_2. The solar surface is approximated adequately by a black surface with the temperature $T = 5800\,\text{K}$. Calculate the fraction of the total incident solar radiation that is absorbed by fresh snow.

10.5 The surface $A_1 = 1\,\text{cm}^2$ shown in Figure P10.5 is black and emits radiation with the total intensity $I_{b,1} = 3000\,\text{W/m}^2 \cdot \text{sr}$. The target area is $A_2 = 2\,\text{cm}^2$.
(a) Calculate the heat current $q_{1 \to 2}$ representing the portion of the radiation emitted by A_1 and intercepted by A_2.
(b) Repeat the preceding calculation by assuming that the surface A_2 is centered at the point P and is parallel to A_1. Comment on what the face-to-face alignment of A_1 and A_2 does to the value of q_{1-2}.

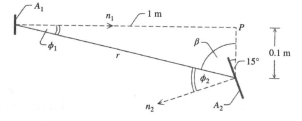

Figure P10.5

10.6 (a) In the small$-\lambda T$ limit, the denominator appearing on the right side of Eq. (10.16) is approximately equal to exp $(C_2/\lambda T)$. Verify that in this limit the radiation function approaches the asymptote listed in Eq. (10.26).

(b) In the large$-\lambda T$ limit, the denominator of the $E_{\lambda,b}$ expression (10.16) approaches $C_2/\lambda T$. Show that the corresponding asymptote of the radiation function is the expression listed in Eq. (10.27).

View Factors

10.7 Determine analytically the view factor F_{12} between the infinitesimal area dA_1 and the parallel disc A_2 shown in the fourth entry in Table 10.2. Verify that in the $H/R \rightarrow 0$ limit the view factor F_{12} approaches 1, and discuss the physical meaning of this limiting configuration.

10.8 Consider the infinitely long enclosure with triangular cross section shown as the third configuration in Table 10.2. This enclosure has three surfaces. Invoke the reciprocity and enclosure relations that exist among the various view factors to derive (and prove the validity of) the F_{12} expression listed in Table 10.2.

10.9 On a clear day, a disc-shaped flat-plate solar collector is oriented so that its axis passes through the center of the sun. The sun's diameter is 1.392×10^6 km, and the average radius of the Earth's orbit is 1.447×10^8 km.
(a) Calculate the geometric view factor F_{21} in which surface 1 is the collector disc and surface 2 is the solar sphere.
(b) Calculate the same view factor by assuming that the solar surface is itself a flat disc of diameter 1.392×10^6 km that is parallel to the disc of the collector. Are the view factors of the coaxial disc–sphere and coaxial disc–disc configurations always equal? Why, then, are the numerical answers to parts (a) and (b) identical?

10.10 The last entry in Table 10.2 suggests that the view factor F_{12} from a sphere to a disc positioned on the same centerline does not depend on the sphere radius (R_1). This result may seem surprising because view factors are geometry dependent, and R_1 is an important part of the sphere–disc geometry. Derive your own version of the F_{12} view factor from the sphere to the disc. (*Hint*: Consider the larger sphere that touches the rim of the disc, and use the view factor information listed for the two-sphere enclosure in Figure 10.15.)

10.11 Figure P10.11 shows two finite-width strips that are infinitely long in the direction normal to the paper. The widths of these strips are L_1 and L_2. Show that the geometric view factor from surface 1 to surface 2 is $F_{12} = [a+b-(c+d)]/(2L_1)$, for which the dimensions a, b, c, and d have been defined directly on Figure P10.11. This method of calculating F_{12} is known as the crossed-strings method [15], in view of the diagonals a and b that cross in the space between L_1 and L_2.

10.12 Consider the two infinitely long parallel plates of width X and spacing H shown in Figure P10.12. This configuration corresponds to the limit $Y/H \rightarrow \infty$ represented by the uppermost curve in Figure 10.11. Rely on the answer to the preceding problem to determine a compact analytical expression for the view factor F_{12}.

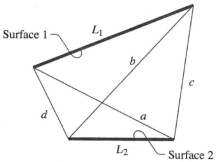

Figure P10.11

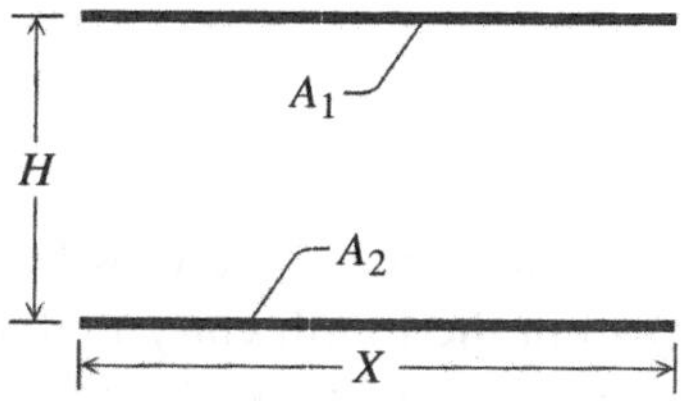

Figure P10.12

Diffuse-Gray Enclosures

10.13 The evacuated space between two infinite parallel plates maintained at different temperatures (T_H, T_L) contains n parallel radiation shields. All the surfaces – the inner surfaces of the end plates T_H and T_L and both sides of each of the radiation shields T_i ($i = 1, 2, ..., n$) – are diffuse-gray surfaces with the same total hemispherical emissivity ϵ.

(a) Determine the radiation heat transfer rate from T_H to T_L.

(b) Show that the heat transfer rate from T_H to T_L is $n + 1$ times smaller than what it would have been if all the shields were absent.

10.14 Figure P10.14 shows a vertical cross section through the center of an enclosed barbecue system. The pile of charcoal ($T_1 = 800\,°C$) is shaped as a disc of diameter 20 cm. The thin-walled metallic enclosure is shaped as a 0.6-m-diameter sphere. The enclosure wall can be modeled as isothermal (T_2). The heat transfer from the outside of the enclosure to the ambient ($T_3 = 25\,°C$) is by radiation and convection. The convection effect is characterized by the overall (forced and natural convection) heat transfer coefficient $h = 20\,\text{W/m}^2 \cdot \text{K}$. The charcoal, enclosure, and ambient can be modeled as black surfaces. Assume that the coarse grill that supports the charcoal is a transparent surface.

(a) Calculate the temperature of the enclosure wall, T_2.

(b) To what degree does radiation contribute to the total heat transfer rate from the outer surface of the enclosure, that is, from T_2 to T_3?

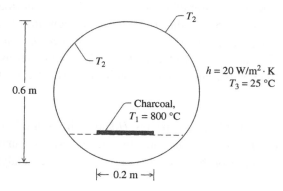

Figure P10.14

10.15 Figure P10.15 shows the cross section through an enclosure that extends to infinity in the direction normal to the paper. The cross section is an isosceles triangle in which $A_2 = A_3 = 10A_1$. Each of the three surfaces can be modeled as black and isothermal. The A_3 surface is insulated with respect to the surroundings, while the net heat current q_{1-2} enters the enclosure through the A_1 surface and exits through the A_2 surface.

  (a) Calculate the numerical values of the view factors F_{12}, F_{13}, F_{21}, and F_{23}.
  (b) Determine the net heat transfer rate q_{1-2}.
  (c) Determine the "floating" temperature of the insulated surface, T_3, as a function of the temperatures of the differentially heated surfaces of the enclosure (T_1, T_2).

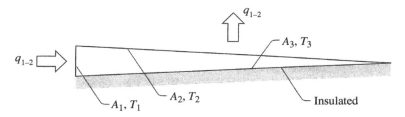

Figure P10.15

10.16 The reason why it is a good idea to keep turning the steak on the grill is that the heat flux on the bottom side of the steak is much larger than the heat flux on the top side. To see this, consider the enclosed barbecue shown in Figur P10.16. The coal pile is shaped as a disc of diameter 20 cm, with the upper area A_1 at $T_1 = 800\,°C$.

   Assume that the underside of the coal pile is insulated by the ash mound collected under it. The enclosure area A_4 is a 0.6-m-diameter sphere with temperature $T_4 = 150\,°C$.

   A fresh steak has just been put on the grill, $T_2 = T_3 = 25\,°C$. Its shape is approximated well by a disc with a diameter of 10 cm. It is positioned at 15 cm directly above the coal. All the surfaces inside the enclosure (A_1, A_2, A_3, A_4) can be modeled as black. Assume further that the coarse grills that hold the steak and the coal are transparent.

   Calculate the average heat flux q_2'' that enters through the bottom of the steak (A_2) and the average heat flux q_3'' that enters through the top (A_3). Compare q_2'' with q_3''.

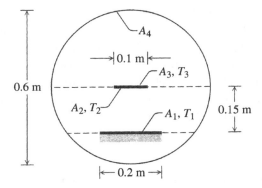

Figure P10.16

10.17 The ceramic wall shown in Figure P10.17 holds an array of flat-plate electrical conductors whose normal operating temperature is $T_2 = 150\,°C$. The emissivity of the conductor surface is $\epsilon_2 = 0.7$. The ceramic wall is a refractory surface. The surroundings can be modeled as black with temperature $T_2 = 25\,°C$. The plates are long in the direction perpendicular to Figure P10.17. Analyze the rectangular three-surface enclosure (T_1, T_2, T_3) formed between two consecutive plates, and calculate the following:

(a) The net heat transfer rate from T_2 to T_3 through the plane labeled A_3 in Figure P10.17.

(b) The temperature of the ceramic wall, T_1.

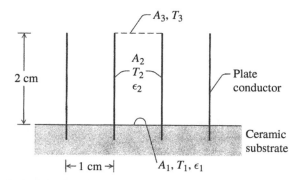

Figure P10.17

10.18 Useful energy is extracted from a geothermal reservoir by using the well shown in Figure 10.18. Cold water is forced to flow downward through the annular space of the coaxial cylinders, and it is heated by contact with the surrounding rock material (the temperature of this material increases almost linearly with depth). The hot water stream returns to the Earth's surface through the inner pipe. A critical component in the design of the well is the wall of the inner pipe. This has to be a good insulator, to prevent the heat transfer that would take place in the radial direction, from the inner hot stream to the cold stream of the annular space.

A possible design is outlined on the right side of Figure P10.18. The dividing wall discussed until now will be made out of two concentric tubes, so that the narrow space created between them can be evacuated. The evacuated annular space has an inner surface of diameter $D_1 = 15$ cm and an outer surface of diameter $D_2 = 16$ cm. The total length (depth) of the well is $L = 1$ km.

Estimate the total heat transfer rate between the hot and cold water streams across the evacuated gap. Assume that this heat leakage is controlled (impeded) primarily by the radiation between the long cylindrical shells D_1 and D_2. The total hemispherical emissivities of these two surfaces are $\epsilon_1 \cong \epsilon_2 \cong 0.6$. The temperature of the inner surface (D_1) is controlled by the upflowing stream of pressurized hot water, $T_1 \cong 220\,°C$. The temperature of the outer surface of the evacuated gap is controlled by the downflowing cold stream and is given by the linear relation

$$T_2(x) = T_2(0) + \left(\frac{dT_2}{dx}\right) x$$

where $T_2(0)$ is the ground-level temperature, $T_2(0) \cong 20\,°C$, and $dT_2/dx \cong 200\,°C/km$.

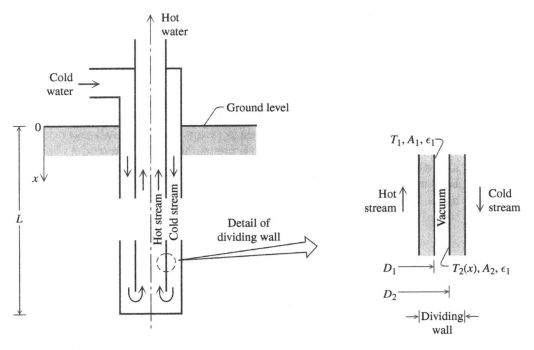

Figure P10.18

10.19 Hot ash from the boiler of a cogeneration plant is stored temporarily in a hopper located in a 5-m-deep pit in the floor of the plant. The hopper is long in the direction perpendicular to the plane of Figure P10.19. The distance from the side wall of the hopper to the vertical wall of the pit is 2.5 m. The temperature of the outer surface of the hopper is 300 °C. This raises some questions concerning the safety of workers who may have to descend into the pit to unclog the flow of ash through the bottom end of the hopper.

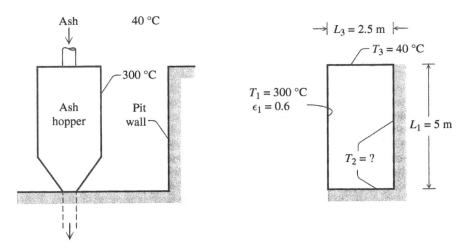

Figure P10.19

A key safety parameter is the temperature of the pit wall and floor. Estimate this temperature by relying on the two-dimensional enclosure model shown on the right side of Figure P10.19. The hopper surface is diffuse-gray, with $\epsilon_1 = 0.6$. The pit surface is adiabatic (reradiating). The ambient is represented by the black surface of temperature 40 °C, which completes the enclosure. Assume that all the heat transfer is due to radiation.

10.20 In the calculation of the pit wall temperature (T_2) in the preceding problem, you were advised to assume that all the heat transfer in the three-surface enclosure model is by radiation. The answer to that calculation turns out to be $T_2 = 232$ °C.

Obtain a more realistic estimate for T_2 by taking into account the natural convection heat transfer from the pit wall to the air that fills the hopper–wall space. Assume that the heated air rises as a boundary layer along the vertical wall of the pit. Assume also that the air temperature midway between the hopper and the pit wall is 40 °C, that is, equal to the ambient air that descends slowly into the 2.5-m-wide space (this ambient air replaces the heated air that rises along the pit wall). Finally, assume that the heat transfer coefficient calculated in this manner applies over the entire surface of the pit wall (5 m side and 2.5 m bottom).

10.21 Another aspect of the ash hopper analyzed in the preceding two problems is that it is designed not only to temporarily store the ash but also to cool it. Hot ash with a temperature of 900 °C falls steadily into the hopper at the rate of 550 kg/h. The specific heat of the ash material is 1 kJ/kg · K. The freshly fallen ash is cooled to the new temperature T_1, as it rests on top of the ash pile.

(a) Estimate the steady-state temperature of the top of the ash pile, T_1, by analyzing the hopper model shown on the right side of Figure P10.21. The top of the ash pile is a diffuse-gray surface with $\epsilon_1 = 0.9$ and $A_1 = 11$ m². The wall of the upper portion of the hopper has the area $A_2 = 27$ m²; this is also modeled as diffuse-gray, with $\epsilon = 0.6$ on both sides. The 40 °C environment is represented by the outermost surface, which is modeled as black. Assume that the heat transfer is due entirely to radiation.

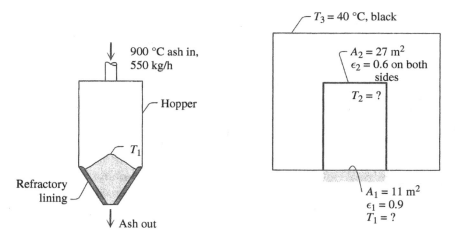

Figure P10.21

(b) Calculate the total heat transfer rate released by the ash flow, that is, from T_1 to T_3 across the two-chamber enclosure model.

(c) Calculate the temperature of the hopper wall, T_2.

10.22 The intensity of the solar irradiation that strikes the Earth is $I = 1360\ \text{W/m}^2$. The Earth's surface absorptivity for solar radiation is $\alpha = 0.7$. At the same time, the Earth loses heat by radiation to the universe of temperature $T_\infty \cong 4\ \text{K}$. The Earth's emissivity is $\epsilon(T) = 0.95$, where T is the Earth's average temperature (T is averaged annually and spatially).

(a) Assume that the atmosphere is absent, and calculate the Earth's average temperature T.

(b) Compare the calculated T value with the actual average temperature of the Earth ($15\,°C$). Comment on the difference and how that difference may be due to the role played by the atmosphere [29] (Figure P10.22).

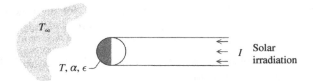

Figure P10.22

Participating Media

10.23 A spherical vessel with a 1-m diameter and $200\,°C$ wall temperature contains a gas at a pressure of 2 atm and temperature of $1000\,°C$. The gas composition on a molar basis is 50% N_2, 30% CO_2, and 20% O_2. The interior surface of the vessel is black. Calculate the following:

(a) The gas emissivity.

(b) The gas absorptivity with respect to the radiation emitted by the surface.

(c) The instantaneous net heat transfer rate from the gas to the wall of the container.

10.24 The equilibrium mixture of H_2, O_2, and H_2O at 1 atm and 3000 K has the following composition in terms of mole fractions [2]: $x_{H_2} = 0.137$, $x_{O_2} = 0.069$, and $x_{H_2O} = 0.794$. The mixture is placed instantly in a long cylinder with a 0.4-m internal diameter. The internal surface of the cylinder may be modeled as black. Estimate the instantaneous cooling rate that must be supplied to the external surface of the cylinder to maintain the wall temperature at 600 K.

10.25 The adiabatic burning of methane with theoretical air generates the ideal gas mixture of products $CO_2 + 2H_2O + 7.52N_2$, in which the numerical coefficients represent moles per one mole of CO_2. The mixture of products exits the combustion chamber at 1 atm and at the adiabatic flame temperature of 2320 K [2]. It passes through a cylindrical duct with a 0.2-m internal diameter and a wall temperature of 500 K. Its flowrate is high enough so that mixture temperature is fairly constant along the duct, this in spite of the radiation cooling effect provided by the duct wall. The latter may be modeled as black. Calculate the heat transfer rate removed from the products, per unit axial length. The emission from the duct wall may be neglected.

10.26 An industrial furnace that bums methane with 400% theoretical air exhausts a gas stream with the following composition: $CO_2 + 2H_2O + 6O_2 + 30.08N_2$. The numerical coefficients represent moles per one mole of CO_2. This mixture of products of combustion is at 700 K and 1 atm and occupies the space between two parallel walls. The distance between the walls is 0.5 m, and the area of one wall is 20 m^2. Both surfaces are at a temperature of 460 K and may be modeled as black. Calculate the net radiation heat transfer rate received by both surfaces. Note that the gas and wall temperatures are comparable; therefore, the radiation emitted by the walls should not be neglected.

10.27 The gap between two large parallel surfaces at different temperatures (T_1, T_2) is occupied by an isothermal gray gas $(T_g, \epsilon_g, \tau_g)$. Each surface has an area A and is diffuse-gray. Their respective emissivities are equal to the emissivity of the gas; their common value happens to be 0.5.
(a) Derive an expression for the net rate of heat transfer between the two surfaces.
(b) Compare your result with the heat transfer rate that occurs when the space between the plates is completely evacuated. Show in this way that the gas has a radiation shielding effect.

10.28 The 10-cm-wide space between the large parallel plates shown in Figure P10.28 is filled with steam at atmospheric pressure. The steam temperature is uniform at a level that could be determined by analyzing the radiation heat transfer in the enclosure formed by the two plates (the method is outlined in the next problem). Model the steam as a gray gas, and determine its absorptivity α_g by assuming that T_g is halfway between T_1 and T_2. Evaluate the goodness of this model by comparing α_g with (i) the absorptivity of steam with respect to radiation arriving from T_1 and (ii) the absorptivity of steam with respect to radiation arriving from T_2.

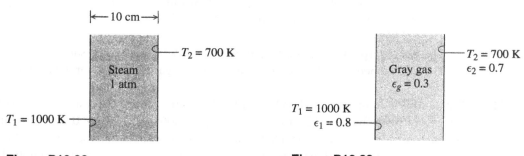

Figure P10.28   **Figure P10.29**

10.29 An isothermal layer of gray gas with the emissivity $\epsilon_g = 0.3$ is sandwiched between two large and parallel walls. The internal surfaces of these walls can be modeled as diffuse-gray, with temperatures and emissivities listed on Figure P10.29.
(a) Calculate the net radiation heat flux from T_1 to T_2.
(b) Determine the gas temperature T_g, and show in this way that T_g is closer to the higher of the two side-wall temperatures.

10.30 The heat transfer process that goes on in a gas furnace can be modeled as shown in Figure P10.30. The surface that is to be heated is cold and diffuse-gray (A_1, T_1, ϵ_1,). The surrounding surface (A_2) is refractory, that is, adiabatic on its back side. The space between the two surfaces is the furnace volume, which is filled by the hot (participating) gas that acts as heat source. The gas is modeled as isothermal and gray (T_g, $\alpha_g = \epsilon_g$). Draw the radiation network for this furnace model, and derive an expression for the net instantaneous heat transfer rate from the gas to the cold surface, q_{g-1}.

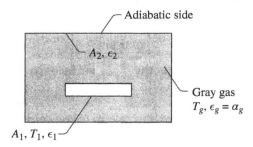

Figure P10.30

10.31 The space between the infinite parallel plates shown on the left side of Figure P10.31 is filled with a gray gas, while the internal parallel surfaces are diffuse-gray. The net heat transfer originates from the gas, which is hot (T_g), and sinks into one of the surfaces, which is cold (T_1). The other surface is refractory, that is, insulated on its back side. The gas emissivity and the emissivity of the cold surface are equal: their common value is 0.5.

(a) Derive an expression for the net instantaneous heat transfer rate from the gas to the cold surface, q_{g-1}.

(b) Consider the alternative shown on the right side of Figure P10.31, where the space has been evacuated and the right wall has assumed the temperature and emissivity of the gas. The transfer of heat is again from T_g to T_1, through vacuum this time. Derive an expression for the net heat transfer rate q_{g-1}, and compare it with the one derived in part (a). In this way you will be able to answer the question of when the heat transfer rate is higher, when the heat source fills the entire volume (gas, part (a)), or when the heat source is flattened into a sheet (part (b)).

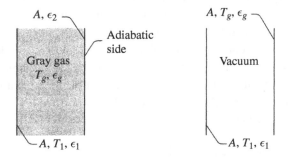

Figure P10.31

Evolutionary Design

10.32 Heat is transferred by radiation from an infinitesimal surface dA_1 to a rectangular area with fixed size $(2a) \times (2b)$ and variable shape a/b. Figure P10.32 shows that the surface dA_1 is situated on the line perpendicular to the center of the rectangular area. The objective is to increase the radiation thermal conductance between dA_1 and the rectangular area by selecting the shape of the rectangular area. What particular shape would you select for the rectangular area? To answer, rely on the formula for the geometric view factor for the elemental configuration shown on the right side of Figure P10.32 [7]:

$$F_{d1-2} = \frac{1}{2\pi} \left\{ \frac{A}{(1+A^2)^{1/2}} \tan^{-1}\left[\frac{B}{(1+A^2)^{1/2}}\right] + \frac{B}{(1+B^2)^{1/2}} \tan^{-1}\left[\frac{A}{(1+B^2)^{1/2}}\right] \right\}$$

where $A = a/c$ and $B = b/c$. Note that the area element A_2 is one quarter of the area shown on the left side of Figure P10.32.

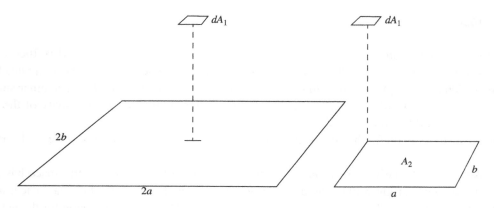

Figure P10.32

10.33 Architects propose to install a small refrigerated panel in the ceiling of a room in order to cool a side wall that is being heated by exposure to sunlight. The wall area is $2A_2$, and the cold panel area is dA_1. The heat transfer between dA_1 and $2A_2$ is by radiation. The degree of freedom in this configuration is the distance (c) between the cold panel and the side wall. The objective is to position the cold panel such that the thermal conductance between it and the side wall is increased. Where should the cold panel be located? One architect argues that the dimension c should be as great as possible such that the cold panel has a broader view of the warm wall. Is this argument correct? Find the answer by analyzing the behavior of the view factor between the two surfaces (Figure P10.33). The view factor between dA_1 and A_2 (not $2A_2$) is given by [7]

$$F_{d1-A_2} = \frac{1}{2\pi} \left[\tan^{-1}\frac{1}{Y} - \frac{Y}{(X^2+Y^2)^{1/2}} \tan^{-1}\frac{1}{(X^2+Y^2)^{1/2}} \right]$$

where $X = qa/b$ and $Y = c/b$.

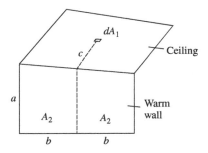

Figure P10.33

10.34 A room is maintained at the temperature T_r above the ambient temperature T_0 by the heat input Q_r received from a reversible heat pump driven by the heat input Q_c from a solar collector of temperature T_c (Figure P10.34). The net solar heat transfer into the collector is fixed, Q^*. The rate of collector-ambient heat loss is $Q_0 = U_c A_c (T_c - T_0)$, where the thermal conductance $U_c A_c$ is fixed, and A_c is the collector surface. Determine the optimal collector temperature such that the heat input to the room Q_r is maximized.

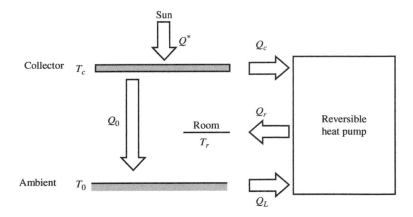

Figure P10.34

Appendix A

Constants and Conversion Factors

Constants.

Universal ideal gas constant	$\bar{R} = 8.314 \text{ kJ/kmol·K}$
	$= 1.9872 \text{ cal/mol} \cdot \text{K}$
	$= 1.9872 \text{ Btu/lbmol} \cdot \text{R}$
	$= 1545.33 \text{ ft} \cdot \text{lbf/lbmol} \cdot \text{R}$
Boltzmann's constant	$k = 1.38054 \times 10^{-23} \text{ J/K}$
Planck's constant	$h = 6.626 \times 10^{-34} \text{ J/s}$
Speed of light in vacuum	$c = 2.99\,8 \times 10^{8} \text{ m/s}$
Avogadro's number	$N = 6.022 \times 10^{23} \text{ molecules/mol}$
Stefan–Boltzmann constant	$\sigma = 5.669 \times 10^{-8} \text{ W/m}^2 \cdot \text{K}^4$
	$= 0.1714 \times 10^{-8} \text{ Btu/h·ft}^2 \cdot \text{R}^4$
Atmospheric pressure	$P_{atm} = 0.101325 \text{ MPa}$
	$= 1.01325 \text{ bar}$
	$= 1.01325 \times 10^5 \text{ N/m}^2$
Ice point at 1 atm	$T_{ice} = 0\,°C = 273.15 \text{ K}$
Gravitational acceleration	$g = 9.807 \text{ m/s}^2$
	$= 32.17 \text{ ft/s}^2$
Mole	1 mol = sample containing 6.022×10^{23} elementary entities (e.g. molecules); abbreviated also as 1 gmol, or
	$1 \text{ mol} = 10^{-3} \text{kmol}$
	$= 10^{-3} \text{kgmol}$
	$= \dfrac{1}{453.6} \text{lbmol}$
Natural logarithm	$\ln x = 2.30258 \log_{10} x$
	$\log_{10} x = 0.4343 \ln x$
Important numbers	$e = 2.71828$
	$\pi = 3.14159$
	$1° = 0.01745 \text{ rad}$

Heat Transfer: Evolution, Design and Performance, First Edition. Adrian Bejan.
© 2022 John Wiley & Sons, Inc. Published 2022 by John Wiley & Sons, Inc.
Companion website: www.wiley.com/go/bejan/heattransfer

Conversion factors.

Acceleration	$1\ \text{m/s}^2 = 4.252 \times 10^7\ \text{ft/h}^2$
Area	$1\ \text{in.}^2 = 6.452\ \text{cm}^2$
	$1\ \text{ft}^2 = 0.0929\ \text{m}^2$
	$1\ \text{yd}^2 = 0.8361\ \text{m}^2$
	$1\ \text{mile}^2 = 2.59\ \text{km}^2$
	$1\ \text{hectare} = (100\ \text{m})^2$
	$1\ \text{acre} = 4047\ \text{m}^2$
	$= 0.405\ \text{hectares}$
Density	$1\ \text{kg/m}^3 = 0.06243\ \text{lbm/ft}^3$
	$1\ \text{lbm/ft}^3 = 16.018\ \text{kg/m}^3$
Energy	$1\ \text{kJ} = 737.56\ \text{ft} \cdot \text{lbf}$
	$= 0.9478\ \text{Btu}$
	$= 3.725 \times 10^{-4}\ \text{hp} \cdot \text{h}$
	$= 2.778 \times 10^{-4}\ \text{kW} \cdot \text{h}$
	$1\ \text{Btu} = 1055\ \text{J}$
	$= 778.16\ \text{ft} \cdot \text{lbf}$
	$= \dfrac{1}{3412.14}\ \text{kw} \cdot \text{h}$
	$= \dfrac{1}{2544.5}\ \text{hp} \cdot \text{h}$
	$1\ \text{cal} = 4.187\ \text{J}$
	$1\ \text{erg} = 10^{-7}\ \text{J}$
Force	$1\ \text{lbf} = 4.448\ \text{N}$
	$= 0.4536\ \text{kgf}$
	$1\ \text{dyne} = 10^{-5}\ \text{N}$
Heat flux	$1\ \text{W/m}^2 = 0.317\ \text{Btu/h} \cdot \text{ft}^2$
	$1\ \text{Btu/h} \cdot \text{ft}^2 = 3.154\ \text{W/m}^2$
Heat transfer coefficient	$1\ \text{W/m}^2 \cdot \text{K} = 0.1761\ \text{Btu/h} \cdot \text{ft}^2 \cdot {}^\circ\text{F}$
	$= 0.8598\ \text{kcal/h} \cdot \text{m}^2 \cdot {}^\circ\text{C}$
	$1\ \text{Btu/h} \cdot \text{ft}^2 \cdot {}^\circ\text{F} = 5.6786\ \text{W/m}^2 \cdot \text{K}$
Heat transfer rate	$1\ \text{Btu/s} = 1055\ \text{W}$
	$1\ \text{Btu/h} = 0.2931\ \text{W}$
	$1\ \text{hp} = 745.7\ \text{W}$
	$1\ \text{ft} \cdot \text{lbf/s} = 1.3558\ \text{W}$
Kinematic viscosity (ν), thermal diffusivity (α)	$1\ \text{m}^2/\text{s} = 10^4\ \text{cm}^2/\text{s}$
	$= 10^4\ \text{stokes}$
	$= 3.875 \times 10^4\ \text{ft}^2/\text{h}$
	$= 10.764\ \text{ft}^2/\text{s}$

Latent heat, specific energy, and specific enthalpy	1 kJ/kg $= 0.4299$ Btu/lbm
	$= 0.2388$ cal/g
	1 Btu/lbm $= 2.326$ kJ/kg
Length	1 in. $= 2.54$ cm
	1 ft $= 0.3048$ m
	1 yd $= 0.9144$ m
	1 mile $= 1.609$ km
Mass	1 lbm $= 0.4536$ kg
	1 kg $= 2.2046$ lbm
	$= 1.1023 \times 10^{-3}$ U.S.ton
	$= 10^{-3}$ tonne
	1 oz $= 28.35$ g
Power	1 Btu/s $= 1055$ W $= 1.055$ kW
	1 Btu/h $= 0.293$ W
	1 W $= 3.412$ Btu/h
	$= 9.48 \times 10^{-4}$ Btu/s
	1 hp $= 0.746$ kW
	$= 0.707$ Btu/s
Pressure and stress	1 Pa $= 1$ N/m^2
	1 psi $= 6895$ N/m^2
	1 atm $= 14.69$ psi
	$= 1.013 \times 10^5$ N/m^2
	1 bar $= 10^5$ N/m^2
	1 Torr $= 1$ mm Hg
	$= 133.32$ N/m^2
	1 psi $= 27.68$ in.H$_2$O
	1 ft H$_2$O $= 0.4335$ psi
Specific heat and specific entropy	1 kJ/kg $\cdot$ K $= 0.2388$ Btu/lbm $\cdot$ °F
	$= 0.2389$ cal/g $\cdot$ °C
	1 Btu/lbm $\cdot$ °F $= 4.187$ kJ/kg $\cdot$ K
Speed	1 mile/h $= 0.447$ m/s
	$= 1.609$ km/h
	1 km/h $= 0.278$ m/s
	$= 0.622$ mi/h
	1 m/s $= 3.6$ km/h
	$= 2.237$ mi/h

Temperature

$$1\,K = 1°C$$
$$1\,K = (9/5)°F$$
$$T(K) = T(°C) + 273.15$$
$$T(°C) = (5/9)\,[T(°F) - 32]$$
$$T(°F) = T(R) - 459.67$$

Temperature difference

$$\Delta T(K) = \Delta T(°C)$$
$$= (5/9)\,\Delta T(°F) = (5/9)\,\Delta T(R)$$

Thermal conductivity

$$1\,W/m \cdot K = 0.5782\,Btu/h \cdot ft \cdot °F$$
$$= 0.01\,W/cm \cdot K$$
$$= 2.39 \times 10^{-3}\,cal/cm \cdot s \cdot °C$$
$$1\,Btu/h \cdot ft\ \ °F = 1.7307\,W/m \cdot K$$

Thermal resistance

$$1\,K/W = 0.5275°F/Btu \cdot h$$
$$1°F/Btu \cdot h = 1.896\,K/W$$

Viscosity

$$1\,N \cdot s/m^2 = 1\,kg/s \cdot m$$
$$= 2419.1\,lbm/ft \cdot h$$
$$= 5.802 \times 10^{-6}\,lbf \cdot h/ft^2$$
$$1\,poise = 1\,g/s \cdot cm$$

Volume

$$1\,liter = 10^{-3}\,m^3 = 1\,dm^3$$
$$1\,in.^3 = 16.39\,cm^3$$
$$1\,ft^3 = 0.02832\,m^3$$
$$1\,yd^3 = 0.7646\,m^3$$
$$1\,gal\,(U.S.) = 3.785\,l$$
$$1\,gal\,(imperial) = 4.546\,l$$
$$1\,pint = 0.5683\,l$$

Volumetric heat
generation rate

$$1\,W/m^3 = 0.0966\,Btu/h \cdot ft^3$$
$$1\,Btu/h \cdot ft^3 = 10.35\,W/m^3$$

Dimensionless groups.[a]

Bejan number	$Be = \Delta P L^2 / \mu \alpha$
Biot number	$Bi = hL/k_s$
Boussinesq number	$Bo = g\beta \Delta T H^3 / \alpha^2$
Eckert number	$Ec = U^2 / c_P \Delta T$
Fourier number	$Fo = \alpha t / L^2$
Graetz number	$Gz = D^2 U / \alpha x = Re_D Pr \dfrac{D}{x}$
Grashof number	$Gr = g\beta \Delta T H^3 / \nu^2$
Nusselt number	$Nu = hL/k_f$

Péclet number	$Pe = UL/\alpha = Re \cdot Pr$
Prandtl number	$Pr = \nu/\alpha = Sc/Le$
Rayleigh number	$Ra = g\beta\Delta T H^3/\alpha\nu$
Rayleigh number based on heat flux	$Ra^* = g\beta q'' H^4/\alpha\nu k$
Reynolds number	$Re = UL/\nu$
Stanton number	$St = h/\rho c_P U = Nu/Re \cdot Pr$
Stefan number	$Ste = c_f \Delta T/h_{sf}$

a) Subscripts $(\ )_s$ = solid, $(\ )_f$ = fluid.

Appendix B

Properties of Solids

Nonmetallic solids.[a]

Material	T (°C)	ρ (kg/m³)	c_p (kJ/kg·K)	k (W/m·K)	α (cm²/s)
Asbestos					
Cement board	20			0.6	
Felt (16 laminations per cm)	40			0.057	
Fiber	50	470	0.82	0.11	0.0029
Sheet	20			0.74	
	50			0.17	
Asphalt	20	2120	0.92	0.70	0.0036
Bakelite	20	1270	1.59	0.230	0.0011
Bark	25	340	1.26	0.074	0.0017
Beef (see meat)					
Brick					
Carborundum	1400			11.1	
Cement	10	720		0.34	
Common	20	1800	0.84	0.38–0.52	0.0028–0.0034
Chrome	100			1.9	
Facing	20			1.3	
Firebrick	300	2000	0.96	0.1	0.00054
Magnesite (50% MgO)	20	2000		2.68	
Masonry	20	1700	0.84	0.66	0.0046
Silica (95% SiO_2)	20	1900		1.07	
Zircon (62% ZrO_2)	20	3600		2.44	
Brickwork, dried in air	20	1400–1800	0.84	0.58–0.81	0.0049–0.0054

Heat Transfer: Evolution, Design and Performance, First Edition. Adrian Bejan.
© 2022 John Wiley & Sons, Inc. Published 2022 by John Wiley & Sons, Inc.
Companion website: www.wiley.com/go/bejan/heattransfer

Material	T (°C)	ρ (kg/m³)	c_p (kJ/kg·K)	k (W/m·K)	α (cm²/s)
Carbon					
Diamond (type llb)	20	3250	0.51	1350	8.1
Graphite (firm, natural)	20	2000–2500	0.61	155	1.02–1.27
Carborundum (SiC)	100	1500	0.62	58	0.62
Cardboard	0–20	~790		~0.14	
Celluloid	20	1380	1.67	0.23	0.001
Cement (Portland, fresh, dry)	20	3100	0.75	0.3	0.0013
Chalk ($CaCO_3$)	20	2000–3000	0.74	2.2	0.01–0.015
Clay	20	1450	0.88	1.28	0.01
Fireclay	100	1700–2000	0.84	0.5–1.2	0.35–0.71
Sandy clay	20	1780		0.9	
Coal	20	1200–1500	1.26	0.26	0.0014–0.0017
Anthracite	900	1500		0.2	
Brown coal	900			0.1	
Bituminous in situ		1300		0.5–0.7	0.003–0.004
Dust	30	730	1.3	0.12	0.0013
Concrete, made with gravel, dry	20	2200	0.88	1.28	0.0066
Cinder	24			0.76	
Cork					
Board	20	150	1.88	0.042	0.0015
Expanded	20	120		0.036	
Cotton	30	81	1.15	0.059	0.0063
Earth					
Coarse-grained	20	2040	1.84	0.59	0.0016
Clayey (28% moisture)	20	1500		1.51	
Sandy (8% moisture)	20	1500		1.05	
Diatomaceous	20	466	0.88	0.126	0.0031
Fat	20	910	1.93	0.17	0.001
Felt, hair	−7	130–200		0.032–0.04	
	94	130–200		0.054–0.051	
Fiber insulating board	20	240		0.048	
Glass					
Borosilicate	30	2230		1.09	
Fiber	20	220		0.035	

Material	T (°C)	ρ (kg/m³)	c_p (kJ/kg·K)	k (W/m·K)	α (cm²/s)
Lead	20	2890	0.68	0.7–0.93	0.0036–0.0047
Mirror	20	2700	0.80	0.76	0.0035
Pyrex	60–100	2210	0.75	1.3	0.0078
Quartz	20	2210	0.73	1.4	0.0087
Window	20	2800	0.80	0.81	0.0034
Wool	0	200	0.66	0.037	0.0028
Granite	20	2750	0.89	2.9	0.012
Gypsum	20	1000	1.09	0.51	0.0047
Ice	0	917	2.04	2.25	0.012
Ivory	80			0.5	
Kapok	30			0.035	
Leather, dry	20	860	1.5	0.12–0.15	~0.001
Limestone (Indiana)	100	2300	0.9	1.1	~0.005
Linoleum	20	535		0.081	
Lunar surface dust, in high vacuum	250	1500 ± 300	~0.6	~0.0006	
Magnezia (85%)	38–204			0.067–0.08	
Marble	20	2600	0.81	2.8	0.013
Meat					
Beef	25				~0.0014
Chuck	43–66	1060			0.0012
Liver	27			0.5	
Eye of loin, parallel to fiber	2–7			0.3	
Ground	6			0.35	
Lean	2–47			0.45	
Chicken					
Muscle, perpendicular to fiber	5–27			0.41	
Skin	5–27			0.03	
Egg, white	33–38			0.55	
Egg, whole	−8			0.46	
Egg, albumen gel, freeze-dried	41			0.04	
Egg, yolk	24–38			0.42	
Fish					
Cod fillets	−19			1.17	
Halibut	43–66	1080			0.0014
Herring	−19			0.8	

Material	T (°C)	ρ (kg/m³)	c_p (kJ/kg·K)	k (W/m·K)	α (cm²/s)
Salmon, perpendicular to fiber	−23			1.3	
	2			0.7	
Salmon, freeze-dried, parallel to fiber	−29			0.04	
Horse	25			0.41	
Lamb, lean	7–57			0.45	
Pork					
Ham, smoked	43–66	1090			0.0014
Fat	25			0.15	
Lean, perpendicular to fiber	27–57			0.52	
Lean, parallel to fiber	7–57			0.45	
Pig skin	25			0.37	
Sausage					
23% fat	25			0.38	
15% fat	25			0.43	
Seal, blubber	(−13)–(−2)			0.21	
Turkey					
Breast, perpendicular to fiber	−3			1.05	
	2			0.7	
Breast, parallel to fiber	−8			1.4	
Leg, perpendicular to fiber	2			0.7	
Whale					
Blubber, perpendicular to fiber	18			0.21	
Meat	−12			1.3	
Mica	20	2900		0.52	
Mortar	20	1900	0.8	0.93	0.0061
Paper	20	700	1.2	0.12	0.0014
Paraffin	30	870–925	2.9	0.24–0.27	~0.001
Plaster	20	1690	0.8	0.79	0.0058
Plexiglas (acrylic glass)	20	1180	1.44	0.184	0.0011
Polyethylene	20	920	2.30	0.35	0.0017
Polystyrene	20	1050		0.157	
Polyurethane	20	1200	2.09	0.32	0.0013
Polyvinyl chloride (PVC)	20	1380	0.96	0.15	0.0011
Porcelain	95	2400	1.08	1.03	0.004
Quartz	20	2100–2500	0.78	1.40	~0.008
Rubber					
Foam	20	500	1.67	0.09	0.0011
Hard (ebonite)	20	1150	2.01	0.16	0.0006

Material	T (°C)	ρ (kg/m³)	c_p (kJ/kg·K)	k (W/m·K)	α (cm²/s)
Soft	20	1100	1.67	~0.2	~0.001
Synthetic	20	1150	1.97	0.23	0.001
Salt (rock salt)	0	2100–2500	0.92	7	0.03–0.036
Sand					
Dry	20			0.58	
Moist	20	1640		1.13	
Sandstone	20	2150–2300	0.71	1.6–2.1	0.01–0.013
Sawdust, dry	20	215		0.07	
Silica stone (85% SiC)	700	2720	1.05	1.56	0.055
Silica aerogel	0	140		0.024	
Silicon	20	2330	0.703	153	0.94
Silk (artificial)	35	100	1.33	0.049	0.0037
Slag	20	2500–3000	0.84	0.57	0.0023–0.0027
Slate					
Parallel to lamination	20	2700	0.75	2.9	0.014
Perpendicular to lamination	20	2700	0.75	1.83	0.009
Snow, firm	0	560	2.1	0.46	0.0039
Soil (see also Earth)					
Dry	15	1500	1.84	1	0.004
Wet	15	1930		2	
Strawberries, dry	−18			0.59	
Sugar (fine)	0	1600	1.25	0.58	0.0029
Sulfur	20	2070	0.72	0.27	0.0018
Teflon (polytetrafluoroethylene)	20	2200	1.04	0.23	0.001
Wood, perpendicular to grain					
Ash	15	740		0.15–0.3	
Balsa	15	100		0.05	
Cedar	15	480		0.11	
Mahogany	20	700		0.16	
Oak	20	600–800	2.4	0.17–0.25	~0.0012
Pine, fir, spruce	20	416–421	2.72	0.15	0.0012
Plywood	20	590		0.11	
Wool					
Sheep	20	100	1.72	0.036	0.0021
Mineral	50	200	0.92	0.042	0.0025
Slag	25	200	0.8	0.05	0.0031

a) Constructed based on data compiled in Refs. [1–9].

Metallic solids.[a), b)]

Metals, alloys	Properties at 20 °C (293 K)				Thermal conductivity k (W/m·K)[a)]						
	ρ (kg/m³)	c_p (kJ/ kg·K)	k (W/ m·K)	α (cm²/s)	−100 °C (173 K)	0 °C (273 K)	100 °C (373 K)	200 °C (473 K)	400 °C (673 K)	600 °C (873 K)	1000 °C (1273 K)
Aluminum											
Pure	2707	0.896	204	0.842	215	202	206	215	249		
Duralumin (94–96% Al, 3–5% Cu, trace Mg)	2787	0.883	164	0.667	126	159	182	194			
Silumin (87% Al, 13% Si)	2659	0.871	164	0.710	149	163	175	185			
Antimony	6690	0.208	17.4	0.125	19.2	17.7	16.3	16.0	17.2		
Beryllium	1850	1.750	167	0.516	126	160	191	215			
Bismuth	9780	0.124	7.9	0.065	12.1	8.4	7.2	7.2			
Cadmium	8650	0.231	92.8	0.464	97	93	92	91			
Cesium	1873	0.230	36	0.836							
Chromium	7190	0.453	90	0.276	120	95	88	85	77		
Cobalt	8900	0.389	70	0.202	21		22.2	26			
Copper											
Pure	8954	0.384	398	1.16	420	401	391	389	378	366	336
Commercial	8300	0.419	372	1.07							
Aluminum bronze	8666	0.410	83	0.233							
Brass	8522	0.385	111	0.341	88		128	144	147		

Metals, alloys	Properties at 20 °C (293 K)				Thermal conductivity k (W/m·K)[a]						
	ρ (kg/m³)	c_P (kJ/ kg·K)	k (W/ m·K)	α (cm²/s)	−100 °C (173 K)	0 °C (273 K)	100 °C (373 K)	200 °C (473 K)	400 °C (673 K)	600 °C (873 K)	1000 °C (1273 K)
Bronze (75% Cu, 25% Sn)	8666	0.343	26	0.086							
Constantan	8922	0.410	22.7	0.061	21						
German silver	8618	0.394	24.9	0.073	19.2		31	40	48		
Gold	19 300	0.129	315	1.27		318		309			
Iron											
Pure	7897	0.452	73	0.205	87	73	67	62	48	40	35
Cast	7272	0.420	52	0.170							
Carbon steel, (0.5% C)	7833	0.465	54	0.148		55	52	48	42	35	29
Chrome steel, (1% Cr)	7865	0.460	61	0.167		62	55	52	42	36	33
Chrome–nickel steel											
(15% Cr, 10% Ni)	7865	0.460	19	0.053							
Invar (36% Ni)	8137	0.460	10.7	0.029							
Manganese steel, (1% Mn)	7865	0.460	50	0.139							

Metals, alloys	Properties at 20 °C (293 K)				Thermal conductivity k (W/m·K)[a]						
	ρ (kg/m³)	c_P (kJ/ kg·K)	k (W/ m·K)	α (cm²/s)	−100 °C (173 K)	0 °C (273 K)	100 °C (373 K)	200 °C (473 K)	400 °C (673 K)	600 °C (873 K)	1000 °C (1273 K)
Nickel–chrome steel											
(20% Ni, 15% Cr)	7865	0.460	14	0.039		14	15.1	15.1	17	19	
Silicon steel, (1% Si)	7769	0.460	42	0.116							
Stainless steel, Type 304	7817	0.460	13.8	0.040			15	17	21	25	
Type 347	7817	0.420	15	0.044	13		16	18	20	23	28
Tungsten steel, (2% W)	7961	0.444	62	0.176		62	59	54	48	45	36
Wrought	7849	0.460	59	0.163		59	57	52	45	36	33
Lead	11 340	0.130	34.8	0.236	36.9	35.1	33.4	31.6	23.3		
Lithium	530	3.391	61	0.340		61	61				
Pure	1746	1.013	171	0.970	178	171	168	163			
Magnesium											
Pure	7300	0.486	7.8	0.022							
Manganese											
Manganin	8400	0.406	21.9	0.064							
Molybdenum	10 220	0.251	123	0.480	138	125	118	114	109	106	99
Monel 505 (at 60 °C)	8360	0.544	19.7	0.043							
Nickel											
Pure	8906	0.445	91	0.230	114	94	83	74	64	69	78

Metals, alloys	Properties at 20 °C (293 K)				Thermal conductivity k (W/m·K)[a]						
	ρ (kg/m³)	c_P (kJ/ kg·K)	k (W/ m·K)	α (cm²/s)	−100 °C (173 K)	0 °C (273 K)	100 °C (373 K)	200 °C (473 K)	400 °C (673 K)	600 °C (873 K)	1000 °C (1273 K)
Nichrome	8666	0.444	17	0.044		17.1	18.9	20.9	24.6		
Niobium	8570	0.270	53	0.230							
Palladium	12 020	0.247	75.5	0.254		75.5	75.5	75.5	75.5		
Platinum	21 450	0.133	71.4	0.250	73	72	72	72	74	77	84
Potassium	860	0.741	103	1.62							
Rhenium	21 100	0.137	48.1	0.166							
Rhodium	12 450	0.248	150,	0.486							
Rubidium	1530	0.348	58.2	1.09							
Silver	10 524	0.236	427	1.72	431	428	422	417	401	386	
Sodium	971	1.206	133	1.14							
Tantalum	16 600	0.138	57.5	0.251		57.4					
Tin	7304	0.220	67	0.417	76	68	63				
Titanium	4540	0.523	22	0.093	26	22	21	20	19	21	22
Tungsten	19 300	0.134	179	0.692		182					
Uranium	18 700	0.116	28	0.129	24	27	29	31	36	41	
Vanadium	6100	0.502	31.4	0.103		31.3					
Wood's metal	1056	0.147	12.8	0.825							
Zinc	7144	0.388	121	0.437	122	122	117	110	100		
Zirconium	6570	0.272	22.8	0.128		23.2					

a) The effect of temperature on thermal conductivity is illustrated further in Figure 1.5.
b) Constructed based on data compiled in Refs. [1–7].

Ice properties.[a]

T (0 °C)	ρ[b] (g/cm^3)	h_{sf} (kJ/kg)	β[b] (K^{-1})
0	0.9164	333.4	
−5		308.5	
−10	0.9187	284.8	1.56×10^{-4}
−15		261.6	
−20	0.9203	241.4	1.38×10^{-4}
−40	0.9228		1.29×10^{-4}
−100	0.9273		1.24×10^{-4}
−200	0.9328		1.1×10^{-5}

a) At atmospheric pressure.
b) Constructed based on data compiled in Ref. [10].

Electrical resistivities: $\rho_e \cong \rho_{e,0}[1 + \alpha_0(T - T_0)]$.

Material	Reference temperature, T_0 (°C)	Reference resistivity, $\rho_{e,0}$ (10^{-8} W·m/A^2)	Temperature coefficient, α_0 (°C^{-1})
Aluminum	20	2.83	0.004
Brass	20	6.2	0.0015
Carbon			
Amorphous	20	3800–4100	–
Graphite	20	720–812	–
Copper			
Pure	20	1.68	0.004
Drawn	20	1.724	0.004
Gold		2.44	0.0034
Iron			
Pure	20	9.61	0.013
Cast	20	75–100	–
Wire	20	97.8	–

Material	Reference temperature, T_0 (°C)	Reference resistivity, $\rho_{e,0}$ (10^{-8} W·m/A^2)	Temperature coefficient, α_0 (°C^{-1})
Lead	20	22	0.0039
Lithium	20	9.28	–
Magnesium	20	4.38	–
Manganese	20	144	–
Monel	20	43.5	0.0019
Mercury	20	96.8	0.0009
Molybdenum	20	5.34	–
Nickel	20	8.54	0.0041
Niobium	0	12.5	–
Palladium	20	10.54	–
Phosphorus, white	11	10^{17}	–
Platinum	20	10.72	0.003
Potassium	20	7.2	–
Silver	20	1.63	0.0038
Sodium	20	4.77	–
Steel			
Soft	20	15.9	0.0016
Transformer	20	11.1	–
Trolley wire	20	12.7	–
Tin	20	11.63	0.0042
Titanium	20	42	–
Tungsten	20	5.51	0.005
Vanadium	20	19.7	–
Ytterbium	25	25	–
Yttrium	25	59.6	–
Zinc	20	5.97	0.0037
Zirconium	20	42.1	–

Porous materials.[a]

Material	Porosity, ϕ	Permeability, K (cm^2)	Contact surface per unit volume (cm^{-1})
Agar–agar		$2 \times 10^{-10} - 4.4 \times 10^{-9}$	
Black slate powder	0.57–0.66	$4.9 \times 10^{-10} - 1.2 \times 10^{-9}$	$7 \times 10^3 - 8.9 \times 10^3$
Brick	0.12–0.34	$4.8 \times 10^{-11} - 2.2 \times 10^{-9}$	
Catalyst (Fischer–Tropsch, granules only)	0.45		5.6×10^5
Cigarette		1.1×10^{-5}	
Cigarette filters	0.17–0.49		
Coal	0.02–0.12		
Concrete (ordinary mixes)	0.02–0.07		
Concrete (bituminous)		$1 \times 10^{-9} - 2.3 \times 10^{-7}$	
Copper powder (hot-compacted)	0.09–0.34	$3.3 \times 10^{-6} - 1.5 \times 10^{-5}$	
Cork board		$2.4 \times 10^{-7} - 5.1 \times 10^{-7}$	
Fiberglass	0.88–0.93		560–770
Granular crushed rock	0.45		
Hair (on mammals)	0.95–0.99		
Hair felt		$8.3 \times 10^{-6} - 1.2 \times 10^{-5}$	
Leather	0.56–0.59	$9.5 \times 10^{-10} - 1.2 \times 10^{-9}$	$1.2 \times 10^4 - 1.6 \times 10^4$
Limestone (dolomite)	0.04–0.10	$2 \times 10^{-11} - 4.5 \times 10^{-10}$	
Sand	0.37–0.50	$2 \times 10^{-7} - 1.8 \times 10^{-6}$	150–220
Sandstone ("oil sand")	0.08–0.38	$5 \times 10^{-12} - 3 \times 10^{-8}$	
Silica grains	0.65		
Silica powder	0.37–0.49	$1.3 \times 10^{-10} - 5.1 \times 10^{-10}$	$6.8 \times 10^3 - 8.9 \times 10^3$
Soil	0.43–0.54	$2.9 \times 10^{-9} - 1.4 \times 10^{-7}$	
Spherical packings (well shaken)	0.36–0.43		
Wire crimps	0.68–0.76	$3.8 \times 10^{-5} - 1 \times 10^{-4}$	29–40

a) Nield and Bejan [11].

Porosity:

$$\phi = \frac{\text{void volume}}{\text{total voulme of sample}}$$

Permeability K of porous medium saturated with one fluid:

$$u = \frac{K}{\mu} \frac{\Delta P}{L} \quad \text{(Darcy's law)}$$

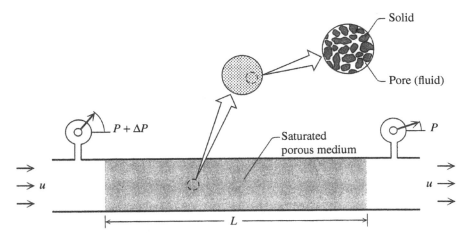

Figure B.1

where

ΔP = Pressure drop across porous column of length L saturated with fluid

μ = Fluid viscosity

u = Fluid velocity averaged over the entire volume (solid + fluid) of the saturated porous medium sample

Darcy's law is valid when the Reynolds number based on u and the pore length scale (D_p) is sufficiently small so that [12],

$$\frac{uD_p}{\nu} \lesssim 10$$

References

1 Eckert, E.R.G. and Drake, R.M. (1972). *Analysis of Heat and Mass Transfer*. New York: McGraw-Hill.

2 Touloukian, Y.S. and Ho, C.Y. (eds.) (1972). *Thermophysical Properties of Matter*. New York: Plenum.

3 Raznjevic, K. (1976). *Handbook of Thermodynamic Tables and Charts*. Washington, DC: Hemisphere.

4 Grigull, U. and Sandner, H. (1984). *Heat Conduction* (trans. J. Kestin). Washington, DC: Hemisphere, Appendix E.

5 Kreith, F. and Black, W.Z. (1980). *Basic Heat Transfer*. New York: Harper and Row, Appendix E.

6 Lienhard, J.H. (1987). *A Heat Transfer Textbook*, 2e. Englewood cliffs, NJ: Prentice-Hall, Appendix A.

7 Witte, L.C., Schmidt, P.S., and Brown, D.R. (1988). *Industrial Energy Management and Utilization*. New York: Hemisphere.

8 Qashou, M.S., Vachon, R.I., and Touloukian, Y.S. (1972). Thermal conductivity of foods. *ASHRAE Trans.* 78 (Pt. 1): 165–183.

9 Dickerson, R. Jr. and Reed, R.B. Jr. (1975). Thermal diffusivity of meats. *ASHRAE Trans.* 81: 356–364.

10 Hobbs, P.V. (1974). *Ice Physics*. Oxford University Press, Chapter 5.

11 Nield, D.A. and Bejan, A. (2017). *Convection in Porous Media*, 5e. New York: Springer-Verlag.

12 Bejan, A. (2013). *Convection Heat Transfer*, 4e. Hoboken: Wiley.

Appendix C

Properties of Liquids

Water at atmospheric pressure.[a]

T (°C)	ρ (g/cm^3)	c_P (kJ/kg·K)	c_v (kJ/kg·K)	h_{fg} (kJ/kg)	β (K^{-1})
0	0.9999	4.217	4.215	2501	-0.6×10^{-4}
5	1	4.202	4.202	2489	$+0.1 \times 10^{-4}$
10	0.9997	4.192	4.187	2477	0.9×10^{-4}
15	0.9991	4.186	4.173	2465	1.5×10^{-4}
20	0.9982	4.182	4.158	2454	2.1×10^{-4}
25	0.9971	4.179	4.138	2442	2.6×10^{-4}
30	0.9957	4.178	4.118	2430	3.0×10^{-4}
35	0.9941	4.178	4.108	2418	3.4×10^{-4}
40	0.9923	4.178	4.088	2406	3.8×10^{-4}
50	0.9881	4.180	4.050	2382	4.5×10^{-4}
60	0.9832	4.184	4.004	2357	5.1×10^{-4}
70	0.9778	4.189	3.959	2333	5.7×10^{-4}
80	0.9718	4.196	3.906	2308	6.2×10^{-4}
90	0.9653	4.205	3.865	2283	6.7×10^{-4}
100[b]	0.9584	4.216	3.816	2257	7.1×10^{-4}

T (°C)	μ (g/cm·s)	ν (cm^2/s)	k (W/m·K)	α (cm^2/s)	Pr	$\dfrac{g\beta}{\alpha\nu} = \dfrac{Ra_H}{H^3 \Delta T}$ (K^{-1}·cm^{-3})
0	0.01787	0.01787	0.56	0.00133	13.44	-2.48×10^3
5	0.01514	0.01514	0.57	0.00136	11.13	$+0.47 \times 10^3$
10	0.01304	0.01304	0.58	0.00138	9.45	4.91×10^3
15	0.01137	0.01138	0.59	0.00140	8.13	9.24×10^3
20	0.01002	0.01004	0.59	0.00142	7.07	14.45×10^3

(Continued)

Heat Transfer: Evolution, Design and Performance, First Edition. Adrian Bejan.
© 2022 John Wiley & Sons, Inc. Published 2022 by John Wiley & Sons, Inc.
Companion website: www.wiley.com/go/bejan/heattransfer

T (°C)	μ (g/cm·s)	ν (cm²/s)	k (W/m·K)	α (cm²/s)	Pr	$\dfrac{g\beta}{\alpha\nu} = \dfrac{Ra_H}{H^3 \Delta T}$ (K⁻¹·cm⁻³)
25	0.00891	0.00894	0.60	0.00144	6.21	19.81×10^3
30	0.00798	0.00802	0.61	0.00146	5.49	25.13×10^3
35	0.00720	0.00725	0.62	0.00149	4.87	30.88×10^3
40	0.00654	0.00659	0.63	0.00152	4.34	37.21×10^3
50	0.00548	0.00554	0.64	0.00155	3.57	51.41×10^3
60	0.00467	0.00475	0.65	0.00158	3.01	66.66×10^3
70	0.00405	0.00414	0.66	0.00161	2.57	83.89×10^3
80	0.00355	0.00366	0.67	0.00164	2.23	101.3×10^3
90	0.00316	0.00327	0.67	0.00165	1.98	121.8×10^3
100	0.00283	0.00295	0.68	0.00166	1.78	142.2×10^3

a) Data collected from Refs. [1–3].
b) Saturated.

Water at saturation pressure.[a]

T (°C)	ρ (g/cm³)	c_P (kJ/kg·K)	μ (g/cm·s)	ν (cm²/s)	k (W/m·K)	α (cm²/s)	Pr
0	0.9999	4.226	0.0179	0.0179	0.56	0.0013	13.7
10	0.9997	4.195	0.0130	0.0130	0.58	0.0014	9.5
20	0.9982	4.182	0.0099	0.0101	0.60	0.0014	7
40	0.9922	4.175	0.0066	0.0066	0.63	0.0015	4.3
60	0.9832	4.181	0.0047	0.0048	0.66	0.0016	3
80	0.9718	4.194	0.0035	0.0036	0.67	0.0017	2.25
100	0.9584	4.211	0.0028	0.0029	0.68	0.0017	1.75
150	0.9169	4.270	0.00185	0.0020	0.68	0.0017	1.17
200	0.8628	4.501	0.00139	0.0016	0.66	0.0017	0.95
250	0.7992	4.857	0.00110	0.00137	0.62	0.0016	0.86
300	0.7125	5.694	0.00092	0.00128	0.56	0.0013	0.98
340	0.6094	8.160	0.00077	0.00127	0.44	0.0009	1.45
370	0.4480	11.690	0.00057	0.00127	0.29	0.00058	2.18

a) Data collected from Refs. [1, 2].

Ammonia, saturated liquid.[a]

T (°C)	ρ (g/cm³)	c_P (kJ/kg·K)	μ (kg/s·m)	ν (cm²/s)	k (W/m·K)	α (cm²/s)	Pr
−50	0.704	4.46	3.06×10^{-4}	4.35×10^{-3}	0.547	1.74×10^{-3}	2.50
−25	0.673	4.49	2.58×10^{-4}	3.84×10^{-3}	0.548	1.81×10^{-3}	2.12
0	0.640	4.64	2.39×10^{-4}	3.73×10^{-3}	0.540	1.82×10^{-3}	2.05
25	0.604	4.84	2.14×10^{-4}	3.54×10^{-3}	0.514	1.76×10^{-3}	2.01
50	0.564	5.12	1.86×10^{-4}	3.30×10^{-3}	0.476	1.65×10^{-3}	2.00

a) Constructed based on data compiled in Ref. [4].

Carbon dioxide, saturated liquid.[a]

T (K)	P (bar)	ρ (g/cm³)	c_P (kJ/kg·K)	μ (kg/s·m)	ν (cm²/s)	k (W/m·K)	α (cm²/s)	Pr
216.6	5.18	1.179	1.707	2.10×10^{-4}	1.78×10^{-3}	0.182	9.09×10^{-4}	1.96
220	6.00	1.167	1.761	1.86×10^{-4}	1.59×10^{-3}	0.178	8.26×10^{-4}	1.93
240	12.83	1.089	1.933	1.45×10^{-4}	1.33×10^{-3}	0.156	7.40×10^{-4}	1.80
260	24.19	1.000	2.125	1.14×10^{-4}	1.14×10^{-3}	0.128	6.03×10^{-4}	1.89
280	41.60	0.885	2.887	0.91×10^{-4}	1.03×10^{-3}	0.102	4.00×10^{-4}	2.57
300	67.10	0.680		0.71×10^{-4}	1.04×10^{-3}	0.081		
304.2	73.83	0.466		0.60×10^{-4}	1.29×10^{-3}	0.074		

a) Constructed based on data compiled in Refs. [5, 6].

Fuels, liquids at $P \cong 1$ atm.[a]

T (°C)	ρ (g/cm³)	c_P (kJ/kg·K)	μ (kg/s·m)	ν (cm²/s)	k (W/m·K)	α (cm²/s)	Pr
Gasoline							
20	0.751	2.06	5.29×10^{-4}	7.04×10^{-3}	0.1164	7.52×10^{-4}	9.4
50	0.721	2.20	3.70×10^{-4}	5.13×10^{-3}	0.1105	6.97×10^{-4}	7.4
100	0.681	2.46	2.25×10^{-4}	3.30×10^{-3}	0.1005	6.00×10^{-4}	5.5
150	0.628	2.74	1.56×10^{-4}	2.48×10^{-3}	0.0919	5.34×10^{-4}	4.6
200	0.570	3.04	1.11×10^{-4}	1.95×10^{-3}	0.0800	4.62×10^{-4}	4.2

(Continued)

T (°C)	ρ (g/cm³)	c_P (kJ/kg·K)	μ (kg/s·m)	ν (cm²/s)	k (W/m·K)	α (cm²/s)	Pr
Kerosene							
20	0.819	2.00	1.49×10^{-3}	1.82×10^{-2}	0.1161	7.09×10^{-4}	25.7
50	0.801	2.14	9.56×10^{-4}	1.19×10^{-2}	0.1114	6.50×10^{-4}	18.3
100	0.766	2.38	5.45×10^{-4}	7.11×10^{-3}	0.1042	5.72×10^{-4}	12.4
150	0.728	2.63	3.64×10^{-4}	5.00×10^{-3}	0.0965	5.04×10^{-4}	9.9
200	0.685	2.89	2.62×10^{-4}	3.82×10^{-3}	0.0891	4.50×10^{-4}	8.5
250	0.638	3.16	2.01×10^{-4}	3.15×10^{-3}	0.0816	4.05×10^{-4}	7.8

Helium, Liquid at $P = 1$ atm.[a]

T (K)	ρ (g/cm³)	c_P (kJ/kg·K)	μ (kg/s·m)	ν (cm²/s)	k (W/m·K)	α (cm²/s)	Pr
2.5	0.147	2.05	3.94×10^{-6}	2.68×10^{-4}	0.0167	5.58×10^{-4}	0.48
3	0.143	2.36	3.86×10^{-6}	2.69×10^{-4}	0.0182	5.38×10^{-4}	0.50
3.5	0.138	3.00	3.64×10^{-6}	2.64×10^{-4}	0.0191	4.62×10^{-4}	0.57
4	0.130	4.07	3.34×10^{-6}	2.57×10^{-4}	0.0196	3.71×10^{-4}	0.69
4.22[b]	0.125	4.98	3.17×10^{-6}	2.53×10^{-4}	0.0196	3.15×10^{-4}	0.80

a) Constructed based on data compiled in Ref. [7].
b) Data collected from Ref. [8].
c) Saturated.

Lithium, saturated liquid.[a]

T (K)	P (bar)	ρ (g/cm³)	c_P (kJ/kg·K)	μ (kg/s·m)	ν (cm²/s)	k (W/m·K)	α (cm²/s)	Pr
600	4.2×10^{-9}	0.503	4.23	4.26×10^{-4}	0.0085	47.6	0.223	0.038
800	9.6×10^{-6}	0.483	4.16	3.10×10^{-4}	0.0064	54.1	0.270	0.024
1000	9.6×10^{-4}	0.463	4.16	2.47×10^{-4}	0.0053	60.0	0.312	0.017
1200	0.0204	0.442	4.14	2.07×10^{-4}	0.0047	64.7	0.355	0.013
1400	0.1794	0.422	4.19	1.80×10^{-4}	0.0043	68.0	0.384	0.011

Nitrogen, liquid at $P = 1$ atm.[a]

T (K)	ρ (g/cm³)	c_P (kJ/kg·K)	μ (kg/s·m)	ν (cm²/s)	k (W/m·K)	α (cm²/s)	Pr
65	0.861	1.988	2.77×10^{-5}	3.21×10^{-3}	0.161	9.39×10^{-4}	3.42
70	0.840	2.042	2.12×10^{-5}	2.53×10^{-3}	0.151	8.77×10^{-4}	2.88
75	0.819	2.059	1.77×10^{-5}	2.17×10^{-3}	0.141	8.36×10^{-4}	2.59
77.3[b]	0.809	2.065	1.64×10^{-5}	2.03×10^{-3}	0.136	8.15×10^{-4}	2.49

a) Constructed based on data compiled in Refs. [5, 6].
b) Interpolated from data in Ref. [9].
c) Saturated.

Potassium, saturated liquid.[a]

T (K)	P (bar)	ρ (g/cm³)	c_P (kJ/kg·K)	μ (kg/s·m)	ν (cm²/s)	k (W/m·K)	α (cm²/s)	Pr
400	1.84×10^{-7}	0.814	0.805	4.13×10^{-4}	0.0051	52.0	0.794	0.0064
600	9.26×10^{-4}	0.767	0.771	2.38×10^{-4}	0.0031	43.9	0.742	0.0042
800	0.0612	0.720	0.761	1.71×10^{-4}	0.0024	37.1	0.677	0.0035
1000	0.7322	0.672	0.792	1.35×10^{-4}	0.0020	31.3	0.589	0.0034
1200	3.963	0.623	0.846	1.14×10^{-4}	0.0018	26.3	0.499	0.0037
1400	12.44	0.574	0.899	0.98×10^{-4}	0.0017	21.5	0.416	0.0041

a) Constructed based on data compiled in Refs. [5, 6].

Mercury, saturated liquid.[a]

T (K)	P (bar)	ρ (g/cm³)	c_P (kJ/kg·K)	μ (kg/s·m)	ν (cm²/s)	k (W/m·K)	α (cm²/s)	Pr	β (K⁻¹)
260	6.9×10^{-8}	13.63	0.141	1.79×10^{-3}	1.31×10^{-3}	8.00	0.042	0.0316	1.8×10^{-4}
300	3.1×10^{-6}	13.53	0.139	1.52×10^{-3}	1.12×10^{-3}	8.54	0.045	0.0248	1.8×10^{-4}
340	5.5×10^{-5}	13.43	0.138	1.34×10^{-3}	1.00×10^{-3}	9.06	0.049	0.0205	1.8×10^{-4}
400	1.4×10^{-3}	13.29	0.137	1.17×10^{-3}	8.83×10^{-4}	9.80	0.054	0.0163	1.8×10^{-4}
500	0.053	13.05	0.135	1.01×10^{-3}	7.72×10^{-4}	10.93	0.062	0.0125	1.8×10^{-4}
600	0.578	12.81	0.136	9.10×10^{-4}	7.10×10^{-4}	11.94	0.071	0.0100	1.9×10^{-4}
800	11.18	12.32	0.140	8.08×10^{-4}	6.56×10^{-4}	13.57	0.079	0.0083	1.9×10^{-4}
1000	65.74	11.79	0.149	7.54×10^{-4}	6.40×10^{-4}	14.69	0.084	0.0076	1.9×10^{-4}

a) Constructed based on data compiled in Refs. [5, 6].

Refrigerant 12 (FREON 12, CCl_2F_2), saturated liquid.[a]

P (bar)	T (K)	ρ (g/cm³)	c_P (kJ/kg·K)	μ (kg/s·m)	ν (cm²/s)	k (W/m·K)	α (cm²/s)	Pr
0.2	211.1	1.579	0.865	5.28×10^{-4}	3.34×10^{-3}	0.101	7.40×10^{-4}	4.52
0.4	223.5	1.554	0.876	4.48×10^{-4}	2.88×10^{-3}	0.097	7.12×10^{-4}	4.05
0.6	231.7	1.522	0.884	4.08×10^{-4}	2.68×10^{-3}	0.094	6.98×10^{-4}	3.84
1.0	243.0	1.488	0.894	3.59×10^{-4}	2.41×10^{-3}	0.089	6.68×10^{-4}	3.61
2.0	260.6	1.435	0.914	2.95×10^{-4}	2.06×10^{-3}	0.083	6.33×10^{-4}	3.25
3.0	272.3	1.392	0.930	2.62×10^{-4}	1.88×10^{-3}	0.079	6.11×10^{-4}	3.08
6.0	295.2	1.321	0.969	2.13×10^{-4}	1.61×10^{-3}	0.070	5.47×10^{-4}	2.95
10.0	314.9	1.247	1.023	1.88×10^{-4}	1.51×10^{-3}	0.063	4.94×10^{-4}	3.05
20.0	346.3	1.099	1.234	1.49×10^{-4}	1.36×10^{-3}	0.053	3.91×10^{-4}	3.47
30.0	367.2	0.955	1.520	1.16×10^{-4}	1.21×10^{-3}	0.042	2.89×10^{-4}	4.20

a) Constructed based on data compiled in Refs. [5, 6].

Refrigerant 22 (FREON 22, $CHClF_2$), saturated liquid.[a]

T (K)	P (bar)	ρ (g/cm³)	c_P (kJ/kg·K)	μ (kg/s·m)	ν (cm²/s)	k (W/m·K)	α (cm²/s)	Pr
180	0.037	1.545	1.058	6.47×10^{-5}	4.19×10^{-4}	0.146	8.93×10^{-4}	0.47
200	0.166	1.497	1.065	4.81×10^{-5}	3.21×10^{-4}	0.136	8.53×10^{-4}	0.38
220	0.547	1.446	1.080	3.78×10^{-5}	2.61×10^{-4}	0.126	8.07×10^{-4}	0.32
240	1.435	1390	1.105	3.09×10^{-5}	2.22×10^{-4}	0.117	7.62×10^{-4}	0.29
260	3.177	1.329	1.143	2.60×10^{-5}	1.96×10^{-4}	0.107	7.04×10^{-4}	0.28
280	6.192	1.262	1.193	2.25×10^{-5}	1.78×10^{-4}	0.097	6.44×10^{-4}	0.28
300	10.96	1.187	1.257	1.98×10^{-5}	1.67×10^{-4}	0.087	5.83×10^{-4}	0.29
320	18.02	1.099	1.372	1.76×10^{-5}	1.60×10^{-4}	0.077	5.11×10^{-4}	0.31
340	28.03	0.990	1.573	1.51×10^{-5}	1.53×10^{-4}	0.067	4.30×10^{-4}	0.36

a) Constructed based on data compiled in Ref. [6].

Sodium, saturated liquid.[a]

T (K)	P (bar)	ρ (g/cm^3)	c_P (kJ/kg·K)	μ (kg/s·m)	ν (cm^2/s)	k (W/m·K)	α (cm^2/s)	Pr
500	7.64×10^{-7}	0.898	1.330	4.24×10^{-4}	0.0047	80.0	0.67	0.0070
600	5.05×10^{-5}	0.873	1.299	3.28×10^{-4}	0.0038	75.4	0.66	0.0057
700	9.78×10^{-4}	0.850	1.278	2.69×10^{-4}	0.0032	70.7	0.65	0.0049
800	0.00904	0.826	1.264	2.30×10^{-4}	0.0028	65.9	0.63	0.0044
900	0.0501	0.802	1.258	2.02×10^{-4}	0.0025	61.4	0.61	0.0041
1000	0.1955	0.776	1.259	1.81×10^{-4}	0.0023	56.7	0.58	0.0040
1200	1.482	0.729	1.281	1.51×10^{-4}	0.0021	54.5	0.58	0.0036
1400	6.203	0.681	1.330	1.32×10^{-4}	0.0019	52.2	0.58	0.0034
1600	17.98	0.633	1.406	1.18×10^{-4}	0.0019	49.9	0.56	0.0033

a) Constructed based on data compiled in Refs. [5, 6].

Unused engine oil.[a]

T (K)	ρ (g/cm^3)	c_P (kJ/kg·K)	μ (kg/s·m)	ν (cm^2/s)	k (W/m·K)	α (cm^2/s)	Pr	β (K^{-1})
260	0.908	1.76	12.23	135	0.149	9.32×10^{-4}	144 500	7×10^{-4}
280	0.896	1.83	2.17	24.2	0.146	8.90×10^{-4}	27 200	7×10^{-4}
300	0.884	1.91	0.486	5.50	0.144	8.53×10^{-4}	6450	7×10^{-4}
320	0.872	1.99	0.141	1.62	0.141	8.13×10^{-4}	1990	7×10^{-4}
340	0.860	2.08	0.053	0.62	0.139	7.77×10^{-4}	795	7×10^{-4}
360	0.848	2.16	0.025	0.30	0.137	7.48×10^{-4}	395	7×10^{-4}
380	0.836	2.25	0.014	0.17	0.136	7.23×10^{-4}	230	7×10^{-4}
400	0.824	2.34	0.009	0.11	0.134	6.95×10^{-4}	155	7×10^{-4}

a) Constructed based on data compiled in Refs. [5, 6].

Critical point data.[a)]

Liquid	Critical temperature		Critical pressure		Critical specific volume (cm³/g)
	K	°C	MPa	atm	
Air	133.2	−140	3.77	37.2	2.9
Alcohol (methyl)	513.2	240	7.98	78.7	3.7
Alcohol (ethyl)	516.5	243.3	6.39	63.1	3.6
Ammonia	405.4	132.2	11.3	111.6	4.25
Argon	150.9	−122.2	4.86	48	1.88
Butane	425.9	152.8	3.65	36	4.4
Carbon dioxide	304.3	31.1	7.4	73	2.2
Carbon monoxide	134.3	−138.9	3.54	35	3.2
Carbon tetrachloride	555.9	282.8	4.56	45	1.81
Chlorine	417	143.9	7.72	76.14	1.75
Ethane	305.4	32.2	4.94	48.8	4.75
Ethylene	282.6	9.4	5.85	57.7	4.6
Helium	5.2	−268	0.228	2.25	14.4
Hexane	508.2	235	2.99	29.5	4.25
Hydrogen	33.2	−240	1.30	12.79	32.3
Methane	190.9	−82.2	4.64	45.8	6.2
Methyl chloride	416.5	143.3	6.67	65.8	2.7
Neon	44.2	−288.9	2.7	26.6	2.1
Nitric oxide	179.3	−93.9	6.58	65	1.94
Nitrogen	125.9	−147.2	3.39	33.5	3.25
Octane	569.3	296.1	2.5	24.63	4.25
Oxygen	154.3	−118.9	5.03	49.7	2.3
Propane	368.7	95.6	4.36	43	4.4
Sulfur dioxide	430.4	157.2	7.87	77.7	1.94
Water	647	373.9	22.1	218.2	3.1

a) Based on a compilation from Ref. [10].

References

1 Bejan, A. (2013). *Convection Heat Transfer*, 4e. Hoboken: Wiley.
2 Raznjevic, K. (1976). *Handbook of Thermodynamic Tables and Charts*. Washington, DC: Hemisphere.
3 Batchelor, G.K. (1967). *An Introduction to Fluid Dynamics*. Cambridge, England: Cambridge University Press.

4 Eckert, E.R.G. and Drake, R.M. (1972). *Analysis of Heat and Mass Transfer*. New York: McGraw-Hill.

5 Green, D.W. and Maloney, J.O. (eds.) (1984). *Perry's Chemical Engineers' Handbook*, 6e, 3-1–3-263. New York: McGraw-Hill.

6 Liley, P.E. (1987). Thermophysical properties. In: *Handbook of Single-Phase Convective Heat Transfer* (eds. S. Kakac, R.K. Shah and W. Aung). New York: Wiley, Chapter 22.

7 Vargaftik, N.B. (1975). *Tables on the Thermophysical Properties of Liquids and Gases*, 2e. Washington, DC: Hemisphere.

8 McCarty, R.D. (1972). *Thermophysical Properties of Helium-4 from 2 to 1500 K with Pressures to 1000 Atmospheres*. Washington, DC: *NBS TN 631*.

9 Jacobsen, R. T., Stewart, R. B., McCarty, R. D., and Hanley, H. J. M. (1973). Thermophysical Properties of Nitrogen from the Fusion Line to 3500 R (1944 K) for Pressures to 150,000 psia (10342×10^5 N/m²). *NBS TN 648*, Washington, DC.

10 Bejan, A. (2016). *Advanced Engineering Thermodynamics*, 4e. Hoboken: Wiley.

Appendix D

Properties of Gases

Air, dry, at atmospheric pressure.[a]

T (°C)	ρ (kg/m³)	c_P (kJ/kg·K)	μ (kg/s·m)	ν (cm²/s)	k (W/m·K)	α (cm²/s)	Pr	$\dfrac{g\beta}{\alpha\nu} = \dfrac{Ra_H}{H^3\Delta T}$ (cm⁻³·K⁻¹)
−180	3.72	1.035	6.50×10^{-6}	0.0175	0.0076	0.019	0.92	3.2×10^4
−100	2.04	1.010	1.16×10^{-5}	0.057	0.016	0.076	0.75	1.3×10^3
−50	1.582	1.006	1.45×10^{-5}	0.092	0.020	0.130	0.72	367
0	1.293	1.006	1.71×10^{-5}	0.132	0.024	0.184	0.72	148
10	1.247	1.006	1.76×10^{-5}	0.141	0.025	0.196	0.72	125
20	1.205	1.006	1.81×10^{-5}	0.150	0.025	0.208	0.72	107
30	1.165	1.006	1.86×10^{-5}	0.160	0.026	0.223	0.72	90.7
60	1.060	1.008	2.00×10^{-5}	0.188	0.028	0.274	0.70	57.1
100	0.946	1.011	2.18×10^{-5}	0.230	0.032	0.328	0.70	34.8
200	0.746	1.025	2.58×10^{-5}	0.346	0.039	0.519	0.68	9.53
300	0.616	1.045	2.95×10^{-5}	0.481	0.045	0.717	0.68	4.96
500	0.456	1.093	3.58×10^{-5}	0.785	0.056	1.140	0.70	1.42
1000	0.277	1.185	4.82×10^{-5}	1.745	0.076	2.424	0.72	0.18

a) Data collected from Refs. [1–3].

Heat Transfer: Evolution, Design and Performance, First Edition. Adrian Bejan.
© 2022 John Wiley & Sons, Inc. Published 2022 by John Wiley & Sons, Inc.
Companion website: www.wiley.com/go/bejan/heattransfer

Steam at $P = 1$ bar.[a]

T (K)	ρ (kg/m³)	c_P (kJ/kg·K)	μ (kg/s·m)	ν (cm²/s)	k (W/m·K)	α (cm²/s)	Pr
373.15	0.596	2.029	1.20×10^{-5}	0.201	0.0248	0.205	0.98
400	0.547	1.996	1.32×10^{-5}	0.241	0.0268	0.246	0.98
450	0.485	1.981	1.52×10^{-5}	0.313	0.0311	0.324	0.97
500	0.435	1.983	1.73×10^{-5}	0.398	0.0358	0.415	0.96
600	0.362	2.024	2.15×10^{-5}	0.594	0.0464	0.633	0.94
700	0.310	2.085	2.57×10^{-5}	0.829	0.0581	0.899	0.92
800	0.271	2.151	2.98×10^{-5}	1.10	0.0710	1.22	0.90
900	0.241	2.219	3.39×10^{-5}	1.41	0.0843	1.58	0.89
1000	0.217	2.286	3.78×10^{-5}	1.74	0.0981	1.98	0.88
1200	0.181	2.43	4.48×10^{-5}	2.48	0.130	2.96	0.84
1400	0.155	2.58	5.06×10^{-5}	3.27	0.160	4.00	0.82
1600	0.135	2.73	5.65×10^{-5}	4.19	0.210	5.69	0.74
1800	0.120	3.02	6.19×10^{-5}	5.16	0.330	9.10	0.57
2000	0.108	3.79	6.70×10^{-5}	6.20	0.570	13.94	0.45

a) Constructed based on data compiled in Refs. [4, 5].

Ideal gas constants and specific heats.[a], [b]

Gas		M (kg/kmol)	R (kJ/kg·K)	c_P (kJ/kg·K)	c_v (kJ/kg·K)
Air, dry	–	28.97	0.287	1.005	0.718
Argon	Ar	39.944	0.208	0.525	0.317
Carbon dioxide	CO_2	44.01	0.189	0.846	0.657
Carbon monoxide	CO	28.01	0.297	1.040	0.744
Helium	He	4.003	2.077	5.23	3.15
Hydrogen	H_2	2.016	4.124	14.31	10.18
Methane	CH_4	16.04	0.518	2.23	1.69
Nitrogen	N_2	28.016	0.297	1.039	0.743
Oxygen	O_2	32.000	0.260	0.918	0.658
Water vapor	H_2O	18.016	0.461	1.87	1.40

a) The c_P and c_v values correspond to the temperature 300 K. This ideal gas model is valid at low and moderate pressures ($P \lesssim 1$ atm).
b) After Ref. [6].

Ammonia, gas at $P = 1$ atm.[a]

T (°C)	ρ (kg/m³)	c_P (kJ/kg·K)	μ (kg/s·m)	ν (cm²/s)	k (W/m·K)	α (cm²/s)	Pr
0	0.793	2.18	9.35×10^{-6}	0.118	0.0220	0.131	0.90
50	0.649	2.18	1.10×10^{-5}	0.170	0.0270	0.192	0.88
100	0.559	2.24	1.29×10^{-5}	0.230	0.0327	0.262	0.87
150	0.493	2.32	1.47×10^{-5}	0.297	0.0391	0.343	0.87
200	0.441	2.40	1.65×10^{-5}	0.374	0.0467	0.442	0.84

a) Constructed based on data compiled in Ref. [7].

Carbon dioxide, gas at $P = 1$ bar.[a]

T (K)	ρ (kg/m³)	c_P (kJ/kg·K)	μ (kg/s·m)	ν (cm²/s)	k (W/m·K)	α (cm²/s)	Pr
300	1.773	0.852	1.51×10^{-5}	0.085	0.0166	0.109	0.78
350	1.516	0.898	1.75×10^{-5}	0.115	0.0204	0.150	0.77
400	1.326	0.941	1.98×10^{-5}	0.149	0.0243	0.195	0.77
500	1.059	1.014	2.42×10^{-5}	0.229	0.0325	0.303	0.76
600	0.883	1.075	2.81×10^{-5}	0.318	0.0407	0.429	0.74
700	0.751	1.126	3.17×10^{-5}	0.422	0.0481	0.569	0.74
800	0.661	1.168	3.50×10^{-5}	0.530	0.0551	0.714	0.74
900	0.588	1.205	3.81×10^{-5}	0.648	0.0618	0.873	0.74
1000	0.529	1.234	4.10×10^{-5}	0.775	0.0682	1.043	0.74

a) Constructed based on data compiled in Refs. [4, 5].

Helium, gas at $P = 1$ atm.[a]

T (K)	ρ (kg/m³)	c_P (kJ/kg·K)	μ (kg/s·m)	ν (cm²/s)	k (W/m·K)	α (cm²/s)	Pr
4.22	16.9	9.78	1.25×10^{-6}	7.39×10^{-4}	0.011	6.43×10^{-4}	1.15
7	7.53	5.71	1.76×10^{-6}	2.34×10^{-3}	0.014	3.21×10^{-3}	0.73
10	5.02	5.41	2.26×10^{-6}	4.49×10^{-3}	0.018	6.42×10^{-3}	0.70
20	2.44	5.25	3.58×10^{-6}	0.0147	0.027	0.0209	0.70
30	1.62	5.22	4.63×10^{-6}	0.0286	0.034	0.0403	0.71
60	0.811	5.20	7.12×10^{-6}	0.088	0.053	0.125	0.70
100	0.487	5.20	9.78×10^{-6}	0.201	0.074	0.291	0.69
200	0.244	5.19	1.51×10^{-5}	0.622	0.118	0.932	0.67
300	0.162	5.19	1.99×10^{-5}	1.22	0.155	1.83	0.67
600	0.0818	5.19	3.22×10^{-5}	3.96	0.251	5.94	0.67
1000	0.0487	5.19	4.63×10^{-5}	9.46	0.360	14.2	0.67

a) Data collected from Ref. [8].

n-Hydrogen, gas at $P = 1$ atm.[a]

T (K)	ρ (kg/m³)	c_P (kJ/kg·K)	μ (kg/s·m)	ν (cm²/s)	k (W/m·K)	α (cm²/s)	Pr
250	0.0982	14.04	7.9×10^{-6}	0.804	0.162	1.17	0.69
300	0.0818	14.31	8.9×10^{-6}	1.09	0.187	1.59	0.69
350	0.0702	14.43	9.9×10^{-6}	1.41	0.210	2.06	0.69
400	0.0614	14.48	1.09×10^{-5}	1.78	0.230	2.60	0.68
500	0.0491	14.51	1.27×10^{-5}	2.59	0.269	3.78	0.68
600	0.0408	14.55	1.43×10^{-5}	3.50	0.305	5.12	0.68
700	0.0351	14.60	1.59×10^{-5}	4.53	0.340	6.62	0.68

a) Constructed based on the data compiled in Refs. [4, 5].

Nitrogen, gas at $P = 1$ atm.[a]

T (K)	ρ (kg/m³)	c_P (kJ/kg·K)	μ (kg/s·m)	ν (cm²/s)	k (W/m·K)	α (cm²/s)	Pr
77.33	4.612	1.123	5.39×10^{-6}	0.0117	0.0076	0.0147	0.80
100	3.483	1.073	6.83×10^{-6}	0.0197	0.0097	0.0261	0.76
200	1.711	1.044	1.29×10^{-5}	0.0754	0.0185	0.103	0.73
300	1.138	1.041	1.78×10^{-5}	0.156	0.0259	0.218	0.72
400	0.854	1.045	2.20×10^{-5}	0.258	0.0324	0.363	0.71
500	0.683	1.056	2.58×10^{-5}	0.378	0.0386	0.535	0.71
600	0.569	1.075	2.91×10^{-5}	0.511	0.0442	0.722	0.71
700	0,488	1.098	3.21×10^{-5}	0.658	0.0496	0.925	0.71

a) Data collected from Ref. [9].

Oxygen, gas at $P = 1$ atm.[a]

T (K)	ρ (kg/m³)	c_P (kJ/kg·K)	μ (kg/s·m)	ν (cm²/s)	k (W/m·K)	α (cm²/s)	Pr
250	1.562	0.915	1.79×10^{-5}	0.115	0.0226	0.158	0.73
300	1.301	0.920	2.07×10^{-5}	0.159	0.0266	0.222	0.72
350	1.021	0.929	2.34×10^{-5}	0.229	0.0305	0.321	0.71
400	0.976	0.942	2.58×10^{-5}	0.264	0.0343	0.372	0.71
500	0.780	0.972	3.03×10^{-5}	0.388	0.0416	0.549	0.71
600	0.650	1.003	3.44×10^{-5}	0.529	0.0487	0.748	0.71
700	0.557	1.031	3.81×10^{-5}	0.684	0.0554	0.963	0.71

a) Constructed based on the data compiled in Refs. [4, 5].

Refrigerant 12 (Freon 12, CCl_2F_2), gas at $P = 1$ bar.[a]

T (K)	ρ (kg/m³)	c_P (kJ/kg·K)	μ (kg/s·m)	ν (cm²/s)	*k* (W/m·K)	α (cm²/s)	*Pr*
300	4.941	0.614	1.26×10^{-5}	0.026	0.0097	0.032	0.80
350	4.203	0.654	1.46×10^{-5}	0.035	0.0124	0.045	0.77
400	3.663	0.684	1.62×10^{-5}	0.044	0.0151	0.061	0.73
450	3.248	0.711	1.75×10^{-5}	0.054	0.0179	0.077	0.70
500	2.918	0.739	1.90×10^{-5}	0.065	0.0208	0.097	0.67

a) Constructed based on the data compiled in Refs. [4, 5].

Refrigerant 22 (Freon 22, $CHCl_2F_2$), gas at $P = 1$ atm.[a]

T (K)	ρ (kg/m³)	c_P (kJ/kg·K)	μ (kg/s·m)	ν (cm²/s)	*k* (W/m·K)	α (cm²/s)	*Pr*
250	4.320	0.587	1.09×10^{-5}	0.025	0.008	0.032	0.80
300	3.569	0.647	1.30×10^{-5}	0.036	0.011	0.048	0.77
350	3.040	0.704	1.51×10^{-5}	0.050	0.014	0.065	0.76
400	2.650	0.757	1.71×10^{-5}	0.065	0.017	0.085	0.76
450	2.352	0.806	1.90×10^{-5}	0.081	0.020	0.105	0.77
500	2.117	0.848	2.09×10^{-5}	0.099	0.023	0.128	0.77

a) Constructed based on the data compiled in Refs. [4, 5].

Additional physical quantities included in the text.

	Figure or table
Figure 1.5	Thermal conductivities of solids, liquids, and gases
Figure 1.12	Heat transfer coefficients
Table 8.2	Surface tension, latent heat of vaporization, and other liquid–vapor properties at saturation
Figure 10.3	The electromagnetic spectrum
Figure 10.7	Monochromatic hemispherical blackbody emissive power
Figure 10.8	Blackbody radiation function
Table 10.1	Blackbody radiation function
Table 10.3	Total hemispherical emissivity: metallic surfaces
Table 10.4	Total hemispherical emissivity: nonmetallic surfaces

References

1 Bejan, A. (2013). *Convection Heat Transfer*, 4e. Hoboken: Wiley.

2 Raznjevic, K. (1976). *Handbook of Thermodynamic Tables and Charts*. Washington, DC: Hemisphere.

3 Batchelor, G.K. (1967). *An Introduction to Fluid Dynamics*. Cambridge, England: Cambridge University Press.

4 Green, D.W. and Maloney, J.O. (eds.) (1984). *Perry's Chemical Engineers' Handbook*, 6e, 3-1–3-263. New York: McGraw-Hill.

5 Liley, P.E. (1987). Thermophysical properties. In: *Handbook of Single-Phase Convective Heat Transfer* (eds. S. Kakac, R.K. Shah and W. Aung). New York: Wiley Chapter 22.

6 Bejan, A. (2016). *Advanced Engineering Thermodynamics*, 4e. Hoboken: Wiley.

7 Eckert, E.R.G. and Drake, R.M. (1972). *Analysis of Heat and Mass Transfer*. New York: McGraw-Hill.

8 McCarty, R.D. (1972). *Thermophysical Properties of Helium-4 from 2 to 1500 K with Pressures to 1000 Atmospheres*. Washington, DC: *NBS TN 631*.

9 Jacobsen, R.T., Stewart, R.B., McCarty, R.D., and Hanley, H.J.M. (1973). *Thermophysical Properties of Nitrogen from the Fusion Line to 3500 R (1944 K) for Pressures to 150,000 psia (10342 × 10^5 N/m^2)*. Washington, DC: *NBS TN 648*.

Appendix E

Mathematical Formulas

Areas and volumes of simple bodies.

Right circular cylinder (Figure E.1)

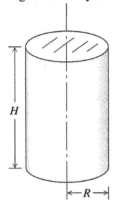

Lateral area $= 2\pi RH$
Total area $= 2\pi R(H + R)$
Volume $= \pi R^2 H$

Truncated right circular cylinder (Figure E.2)

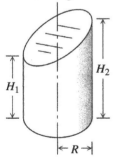

Lateral area $= \pi R(H_1 + H_2)$

Total area $= \pi R \left\{ H_1 + H_2 + R + \left[R^2 + \frac{1}{4}(H_2 - H_1)^2 \right]^{1/2} \right\}$

Volume $= \frac{\pi}{2} R^2 (H_1 + H_2)$

Heat Transfer: Evolution, Design and Performance, First Edition. Adrian Bejan.
© 2022 John Wiley & Sons, Inc. Published 2022 by John Wiley & Sons, Inc.
Companion website: www.wiley.com/go/bejan/heattransfer

Right circular cone (Figure E.3)

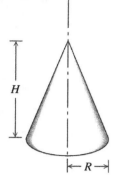

Later area $=\ \pi R(R^2 + H^2)^{1/2}$

Total area $=\ \pi R[R + (R^2 + H^2)^{1/2}]$

Volume $\quad =\ \dfrac{\pi}{3}R^2 H$

Frustum of right circular cone (Figure E.4)

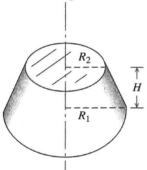

Lateral area $=\ \pi[H^2 + (R_1 - R_2)^2]^{1/2}(R_1 + R_2)$

Volume $\quad\ \ =\ \dfrac{\pi}{3}H(R_1^2 + R_2^2 + R_1 R_2)$

Sphere of radius R

Area $\quad =\ 4\pi R^2 = \pi D^2$

Volume $=\ \dfrac{4}{3}\pi R^3$

Spherical sector (Figure E.5)

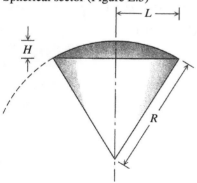

Total area $=\ \pi R(L + 2H)$

Volume $\quad\ =\ \dfrac{2}{3}\pi R^2 H$

Spherical segment (Figure E.6)

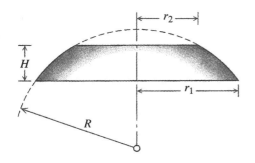

Lateral = $2\pi RH$
(spherical) area
Total area $= 2\pi RH + \pi(r_1^2 + r_2^2)$
Volume $= \dfrac{\pi}{6}H(3r_1^2 + 3r_2^2 + H^2)$

Error Function

Definition and properties:

$$\text{erf}(x) = \frac{2}{\pi^{1/2}} \int_0^x e^{-m^2}\,dm$$

$$\text{erf}(-x) = -\text{erf}(x)$$

$$\text{erfc}(x) = 1 - \text{erf}(x) \quad \text{(complimentary error function)}$$

$$\frac{d}{dx}[\text{erf}(x)]_{x=0} = \frac{2}{\pi^{1/2}} = 1.12838$$

The usual error function table: finding $\text{erf}(x)$ when x is specified:

x	$\text{erf}(x)$	x	$\text{erf}(x)$
0	0	0.9	0.79691
0.01	0.01128	1	0.8427
0.1	0.11246	1.2	0.91031
0.2	0.2227	1.4	0.95229
0.3	0.32863	1.6	0.97635
0.4	0.42839	1.8	0.98909
0.5	0.5205	2	0.99532
0.6	0.60386	2.5	0.99959
0.7	0.6778	3	0.99998
0.8	0.74210	∞	1

The inverted table: finding x when erf(x) is specified:

erf(x)	x
0	0
0.1	0.08886
0.2	0.17914
0.25	0.22531
0.3	0.27246
0.4	0.37081
0.5	0.47694
0.6	0.59512
0.7	0.73287
0.75	0.81342
0.8	0.90619
0.9	1.16309
1	∞

Closed-form approximate expressions for erf(x) and erfc(x), accurate within 0.42% [1]:

$$\text{erf}(x) \cong 1 - A \exp[-B(x + C)^2]$$
$$\text{erfc}(x) \cong A \exp[-B(x + C)^2]$$

where A, B, and C are three constants

$$A = 1.5577\ B = 0.7182\ C = 0.7856$$

Exponential integral function.

x	$\int_x^\infty \frac{e^{-u}}{u} du$	x	$\int_x^\infty \frac{e^{-u}}{u} du$
0.00	$+\infty$		
0.01	4.0379	1.60	0.08631
0.02	3.3547	1.70	0.07465
0.04	2.6813	1.80	0.06471
0.07	2.1508	1.90	0.05620
0.10	1.8229	2.00	0.04890
0.15	1.4645	2.20	0.03719
0.20	1.2227	2.40	0.02844
			(Continued)

x	$\int_x^\infty \frac{e^{-u}}{u}du$	x	$\int_x^\infty \frac{e^{-u}}{u}du$
0.30	0.9057	2.60	0.02185
0.40	0.7024	2.80	0.01686
0.50	0.5598	3.00	0.01305
0.60	0.4544	3.20	0.01013
0.70	0.3738	3.40	0.00789
0.80	0.3106	3.60	0.00616
0.90	0.2602	3.80	0.00482
1.00	0.2194	4.00	0.00378
1.10	0.18599	4.20	0.00297
1.20	0.15841	4.40	0.00234
1.30	0.13545	4.60	0.00184
1.40	0.11622	4.80	0.00145
1.50	0.10002	5.00	0.00115

Leibniz's Formula for Differentiating an Integral

$$\frac{d}{dx}\left[\int_{a(x)}^{b(x)} F(x,m)dm\right] = \int_{a(x)}^{b(x)} \frac{\partial F(x,m)}{\partial x}dm + F(x,b)\frac{db}{dx} - F(x,a)\frac{da}{dx}$$

Hyperbolic Functions

Definitions:

$$\sinh u = \frac{1}{2}(e^u - e^{-u}) \qquad \cosh u = \frac{1}{2}(e^u + e^{-u})$$

$$\tanh u = \frac{\sinh u}{\cosh u} = \frac{e^u - e^{-u}}{e^u + e^{-u}}$$

$$e^u = \cosh u + \sinh u \qquad e^{-u} = \cosh u - \sinh u$$

Derivatives:

$$\frac{d}{dx}(\sinh u) = (\cosh u)\frac{du}{dx} \qquad \frac{d}{dx}(\cosh u) = (\sinh u)\frac{du}{dx}$$

$$\frac{d}{dx}(\tanh u) = \frac{1}{\cosh^2 u}\frac{du}{dx}$$

Identities:

$$\sinh(u + v) = \sinh u \ \cosh v + \cosh u \ \sinh v$$
$$\sinh(u - v) = \sinh u \ \cosh v - \cosh u \ \sinh v$$
$$\cosh(u + v) = \cosh u \ \cosh v + \sinh u \ \sinh v$$
$$\cosh(u - v) = \cosh u \ \cosh v - \sinh u \ \sinh v$$
$$\cosh^2 u - \sinh^2 u = 1$$
$$\sinh 2u = 2 \sinh u \ \cosh u$$
$$\cosh 2u = \cosh^2 u + \sinh^2 u$$
$$\sinh iu = i \ \sin u \quad \cosh iu = \cos u$$
$$\sinh u = -i \ \sin iu \quad \cosh u = \cos iu$$

Inverse Hyperbolic Functions:

$$\sinh^{-1} u = \ln[u + (u^2 + 1)^{1/2}]$$
$$\cosh^{-1} u = \ln[u \pm (u^2 - 1)^{1/2}] \quad (u \geq 1)$$
$$\tanh^{-1} u = \frac{1}{2} \ln\left(\frac{1 + u}{1 - u}\right) \quad (u^2 < 1)$$

x	sinh x	cosh x	tanh x
00.0	0.0000	1.0000	0.0000
0.1	0.1002	1.0050	0.0997
0.2	0.2013	1.0201	0.1974
0.3	0.3045	1.0453	0.2913
0.4	0.4108	1.0811	0.3800
0.5	0.5211	1.1276	0.4621
0.6	0.6367	1.1855	0.5370
0.7	0.7586	1.2552	0.6044
0.8	0.8881	1.3374	0.6640
0.9	1.0265	1.4331	0.7163
1.0	1.1752	1.5431	0.7616
1.1	1.3356	1.6685	0.8005
1.2	1.5095	1.8107	0.8337
1.3	1.6984	1.9709	0.8617
1.4	1.9043	2.1509	0.8854
1.5	2.1293	2.3524	0.9052
1.6	2.3756	2.5775	0.9217
1.7	2.6456	2.8283	0.9354
1.8	2.9422	3.1075	0.9468
1.9	3.2682	3.4177	0.9562
2.0	3.6269	3.7622	0.9640

(Continued)

x	sinh x	cosh x	tanh x
2.1	4.0219	4.1443	0.9705
2.2	4.4571	4.5679	0.9757
2.3	4.9370	5.0372	0.9801
2.4	5.4662	5.5569	0.9837
2.5	6.0502	6.1323	0.9866
2.6	6.6947	6.7690	0.9890
2.7	7.4063	7.4735	0.9910
2.8	8.1919	8.2527	0.9926
2.9	9.0596	9.1146	0.9940
3.0	10.018	10.068	0.9951
3.5	16.543	16.573	0.9982
4.0	27.290	27.308	0.9993
4.5	45.003	45.014	0.9998
5.0	74.203	74.210	0.9999
6.0	201.71	201.72	1.0000
7.0	548.32	548.32	1.0000
8.0	1490.5	1490.5	1.0000
9.0	4051.5	4051.5	1.0000
10.0	11 013	11 013	1.0000

Bessel functions.

x	$J_0(x)$	$J_1(x)$	x	$J_0(x)$	$J_1(x)$
0.0	1.0000	0.0000	1.3	0.6201	0.5220
0.1	0.9975	0.0499	1.4	0.5669	0.5419
0.2	0.9900	0.0995	1.5	0.5118	0.5579
0.3	0.9776	0.1483	1.6	0.4554	0.5699
0.4	0.9604	0.1960	1.7	0.3980	0.5778
0.5	0.9385	0.2423	1.8	0.3400	0.5815
0.6	0.9120	0.2867	1.9	0.2818	0.5812
0.7	0.8812	0.3290	2.0	0.2239	0.5767
0.8	0.8463	0.3688	2.1	0.1666	0.5683
0.9	0.8075	0.4059	2.2	0.1104	0.5560
1.0	0.7652	0.4400	2.3	0.0555	0.5399
1.1	0.7196	0.4709	2.4	0.0025	0.5202
1.2	0.6711	0.4983			

Reference

1 Greene, P.R. (1989). A useful approximation to the error function: applications to mass, momentum and energy transport in shear layers. *J. Fluids Eng.* 111: 224–226.

Appendix F

Turbulence Transition

One of the most critical aspects of fluid flow calculations is knowing the regime in which a specified flow is most likely to exist. This aspect is particularly important in convection transfer calculations, as the laminar flow formulas predict transfer rates that differ significantly from the rates predicted by turbulent flow correlations.

Over more than 100 years since Osborne Reynolds' demonstration of the transition to turbulence in pipe flow, we have accumulated a long listing of empirical observations of transition. The most frequently used observations of this kind are summarized in the first column of Table F.1.

The laminar–turbulent transition of each flow is marked by a critical range of values of a characteristic dimensionless group of the flow. In laminar boundary layer flow over a flat plate, for example, the Reynolds number based on the distance x from the leading edge, $U_\infty x/\nu$, is as low as 2×10^4 when the free stream is highly disturbed and as high as 10^6 and even higher when the free stream is without disturbances. A single, representative value was adopted in Eq. (5.85) only for the sake of simplicity and convenience.

Continuing the reading of Table F.1, we see that the natural convection flow along a vertical wall undergoes transition at a Rayleigh number based on local altitude y and wall–fluid temperature difference of order 10^9. This value refers to common fluids such as air and water, that is, to fluids with Prandtl numbers of order 1.

The transition to turbulence in a round jet that discharges freely into a reservoir is characterized by a considerably smaller number (roughly 30), where the number is the Reynolds number based on nozzle diameter and cross section-averaged velocity.

In summary, the traditional record of transition observations consists of a collection of special "critical" numbers, which can be as low as 30 and as high as 10^{12}, depending on the particular flow configuration.

The objective of this appendix is to draw attention to a theory of physics that predicts all such numbers and unifies them. The various "critical" numbers are manifestations of a unique phenomenon of evolution in nature (constructal law [2, 3]), which is the basis for the numbers assembled in the second column of Table F.1. These new numbers represent the *local Reynolds number* of the laminar–turbulent transition region of the flow. The local Reynolds number Re_l is defined as

$$Re_l = \frac{VD}{\nu} \tag{F.1}$$

in which V is the *local longitudinal* velocity scale of the flow and D is the *local transversal* length scale.

In a relatively narrow order-of-magnitude band, all the traditional critical numbers from the first column of Table F.1 correspond to a single local Reynolds number:

$$Re_l \sim 10^2 \quad \text{(at transition)} \tag{F.2}$$

Heat Transfer: Evolution, Design and Performance, First Edition. Adrian Bejan.
© 2022 John Wiley & Sons, Inc. Published 2022 by John Wiley & Sons, Inc.
Companion website: www.wiley.com/go/bejan/heattransfer

Table F.1 Traditional critical numbers for transition in several key flows and the corresponding local Reynolds number scale.

Flow	Traditional critical number	Local Reynolds number
Boundary layer flow over flat plate	$Re_x \sim 2 \times 10^4 – 10^6$	$Re_l \sim 94$–660
Natural convection boundary layer along vertical wall with uniform temperature ($Pr \sim 1$)	$Ra_y \sim 10^9$	$Re_l \sim 178$
Natural convection boundary layer along vertical wall with constant heat flux ($Pr \sim 1$)	$Ra_y^* \sim 4 \times 10^{12}$	$Re_l \sim 330$
Round jet	$Re_{nozzle} \sim 30$	$Re_l \gtrsim 30$
Wake behind long cylinder in cross-flow	$Re \sim 40$	$Re_l \gtrsim 40$
Pipe flow	$Re \sim 2000$	$Re_l \sim 500$
Film condensation on a vertical wall	$Re \sim 450$	$Re_l \sim 450$

Source: After Ref. [1].

For example, the critical local Reynolds number of laminar boundary layer flow over a flat plate is nearly the same as that of the buoyancy-driven jet along a heated vertical wall. It can be argued that the Re_l range for transition in round jet flow is actually higher than the listed nozzle Reynolds number, because the laminar jet expands rapidly outside the nozzle (i.e. the D scale of the jet is larger than the nozzle diameter). The same observation applies to the transition in the wake behind a long cylinder, where the listed Reynolds number is based on diameter of the cylinder, not on the transversal length scale of the wake.

On the high side of the common transition criterion $Re_l \sim 10^2$, the transition in pipe flow occurs at diameter-based Reynolds numbers of order 2000. The actual thickness of the centerline flow "fiber" that exhibits the sinuous motion is considerably smaller than the pipe diameter; therefore the *local* Reynolds number is correspondingly smaller than 2000. This is why a smaller value ($Ra_l \sim 500$) is listed in the second column of the table.

Figure F.1 reviews the local Reynolds numbers that correspond to the transitions considered in Table F.1. This figure condenses the transition observations to a relatively narrow band of values centered around $Re_l \sim 10^2$. Compare this narrow band with the 10–10^{12} range covered by the traditional critical numbers.

Figure F.1 also shows the flow-straightening effect that solid walls have on transition. Flows without solid walls (jets, wakes, plumes) exhibit Re_l values that are on the low side of 10^2. Flows stiffened by one solid wall have somewhat higher local Reynolds numbers at transition. The pipe flow is straightened by solid surfaces from all sides, and, consequently, its transition Re_l value is on the high side of 10^2.

The earliest theoretical derivation of the $Re_l \sim 10^2$ criterion was reported in 1981, as a consequence of the theory on how to predict the natural meandering (buckling) wavelength of any flow at transition, $\lambda = 1.814D$ [1, 4]. The theory is also in Refs. [5, 6], as an example of the manifestation of the constructal law.

The local Reynolds number criterion is unambiguous with respect to the wavelength λ that prevails at transition. This wavelength is roughly twice the thickness of the flow, i.e. unique. In addition, the theory of the preceding paragraph anticipates analytically the observed fact that the original deformation of the flow is a *sinusoid*. These attributes distinguish fundamentally the Re_l criterion from classical hydrodynamic stability analyses in which the shape of the initial deformation of the flow (the "disturbance") is arbitrary, and it must be assumed because it is observed, i.e. empirical.

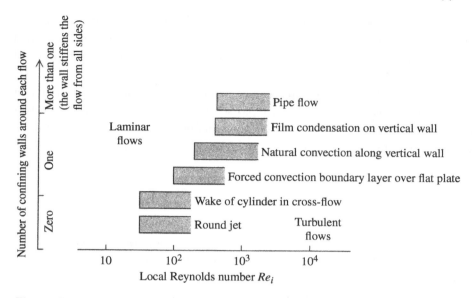

Figure F.1

The local Reynolds number criterion and the constructal theory behind it have several important consequences. One is that the Reynolds number of the smallest eddy in the ensuing flow field is of order 10^2. This discovery contradicts the established view in modern fluid mechanics [7, 8] that holds that the smallest eddy Reynolds number is of order 1.

The same theory and Eq. (F.2) were used to predict (on the back of an envelope) the pulsating frequency of the flames of fires [9]. In addition, the theory predicted that the fire should pulsate only when its base D is greater than approximately 2 cm.

Another consequence of the Re_l criterion is that in turbulent flow near a wall it predicts the existence of a viscous sublayer. The thickness of this sublayer must be of order 10 in the classical y^+ coordinate [5], This y^+ thickness can be verified by intersecting the two u^+ expressions listed on the right side of Eq. (5.116).

The same theory predicts Colburn's analogy between heat transfer and friction in turbulent flow near a wall [5]:

$$St_x Pr^{2/3} = \frac{1}{2} C_{f,x} \tag{F.3}$$

where St_x and $C_{f,x}$ are the local Stanton number and skin friction coefficient, respectively. An integral part of that analysis is the proof that Colburn's analogy is valid only for fluids with Prandtl numbers of order 1 or greater.

The length and time scales of the unstable flow and the process of eddy formation can be used to anticipate the growth (mixing) of turbulent jets, shear layers, and plumes. The time-averaged version of these mixing regions turns out to be one that increases linearly in the downstream direction. The kinematic analysis based on the eddy length and time scales, however, reveals the instantaneous structure of the mixing region, that is, the main eddy sizes and their distribution over the mixing region [5].

Finally, the predicted sinusoidal shape and the proportionality between wavelength and flow thickness during the first stages of inviscid-stream deformation provide a single theoretical explanation for the many "meandering" flows that surround us [2–6]. Examples of such flows are the river meanders, the waving of flags, the fall of paper ribbons, the deformation of fast liquid jets shot through the air, and the sinuous structure of all turbulent plumes [10–18]. The same scales are the basis for the geometric similarity that exists between the laminar sections of straight boundary layer-type flows [5, 19] and between the shapes of the cross sections of rivers of all sizes [20].

The physics law of evolution is also the reason why the cross sections of all jets and plumes (laminar or turbulent) morph toward becoming round [21]. It is why rapid solidification (e.g. snowflakes) evolves through several transitions toward a tree-shaped solid structure whose volume must grow slow–fast–slow, in S-curve fashion [22]. More manifestations of the physics law of evolution everywhere are reviewed in Refs. [2, 3, 23, 24].

References

1 Bejan, A. (1987). Buckling flows: a new frontier in fluid mechanics. In: *Annual Review of Numerical Fluid Mechanics and Heat Transfer*, vol. 1 (ed. T.C. Chawla), 262–304. Washington, DC: Hemisphere.

2 Bejan, A. and Zane, J.P. (2012). *Design in Nature*. New York: Doubleday.

3 Bejan, A. (2016). *The Physics of Life*. New York: St. Martin's Press.

4 Bejan, A. (1981). On the buckling property of inviscid jets and the origin of turbulence. *Lett. Heat Mass Transfer* 8: 187–194.

5 Bejan, A. (2013). *Convection Heat Transfer*, 4e. Hoboken, NJ: Wiley.

6 Bejan, A. (1982). *Entropy Generation Through Heat and Fluid Flow*, 64–97. Wiley.

7 Tennekes, H. and Lumley, J.L. (1972). *A First Course in Turbulence*, 20. Cambridge, MA: MIT Press.

8 Bradshaw, P. (ed.) (1978). Turbulence. In: *Topics in Applied Physics*, 2e, vol. 12, 22. Berlin: Springer-Verlag.

9 Bejan, A. (1991). Predicting the pool fire vortex shedding frequency. *J. Heat Transfer* 113: 261–263.

10 Bejan, A. (1982). Theoretical explanation for the incipient formation of meanders in straight rivers. *Geophys. Res. Lett.* 9 (8): 831–834.

11 Anand, A. and Bejan, A. (1986). Transition to meandering rivulet flow in vertical parallel-plate channels. *J. Fluids Eng.* 108: 269–272.

12 Bejan, A. (1982). The meandering fall of paper ribbons. *Phys. Fluids* 25 (5): 741–742.

13 Kimura, S. and Bejan, A. (1983). Mechanism for transition to turbulence in buoyant plume flow. *Int. J. Heat Mass Transfer* 26: 1515–1532.

14 Kimura, S. and Bejan, A. (1983). The buckling of a vertical liquid column. *J. Fluids Eng.* 105: 469–473.

15 Blake, K.R. and Bejan, A. (1984). Experiments on the buckling of thin fluid layers undergoing end-compression. *J. Fluids Eng.* 106: 74–78.

16 Stockman, M.G. and Bejan, A. (1982). The nonaxisymmetric (buckling) flow regime of fast capillary jets. *Phys. Fluids* 25 (9): 1506–1511.

17 Anderson, R. and Bejan, A. (1983). Buckling of a turbulent jet surrounded by a highly flexible duct. *Phys. Fluids* 26 (11): 3193–3200.

18 Bejan, A. (1985). The method of scale analysis: natural convection in fluids. In: *Natural Convection: Fundamentals and Applications* (eds. S. Kakac, W. Aung and R. Viskanta), 75–94. Washington, DC: Hemisphere.

19 Gore, R.A., Crowe, C.T., and Bejan, A. (1990). The geometric similarity of the laminar sections of boundary layer-type flows. *Int. Commun. Heat Mass Transfer* 17: 465–475.

20 Bejan, A. (2016). *Advanced Engineering Thermodynamics*, 4e, 657. Hoboken, NJ: Wiley.

21 Bejan, A., Ziaei, S., and Lorente, S. (2014). Evolution: why all plumes and jets evolve to round cross sections. *Sci. Rep.* 4 (4730).

22 Bejan, A., Lorente, S., Yilbas, B.S., and Sahin, A.Z. (2013). Why solidification has an S-shaped history. *Sci. Rep.* 3 (1711).

23 Walsh, E.J., Hernon, D.H., McEligot, D.M. et al. (2006). Application of constructal theory to prediction of boundary layer transition onset. Article GT2006-91166. *Proceedings GT2006, ASME Turbo Expo 2006: Power for Land, Sea and Air*, Barcelona (8–11 May 2006).

24 Bejan, A. (2020). *Freedom and Evolution: Hierarchy in Nature, Society and Science*. New York: Springer Nature.

Appendix G

Extremum Subject to Constraint

The operation of determining the extremum of a function $F(x, y)$ that obeys the constraint

$\qquad G(x, y) = C$, constant

is equivalent to determining the extremum of the aggregate function

$\qquad \Phi = F + \lambda G$

where λ is an undetermined coefficient called Lagrange multiplier. The extremum of $\Phi(x, y)$ is located by solving the system

$$\frac{\partial \Phi}{\partial x} = 0, \qquad \frac{\partial \Phi}{\partial y} = 0$$

The (x, y) solution of this system will depend on λ. Finally, λ is determined by substituting the $(x(\lambda), y(\lambda))$ solution into the constraint $G(x, y) = C$.

Heat Transfer: Evolution, Design and Performance, First Edition. Adrian Bejan.
© 2022 John Wiley & Sons, Inc. Published 2022 by John Wiley & Sons, Inc.
Companion website: www.wiley.com/go/bejan/heattransfer

Author Index

Heat Transfer: Evolution, Design and Performance, First Edition. Adrian Bejan.
© 2022 John Wiley & Sons, Inc. Published 2022 by John Wiley & Sons, Inc.
Companion website: www.wiley.com/go/bejan/heattransfer

Subject Index

Heat Transfer: Evolution, Design and Performance, First Edition. Adrian Bejan.
© 2022 John Wiley & Sons, Inc. Published 2022 by John Wiley & Sons, Inc.
Companion website: www.wiley.com/go/bejan/heattransfer